Polymer Durability

ADVANCES IN CHEMISTRY SERIES 249

Polymer Durability

Degradation, Stabilization, and Lifetime Prediction

Roger L. Clough, EDITOR
Sandia National Laboratories

Norman C. Billingham, EDITOR
University of Sussex

Kenneth T. Gillen, EDITOR
Sandia National Laboratories

Developed from a symposium sponsored
by the Division of Polymer Chemistry, Inc.,
at the 206th National Meeting
of the American Chemical Society,
Chicago, Illinois,
August 22–27, 1993

American Chemical Society, Washington, DC 1996

Library of Congress Cataloging-in-Publication Data

Polymer durability: degradation, stabilization, and lifetime prediction / Roger
L. Clough, editor, Norman C. Billingham, editor, Kenneth T. Gillen, editor.

   p.   cm.—(Advances in chemistry series, ISSN 0065–2393; 249)

"Developed from a symposium sponsored by the Division of
Polymer Chemistry, Inc., at the 206th National Meeting
of the American Chemical Society, Chicago, Illinois, August
22–27, 1993."

  Includes bibliographical references and indexes.

  ISBN 0–8412–3134–6 (hardbound)

  1. Polymers—Deterioration—Congresses
  2. Stabilizing agents—Congresses

  I. Clough, Roger L. (Roger Lee), 1949–   . II. Billingham, N. C.
III. Gillen, Kenneth T. IV. American Chemical Society. Division of Polymer
Chemistry. V. American Chemical Society. Meeting (206th: 1993: Chicago, Ill.).
VI. Series.

QD1.A355 no. 249
[QD381.9.D47]
540 s—dc20
[668.9]   95–42163
                                                    CIP

The paper used in this publication meets the minimum requirements of American National
Standard for Information Sciences—Permanence of Paper for Printed Library Materials, ANSI
Z39.48–1984.   ∞

PRINTED IN THE UNITED STATES OF AMERICA

FOREWORD

The ADVANCES IN CHEMISTRY SERIES was founded in 1949 by the American Chemical Society as an outlet for symposia and collections of data in special areas of topical interest that could not be accommodated in the Society's journals. It provides a medium for symposia that would otherwise be fragmented because their papers would be distributed among several journals or not published at all.

Papers are reviewed critically according to ACS editorial standards and receive the careful attention and processing characteristic of ACS publications. Volumes in the ADVANCES IN CHEMISTRY SERIES maintain the integrity of the symposia on which they are based; however, verbatim reproductions of previously published papers are not accepted. Papers may include reports of research as well as reviews, because symposia may embrace both types of presentation.

ABOUT THE EDITORS

ROGER L. CLOUGH is Manager of the Organic Materials Synthesis and Degradation Department at Sandia National Laboratories. He received his Ph.D. degree in chemistry from the California Institute of Technology in 1975 under the supervision of J. D. Roberts. He then spent a year as a North Atlantic Treaty Organization Postdoctoral Fellow at the Technical University in Munich, with E. O. Fischer. This fellowship was followed by a year as a National Science Foundation Fellow at the University of California at Los Angeles with Chris Foote. Clough has authored more than 100 research papers, holds five patents, has written numerous review articles on stability and stabilization of polymers to ionizing radiation, and has edited two books on the subject of radiation effects on polymers. His research activities include polymer oxidation mechanisms, lifetime prediction methodologies, oxygen diffusion effects in materials degradation, development of analytical techniques for monitoring materials degradation, and understanding changes in mechanical properties and color that occur as polymers undergo aging. The department that Clough manages has responsibility for developing a variety of high-performance materials and for investigating a wide range of polymer degradation problems involving environments of elevated temperature, UV light, or ionizing radiation.

NORMAN C. BILLINGHAM is Reader in Chemistry in the School of Chemistry and Molecular Sciences at the University of Sussex, Brighton, United Kingdom. He received his B.Sc. and Ph.D. degrees from the University of Birmingham. In 1967 he moved to Sussex as an ICI Research Fellow, and he has spent his research career in Sussex, moving from Lecturer to Reader in 1988. His research interests in degradation and stabilization of polymers include development of new methods of study, especially chemiluminescence and optical microscopy. He has worked on solubility, diffusion, and loss of stabilizing additives and on application of UV and fluorescence microscopy to study additive location and migration. Other research interests are in polymer synthesis, with an emphasis on use of group-transfer and related polymerization methods to prepare monodisperse block copolymers for evaluation as dispersants and binders in ceramics processing and in membrane applications. A particular interest is water-soluble and amphiphilic polymers and their solution behavior. Billingham has published more than 100 papers and reviews. He was awarded the 1994 Royal Society of Chemistry Industrial Award in Macromolecules and Polymers for his work in degradation and stabilization of polymers. He is editor of the Elsevier journal *Polymer Degradation and Stability*.

KENNETH T. GILLEN received his B.S. degree in chemistry from the University of California at Berkeley and his Ph.D. degree in chemistry from the University of Wisconsin at Madison in 1970. After a two-year postdoctorate at Bell Telephone Laboratories in Murray Hill, New Jersey, and a one-year position at JEOL, Ltd., he joined Sandia National Laboratories in 1974. Since 1991, Gillen has been a Distinguished Member of the Technical Staff in the Properties of Organic Materials Department. His research interests focus on aging effects of polymeric materials, where he develops experimental techniques and creates and verifies models for various aging environments. He has coauthored approximately 100 publications and eight invited book chapters. Gillen serves as the U.S. representative on the International Electrotechnical Commission (IEC) Radiation Effects Working Group, which creates and rewrites IEC radiation standards on endurance of insulating materials. For the past eight years, he has been the organizer of the Department of Energy Conferences on Compatibility, Aging, and Service Life.

CONTENTS

PREFACE

Polymers were originally introduced into commercial use as bulk structural materials primarily because they were inexpensive and easy to form compared with existing alternatives such as metal and wood. They were, however, subject to deterioration due to a variety of environmental factors, including light, mechanical stress, temperature, pollutants, and others. In particular, because most common macromolecules consist primarily of carbon and hydrogen in a reduced state, and because oxygen is abundant in most terrestrial applications, polymers are highly susceptible to degrading oxidation reactions. Because of this susceptibility, polymers quickly gained a reputation for low reliability; indeed, during the 1950s and 1960s, the word "plastic" became commonly associated with the concept of "cheap" or "poor quality".

Today, polymers are used in ever-widening applications, and there is a growing emphasis on their durability and reliability. This emphasis results partly from increased consumer demand for quality in products. In addition, the existing paradigm of easy disposal and replacement of polymeric products is becoming much less attractive because of emerging environmental concerns: there is a growing awareness that enhanced lifetimes of polymeric materials could reduce the energy consumption needed in the manufacturing of materials and could also help alleviate the burden of solid waste. Moreover, there are now many instances in which molecular engineering of polymer-based materials has provided properties uniquely suited to meet a wide range of "high-tech" applications, including structural, surface-protective, electronic, and optical uses. These applications frequently have very high performance demands and/or high cost, which makes material stability of paramount importance.

Polymers have come into widespread use primarily during the latter half of this century. As a result, work on their long-term aging characteristics is still in the early stages compared with, for example, metals, for which stability optimization through compositional and processing changes has been under experimentation over several millennia. The stability of polymers is an exceedingly complex problem, and this complexity remains an impediment to scientific progress in this area. Controlling degradation requires understanding of many different phenomena, including the diverse chemical mechanisms underlying structural change in macromolecules, the influence of polymer morphology, the complexities of oxidation chemistry, the intricate reaction pathways of stabilizer additives, the interactions of fillers and other ingredients, as well as impurities, and the reactive–diffusion processes that often take place (oxygen or other reactants coming in, and additives going out). Finally comes the difficulty of understanding the relationship between the numerous changes in material composition that occur upon aging (which may be inhomogeneous on both micro- and macroscopic scales), and the observed changes in the physical properties and/or failure mechanisms of interest for the material (e.g., mechanical strength, cracking, color, or electrical properties).

This book provides an overview of the state of the art in the science of polymer durability. Its 39 chapters are written by internationally recognized experts in the major technology areas of the field. The book is organized into three main sections covering degradation, stabilization, and lifetime prediction. The degradation section discusses fundamentals of the molecular mechanisms by which polymers undergo aging and deterioration, particularly under environments of UV light (as in outdoor exposure), or of thermal exposure. This section also describes a variety of important analytical techniques used for studying degradation, with special emphasis on the very sensitive technique of chemiluminescence. The second section of the book covers major types of additives used for polymer stabilization during processing, long-term (elevated temperature or room temperature) applications, and UV exposure. Various chapters describe the action mechanisms of different stabilizer types, and the effects of these additives on physical property retention. Several chapters discuss the problem of migration and retention of stabilizer additives in materials. Other stabilization-related topics include the development of specialized protective coatings and the influence of blends, copolymers, and fillers on stability. The final part of the book discusses progress in developing methods for predicting the aging rate and lifetime of a material in a particular application. These methods can involve the design of effective accelerated aging tests combined with modeling of aging processes. Also included in this section is a chapter on modeling of degradation during melt processing.

ROGER L. CLOUGH
Sandia National Laboratories
MS 1407, P.O. Box 5800
Albuquerque, NM 87185–1407

NORMAN C. BILLINGHAM
School of Chemistry and Molecular Sciences
University of Sussex
Falmer, Brighton
BN1 9QJ United Kingdom

KENNETH T. GILLEN
Sandia National Laboratories
MS 1407, P.O. Box 5800
Albuquerque, NM 87185–1407

February 14, 1995

DEGRADATION

To achieve the goal of enhanced polymer durability, one must first understand the underlying degradation mechanisms. Material degradation can be exceedingly complex because of the diversity of chemical reaction mechanisms that may operate, together with numerous complicating factors including morphology, complex stabilizer chemistries, reactive diffusion processes, interactions of fillers or other ingredients, and the complex relationship between molecular-level changes and macroscopic properties. Nevertheless, the past decade has seen significant progress in understanding degradation processes.

The first section of this book is devoted to polymer degradation mechanisms. Chapters 1–10 cover thermal and photochemical degradation mechanisms, with early chapters describing changes in macromolecular structure that occur upon thermal exposure. The first two chapters include in-depth studies of the longstanding issue concerning the existence of imperfections in the polymer structure, introduced during the polymerization process, that may serve as "weak link" sites for degradation chemistry. A series of chapters then discusses photodegradation mechanisms in a wide variety of different polymer types, and one chapter reviews recent results on the behavior of photogenerated singlet oxygen produced from O_2 dissolved within bulk polymers. The section ends with a group of chapters that describe analytical techniques applied to monitoring of polymer degradation. The very sensitive technique of chemiluminescence is emphasized, including the use of this technique to visualize the phenomenon of "infectious" spreading of oxidative degradation both within a solid polymer sample and also through space to adjacent polymer samples. Other important analytical techniques are described, including the emerging method of X-ray tomography for mapping heterogeneous degradation through the interior of polymer materials and the use of GC-mass spectroscopy to characterize volatile degradation products.

Auxiliary Mechanism for Transfer to Monomer during Vinyl Chloride Polymerization

Implications for Thermal Stability of Poly(vinyl chloride)

W. H. Starnes, Jr.[1], Haksoo Chung[1], B. J. Wojciechowski[1], D. E. Skillicorn[2], and G. M. Benedikt[3]

[1]Applied Science Ph.D. Program, Department of Chemistry, College of William and Mary, Williamsburg, VA 23187–8795
[2]The Geon Company Technical Center, Avon Lake, OH 44012
[3]BFGoodrich Company Research and Development Center, Brecksville, OH 44141

Chain transfer to monomer during the free-radical polymerization of vinyl chloride is shown to occur, in part, by a mechanism that begins with the abstraction of methylene hydrogen from the polymer by a propagating macroradical. The resultant radical then donates a chlorine atom to the monomer to form an allylic structure that can contribute to the thermal instability of poly(vinyl chloride). Double-bond contents found by NMR spectroscopy and their correlation with molecular weights confirm the operation of this transfer process and verify a new theory for transfer to monomer. The theory involves two transfer constants and allows their values to be obtained.

W HEN POLY(VINYL CHLORIDE) (PVC) IS PREPARED AT TEMPERATURES within the usual commercial range, most if not all of its ethyl-branch segments have the EB structure shown in Scheme I (*1*). This arrangement results from a process that starts with head-to-head addition of monomer and involves the head-to-head radical, **1**, and the rearranged radicals, **2** and **3** (*1*). At a given temperature of polymerization, the sum of the EB and chloromethyl (MB) branch concentrations (*2*) is independent of the pressure (or molar concentration) of vinyl chloride (VC) (*3*). These observations require operation of the

0065–2393/96/0249–0003$12.00/0

entire mechanism in Scheme I (3). They also show that this scheme, as drawn, incorporates all reactions occurring after head-to-head emplacement that have significant structural or kinetic implications (3). Moreover, they lead directly to the following major conclusions (3):

1. Radicals **2** and **3** do not undergo unimolecular β-scission to generate chlorine atoms that become kinetically free.

2. In bulk or suspension polymerizations of VC, rates of propagation are not controlled by diffusion, even up to conversions of about 90%.

3. The monomer transfer constant, $C_{M,HH}$, that pertains to transfer occurring after head-to-head emplacement (*see* Scheme I) is a true constant that is independent of the VC concentration (eq 1).

$$C_{M,HH} = k_p'k_4/(k_p + k_p')(k_1 + k_4) = k_p'k_5/(k_p + k_p')(k_3 + k_5) \qquad (1)$$

When extrinsic transfer agents are absent, transfer to monomer produces most of the long-chain ends in PVC (4, 5). Yet the instantaneous number-average molecular weight ($\overline{M}_n$) of PVC decreases rapidly with decreasing VC

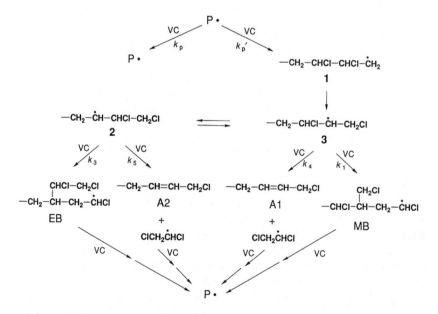

Scheme I. Mechanistic sequelae of head-to-head addition during the free-radical polymerization of vinyl chloride (VC), where P· is the ordinary head-to-tail macroradical, and the ks are rate constants.

concentration (2), despite the constancy of $C_{M,HH}$. These findings necessitate the intervention of an auxiliary monomer-transfer process that increases in importance as the VC content declines. Such a mechanism was described and discussed heretofore (3, 6). This mechanism is depicted in Scheme II, and its first step can be either inter- or intramolecular, as shown. Addition of radical **4** to the monomer would start the growth of a long-branch structure (**5**), whose presence was supported by NMR observations (7). Alternatively, **4** might undergo a transfer reaction with VC that exhibits second-order kinetics and is, therefore, analogous to the transfers in Scheme I (3). The resultant internal allylic (IA) structure would be a thermally labile site (8, 9). Thus, this auxiliary transfer pathway has potential implications for both the stability of PVC and the overall mechanism for the polymerization of VC. We examine these topics in some detail and present conclusive evidence for the occurrence of chain transfer by the auxiliary route.

Experimental Section

Subsaturation polymerizations were performed according to a published procedure (2, 10), starting with 296 g of water, 0.164 g of ammonium persulfate, and 1.89 g of a PVC latex resin that was required for use as seed. Polymerization-grade VC was introduced continuously from a colder reservoir (2, 10), and the reaction mixtures were agitated with an efficient stirrer that was operated at the highest possible speed. Polymerizations were stopped before the agglomeration of primary

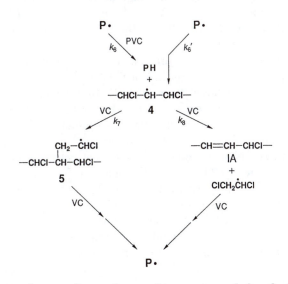

Scheme II. Auxiliary mechanism for transfer to monomer during the free-radical polymerization of vinyl chloride (VC), where P· is the ordinary head-to-tail macroradical, and the ks are composite rate constants as defined in the text.

particles occurred. The polymers were isolated by filtration, washed three times with water, and then dried thoroughly under vacuum at ca. 40 °C.

Number-average molecular weights of PVC specimens were obtained from the equation, $\log \overline{M}_n = 4.6549 + 1.2385[\log(I.V.)]$, where $I.V.$ is inherent viscosity measured in cyclohexanone, $(0.2 \text{ g})/(100 \text{ mL})$, at 30 °C, according to ASTM Method D1243-79 (11). Proton NMR measurements were made with a Bruker AMX500 instrument at 50 °C by using dilute solutions of the polymers in tetrahydrofuran-d_8.

Results and Discussion

Double-Bond Concentrations. One chloroallylic terminus (A1 or A2) will occur in every number-average polymer molecule formed by the transfers in Scheme I. Moreover, one IA group per molecule will result from the transfer of Scheme II. Thus, when other transfer agents have not been added, the total double-bond content of PVC should be close to one per molecule regardless of the relative importance of the different monomer-transfer routes.

Many researchers have tried to determine quantitatively the number of alkene linkages in undegraded PVC. Chemical techniques used for this purpose have included the addition of Br_2 or ICl. In one study involving bromine (12), a total double-bond content of about one per molecule was obtained for a number of PVC specimens made at 55 °C. However, other investigations involving the bromination of various "industrial" (13) or "commercial" (14) PVCs have given values per molecule that range from 0.7 to 1.1 [as calculated by us from the published data (13)] or from 1.0 to 1.95 (14). Values per molecule of about 0.9 (15) and 1.0–1.4 (16) were found for several PVC samples via the addition of ICl.

The International Union of Pure and Applied Chemistry, Working Party on Defects in the Molecular Structure of Poly(vinyl chloride), conducted the most extensive study of unsaturation in PVC. In a round-robin protocol involving several laboratories, the addition of Br_2 or ICl and IR and ^1H NMR spectroscopy were used to measure total unsaturation in a variety of specimens. Double-bond (17, 18) and $\overline{M}_n$ (19) data led to average total double-bond contents per molecule (computed by us) that ranged from 1.0 to 1.3 for bromination (17), from 0.9 to 1.3 for ICl addition (17) [ignoring the very low value found (17) for a polymer made at −30 °C (19)], and from 1.0 to 1.45 for IR analysis, which was performed by only one group (17). All of these ranges apply to different subsets of polymers. From ^1H NMR measurements, one Working Party laboratory found total unsaturation values that can be shown from the published data (17–19) to range from 1.0 to 1.7 per molecule. Other NMR values were obtained by a second laboratory that was designated as "E" in one article (17) and apparently as "B" in another (18). However, these values are uninterpretable, because, inexplicably, they are twice as large in one report (18) as they are in the other (17).

The ^{1}H NMR method avoids the calibration problems that the other techniques present (*17, 18*). Thus, it should be highly reliable. Even so, the NMR observations made heretofore, including those just described, do not support the consistent occurrence of 1 double bond *in toto* per number-average PVC molecule. One research group (*20*) reported total ^{1}H NMR double-bond values of 0.8, 2.8, and 2 per molecule for three of the Working Party samples. These results do not agree satisfactorily with the values of 1.0, 1.7, and 1.0, respectively, that can be calculated for those polymers from the NMR (*17, 18*) and $\overline{M}_n$ (*19*) data of the Working Party. Another publication (*21*) presented ^{1}H NMR and $\overline{M}_n$ observations for some suspension PVCs that indicated total double-bond contents ranging from 1.0 to as high as 4.9 per molecule.

Possible causes of these apparent inconsistencies are inadequate sensitivity and resolution in the NMR analyses, errors in the $\overline{M}_n$ values, and partial dehydrochlorination of some of the polymers studied. In an attempt to avoid these problems, we used ^{1}H NMR spectroscopy at a field strength of 500 MHz to determine the unsaturation in some virgin PVC specimens that were protected from exposure to environments that might have caused them to degrade. The $\overline{M}_n$s of these polymers were obtained from viscosity data by a well-established procedure (*11*), and their double-bond contents were derived in the following manner (*20*) from the areas of the resonances that Figure 1 exemplifies.

First, the total concentration per monomer unit of the chloroallylic chain end (A1 and A2), which also equals $C_{M,HH}$, was obtained. This value was determined from the total area, divided by 2, of the terminal methylene resonances centered at 4.06 and 4.13δ, and the area of the principal CHCl multiplet (not shown), which is centered near 4.5δ. Then the internal double-bond content, [IA], per VC unit was deduced by subtracting the (A1 and A2) chloromethyl proton area from that of the complex multiplet at about 5.5–6.0δ and dividing the result by twice the area of the principal CHCl envelope. From the relative areas of the 4.06 and 4.13δ resonances, the cis:trans isomer ratio of the chloroallylic chain end was about 1:10 for all of the polymers studied.

Results of the NMR study are presented in Table I, which includes data for two sets of polymers. One set was prepared at 80 °C or 55 °C under the constant VC pressures that are represented by the P/P_o ratios given in footnotes, where P is the actual VC pressure and P_o is the VC pressure required to saturate the polymer phase with monomer at a given temperature. This group of polymers was made by adapting a published procedure (*2, 10*) that avoids diffusion control of the rate of polymerization under subsaturation conditions. The other polymers were prepared in aqueous suspension by using a variety of recipes and procedures that were typical of those encountered in commercial operations. In these polymerizations, conversions after the pressure drop amounted to only a few percent, at most. Thus, the average P/P_o value was very close to 1 for all of the polymers in this group.

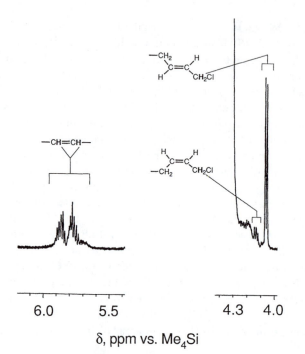

Figure 1. Partial 500-MHz ^{1}H NMR spectra of a suspension PVC sample made at 82 °C.

All of the $[CH=CH]_{total}$ concentrations in Table I are about 1 per molecule. They can, in fact, be regarded as having that value exactly when the experimental uncertainties in the $\overline{M}_n$s of the polymers (Perkins, G., The Geon Company, unpublished observations) are taken into account. Moreover, the data show that, as predicted, changes in the relative importance of the head-to-head and auxiliary transfers (as measured by $C_{M,HH}$ and [IA], respectively) have no effect at all on the $[CH=CH]_{total}$ value. However, in the case of the polymers made with VC subsaturation, the tabulated values of $C_{M,HH}$ and [IA] are too low and too high, respectively, because some of the chloroallylic chain ends are converted into internal alkene structures in a secondary side reaction. This complication is examined in a following section.

The reactions of kinetically free chlorine atoms are highly unselective (22). Consequently, if such atoms were formed from radical **4** by unimolecular β-scission (eq 2), they should have reacted, in part, with PVC monomer units to re-create their progenitor (eq 3).

$$4 \rightarrow IA + Cl^{\bullet} \tag{2}$$

**Table I. Double-Bond Concentrations in PVC Determined
by 500-MHz ^{1}H NMR Spectroscopy**

Polym Temp (°C)	$\overline{M}_n \times 10^{-3}$	$C_{M,HH} \times 10^3$	[IA] per 1000 VC	$[CH{=}CH]_{total}$ per Molecule
82	20.0	2.1	0.6	0.9
80[a]	24.2	1.8	0.6	0.9
80[b]	19.6	1.05[c]	2.3[c]	1.05
70	28.0	1.7	0.5	1.0
61	36.5	1.1	0.5	0.9
57	41.0	1.2	0.0	0.8
56	42.0	0.9	0.6	1.0
55[d]	47.3	0.65[c]	0.5[c]	0.9
55[e]	42.3	0.6[c]	0.85[c]	1.0
55[b]	31.8	0.5[c]	1.6[c]	1.0
53	46.5	1.0	0.1	0.8
52	48.0	0.8	0.3	0.8
49	52.5	0.8	0.5	1.1
40	68.5	0.6	0.2	0.9
36	81.0	0.6	0.0	0.8
36	81.0	0.5	0.3	1.1
32	93.5	0.4	0.2	1.0

NOTE: Polym Temp is polymerization temperature.
[a] Value of P/P_o is 1.00.
[b] Value of P/P_o is 0.59.
[c] Unreliable value; *see* text for discussion.
[d] Value of P/P_o is 0.92.
[e] Value of P/P_o is 0.77.

$$-CHCl-CH_2-CHCl- \text{ (PVC)} + Cl^\bullet \rightarrow HCl + \mathbf{4} \tag{3}$$

Equations 2 and 3 are the propagation steps of an ancillary chain reaction that would have produced additional IA moieties (23) and thus would have caused $[CH{=}CH]_{total}$ to be greater than 1 per molecule. Hence, our failure to find that result argues strongly against the formation of a significant number of free chlorine atoms from **4**. Instead, we can conclude that **4**, like radicals **2** and **3** (3), transfers Cl directly to monomer in a process whose overall kinetics are second-order.

Total Monomer-Transfer Constants.   The total monomer-transfer constant ($C_{M,total}$) is defined by eq 4, where $C_{M,aux}$ refers to the auxiliary transfer of Scheme II and thus is equal to the IA concentration per monomer unit.

$$C_{M,total} = C_{M,HH} + C_{M,aux} \tag{4}$$

From this equation and our previous arguments, the total double-bond content per VC unit should be equivalent to the $C_{M,total}$ value deduced from the $\overline{M}_n$ of the polymer.

The literature contains several Arrhenius expressions that are based on $\overline{M}_n$ data and can be used to calculate $C_{M,total}$ for conventional polymerizations performed at various temperatures. Four equations of this type, including three reported previously, appear in a paper by Carenza et al. (24). We used these equations to calculate four values of $C_{M,total}$ for each of the temperatures in Table I. Table II compares the averages of these values with the $C_{M,total}$ values obtained with eq 4 from the double-bond concentrations in the third and fourth columns of Table I (data for the subsaturation polymers are irrelevant here and thus were omitted). The agreement is excellent in every case, and this result can be regarded as an additional verification of the mechanistic theory we propose.

Determination of Head-To-Head and Auxiliary Transfer Constants from Molecular Weights.

In the transfer process of Scheme II, radical **4** must occur in several microenvironments that differ with respect to local tacticity and the presence or absence of defect structures. Nevertheless, single rate constants can be defined for all of the reactions by which **4** is formed or destroyed. For example, k_6 can be described by eq 5, in which k_{6a}, k_{6b}, ... k_{6n} are the rate constants for reaction of the various structures whose mole fractions are denoted by f_a, f_b, ... f_n, where $f_a + f_b + ... + f_n = 1$.

Table II. Comparison of $C_{M,total}$ Values

Polym Temp (°C)	Literature[a]	¹H NMR[b]
82	3.1 ± 0.5	2.7 ± 0.4
80	2.9 ± 0.5	2.4 ± 0.4
70	2.2 ± 0.3	2.2 ± 0.3
61	1.7 ± 0.3	1.6 ± 0.2
57	1.5 ± 0.2	1.2 ± 0.2
56	1.4 ± 0.2	1.5 ± 0.2
53	1.3 ± 0.2	1.1 ± 0.2
52	1.3 ± 0.2	1.1 ± 0.2
49	1.1 ± 0.1	1.3 ± 0.2
40	0.8 ± 0.1	0.8 ± 0.1
36	0.7 ± 0.1	0.6 ± 0.1
36	0.7 ± 0.1	0.8 ± 0.1
32	0.6 ± 0.1	0.6 ± 0.1

NOTE: Values are $C_{M,total} \times 10^3$.
[a] Average values obtained from the four Arrhenius equations in ref. 24; deviations are average deviations from the mean.
[b] Sum of the values in columns 3 and 4 of Table I; deviations are the estimated experimental errors.

$$k_6 = k_{6a} f_a + k_{6b} f_b + \ldots + k_{6n} f_n \tag{5}$$

Rate constants k_7 and k_8 can be defined in a similar way, and the first-order constant, k_6', can be expressed as a sum of specific rate constants for intramolecular hydrogen abstraction via cyclic transition states having rings of various sizes (eq 6).

$$k_6' = k_{6a}' + k_{6b}' + \ldots + k_{6n}' \tag{6}$$

Because $C_{M,aux}$ is equal to the IA content per monomer unit, it is given by eq 7:

$$C_{M,aux} = \frac{d[IA]/dt}{d[PVC]/dt} = \frac{k_8[4][VC]}{k_p[P^\bullet][VC]} = \frac{k_8[4]}{k_p[P^\bullet]} \tag{7}$$

where the bracketed terms are concentrations and t is reaction time. Here, [PVC] refers to polymerized monomer units rather than polymer molecules, and head-to-tail propagation is the only reaction of $P^\bullet$ that was taken into account. The other reactions of $P^\bullet$ can be ignored because the sum of their rates is quite small. Under steady-state conditions, the formation rate of **4** will equal its rate of disappearance. Hence, eq 8 will apply. The combination of eq 8 with eq 7, so as to eliminate $[4]/[P^\bullet]$, produces eq 9. Because the reciprocal of the number-average degree of polymerization, $(\overline{DP})_n^{-1}$, is equal to $C_{M,total}$, eq 10 follows directly from eq 4. Equation 11 then results from the substitution, into eq 10, of the expression for $C_{M,aux}$ from eq 9.

$$(k_6[PVC] + k_6')[P^\bullet] = (k_7 + k_8)[VC][4] \tag{8}$$

$$C_{M,aux} = \frac{k_8(k_6[PVC] + k_6')}{k_p(k_7 + k_8)[VC]} \tag{9}$$

$$(\overline{DP})_n^{-1} = C_{M,HH} + C_{M,aux} \tag{10}$$

$$(\overline{DP})_n^{-1} = C_{M,HH} + \frac{k_8(k_6[PVC] + k_6')}{k_p(k_7 + k_8)[VC]} \tag{11}$$

When intermolecular hydrogen abstraction by $P^\bullet$ is much faster than the analogous intramolecular process (i.e., when $k_6[PVC] \gg k_6'$), eq 11 reduces into eq 12.

$$(\overline{DP})_n^{-1} = C_{M,HH} + \frac{k_6 k_8}{k_p(k_7 + k_8)} \frac{[PVC]}{[VC]} \tag{12}$$

This equation can be used in a number of ways to test for the credibility of the dual transfer-process proposal. If the proposal were correct, plots of $(\overline{DP})_n^{-1}$ vs. [PVC]/[VC] should produce straight lines whose intercepts are equal to the $C_{M,HH}$ values determined by NMR spectroscopy. Moreover, calculations based on the slopes of such lines should yield values of $C_{M,aux}$ that also are equivalent to the corresponding NMR values.

An important paper by Hjertberg and Sörvik (2) contains the best data that are available for use in plots of eq 12. Most of these data appear in Table III. Because the composite activation energy for transfer to monomer is higher than that for normal chain propagation (24), the molecular weight of PVC should increase when the polymerization temperature is decreased under controlled conditions that otherwise are the same. This trend is observed for all of the $\overline{M}_n$ values in Table III except those of the 45 °C polymers that were made at P/P_o ratios of 0.76, 0.70, and 0.61. Calculations based on VC solubility data (2, 25) show that the low $\overline{M}_n$ values of these specimens could not have been caused by major increases in the equilibrium [PVC]/[VC] values upon going from 55 °C to 45 °C. These lower $\overline{M}_n$ values might have resulted, instead, from excessive VC starvation brought about by diffusion control, attempts to prevent such control (2, 10) notwithstanding. Regardless, these three $\overline{M}_n$ values obviously are in error, and therefore we have not used them for plotting.

Figure 2 shows plots of eq 12 based on the other data in Table III [the line for 55 °C also includes a point for $P/P_o = 0.53$ (2)]. The [PVC]/[VC] ratios used here were obtained from reported values of (g VC)/(100 g PVC) (2) that were derived from information published previously (25). In all cases, the double-regression fits are quite good, and the $C_{M,HH}$ values obtained from their intercepts agree superbly with the values found by NMR spectroscopy (Table IV). An analogous plot (not shown) was constructed from data given in Table I for the polymers made by us at 55 °C. This plot also was linear ($R^2 = 0.97$), and its slope (0.076) was similar to that of the 55 °C line in Figure 2 (0.10). Moreover, Table IV shows that the intercepts of these two lines were identical.

Activation energy differences might be expected to cause the k_e/k_p ratio to increase as the temperature is raised. This change would increase the slopes of the plots of eq 12. However, that trend is not followed by the lines for

Table III. Molecular Weights of Subsaturation Polymers

Polym Temp (°C)	P/P_o					
	0.97	0.92	0.85	0.76	0.70	0.61
80	26.7	25.3	26.0	20.9	20.2	19.2
65	37.1	35.2	34.1	31.6	30.3	25.2
55	48.6	45.0	41.0	37.7	33.1	29.3
45	59.8	50.4	46.8	32.0[a]	26.5[a]	24.1[a]

NOTE: Values are $M_n \times 10^{-3}$. Data are from ref. 2.
[a] Validity doubtful; see text for discussion.

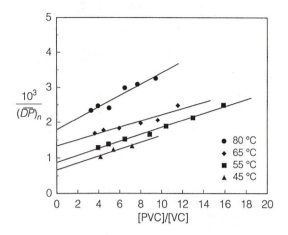

Figure 2. Double-regression plots of eq 12 based on data obtained by Hjertberg and Sörvik (2) for subsaturation PVCs. The R² values for these plots are as follows: 80 °C, 0.91; 65 °C, 0.90; 55 °C, 0.99; and 45 °C, 0.92

55 °C and 65 °C. It may be opposed by an increase in the k_7/k_8 ratio with increasing temperature. The incursion of the process associated with k_6' is another possible complication in this regard (eq 11).

When $k_6' \gg k_6[\text{PVC}]$, eq 11 reduces to eq 13:

$$(\overline{DP})_n^{-1} = C_{\text{M,HH}} + \frac{k_6' \, k_8}{k_p \, (k_7 + k_8)[\text{VC}]} \tag{13}$$

Plots of the latter equation (not shown) were made from the data on which the lines in Figure 2 are based by using VC molar concentrations that were determined in a manner described elsewhere (3). These plots also were linear, and their R^2 values were virtually identical to those of the corresponding plots in Figure 2. However, the $C_{\text{M,HH}}$ values found from their intercepts (0.5, 0.7, 1.2, and 1.5 for 45 °C, 55 °C, 65 °C, and 80 °C, respectively) were somewhat lower than the NMR values of $C_{\text{M,HH}}$. Furthermore, the linearity, per se, of plots of $(\overline{DP})_n^{-1}$ versus $[\text{VC}]^{-1}$ does not necessarily mean that k_6' is much larger than $k_6[\text{PVC}]$. The reason is that, at a given temperature, the sum of the molar concentrations of VC and polymerized VC units remains approximately constant $[\pm(2-5)\%]$ in the concentration ranges of interest to us. [The concentrations of polymerized units upon which this conclusion is based were obtained by assuming volume additivity and using a value of 1.4 g/mL for the density of the polymer (5).] In other words, eq 14 applies when the "constant" total concentration is denoted as A. If eq 11 and 14 are combined to eliminate [PVC], the result is eq 15.

Table IV. Comparison of $C_{M,HH}$ Values

Polym Temp (°C)	By ^1H NMR[a]	From M_n Values[b]
80	1.8 ± 0.3	1.8 ± 0.3
65	1.4 ± 0.2	1.3 ± 0.2
55	0.95 ± 0.1	0.9 ± 0.1
		(0.9 ± 0.1)[c]
45	0.7 ± 0.1	0.7 ± 0.1

NOTE: Values are $C_{M,HH} \times 10^3$.
[a] Values based on data in Table I by taking averages where appropriate. Deviations are the estimated experimental errors.
[b] Values are intercepts of the plots in Figure 2 unless noted otherwise; *see* text for discussion. Deviations are the estimated experimental errors.
[c] Value based on data from the present work; *see* text.

$$[PVC] + [VC] \approx A \qquad (14)$$

$$(\overline{DP})_n^{-1} \approx C_{M,HH} - \frac{k_6 k_8}{k_p (k_7 + k_8)} + \frac{k_8(k_6 A + k_6')}{k_p(k_7 + k_8)[VC]} \qquad (15)$$

This equation shows that plots of $(\overline{DP})_n^{-1}$ versus $[VC]^{-1}$ must be linear, regardless of the relative magnitudes of k_6A and k_6'. Equation 15 also shows that $C_{M,HH}$ can be obtained from the slopes and intercepts of such plots, according to eq 16, when k_6' is insignificant.

$$C_{M,HH} \approx intercept + slope/A \qquad (16)$$

When one introduces mean A values obtained from eq 14 for the P/P_o ranges of interest, eq 16 yields $C_{M,HH}$ values of 0.6, 0.8, 1.3, and 1.7 for 45 °C, 55 °C, 65 °C, and 80 °C, respectively. The agreement of these values with the NMR values in Table IV is better than that of the values based on eq 13. From this result and our previous arguments, we conclude that k_6' probably is indeed small and that eq 12 is at least a reasonable (and useful) approximation. However, minor amounts of intramolecular hydrogen abstraction by $P^{\bullet}$ were observed under some conditions (26).

By comparing eq 9 and 12, we learn that slopes of the plots of eq 12 can be used to calculate $C_{M,aux}$ values when k_6' is ignored. Unfortunately, such values cannot be compared directly with the internal double-bond contents that were reported by Hjertberg and Sörvik (2) for the resins they prepared. Those contents were determined by ozonolysis, a method that now is known to yield values that are much too low when they pertain to PVC molecules

Table V. Comparison of $C_{M,aux}$ Values

Polym Temp (°C)	P/P$_o$	By ^1H NMR[a]	From M_n Values[a]
80	1.00	0.6 ± 0.1	0.5 ± 0.1
80	0.59	1.55 ± 0.2	1.5 ± 0.2
55	0.92	0.2 ± 0.1	0.4 ± 0.1
55	0.77	0.5 ± 0.1	0.65 ± 0.1
55	0.59	1.15 ± 0.2	1.0 ± 0.2

NOTE: Values are $C_{M,aux} \times 10^3$.
[a] *See* text for discussion. Deviations are the estimated experimental errors.

with rather low molecular weights (27). On the other hand, $C_{M,aux}$ values found from eq 9 can be compared with those determined by NMR spectroscopy for the polymers made in our work under constant VC pressures. The latter values require discussion, however, as we now will show.

The NMR data in Table I that relate to our 80 °C and 55 °C polymers reveal that $[CH=CH]_{total}$ (per molecule) remained unchanged, whereas the (A1 and A2) content (per monomer unit) decreased with decreasing VC pressure. These observations can be reconciled in terms of a secondary free-radical process that converts a chloroallylic chain end into an internal alkene array (26, 28). The internal alkene array seems very likely to be a supplemental IA moiety that results from VC addition to a $-CH_2CH=CHC^{\cdot}HCl$ radical formed from A1/A2 by hydrogen abstraction (28).

In any case, whatever the exact nature of the secondary process may be, its occurrence obviously requires subtraction of the number of internal double bonds that it forms from the NMR internal double-bond content, in order to obtain the correct value of $C_{M,aux}$ when P/P_o is less than 1. In the case of the polymer made at 55 °C and with a P/P_o ratio of 0.59, for example, we note that the true value of $C_{M,HH}$ at 55 °C is about 0.95×10^{-3}. (This value, which is the one found by NMR analysis when $P/P_o = 1$, was obtained by averaging the values for 53 °C and 56 °C in Table I). The measured $C_{M,HH}$ value of our prototypical polymer is only 0.5×10^{-3}. Thus, 0.45×10^{-3}/(VC unit) of its allylic ends formed by transfer were converted later into internal double bonds. Subtraction of that value from the total internal double-bond content of 1.6×10^{-3}/(VC unit) yields a value of 1.15×10^{-3} for $C_{M,aux}$. This value and the others found from the NMR data for our polymers are listed in the third column of Table V. The fourth column of Table V contains the $C_{M,aux}$ values obtained from eq 9 when k_6' is neglected. The slopes of eq 12 used in eq 9 were taken from the plot for 80 °C in Figure 2 and the plot (not shown) of the data for our 55 °C polymers. Agreement between the two sets of $C_{M,aux}$ values is satisfactory and thus further confirms the validity of the mechanisms we have suggested for chain transfer to VC.

Implications for PVC Thermal Stability. Stability data obtained for several model substances (8, 9, 29) indicate that the EB and MB structures are unlikely to initiate the thermal degradation of PVC. The chloroallylic chain ends (A1 and A2) also are not so unstable that one needs to regard them as thermally labile sites (30, 31). Thus, the head-to-head emplacement of VC has little, if anything, to do with the thermal instability of the polymer. The auxiliary transfer process, in contrast, is important in this respect, because the IA group that it creates loses HCl quickly when heated (8, 9, 31).

Other PVC structural defects whose thermal dehydrochlorination is swift are those in which chlorine is bonded to a branch-point carbon atom (8, 9, 31). The most abundant of these defects (2, 6–8) is the dichlorobutyl branch structure (6) whose mechanism of formation is shown in Scheme III. Another group bearing tertiary chloride is the long-branch-point arrangement, 7, which is made in analogous fashion according to Scheme IV (6). Furthermore, we have found very recently that 1,3-diethyl- and/or 2-ethyl-n-hexyl branch structures are present in small amounts when the polymer is synthesized under low monomer pressures (26). Both of those arrangements should possess two tertiary chlorines (6), and both of them should be formed by mechanisms that require two intramolecular hydrogen abstractions by propagating macroradicals (6).

tert-Chloride and/or IA structures are believed by most workers to be the principal defects that contribute to the thermal instability of PVC (31). All of these groups result from ancillary processes that begin with inter- or intramolecular hydrogen abstraction by a growing macroradical. Thus we can now reasonably propose that these abstractions are the major source of the PVC stability problem. We recognize, of course, that degradation starts, to some extent, from the ordinary VC units as well (8, 9, 31).

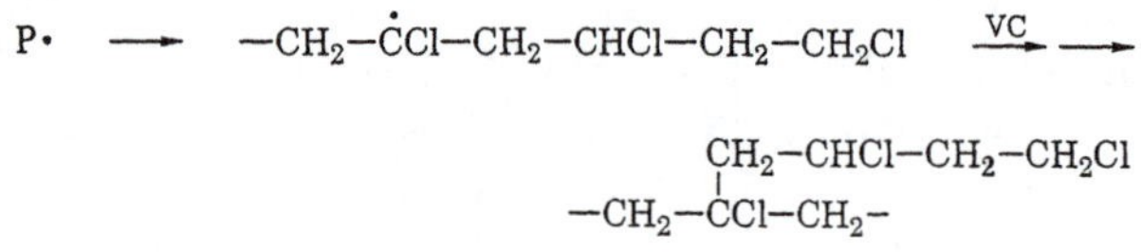

$$P\cdot \longrightarrow -CH_2-\overset{\bullet}{C}Cl-CH_2-CHCl-CH_2-CH_2Cl \overset{VC}{\longrightarrow} \longrightarrow$$

$$\begin{array}{c} CH_2-CHCl-CH_2-CH_2Cl \\ | \\ -CH_2-\overset{}{C}Cl-CH_2- \\ \mathbf{6} \end{array}$$

Scheme III.

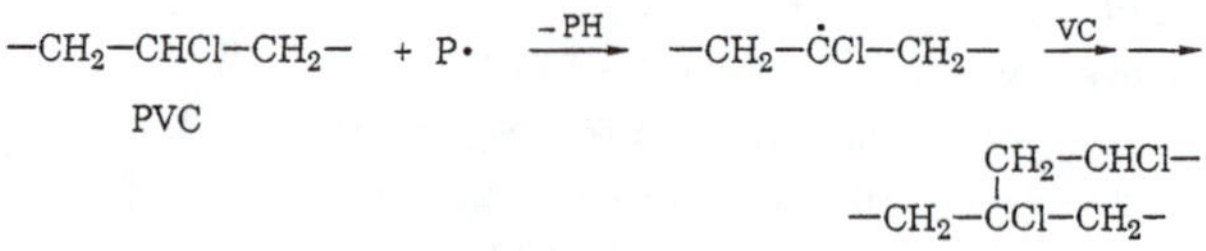

$$-CH_2-CHCl-CH_2- + P\cdot \overset{-PH}{\longrightarrow} -CH_2-\overset{\bullet}{C}Cl-CH_2- \overset{VC}{\longrightarrow}$$
$$\text{PVC}$$

$$\begin{array}{c} CH_2-CHCl- \\ | \\ -CH_2-\overset{}{C}Cl-CH_2- \\ \mathbf{7} \end{array}$$

Scheme IV.

All of the hydrogen abstractions compete directly with ordinary chain propagation. Hence, one way to reduce their importance is to keep the VC concentration at the highest possible level. Decreasing the temperature of polymerization also should help in this regard, because the activation energies of the abstractions probably are appreciably higher than that for propagation by addition to monomer (26). This supposition was, in fact, verified for the abstraction that leads to **6** (32).

Acknowledgments

The research performed at the College of William and Mary was supported by a grant from the National Science Foundation (DMR-8996253) and by the Geon Vinyl Division of the BFGoodrich Company.

References

1. Starnes, W. H., Jr.; Wojciechowski, B. J.; Velazquez, A.; Benedikt, G. M. *Macromolecules* **1992,** *25,* 3638; correction: *Macromolecules* **1992,** *25,* 7080.
2. Hjertberg, T.; Sörvik, E. M. In *Polymer Stabilization and Degradation;* Klemchuk, P. P., Ed.; ACS Symposium Series No. 280; American Chemical Society: Washington, DC, 1985; p 259.
3. Starnes, W. H., Jr.; Wojciechowski, B. J. *Makromol. Chem. Macromol. Symp.* **1993,** *70/71,* 1.
4. Langsam, M. In *Encyclopedia of PVC*; Nass, L. I.; Heiberger, C. A., Eds.; Dekker: New York, 1986; Vol. 1, p 47.
5. Xie, T. Y.; Hamielec, A. E.; Wood, P. E.; Woods, D. R. *Polymer* **1991,** *32,* 537.
6. Starnes, W. H., Jr.; Schilling, F. C.; Plitz, I. M.; Cais, R. E.; Freed, D. J.; Hartless, R. L.; Bovey, F. A. *Macromolecules* **1983,** *16,* 790.
7. Hjertberg, T.; Sörvik, E. M. *Polymer* **1983,** *24,* 673.
8. Hjertberg, T.; Sörvik, E. M. In *Degradation and Stabilisation of PVC*; Owen, E. D., Ed.; Elsevier Applied Science: New York, 1984; p 21, and references cited therein.
9. Iván, B.; Kelen, T.; Tüdős, F. In *Degradation and Stabilization of Polymers*; Jellinek, H. H. G.; Kachi, H., Eds.; Elsevier: New York, 1989; Vol. 2, p 483, and references cited therein.
10. Hjertberg, T. *J. Appl. Polym. Sci.* **1988,** *36,* 129.
11. Skillicorn, D. E.; Perkins, G. G. A.; Slark, A.; Dawkins, J. V. *J. Vinyl Technol.* **1993,** *15,* 105.
12. Hjertberg, T.; Sörvik, E. M. *J. Macromol. Sci. Chem.* **1982,** *17,* 983.
13. Guyot, A.; Bert, M.; Burille, P.; Llauro, M.-F.; Michel, A. *Pure Appl. Chem.* **1981,** *53,* 401.
14. Boissel, J. *J. Appl. Polym. Sci.* **1977,** *21,* 855.
15. Hildenbrand, P.; Ahrens, W.; Brandstetter, F.; Simak, P. *J. Macromol. Sci. Chem.* **1982,** *17,* 1093.
16. Park, G. S.; Saremi, A. H. *Eur. Polym. J.* **1989,** *25,* 665.
17. Hjertberg, T.; Sörvik, E. *J. Vinyl Technol.* **1985,** *7,* 53.
18. Maddams, W. F. *J. Vinyl Technol.* **1985,** *7,* 65.
19. Michel, A. *J. Vinyl Technol.* **1985,** *7,* 46.

20. Darricades-Llauro, M. F.; Michel, A.; Guyot, A.; Waton, H.; Pètiaud, R.; Pham, Q. T. *J. Macromol. Sci. Chem.* **1986,** *23,* 221.
21. Llauro-Darricades, M.-F.; Bensemra, N.; Guyot, A.; Pètiaud, R. *Makromol. Chem. Macromol. Symp.* **1989,** *29,* 171.
22. Hendry, D. G.; Mill, T.; Piskiewicz, L.; Howard, J. A.; Eigenmann, H. K. *J. Phys. Chem. Ref. Data* **1974,** *3,* 937.
23. Starnes, W. H., Jr.; Edelson, D. *Macromolecules* **1979,** *12,* 797.
24. Carenza, M.; Palma, G.; Tavan, M. *J. Polym. Sci. Polym. Symp.* **1973,** *42,* 1031.
25. Nilsson, H.; Silvegren, C.; Törnell, B. *Eur. Polym. J.* **1978,** *14,* 737.
26. Starnes, W. H., Jr.; Chung, H.; Wojciechowski, B. J.; Skillicorn, D. E.; Benedikt, G. M. *Polym. Prepr. (Am. Chem. Soc. Div. Polym. Chem.)* **1993,** *34 (2),* 114.
27. Rogestedt, M.; Hjertberg, T. *Macromolecules* **1993,** *26,* 60.
28. Starnes, W. H., Jr.; Chung, H.; Pike, R. D.; Wojciechowski, B. J.; Zaikov, V. G.; Benedikt, G. M.; Goodall, B. L.; Rhodes, L. F. *Polym. Prepr. (Am. Chem. Soc. Div. Polym. Chem.)* **1995,** *36(2),* 404.
29. Starnes, W. H., Jr.; Plitz, I. M. *Macromolecules* **1976,** *9,* 633; references cited therein.
30. van den Heuvel, C. J. M.; Weber, A. J. M. *Makromol. Chem.* **1983,** *184,* 2261.
31. Starnes, W. H., Jr.; Girois, S. *Polym. Yearbook* **1995,** *12,* 105.
32. Starnes, W. H., Jr.; Wojciechowski, B. J.; Chung, H.; Benedikt, G. M.; Park, G. S.; Saremi, A. H. *Macromolecules* **1995,** *28,* 945.

RECEIVED for review December 6, 1993. ACCEPTED revised manuscript January 18, 1995.

Thermal Stability, Degradation, and Stabilization Mechanisms of Poly(vinyl chloride)

Béla Iván[1]

Central Research Institute for Chemistry of the Hungarian Academy of Sciences, H-1525 Budapest, Pusztaszeri u. 59-67, Hungary

This survey concerns the major features of stability, thermal degradation, and stabilization of poly(vinyl chloride) (PVC). The effect of microstructure and chain defects of PVC on the stability of the resin is analyzed and reviewed in the light of recent findings. New experimental results of thermal degradation of PVC in dilute solution in the presence of stabilizers are also summarized. These results indicate that the most important role of PVC stabilizers is not preventing initiation by labile chlorine substitution but blocking the fast zip-elimination of HCl during degradation. After stabilizer consumption, the blocked structures undergo reinitiation (reversible blocking mechanism). The possible reaction of blocking is the attachment of ester or thioglycolate groups to the propagating polyenes. Reinitiation is explained by the rapid acid-catalyzed thermolysis of these structures by the appearance of free HCl after consumption of stabilizers. Comparison of experimental findings with degradation kinetics expected on the basis of the Frye–Horst, the Minsker, and the Michell stabilization mechanisms indicates that none of these mechanisms is able to explain stabilization of PVC.

POLY(VINYL CHLORIDE) (PVC) is still one of the major commodity polymers produced and consumed worldwide, despite some recent concerns. Because of the attractive economy of production and processing and the ease of property variations from elastomeric to hard final products, this polymer is expected to be one of the significant members of the commercial plastics arena

[1]Current address: University of Mainz, Institute of Physical Chemistry, Welderweg 11-15, D-55099 Mainz, Germany

0065–2393/96/0249–0019$12.00/0

for the foreseeable future. Despite its advantageous properties, PVC exhibits relatively low thermal, thermooxidative, and light stability (*see* reference 1 for a comprehensive review). Therefore, this resin cannot be used without efficient stabilization; processing and use also require the suppressing of degradation. As a consequence, the stability, degradation, and stabilization of this resin have been the subject of research and development projects since its appearance on the market more than half a century ago.

Even though the identification and use of suitable stabilizers has occurred, the fine details of the mechanism and kinetics of PVC degradation and stabilization have remained largely unknown. In the last 10 years, increased interest and research efforts have revealed several new aspects of these processes. The focus of these new efforts include studying the microstructure of PVC with the newest powerful analytical techniques (*1–6*), the relation between chain defect structures and stability (*1, 2, 7–15*), the major characteristics of the kinetics of thermal dehydrochlorination of PVC (*1, 8, 15–23*), degradation of dilute PVC solutions in the presence of stabilizers (*1, 19, 24–29*), and the formulation of the reversible blocking mechanism (*15, 24–27*) of PVC stabilization.

This chapter surveys thermal stability, degradation, and stabilization of PVC in light of these recent results. An overview of the Frye–Horst (*30, 31*), Minsker (*32*), and Michell (*33, 34*) stabilization mechanisms and the reversible blocking (*15, 24–27*) mechanism will also be included.

Microstructure and Thermal Stability of PVC

Irrespective of the type of degradation (thermal, thermooxidative, or light), the major chain degradation process of PVC is zip-elimination of HCl and the simultaneous formation of conjugated double-bond-containing sequences (polyenes) in the polymer chain (Scheme I). The main dehydrochlorination

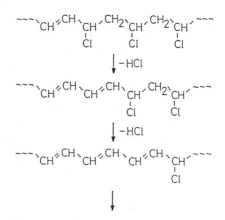

Scheme I. Zip-elimination of HCl from PVC and simultaneous formation of sequences containing conjugated double bonds (polyenes) in the chain during degradation.

process of PVC degradation yields a corrosive gas, HCl, and leads to highly reactive polyenes absorbing light in the UV and visible region. This absorption results in undesired coloration from yellow to nearly black in some cases. The oxygen sensitivity of polyenes and their secondary processes such as cross-linking (35) result in further undesired deterioration and changes in chemical resistance and physical properties of the resin.

Because heat is one of the major causes of degradation during processing, thermal degradation and its prevention have attracted the most attention. The primary dehydrochlorination process during thermal degradation of this polymer involves three steps (1):

- initiation of HCl loss
- rapid zip-elimination of HCl and simultaneous formation of polyenes in the PVC chain
- termination of the zipping process

On the basis of experiments with low-molecular weight model compounds, allylic and tertiary chlorine containing defects in the polymer chain were thought to be exclusively responsible for the low thermal stability of PVC (1). However, kinetic studies (36) and some mechanistic considerations (2) indicated random initiation by regular $-CH_2CHCl-$ repeat units. Thermal degradation experiments (8) with PVC enriched with allyl chlorine (37) proved that random initiation of HCl loss also occurs during thermal degradation of PVC. The rate constant of random initiation was four orders of magnitude lower than that of initiation of degradation by defect structures such as allylic or tertiary chlorines during degradation of dilute PVC solutions at 200 °C (8). One of the most interesting aspects of these results is that the difference between the concentrations of labile and regular secondary chlorines in PVC yields similar dehydrochlorination rates either by random or labile site initiations. In a practical sense, the low thermal stability of PVC can be viewed as its inherent property because of the existence of random initiation. Therefore, PVCs having high thermal stability probably cannot be prepared by free-radical polymerization of vinyl chloride (1, 8, 15).

Degradation studies were performed with PVCs containing increased concentrations of tertiary (7, 11, 12) or allylic (8–12) chlorines and with polymers having at least one of the labile structures: polychloroprene (38–40), chlorinated butyl rubber (39), chlorinated ethylene–propylene copolymers (39), and tertiary chlorine-terminated polyisobutylenes (41). These studies have undoubtedly proved the importance of chain defects of PVC containing allylic and tertiary chlorine and their effect on thermal dehydrochlorination. Substitution of labile chlorines with alkyl groups by treatment with alkylaluminum compounds such as trimethylaluminum (39, 42–44) and dimethylcyclopentadienyl aluminum (7, 38, 40, 43–46) yielded significant decreases in the rate of HCl loss in every case. These results provided direct additional evidence for the thermal instability of moieties containing allylic and tertiary chlorine.

Detailed microstructure investigations (1, 4, 12) revealed that allylic chlorines can be present at chain ends (**I** in Chart I) and in the main chain (**II** in Chart I). Recent studies (4, 47) indicated that the majority of allylic chlorines are located at chain ends. However, on the basis of ¹H NMR studies with low-molecular weight PVC fractions, chain end **I** in Chart I is thermally stable (48). Tertiary chlorines can be present at branching points for short- and long-chain branches in PVC. Ethyl branches, considered previously as one of the major sources of tertiary chlorines, have tertiary hydrogens at the branch points (6, 47). Thus, tertiary chlorines are mainly located at the starting points of *n*-butyl (**III**, Chart I) and long-chain branches (**IV**, Chart I) (1).

Minsker and co-workers (13, 14) claimed that keto-allyl groups (**V** in Chart I) are the most important labile structures in PVC, and these chain defects are solely responsible for the low thermal stability of the resin. However, conclusive experimental evidence for the presence of keto-allyl groups in PVC has not been reported to date. Data obtained by studying the thermal decomposition of model compounds containing keto-allyl groups and their derivatives (49) indicated that only the cis isomer of **V** in Chart I leads to rapid dehydrochlorination, whereas the trans isomer is more stable than the secondary chlorine-containing regular monomer units of the polymer.

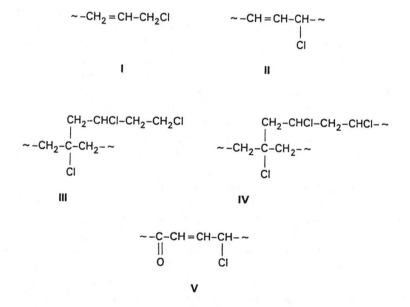

Chart I.

The source of random initiation of HCl loss has not been satisfactorily revealed. On the basis of model experiments with secondary chloroalkanes, one might conclude that spontaneous unimolecular elimination of HCl from regular –CH$_2$–CHCl– repeat units by heat occurs. This process yields an allylic chlorine moiety and leads to subsequent fast zip-elimination of HCl. Formation of cyclic chloronium ions by two neighboring monomer units resulting in initiation of HCl loss was also proposed for random initiation (2).

The effects of tacticity on the thermal stability of PVC were studied intensively by Millán and co-workers (50, 51). The rate of HCl loss increased with increasing concentrations of GTTG$^-$ isotactic triads (50, 51). The length of polyenes in thermally degraded PVCs also depends on the tacticity. On the basis of these findings the stereostructure of the polymer can be considered as one of the potential sources of random initiation. Although a double logarithmic plot exhibited correlation between the rate of dehydrochlorination and the concentration of isotactic triads in PVC (12), several effects that occur simultaneously, such as initiation by chain defects, effects induced by the morphology of the resins and HCl-catalysis, should be separated to draw reliable conclusions on the role of tacticity in the thermal stability of PVC.

Stabilization Mechanisms

The Reversible Blocking Mechanism. Primary heat stabilizers of PVC, such as metal carboxylates and organotin compounds, usually possess two major roles: scavenging HCl and reacting with the polymer to prevent HCl loss and simultaneous discoloration (1). The major problem of PVC stabilization mechanisms is determining how the stabilizers influence primary degradation processes such as initiation of HCl loss, fast zip-elimination of HCl, and termination of unzipping. However, most of the stabilization studies with PVC have been carried out in bulk (with polymer melts, films, and powders), and the reaction conditions were far from ideal for investigating the kinetic and mechanistic details of the stabilization process. The effects of sample preparation, sample inhomogeneity, stabilizer distribution in the matrix, morphological differences and changes, diffusion problems, HCl-catalysis, and the influence of oxygen in air cannot be excluded when using solid samples.

To overcome the problems connected with solid PVC samples, systematic thermal (1, 15, 24, 25) and thermooxidative (1, 15, 25, 27) experiments with dilute PVC solutions in the presence of a variety of stabilizers were performed in our laboratories during the last 10 years. PVC solutions (1%) in 1,2,4-trichlorobenzene were degraded at 200 °C. The kinetics of free HCl evolution, formation of double bonds in the main chain, the average length of polyenes, and the initiation of zip-elimination were investigated in the absence and presence of industrial stabilizers such as lead stearate, barium stearate, cadmium stearate, calcium stearate, zinc stearate, dibutyltin distearate, dibutyltin maleate, and dioctyltin bisthioglycolate.

Our stabilization studies with dilute PVC solutions have allowed us to gain new information on the kinetics and mechanism of PVC stabilization. The following major experimental findings were obtained (*see* Figures 1–4 for the characteristics of these results):

1. All the stabilizers yielded induction periods for the evolution of free HCl.

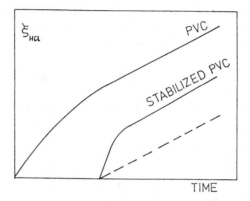

Figure 1. Extent of free HCl evolution as a function of time obtained in thermal degradation experiments for unstabilized PVC and stabilized PVC (solid lines) and constructed by Frye–Horst, Minsker, and Michell mechanisms (dashed line) for stabilized PVC. Axes are in arbitrary units.

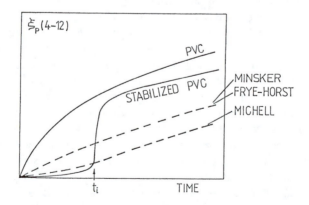

Figure 2. Number of double bonds in polyenes with length 4–12 [$\xi_p(4-12)$] as a function of time in experiments (solid lines) and constructed on the basis of different stabilization mechanisms (dashed lines). Axes are in arbitrary units.

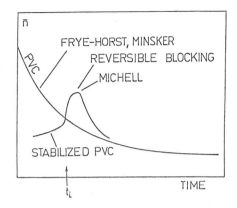

Figure 3. Average length of polyenes as a function of time obtained by experiments and constructed by the different stabilization mechanisms. Axes are in arbitrary units.

2. The initial rates of free HCl evolution at the end of the induction periods were significantly higher in the presence of stabilizers.
3. Detectable amounts of short polyenes were formed during the induction periods.
4. The formation of double bonds and longer polyenes increased sharply at the end of the induction periods.
5. The average length of polyenes as a function of time exhibited a maximum in the vicinity of the end of the induction periods.
6. In some cases the concentration of labile structures in PVC estimated by kinetic analysis was higher at the end of the induction period than that in virgin PVC.
7. Initiation of HCl loss was not prevented during the induction periods, and the rate of initiation did not change at the end of the induction periods.

This unexpected and strange set of observations—the rapid increase in the rates of free HCl evolution and in the corresponding double bonds (polyenes) but no change in the rate of formation of internal segments containing double bonds (new initiating sites) at the end of the induction periods—cannot be explained by the widely cited Frye–Horst (*30, 31*), Minsker (*32*), or Michell (*33, 34*) mechanisms. The unchanged rate of initiation at the end of the induction period indicates that the sudden increase in the rate of HCl loss and double bond formation is due to reinitiation at already existing polyenes formed during the induction periods. During the induction periods, initiation of HCl loss is not inhibited by the stabilizers. However, shortly after initiation the propagating polyenes formed by unzipping of HCl are blocked by the

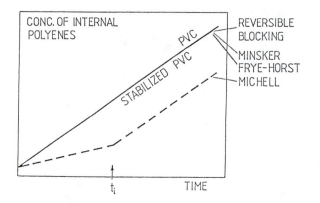

Figure 4. Concentration of internal polyenes (initiating sites) as a function of time obtained by experiments and constructed by the different stabilization mechanisms. Axes are in arbitrary units.

stabilizers. Therefore, only short sequences containing conjugated double bonds are formed until stabilizer is present in the degrading systems. This mechanism, the *reversible blocking mechanism*, together with the primary process of thermal degradation of PVC is exhibited in Scheme II. For comparison, the Frye–Horst and Minsker mechanisms (*see* next section) of replacement by stabilizers of labile chlorines originally present in the resin are also shown in Scheme II.

The most plausible explanation for the blocking reaction is the attachment of ester (or thioglycolate) groups to the propagating polyenes. Indeed, some literature data indicate that the organic groups of stabilizers incorporate into the PVC chain in the course of degradation in dilute solution (*28, 29, 52*) and in bulk samples (*30, 31*). The rapid degradation (dehydrochlorination and simultaneous double bond formation in the PVC chain) obtained at the end of the induction periods is due to the rapid acid-catalyzed thermolysis (E_i elimination) of the allylic ester during the appearance of free HCl (Scheme III). This reaction yields a polyene terminated with an allylic chlorine; the thermal stability of this polyene is significantly lower than that of the corresponding ester in the absence of an acid. Interestingly, the acid-catalyzed thermolysis of labile (tertiary, allylic, and benzylic) esters and carbonates was used to develop a large variety of photoresists in which the proton (acid) was generated by photochemical means (*53–55*).

Critical Evaluation of PVC-Stabilization Mechanisms. The widely accepted (*56, 57*) Frye–Horst hypothesis (*30, 31*) assumes that primary PVC stabilizers (metal carboxylates and organotin compounds) act simply by substituting labile (tertiary and allylic) chlorines (**I–IV** in Chart I) in the resin

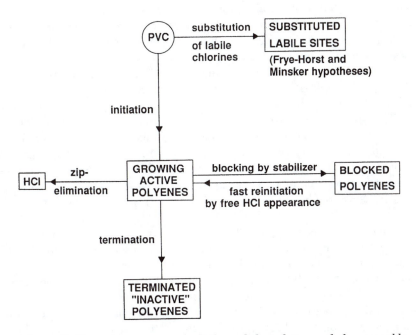

Scheme II. The primary processes of thermal degradation and the reversible blocking mechanism of PVC stabilization (Frye–Horst and Minsker mechanisms are also shown).

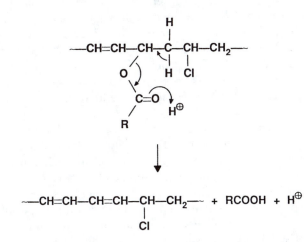

Scheme III. Acid catalyzed E_i elimination of allylic ester from PVC chain after appearance of free HCl.

for more stable groups (ester or thioglycolate). The Minsker mechanism is analogous to the Frye–Horst proposition: Minsker and co-workers (32) claimed that the sole function of stabilizers is an exchange reaction between the ketoallyl groups (V in Chart I) and the ester or thioglycolate groups of stabilizers. Thus, suppressing or preventing degradation by stabilizers is due to the fast substitution of labile chlorines in PVC. This process subsequently inhibits initiation of HCl loss during degradation according to the Frye–Horst and Minsker mechanisms.

The Michell mechanism (33, 34) assumes radical chain dehydrochlorination of PVC in which chlorine radicals are the chain carriers. In the presence of stabilizers, the chlorine atoms are thought to react with the stabilizer and to yield metal chlorides and stearate free radicals. This reaction terminates the chain propagation of HCl loss by trapping chlorine atoms and by esterifying PVC macroradicals that have unpaired electrons with stearate free radicals.

Experimental results obtained in the course of our systematic studies concerning degradation of dilute PVC solutions in the presence of stabilizers offer a unique opportunity for a critical evaluation of these different stabilization mechanisms. On the basis of these stabilization mechanisms, kinetic curves (i.e., the expected shape of free HCl evolution), concentration of double bonds, average polyene length, and number of internal polyene sequences (initiating sites of degradation) as a function of time can be constructed and compared to experimental findings. Although random initiation was not known at the time the Frye–Horst mechanism (30, 31) was proposed, this process is also considered in constructing the kinetic curves. Figures 1–4 summarize the results of comparing the expected kinetics of the various stabilization mechanisms with experimental findings obtained during degradation of dilute PVC solutions in the presence of stabilizers (1, 15, 19, 24–27). In these curves, experimental findings are drawn by solid curves, whereas the kinetic curves expected on the basis of the Frye–Horst, Minsker, and Michell mechanisms are displayed by dotted lines. When the experimental results and the predictions according to one or other mechanisms are in agreement, solid lines are used.

As shown in Figure 1, the Frye–Horst, Minsker, and Michell mechanisms do not lead to increased rates of free HCl evolution at the end of the induction periods. According to all of these mechanisms, dehydrochlorination should occur by a constant rate (by random initiation on the basis of the Frye–Horst and Minsker mechanisms) after consumption of stabilizers.

The concentration of double bonds in polyenes with conjugation length 4–12 [$\xi_p(4$–$12)$] versus time plots are exhibited in Figure 2. Because the exclusive role of stabilizers is substitution of labile chlorines originally present in PVC by more stable groups according to the Frye–Horst (30, 31) and Minsker (32) mechanisms, the concentration of double bonds as a function of time should increase monotonously because of random initiation. A rate increase and thus a break point is expected by the Michell mechanism at the end of the induction period in the $\xi_p(4–12)$ versus time plot. This increase is due to the raise of the kinetic chain length of dehydrochlorination in the radical mechanism proposed by Michell (*33, 34*) after the consumption of stabilizers. As this analysis indicates, neither of these mechanisms can provide an explanation for the sharp increase of $\xi_p(4–12)$ at the end of the induction periods.

According to the trends in Figure 3, only the Michell mechanism would yield maximum curve for the average length of polyenes (*n*) as a function of time. However, because of the absence of inhibition of random initiation by stabilizers, the shape of the *n* versus time plots should be similar to that of the unstabilized PVC on the basis of the Frye–Horst and Minsker mechanisms.

As discussed in the previous section, estimation of the concentration of labile sites leading to fast HCl loss at the end of the induction periods (t_i) gave higher concentrations in some cases than the labile chlorine concentration in the virgin PVC. However, the concentrations of any labile sites at t_i should not be higher than that in the starting material according to the Frye–Horst, Minsker, and Michell mechanisms.

The concentration of internal polyenes (initiating sites) as a function of time is shown in Figure 4. By considering random initiation, the Frye–Horst and Minsker mechanisms give the same plot as observed for unstabilized and stabilized PVC. However, if the concentration of tertiary chlorine in PVC is not negligibly low (which is the case in most PVCs), then rates of polyene formation in stabilized PVC are expected to be lower than the rates of formation in the unstabilized resin during the induction period according to the Frye–Horst mechanism. This difference is due to substitution of these labile chlorines. According to the Michell mechanism, the initiation should be lower for the stabilized PVC during the induction period. However, this mechanism would result in an increase of initiation rate after the end of the induction period, and this result is contrary to experimental findings.

Concluding Remarks

According to our analysis, the Frye–Horst, Minsker, and Michell mechanisms do not fully explain the currently available experimental results concerning the stabilization of PVC. The grounds of the Michell mechanism—a radical degradation and corresponding stabilization—are entirely questionable. One shortcomings of this mechanism is that it is still not completely clear whether thermal degradation of PVC occurs by molecular, ionic, or radical mechanisms (*1*). On the other hand, most of the primary PVC stabilizers have ionic character, and the exclusive formation of carboxylate radicals in a simple exchange reaction between chlorine atoms and stabilizers composed of ionic bonds is hardly believable. Chlorine radicals are highly reactive species possessing low selectivity in their reactions.

As the experimental data of our systematic studies indicate, the main role of stabilizers is not the substitution of labile chlorines present in PVCs (as claimed by the Frye–Horst and Minsker mechanisms). Instead, the main role is the blocking of the propagating ("active") polyenes, as concluded by the reversible blocking mechanism (*see* Schemes II and III). In a recent paper, Starnes (58) argued that the reversible blocking mechanism should be viewed as the Frye–Horst hypothesis. He claimed that the reversible blocking mechanism should also result in replacement of allylic chlorines adjacent to polyenes (58). However, this replacement would occur only if degradation of PVC proceeded exclusively by a step-by-step molecular mechanism (59).

As analyzed in a recent review (1), experimental findings exist to support either a radical or ionic mechanism of thermal dehydrochlorination of PVC. Allylic chlorines adjacent to polyenes would not exist in the extremely fast zip-elimination process by either of these mechanisms. As shown earlier by thermal degradation of PVCs containing increased concentrations of allylic chlorines adjacent to short polyenes, the activity of these labile chlorines is significantly lower than that of the unzipping (propagating) active polyenic structures (8). This difference also indicates that allylic chlorines most likely do not exist in the zip-elimination process at all. Therefore, the blocking reaction by thermal stabilizers of PVC cannot be viewed as a simplified displacement of labile allylic chlorines adjacent to polyenes in degrading PVC, as claimed by the Frye–Horst mechanism (59).

As exhibited in Scheme II, the reversible blocking mechanism (15, 24–27) also leads to a significant shift in our views concerning stabilization of PVC. Therefore, the focus of PVC stabilization is changed from preventing initiation to the second major step of the primary degradation of PVC, that is, to the unzipping of HCl loss and thus to the fast interaction between the propagating polyenes and stabilizers according to the reversible blocking mechanism.

Acknowledgments

Support by the Hungarian Scientific Fund (OTKA 4453) is gratefully acknowledged.

References

1. Iván, B.; Kelen, T.; Tüdös, F. In *Degradation and Stabilization of Polymers*; Jellinek, H. H. G.; Kachi, H., Eds.; Elsevier Science: Amsterdam, Netherlands, 1989; Vol. 2, pp 483–714.
2. Starnes, W. H., Jr. *Dev. Polym. Degrad.* **1981**, *3*, 35.
3. Hjertberg, T.; Wendel, A. *Polymer* **1982**, *23*, 1641.
4. Llauro-Darricades, M. F.; Michel, A.; Guyot, A.; Waton, A.; Petiaud, R.; Pham, Q. T. *J. Macromol. Sci. Chem.* **1986**, *A23*, 221.
5. Benedikt, G. M. *J. Macromol. Sci. Pure Appl. Chem.* **1992**, *129*, 85.

6. Starnes, W. H., Jr.; Wojciechowski, B. J. *Makromol. Chem. Macromol. Symp.* **1993**, *70/71*, 1.
7. Iván, B.; Kennedy, J. P.; Kende, I.; Kelen, T.; Tüdös, F. *J. Macromol. Sci. Chem.* **1981**, *A16*, 1473.
8. Iván, B.; Kennedy, J. P.; Kelen, T.; Tüdös, F.; Nagy, T. T.; Turcsányi, B. *J. Polym. Sci. Part A: Polym. Chem.* **1983**, *21*, 2177.
9. Braun, D.; Böhringer, B.; Iván, B.; Kelen, T.; Tüdös, F. *Eur. Polym. J.* **1986**, *22*, 1.
10. Braun, D.; Böhringer, B.; Iván, B.; Kelen, T.; Tüdös, F. *Eur. Polym. J.* **1986**, *22*, 299.
11. Hjertberg, T.; Sörvik, E. M. *Polymer* **1983**, *24*, 685.
12. Rogestedt, M.; Hjertberg, T. *Macromolecules* **1993**, *60*, 26.
13. Minsker, K. S.; Berlin, A. A.; Lisitskii, V. V.; Kolesov, S. V. *Vysokomol. Soedin. Ser. A* **1977**, *19*, 32.
14. Minsker, K. S.; Kolesov, S. V.; Yanborisov, V. M.; Berlin, A. A.; Zaikov, G. E. *Polym. Degrad. Stab.* **1986**, *16*, 99.
15. Iván, B.; Kelen, T.; Tüdös, F. *Makromol. Chem., Macromol. Symp.* **1989**, *29*, 59.
16. Simon, P. *Polym. Degrad. Stab.* **1992**, *35*, 45.
17. Simon, P. *Polym. Degrad. Stab.* **1992**, *35*, 157.
18. Simon, P.; Valko, L. *Polym. Degrad. Stab.* **1992**, *35*, 249.
19. Iván, B.; Turcsányi, B.; Kelen, T.; Tüdös, F. *Izv. Chim.* **1985**, *18*, 287.
20. Danforth, J. D. *J. Am. Chem. Soc.* **1971**, *93*, 6843.
21. Danforth, J. D.; Takeuchi, T. *J. Polym. Sci. Part A: Poly. Chem.* **1973**, *11*, 2091.
22. Danforth, J. D.; Spiegel, J.; Bloom, J. *J. Macromol. Sci. Chem.* **1982**, *A17*, 1107.
23. Danforth, J. D.; Indiveri, J. *J. Phys. Chem.* **1983**, *87*, 5376.
24. Iván, B.; Kelen, T.; Tüdös, F. *Polym. Mater. Sci. Eng.* **1988**, *58*, 548.
25. Iván, B.; Kelen, T.; Tüdös, F. *Polym. Prepr.* **1988**, *29(1)*, 132.
26. Iván, B.; Turcsányi, B.; Kelen, T.; Tüdös, F. *J. Vinyl Technol.* **1990**, *12*, 126.
27. Iván, B.; Turcsányi, B.; Kelen, T.; Tüdös, F. *Angew. Makromol. Chem.* **1991**, *189*, 35.
28. Ayrey, G.; Hsu, S. Y.; Poller, R. C. *J. Polym. Sci. Part A: Polym. Chem.* **1984**, *22*, 2871.
29. Bartholin, M.; Bensemra, M.; Hoang, V. T.; Guyot, A. *Polym. Bull.* **1990**, *23*, 425.
30. Frye, A. F.; Horst, R. W. *J. Polym. Sci.* **1959**, *40*, 419.
31. Frye, A. F.; Horst, R. W. *J. Polym. Sci.* **1960**, *45*, 1.
32. Kolesov, S. V.; Berlin, A. A.; Minsker, K. S. *Vysokomol. Soedin., Ser. A* **1977**, *A19*, 381.
33. Michell, E. W. J. *Br. Polym. J.* **1972**, *4*, 343.
34. Michell, E. W. J. *J. Mater. Sci.* **1985**, *20*, 3816.
35. Kelen, T.; Iván, B.; Nagy, T. T.; Turcsányi, B.; Tüdös, F.; Kennedy, J. P. *Polym. Bull.* **1978**, *1*, 79.
36. Tüdös, F.; Kelen, T. *Macromol. Chem.* **1973**, *8*, 393.
37. Iván, B.; Kennedy, J. P.; Kelen, T.; Tüdös, F. *J. Polym. Sci., Part A: Polym. Chem.* **1981**, *19*, 679.
38. Iván, B.; Kennedy, J. P.; Planthottam, S. S.; Kelen, T.; Tüdös, F. *J. Polym. Sci., Part A: Polym. Chem.* **1980**, *18*, 1685.
39. Iván, B.; Kennedy, J. P.; Kelen, T.; Tüdös, F. *Polym. Bull.* **1980**, *2*, 461.
40. Iván, B.; Kennedy, J. P.; Kelen, T.; Tüdös, F. *Polym. Bull.* **1980**, *3*, 45.
41. Iván, B.; Kennedy, J. P.; Kelen, T.; Tüdös, F. *J. Macromol. Sci. Chem.* **1981**, *A16*, 533.
42. Rogestedt, M.; Hjertberg, T. *Macromolecules* **1992**, *23*, 6332.
43. Iván, B.; Kennedy, J. P.; Kelen, T.; Tüdös, F. *J. Polym. Sci. Part A: Polym. Chem.* **1981**, *19*, 9.

44. Tüdös, F.; Iván, B.; Kelen, T.; Kennedy, J. P. *Dev. Polym. Degrad.* **1985**, *6*, 147.
45. Iván, B.; Kennedy, J. P. ; Kelen, T.; Tüdös, F. *Polym. Bull.* **1981**, *6*, 147.
46. Iván, B.; Kennedy, J. P. ; Kelen, T.; Tüdös, F. *J. Macromol. Sci. Chem.* **1982**, *A17*, 1033.
47. Starnes, W. H., Jr.; Chung, H.; Wojciechowski, B. J. *Polym. Prepr.* **1993**, *34(2)*, 114.
48. Van den Heuvel, C. J. M.; Weber, A. J. M. *Makromol. Chem.* **1983**, *184*, 2261.
49. Panek, M. G.; Villacorta, G. M.; Starnes, W. H., Jr.; Plitz, I. M. *Macromolecules* **1985**, *18*, 1040.
50. Millán, J.; Martínez, G.; Jimeno, M. L.; Tiemblo, P.; Mijangos, C.; Gómez-Elvira, J. M. *Makromol. Chem. Macromol. Symp.* **1991**, *48/49*, 403.
51. Martínez, G.; Gómez-Elvira, J. M.; Millán, J. *Polym. Degrad. Stab.* **1993**, *40*, 1.
52. Starnes, W. H., Jr.; Plitz, I. M. *Macromolecules* **1976**, *9*, 633.
53. Wilson, C. G.; Ito, H.; Fréchet, J. M. J.; Tessier, T. G.; Houlihan, F. M. *J. Electrochem. Soc.* **1986**, *133*, 181.
54. Fréchet, J. M. J.; Bouchard, F.; Eichler, E.; Houlihan, F. M.; Iizawa, T.; Kryczka, B.; Wilson, C. G. *Polym. J.* **1987**, *19*, 31.
55. Shirai, M.; Kinoshita, H.; Tsunooka, M. *Eur. Polym. J.* **1992**, *28*, 379.
56. Andreas, H. In *Plastics Additives Handbook;* Gachter, R.; Müller, H., Eds.; Hanser: Munich, Germany, 1985; p 193.
57. Jennings, T. C.; Fletcher, C. W. In *Encyclopedia of PVC;* Nass, L. I.; Heiberger, C. A., Eds.; Dekker: New York, 1988; Vol. 2, 2nd ed., p 45.
58. Starnes, W. H., Jr. *Polym. Prepr.* **1994**, *35(1)*, 425.
59. Iván, B., submitted for publication.

RECEIVED for review December 6, 1993. ACCEPTED revised manuscript November 29, 1994.

Thermolysis of Poly(chloroethyl methacrylates) and Poly(chloroethyl acrylates)

Wolfram Schnabel

Hahn-Meitner-Institut Berlin GmbH, Bereich C, Glienicker Str. 100, D-14109 Berlin, Germany

Poly(ethyl acrylates) and poly(ethyl methacrylates), chlorinated in the side groups, release chlorine and undergo intermolecular cross-linking if heated to temperatures up to 500 K. These processes are most significant in the cases of the dichloroethyl esters. At T > 500 K both acrylate and methacrylate polymers decompose into low molecular fragments. In methacrylate polymers this process predominantly consists of depolymerization (formation of monomer). As far as the onset temperature for mass loss is concerned, the chlorinated polymers definitely exhibit a lower stability than the nonchlorinated ones. Oxygen enhances cross-linking of the mono- and dichloroethyl methacrylate polymers but retards or prevents cross-linking in all other cases. The simultaneous action of both heat and UV light reduces the thermal stability. This effect is quite significant in the cases of the methacrylate polymers that are prone to unzipping.

POLY(CHLOROETHYL METHACRYLATES) readily undergo intermolecular cross-linking during heating to temperatures exceeding about 430 K [*see* Chart I for structures for poly(mono-, di-, and trichloroethyl methacrylates) (PMCMA, PDCMA, and PTCMA, respectively)]. For this reason, these polymers became interesting resist materials for applications in X-ray and electron-beam lithography. Actually, high-energy radiation induces main-chain degradation of these polymers and thus improves their solubility in appropriate fluid developers; the polymers act as positive tone resists. Thermal cross-linking before exposure to high-energy radiation significantly improves the development of fine line structures of high contrast. Because little was known of the mechanism

0065–2393/96/0249–0033$12.00/0

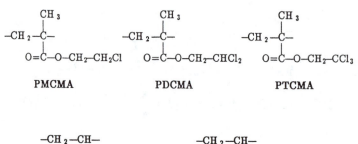

Chart I.

of the chemical reactions induced thermally in this class of polymers, systematic studies have been performed in our laboratory. These studies were extended to the thermolysis of corresponding acrylate polymers [poly(di- and trichloroethyl acrylate) (PDCEA and PTCEA, respectively); *see* Chart I].

In this chapter results obtained with these polymers will be reviewed (*1–6*). From earlier work (*7, 8*) it is well known that different mechanisms become operative in the thermal degradation of nonchlorinated methyl and ethyl esters of polyacrylic and polymethacrylic acid; the most significant difference is that the polymethacrylates, in contrast to the polyacrylates, readily undergo depolymerization. This difference was explained in terms of inter- and intramolecular hydrogen abstraction reactions that undergo terminal carbon-centered radicals of polyacrylate radicals (*8–11*).

In the case of polymethacrylates, hydrogen abstraction reactions are much less likely, mainly for steric reasons. Therefore, depolymerization (unzipping) is favored. Another difference applies to the capability of unsubstituted polyacrylates such as poly(ethyl acrylate) (PEA) to form cross-linked networks during heating. By contrast, unsubstituted polymethacrylates do not cross-link. Cross-linking of polyacrylates occurs via the combination of lateral macroradicals sited on the backbone or via intermolecular reesterification (*10*).

Principally, the characteristic differences in the thermal behavior of the two families of polymers also exist between chloroethyl acrylate and methacrylate polymers. However, in all cases thermally induced initiation processes in pendant groups were dominant at all temperatures. This phenomenon is due to the readiness at which C–Cl bonds are cleaved; interestingly, this cleavage is most pronounced in the case of the dichloroethyl esters. Molecular oxygen plays a peculiar role in cross-linking and depolymerization, which are the major chemical reactions leading to important alterations in the polymers. For instance, gel formation (intermolecular cross-linking) is enhanced in the cases of PMCMA and PDCMA, whereas it is prevented in the case of

PTCMA. For this reason, the first part of this chapter is divided into two subsections devoted to processes occurring in the absence and presence of O_2. The second part deals with the influence of UV light on the thermal degradation of the polymers.

Thermolysis in the Absence of Oxygen

Poly(chloroethyl methacrylates).   Elimination of chlorine, gel formation (intermolecular cross-linking), and depolymerization (unzipping) are the major chemical processes that determine the fate of the poly(chloroethyl methacrylates) during heating to elevated temperatures. For the initiation of the thermolysis at relatively low temperatures (below about 500 K), chlorine elimination as the cleavage of C–Cl bonds is the major process. With rising temperature scissions of other bonds in the pendant groups and also in the backbone of the polymers become more likely. Table I summarizes the major effects observed with PMCMA, PDCMA, PTCMA, and poly(ethyl methacrylate) (PEMA).

With respect to the chlorine-containing polymers, gel formation is the major process at temperatures below about 500 K. At temperature $(T) > $ ~500 K the polymers decompose readily, mainly via depolymerization. PEMA

Table I. Thermolysis of Poly(chloroethyl methacrylates)

Polymer	Cross-linking	Depolymerization	Elimination of Cl	Oxygen Effects
PMCMA	Major process at $T < 500$ K; unimportant at $T > 500$ K	Not occurring at $T < 500$ K; major process at $T > 500$ K	Minor process at all T	Cross-linking enhanced, rate of mass loss and unzipping retarded at $T < 500$ K
PDCMA	Major process at $T < 500$ K; unimportant at $T > 500$ K	Not occurring at $T < 500$ K; major process at $T > 500$ K	Important but not dominant at all T	Cross-linking enhanced, rate of mass loss and unzipping retarded at $T < 500$ K
PTCMA	Major process at $T < 500$ K; unimportant at $T > 500$ K	Not occurring at $T < 500$ K; major process at $T > 500$ K	Minor process at all T	Cross-linking prevented, rate of mass loss and unzipping retarded
PEMA	Not occurring at all T	Very important at all T	Not applicable	Rate of mass loss accelerated, main-chain scission very effective

does not cross-link, but it depolymerizes readily. Notably, PDCMA cross-links with a much higher rate than PMCMA and PTCMA. This result is paralleled by the rate of chlorine elimination, which is much higher in the case of PDCMA than in the cases of PMCMA and PTCMA. Typical results concerning gel formation and mass loss are shown in Figures 1 and 2, respectively. Table II shows a typical result of the composition of the condensate of volatiles; monomer is the major product. The thermal behavior of the methacrylate polymers is compared in Table III. Obviously, PEMA, the nonchlorinated polymer, exhibits the highest thermal stability; and PDCMA, the polymer containing dichloroethyl groups, is least stable. Generally, the stability decreases in the series PEMA > PMCMA > PTCMA > PDCMA.

Poly(chloroethyl acrylates). Table IV summarizes the observations made during heating acrylate polymers. The chlorine-containing polymers do not exhibit a pronounced tendency to undergo depolymerization and, in this way, resemble PEA, the nonchlorinated polymer. However, PTCMA decomposes at $T > 510$ K to some extent by unzipping. At $T > {\sim}500$ K main-chain scission and decomposition into low-molecular fragments are the major chemical processes.

The chlorine-containing polymers PDCEA and PTCEA release chlorine during heating. Interestingly, this process is very important in the case of PDCEA and plays only a minor role in the case of PTCEA. The release of chlorine is paralleled by gel formation and indicates that intermolecular cross-linking is related to the cleavage of C–Cl bonds and the subsequent combination of lateral radicals generated in this way. Notably, the insolubilization of PDCEA occurs very rapidly. As can be seen from Figure 3, gel formation to ~100% conversion is completed within some minutes after the sample has been warmed up to the set temperature. By contrast, PTCEA cross-links only to a small extent over a heating period of many hours (4). The thermal behavior of the acrylate polymers is compared in Table III. Obviously, PEA

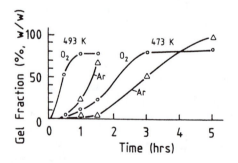

Figure 1. Thermal cross-linking of PDCMA in Ar ($\triangle$) and O_2 ($\bigcirc$) at 493 K and 473 K as plot of gel fraction of residual polymer vs. time of heat treatment. (Reproduced with permission from reference 2. Copyright 1993 Elsevier.)

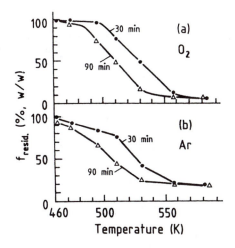

*Figure 2. Thermolysis of PDCMA as isothermal mass loss at 30- (●) and 90-
(△) min heat treatment in O_2 (a) and Ar (b). (Reproduced with permission from
reference 2. Copyright 1993 Elsevier.)*

exhibits the highest thermal stability and PDCEA is least stable. Generally,
the stability decreases in the series PEA > PTCEA > PDCEA.

Thermolysis in the Presence of Oxygen

Poly(chloroethyl methacrylates).   The effect of molecular oxygen
on the thermal reactions of poly(chloroethyl methacrylates) strongly depends
on the degree of chlorination. Most strikingly, in the cases of PMCMA and
PDCMA, cross-linking is enhanced by O_2. Typical results obtained in the case
of PDCMA are presented in Figure 1. The gel fraction representing the cross-
linked portion of the polymer sample increases more rapidly in the presence
of O_2 than in its absence. By contrast, PTCMA does not cross-link at all in
the presence of O_2. As far as mass loss is concerned the thermal stability of
all three chlorinated polymers is increased by O_2. Typical PDCMA results
(Figure 2) demonstrate a significant shift in the onset temperature of mass
loss to higher temperatures and an increase in the rate of mass loss in the
presence of O_2. On the other hand, PEMA exhibits a quite different behavior:
O_2 accelerates the rate of mass loss, and main-chain scission is more effective
in the presence of O_2 than in its absence. These results are summarized in
the last column of Table I.

Poly(chloroethyl acrylates).   In the case of PDCEA O_2 retards
cross-linking, but the release of chlorine is not affected. The rate of mass loss

Table II. Components of Condensate Collected during 60-min Thermal Treatment of PMCMA at 541 K under Argon and at 505 K under O$_2$

Structure	Ar	O$_2$
CH$_3$–CO–H	+	+
ClCH$_2$–CO–H	−	0.3
ClCH$_2$–CH$_2$Cl	0.06	0.3
ClCH$_2$–CH$_2$OH	−	0.2
CH$_2$=C(CH$_3$)–COOCH$_3$	−	+
CH$_2$=C(CH$_3$)–COOCH=CH$_2$	+	+
CH$_3$–CO–CO–Cl	−	+
CH$_2$=C(CH$_3$)–COOH	−	0.1
Cl–CO–OCH$_2$CH$_2$Cl	−	+
CH$_2$=C(CH$_3$)–COOCH$_2$CH$_3$	+	−
CH$_2$=C(CH$_3$)–COOCH$_2$CH$_2$Cl	99.0	95.6
CH$_3$–CO–CO–OCH$_2$CH$_2$Cl	−	1.0
H$_2$C–C(CH$_3$)(epoxide)–COOCH$_2$CH$_2$Cl	−	2.0
ClCH$_2$–C(CH$_3$)(OH)–COOCH$_2$CH$_2$Cl	−	0.3

NOTE: Values are percent concentration (w/w); + means traces detected, − means not detected. Results for first structure (CH$_3$CHO) were determined in a separate run.

**Table III. Anoxic Thermolysis of
Poly(chloroethyl acrylates) and
Poly(chloroethyl methacrylates)**

Polymer	T(onset) (30 min)	T(50%) (30 min)	T(50%) (90 min)
PEA	580	640	620
PDCEA	505	550	530
PTCEA	520	575	565
PEMA	490	600	590
PMCMA	470	590	570
PDCMA	460	530	510
PTCMA	470	540	520

NOTE: Values are decomposition temperature (K) for onset of mass loss and 50% mass loss after 30- and 90-min heat treatment.

Table IV. Thermolysis of Poly(chloroethyl acrylates)

Polymer	Cross-linking	Depolymerization	Elimination of Cl	Oxygen Effects
PDCEA	Major process at $T < 500$ K; unimportant at $T > 500$ K	Negligible at all T	Important at all T	Cross-linking retarded, rate of mass loss enhanced, and release of Cl not affected
PTCEA	Minor process at $T < 500$ K; polymer chars at $T > 500$ K	Not occurring at $T < 500$ K; important but not dominant at $T > 500$ K	Minor process at all T	Cross-linking retarded, rate of mass loss enhanced
PEA	Minor process at all T	Not occurring at all T	Not applicable	

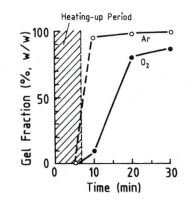

Figure 3. Thermal cross-linking of PDCEA in Ar (○) and O₂ (●) at 494 K as plot of gel fraction of residual polymer vs. time of heat treatment. (Reproduced with permission from reference 4. Copyright 1994 Elsevier.)

is enhanced compared with that observed in the absence of O_2. Similar oxygen effects were found in the case of PTCEA.

Mechanistic Aspects

The thermal decomposition of pendant groups essentially determines the initiation of chemical processes occurring at $T < 500$ K. Among the various types of bond breakage that can be envisaged, the cleavage of C–Cl bonds seems to be an important elementary reaction. An initiation mechanism based on this reaction is discussed below for the case of PDCMA (reaction 1).

$$
\begin{array}{ccc}
\text{CH}_3 & & \text{CH}_3 \\
| & & | \\
-\text{CH}_2-\text{C}- & \xrightarrow{\Delta} & -\text{CH}_2-\text{C}- \quad + \text{Cl}\bullet \\
| & & | \\
\text{O}=\text{C}-\text{O}-\text{CH}_2-\text{CHCl}_2 & & \text{O}=\text{C}-\text{O}-\text{CH}_2-\overset{\bullet}{\text{CHCl}} \\
& & (\text{I}\bullet)
\end{array} \tag{1}
$$

Radicals of type **I·** undergo different reactions: cross-linking according to reaction 2 and decomposition according to reactions 3 and 4.

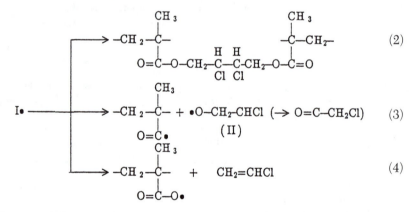

The intermediate product **II** rearranges to chloroacetaldehyde.

In the presence of O_2 most of the free radicals generated by the thermolysis of the polymers are prone to react with O_2 to form peroxyl radicals. A portion of these free radicals will be converted into oxyl radicals according to the general mechanism described by reactions 5 and 6.

$$
\text{R}\bullet + \text{O}_2 \longrightarrow \text{R}-\text{O}-\text{O}\bullet
\begin{cases}
\xrightarrow{\text{ROO}\bullet} 2\ \text{RO}\bullet + \text{O}_2 \\
\xrightarrow{\text{RH}} \text{ROOH} + \text{R}\bullet
\end{cases} \tag{5}
$$

$$
\text{ROOH} \longrightarrow \text{RO}\bullet + \bullet\text{OH} \tag{6}
$$

Oxyl radicals appear to play a prominent role in the enhancement of cross-linking by molecular oxygen as was observed for PMCMA and PDCMA. At elevated temperatures lateral oxyl radicals (RO•) probably undergo a kind of transesterification that results in the immediate formation of intermolecular linkages and the release of low-molecular oxyl radicals. The reaction of oxyl radicals with ester groups is depicted by reaction 7. Note that RO• denotes a macroradical.

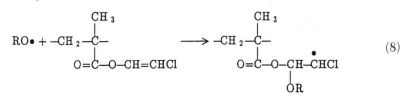

$$RO\bullet + -CH_2-\overset{\overset{\displaystyle CH_3}{|}}{\underset{\underset{\displaystyle O=C-O-CH_2CHCl_2}{|}}{C}}- \longrightarrow -CH_2-\overset{\overset{\displaystyle CH_3}{|}}{\underset{\underset{\displaystyle O=C-OR}{|}}{C}}- + \bullet O-CH_2CHCl_2 \qquad (7)$$

The occurrence of a similar reaction of oxyl radicals was suggested some time ago in a mechanism concerning the oxidative degradation of poly(methyl acrylate) (12). An additional route for the formation of linkages between macromolecules would be the addition of oxyl radicals to C=C double bonds according to reaction 8.

$$RO\bullet + -CH_2-\overset{\overset{\displaystyle CH_3}{|}}{\underset{\underset{\displaystyle O=C-O-CH=CHCl}{|}}{C}}- \longrightarrow -CH_2-\overset{\overset{\displaystyle CH_3}{|}}{\underset{\underset{\displaystyle O=C-O-CH-CHCl}{|}}{C}}- \qquad (8)$$
$$\underset{OR}{|}$$

Double bond formation by elimination of HCl is feasible according to reaction 9.

$$-CH_2-\overset{\overset{\displaystyle CH_3}{|}}{\underset{\underset{\displaystyle O=C-O-CH_2-CHCl_2}{|}}{C}}- \xrightarrow{\Delta} -CH_2-\overset{\overset{\displaystyle CH_3}{|}}{\underset{\underset{\displaystyle O=C-O-CH=CHCl}{|}}{C}}- + HCl \qquad (9)$$

Notably, the processes according to reactions 7–9 are insensitive toward O_2. Evidence for the occurrence of reaction 7 in the presence of O_2 is provided by the formation of 2-chloroethanol and 2,2-dichloroethanol in the cases of PMCMA and PDCMA, respectively. The alcohols are formed when alkoxyl radicals abstract hydrogens from surrounding molecules. This reaction is exemplified by reaction 10 for the case of PDCMA.

$$RH + \bullet O-CH_2CHCl_2 \longrightarrow R\bullet + HO-CH_2CHCl_2 \qquad (10)$$

Notably, 2,2-dichloroethanol and 2-chloroethanol are formed only in traces in the absence of O_2. In connection with the effect of O_2 on the thermal cross-linking of PMCMA and PDCMA, the quite different thermolysis course observed in the case of PTCMA should be pointed out. The thermal cross-linking of PTCMA is completely prevented by O_2. This apparently contradictory behavior can be understood by considering that reaction 7 cannot take place in the case of PTCMA because steric hindrance significantly reduces the reactivity of radical $R^2O^\cdot$.

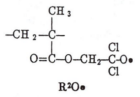

$$R^2O\bullet$$

Thermolysis under the Influence of UV Light

Under the influence of UV light the thermal degradation of poly(chloroethyl methacrylates) occurs overwhelmingly by depolymerization and commences at lower temperatures than in the dark (350 K) (3). At ambient temperature irradiation at $\lambda = 253.7$ nm of PTCMA results in the formation of an insoluble gel due to cross-linking, whereas PMCMA and PDCMA undergo predominantly main-chain scission. Cross-linking of PTCMA occurs because CCl_3 groups, in contrast to CH_2Cl and $CHCl_2$ groups, absorb light relatively strongly at $\lambda = 253.7$ nm. In this way, $R-CCl_2^\cdot$ radicals capable of forming cross-links by combination are generated. UV irradiation of the three chlorinated poly(ethyl methacrylates) at temperatures between 430 and 470 K induces main-chain scission. Terminal radicals thus formed initiate the depolymerization; that is, under the influence of UV irradiation the onset temperature for mass loss is significantly lowered. The results can be interpreted on the basis of reactions 11 and 12, which show that the chemical deactivation of excited carbonyl groups involves two different chemical routes.

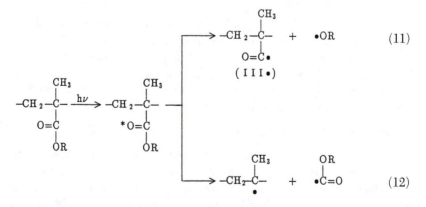

Radical **III**$^\cdot$ can either abstract hydrogen inter- or intramolecularly according to reaction 13 or decompose according to reaction 14.

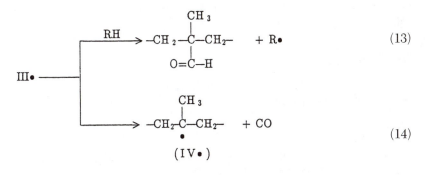

Notably, reaction 13 results in the formation of aldehyde groups. These groups were identified (*13*) in photolyzed poly(acrylates) and poly(methacrylates) via the conversion with 2,4-nitrophenylhydrazine and the separation of the polymer containing hydrazone groups from the reagent mixture by gel permeation chromatography (*13*). The new absorption band formed during the irradiation (Figure 4) is attributed to aldehyde groups. Reaction 14 yields carbon monoxide and radical **IV·**. Decomposition of **IV·** according to reaction 15 results in the scission of a bond in the polymer backbone.

$$
\begin{array}{c}
\text{CH}_3 \qquad \text{CH}_3 \\
-\text{CH}_2-\overset{|}{\underset{\bullet}{\text{C}}}-\text{CH}_2-\overset{|}{\text{C}}-\text{CH}_2- \longrightarrow -\text{CH}_2-\text{C}=\text{CH}_2 + \quad \bullet\overset{|}{\text{C}}-\text{CH}_2- \\
(\text{IV}\bullet) \qquad \text{O}=\text{C} \qquad\qquad\qquad\qquad \text{O}=\text{C} \\
\qquad\qquad \overset{|}{\text{OR}} \qquad\qquad\qquad\qquad\qquad \overset{|}{\text{OR}}
\end{array} \tag{15}
$$

In this way photolysis gives rise to the formation of terminal macroradicals capable of initiating unzipping. Thus, an explanation is provided for the sig-

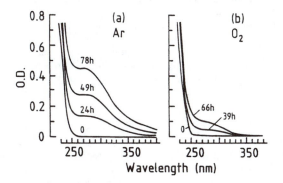

Figure 4. Change in optical absorption spectrum of PDCMA during irradiation at λ = 253.7 nm in Ar (a) and O₂ (b) at 297 K: I = 6 × 10¹⁴ photons/(cm² s). (Reproduced with permission from reference 3. Copyright 1994 Elsevier.)

nificant lowering of the onset temperature for mass loss if the polymers are subjected to the simultaneous impact of heat and UV light.

Experimental

Monomers. Chlorine-substituted ethyl methacrylates and acrylates were synthesized by esterification of methacrylic acid and acrylic acid with chlorine-substituted ethanols. The compounds were obtained from either E. Merck or Aldrich.

Polymers. The polymers were prepared by free-radical polymerization of the monomers to a conversion not exceeding 15% by using 2,2-azo-bis(2-methyl-propionitrile) as initiator. After several reprecipitations from solution the polymers were dried *in vacuo* for several days. Appropriate solvent–precipitant systems were methyl ethyl ketone–*n*-hexane or toluene–methanol for polymethacrylates and acetone–water for polyacrylates. Gel permeation chromatography measurements based on PMMA calibration yielded the following weight average molar masses: 9×10^5 (PMCMA), 3.2×10^5 (PDCMA), 3.4×10^5 (PTCMA), 7.2×10^5 (PDCEA), and 5.5×10^5 (PTCEA).

Thermal Degradation Experiments. The isothermal mass loss of polymer samples was determined at temperatures between 373 K and 573 K by using 60-mg samples in glass tubes placed in a small furnace. During heating the tubes containing the polymer were flushed constantly either with oxygen or argon at a rate of 50 mL/min. In addition, differential scanning calorimetric tests were performed with 3–4-mg polymer samples by using a Perkin-Elmer apparatus (model DSC 2C). The scan rate was varied from 1 to 5 K/min at a gas flow rate (N_2 or O_2) of 30 mL/min.

Analysis of Condensable Volatile Products and Residual Polymer. Volatile products generated in the presence and absence of O_2 were condensed in a cold trap at -78 °C. For the determination of HCl the gas stream was conducted from the cold trap into an aqueous solution of $AgNO_3/HNO_3$. The amount of HCl formed was calculated on the basis of precipitated AgCl. Similarly, acetaldehyde was determined by conducting the gas stream into an aqueous 2 N HCl solution of 2,4-nitrohydrazine. Gas chromatography–mass spectrometry techniques were applied for volatile product analysis. Details were described elsewhere (1). The chlorine content of the residual polymer was determined by elemental analysis by Mikroanalytisches Labor Pascher. The gel content of cross-linked samples was determined by extraction with methyl ethyl ketone with the aid of a Soxhlet extractor.

References

1. Popovic, I.; Song, J.; Fischer, Ch.-H.; Katsikas, L.; Hohne, G.; Velickovic, I.; Schnabel, W. *Polym. Degrad. Stab.* **1991**, *32*, 265–283.
2. Song, J.; Fischer, Ch.-H.; Schnabel, W. *Polym. Degrad. Stab.* **1993**, *41*, 141–147.
3. Song, J.; Fritz, P. M., Fischer, Ch.-H.; Schnabel, W. *Polym. Degrad. Stab.* **1994**, *43*, 177–185.
4. Song, J.; Schnabel, W. *Polym. Degrad. Stab.* **1994**, *43*, 335–342.

5. Popovic, I.; Katsikas, L.; Voloschuk, K. A.; Velickovic, I.; Schnabel, W. *J. Therm. Anal.* **1992,** *38,* 267–275.
6. Popovic, I.; Katsikas, L.; Velickovic, I.; Schnabel, W. In *The Thermal Degradation of Poly(2–mono, 2,2-di- and 2,2,2–trichloroethyl methacrylate): Kinetics and Mechanism;* Scientific Series of the International Bureau; Forschungszentrum Julich GmbH, Julich, Vol. 8.
7. Grassie, N.; Scott, G. *Polymer Degradation and Stabilization; Cambridge University Press:* **1985**.
8. Madorski, S. L. *Thermal Degradation of Organic Polymers;* Interscience: New York, 1964.
9. Cameron, G. G.; Kane, D. R. *Makromol. Chem.* **1967,** *109,* 194.
10. Cameron, G. G.; Kane, D. R. *J. Polym. Sci., Polym. Lett.* **1964,** *2,* 693.
11. Grassie, N.; Speakman, J. G. *J. Polym. Sci. A–I,* **1971,** *9,* 919.
12. Cameron, G. G.; Davie, F. *Chem. Zvesti.* **1972,** *26,* 200.
13. Fritz, P. M.; Zhu, Q. Q.; Schnabel, W. *Eur. Polym. J.* **1994,** *30,* 1335–1338.

RECEIVED for review January 26, 1994. ACCEPTED revised manuscript September 23, 1994.

Thermal Degradation of Automotive Plastics: A Possible Recycling Opportunity

M. Day, J. D. Cooney, C. Klein, and J. Fox

National Research Council Canada, Institute for Environmental Chemistry, Ottawa, Ontario K1A 0R6, Canada

The thermal degradation of a series of 50/50 mixtures of four automotive plastics [polypropylene, acrylonitrile–butadiene–styrene, poly-(vinyl chloride), and polyurethane] was studied by using dynamic thermogravimetry in nitrogen. Comparison of the weight-loss data obtained experimentally with data predicted from the behavior of the individual polymers along with kinetic data obtained by using isoconversional techniques indicates that interactions can occur between the polymers that make up a polymer mix. Both stabilization and destabilization of weight-loss degradation processes were noted depending on the polymeric composition of the mixture. The presence of poly-(vinyl chloride) in the polymer mixtures had significant effects on the weight-loss processes. The results could have important implications for the recovery of chemicals from mixed-plastic waste streams such as those produced by automobile shredding operations.

THE DISPOSAL OF PLASTIC WASTE INTO LANDFILL SITES is well recognized as both a waste of a valuable, nonrenewable resource and a waste of a source of energy. Pyrolysis or tertiary recycling, meanwhile, represents an opportunity to preserve these hydrocarbons to produce valuable petrochemicals. Over the years a number of projects were developed to produce marketable products from plastic wastes, and these projects had varying success (1–4). This type of recycling has been projected to grow 10-fold in the next 10 years (5) because of the anticipated increase in the price of oil and the increased costs associated with current conventional disposal options.

0065–2393/96/0249–0047$12.00/0

The mixed-waste stream produced when junked automobiles are shredded to recover ferrous and nonferrous metals represents a particularly attractive stream for converting organic wastes to hydrocarbons by pyrolysis (6, 7). This material, known as auto-shredder residue (ASR), is produced when old, discarded automobiles are shredded in a hammer mill. The waste is a complex mix of plastics, rubber, textiles, glass, foam, dirt, rust, etc., contaminated with automobile fluids and lubricants. Analysis and characterization of the material in our laboratories (8) has shown the material to be highly variable in composition, as would be expected based on the variability of materials used in the construction of automobiles. Whereas much information exists on the thermal degradation of single polymer systems, the thermolysis of mixed polymer systems has received little attention. Most of the work on mixed polymer systems has centered on thermal stability studies of polymer blends, usually containing poly(vinyl chloride) (PVC) (9). These studies revealed that the blending of two polymers can either stabilize or destabilize the polymer components present depending on the polymers and their degradation products. In this chapter, studies of the thermal degradation of 50/50 blends of the four major plastics used in automobiles will be presented. The data will then be analyzed to determined the types of interaction to be expected when mixed-polymer systems such as those found in ASR are subjected to thermal recycling processes.

Experimental

The four automotive plastics used in this study were obtained from major resin suppliers to the automotive industry. They included:

- A clean, flexible polyurethane (PU) foam used in automotive upholstery
- Himont's PRO-FAX SV-152 impact-resistant polypropylene (PP)
- BF Goodrich Geon PVC
- Dow Magnum acrylonitrile–butadiene–styrene resin (ABS)

Each polymer was ground cryogenically to a fine powder (less than 20 mesh) by using a Wiley Laboratory Mill. The 50/50 mixtures were prepared by weighing equal proportions of the polymers into a container, which was then agitated for 1 h to ensure thorough mixing.

The thermogravimetry (TG) studies were conducted on a 2100 system (TA Instruments, Inc.) employing a 951 TG balance. A dynamic nitrogen atmosphere (50 mL/min) and programmed heating rates of 0.1–50 °C/min were used. Sample weights ranged from 11 to 13 mg.

The kinetic parameters for the thermal degradation processes were determined by the isoconversional method of Ozawa (10) and Flynn and Wall (11). In addition, apparent activation energies were estimated by using Kissinger's peak maximum temperature technique (12). To facilitate the determination of the peak maximum temperature from the derivative weight-loss curve, Jandel's Scientific Peak Fit software program version 3.1 was employed to deconvolute the curve.

Results and Discussion

PP/ABS.   The thermograms obtained from the PP/ABS mix are shown
in Figure 1. The four curves shown in this and the following figures correspond to the weight losses for the two pure polymers as well as those obtained
with the experimental mixture and the anticipated behavior based on the individual polymers in the absence of any interactions. The rate of weight loss
appears to occur less rapidly in the mixture than would be anticipated. For
example, at a heating rate of 5 °C/min at a temperature of 430 °C, the mix
lost approximately only 41.1%, whereas the anticipated loss was 51.7%.

Calculation of the kinetic parameters for the thermal degradation by using
the Flynn-Wall approach gave the apparent activation energies (E) presented
in Figure 2. Interestingly, there are significant differences between the values
determined from the experimental data points and those obtained from the
calculated data points based on the principle of additivity. The rationale for
these differences are, at the moment, unclear, because the Kissinger method
value for ABS of 163 kJ/mol is very close to the anticipated value of 159 kJ/
mol, whereas the value for PP is increased slightly from an anticipated value
of 161 kJ/mol to one of 198 kJ/mol found experimentally. However, based on
the data presented in Figures 1 and 2, it would appear that some interactions
are taking place between these two polymers when subjected to thermal deg-

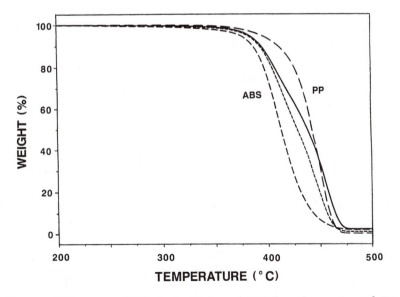

*Figure 1. TG curves at 5 °C/min for PP (— —) ABS (– – –), experimental 50/
50 mixture (———), and calculated for the mixture in the absence of interaction
(----).*

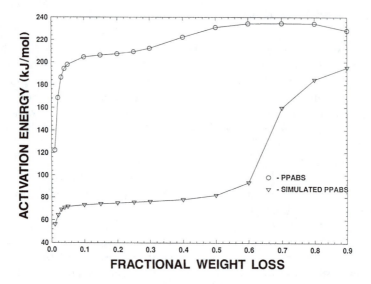

Figure 2. Activation energies as a function of fractional weight loss for 50/50 mixture of PP/ABS derived from experimental (O) and calculated (∇) weight-loss data.

radation, especially in terms of the weight loss associated with the degradation of the ABS.

PP/PVC.   The experimental and predicted TG thermograms for the mix of PP and PVC and those of the pure polymers are shown in Figure 3. Once again, the weight loss with the mix is less pronounced than would be predicted. For example, the onset of dehydrochlorination appears retarded, and the weight loss associated with the process appears much less than anticipated (i.e., 17.0% as opposed to 28.6% at a temperature of 350 °C). However, when the kinetic apparent E values are examined (Figure 4), an interesting observation is noted. For the initial 20% weight loss associated with the dehydrochlorination reaction, the experimentally observed and anticipated E values are almost identical and very similar to the Kissinger values of 133 and 130 kJ/mol determined at the peak maximum. However, when the region associated with the degradation of PP and the polyenes produced from the PVC are considered (fractional weight loss $\propto$ 0.3–0.9), there is a marked difference in the E values. Therefore, the incorporation of PP with PVC could apparently lead to a certain amount of stabilization of the initial degradation of PVC, whereas the char formation action of PVC could be responsible for a retardation of the weight loss associated with the degradation of PP.

PP/PU.   The TG curves obtained for the PP/PU mix are presented in Figure 5. These polymers show their main decomposition processes in very

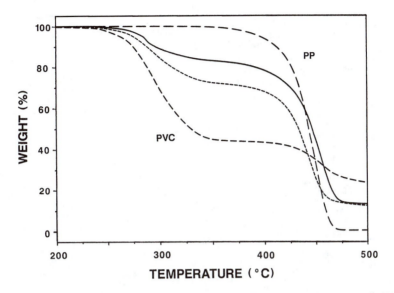

Figure 3. TG curves at 5 °C/min for PP (— —), PVC (– – –), experimental 50/50 mixture (——), and calculated for the mixture in the absence of interactions (---).

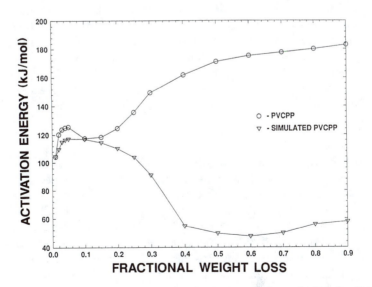

Figure 4. Activation energies as a function of fractional weight loss for 50/50 mixtures of PP/PVC derived from experimental (O) and calculated (∇) weight-loss data.

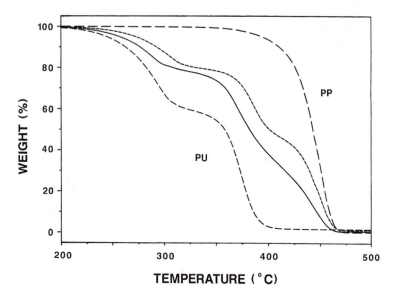

Figure 5. TG curves at 5 °C/min for PP (— —), PU (– – –), experimental 50/50 mixture (——), and calculated for the mixture in the absence of interactions (---).

clearly defined temperature regions that have only a slight overlap. In the case of this polymer mix, the weight loss occurs more rapidly in the mix than would be expected. For example, at 300 °C the mix lost approximately 6.2% more than predicted, whereas at 400 °C the loss was approximately 12.1% more. These results suggest the degradation processes in this particular blend are accelerated due to the presence of the other polymer. The kinetic parameters presented in Figure 6 appear to confirm this interaction for the PP/PU systems, although the actual E values are only slightly smaller than the predicted values.

ABS/PVC. The weight-loss behavior of the ABS/PVC mix (Figure 7) in many ways resembles that of the PP/PVC mix. For example, the weight loss associated with the dehydrochlorination reaction is much more pronounced than expected, and the weight loss associated with the process is much less (i.e., 18.1% in comparison to 29.5% at 350 °C). In a similar manner to that observed with PP/PVC, the apparent E for the dehydrochlorination process (Figure 8) was relatively constant and similar for the experimental and simulated data over the first 25% weight loss. However, differences are noted between the E values at higher conversion values, and the experimental values are consistently higher than predicted values. Therefore, like PP, ABS is bestowing a certain amount of stability to the initial dehydrochlorination of PVC.

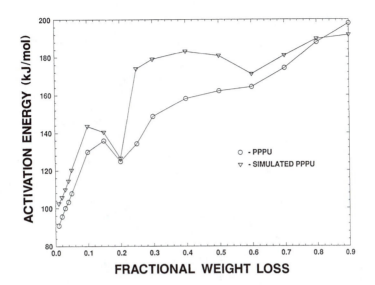

Figure 6. Activation energies as a function of fractional weight loss for 50/50 mixtures of PP/PU derived from experimental (O) and calculated (∇) weight-loss data.

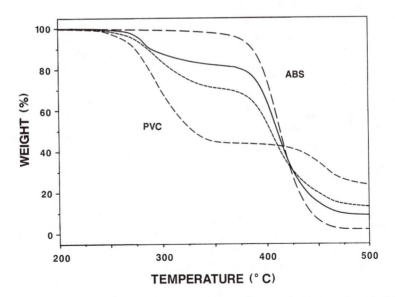

Figure 7. TG curves at 5 °C/min for ABS (— —), PVC (- - -), experimental 50/50 mixture (——), and calculated for the mixture in the absence of interactions (----).

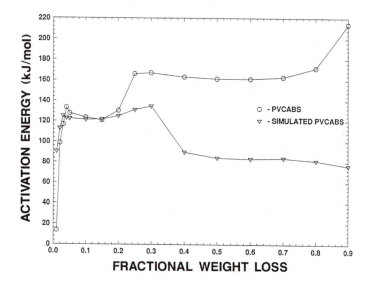

Figure 8. Activation energies as a function of fractional weight loss for 50/50 mixtures of ABS/PVC derived from experimental (O) and calculated (▽) weight-loss data.

ABS/PU. The TG curves for the ABS/PU mixture are presented in Figure 9. Once again, as was noted from the PP/PU mixture, three degradation regions have been identified. In all cases the peak degradation temperatures of each region were noted to occur at lower temperatures than predicted. For example, in the case of the PU they occurred at 287 °C and 366 °C (predicted values were 295 °C and 374 °C), whereas ABS peaked at 400 °C when it was predicted to peak at 412 °C. In terms of measured E values (Figure 10), the experimentally determined values were consistently higher than expected, especially for the second stage of the PU degradation. This kinetic data further suggests that the presence of ABS is retarding the degradation of PU due to diffusion control in addition to chemical interaction processes.

PU/PVC. In the case of the PU/PVC mixture, it is clearly evident from the weight-loss curves shown in Figure 11 that interactions are occurring between the two polymers. For example, the experimentally determined temperature of dehydrochlorination is much lower (273 °C) than was expected based on the additivity rule (295 °C). It is possible that the degradation of the PU is influenced by the accumulation of HCl within the polymer mix that may catalyze the weight-loss processes and cause an acceleration in the initial weight-loss processes of PU. However, as degradation proceeds, possible cross-linking reactions involving PVC may retard the diffusion of liberated

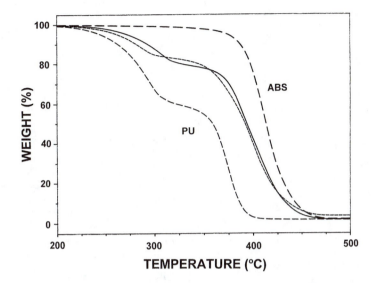

Figure 9. TG curves at 5 °C/min for ABS (— —), PU (- - -), experimental 50/ 50 mixture (———), and calculated for the mixture in the absence of interactions (---).

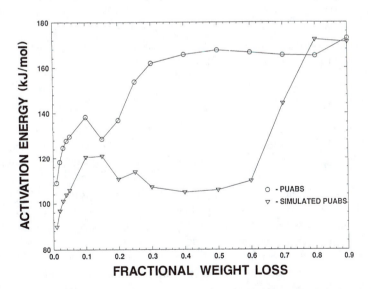

Figure 10. Activation energies as a function of fractional weight loss for 50/50 mixtures of ABS/PU derived from experimental (O) and calculated (∇) weight-loss data.

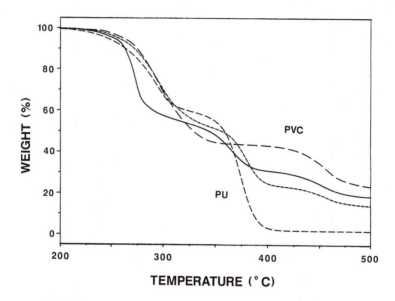

*Figure 11. TG curves at 5 °C/min for PVC (— —), PU (– – –), experimental 50/
50 mixture (——), and calculated for the mixture in the absence of interactions
(----).*

products at temperatures above 380 °C, where a reduction in rate of weight loss is noted. Interestingly, examination of the kinetic parameters for the thermal degradation processes as shown in Figure 12 reveal that the E values determined from the experimental data very closely resemble those predicted for the behavior of the individual polymers.

Conclusion

The thermal degradation of mixed polymer systems cannot be predicted on the basis of the behavior of individual polymers. Both diffusion and chemical interactions appear possible when 50/50 polymer mixtures are being considered. The effects may either stabilize or destabilize the component polymers present in a polymer mix and influence the kinetics and mechanism of the degradation processes. In the case of PVC-containing mixtures, the liberation of HCl or chlorine radicals is of particular concern, because they could have a significant effect on pyrolytic degradation processes and subsequent product yields. The data provided in this study identify some of the concerns that must be addressed if pyrolysis is to play a key role in the recycling of automotive plastics produced by automobile shredding operations.

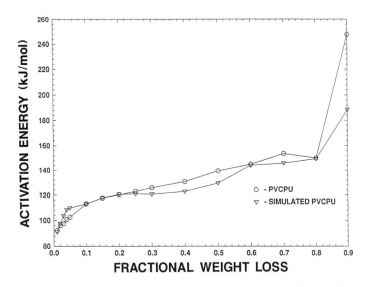

Figure 12. Activation energies as a function of fractional weight loss for 50/50 mixtures of PVC/PU derived from experimental (O) and calculated (∇) weight-loss data.

References

1. Kaminsky, W. *J. Anal. Appl. Pyrolysis* **1985,** *8,* 439–448.
2. Scott, D. S.; Czernik, S. R.; Piskorz, J.; Radlein, D. St. A. G. *Energy Fuels* **1990,** *4,* 407–411.
3. Bertolini, G. E.; Fontaine, J. *Conserv. Recycl.* **1987,** *10,* 331–343.
4. Kaminsky, W. *Makromol. Chem., Macromol. Symp.* **1992,** *57,* 145–160.
5. Anon. *Reuse Recycle Wastes* **1992,** *22(9),* 68.
6. Braslaw, J.; Melotik, D. J.; Gealer, R. L.; Wingfield, R. C. *Thermochim. Acta* **1991,** *186,* 1–18.
7. Roy, C.; deCaumia, B.; Mallette, P. In *Proceedings of the 16th Institute of Gas Technology Conference on Energy from Biomass and Wastes;* Institute of Gas Technology: Orlando, FL, 1992; p 1.
8. Day, M.; Graham, J.; Lachmansingh, R.; Chen, E. *Resources Conserv. Recycl.* **1993,** *9,* 255–279.
9. McNeill, I. C.; Gorman, J. C. *Polym. Degrad. Stab.* **1991,** *22,* 263–276.
10. Ozawa, T. *Bull. Chem. Soc. Jpn.* **1965,** *38,* 1881–1886.
11. Flynn, J. H.; Wall, L. A. *Polym. Lett.* **1966,** *4,* 323–528.
12. Kissinger, H. E. *Anal. Chem.* **1967,** *21,* 1702–1706.

RECEIVED for review December 6, 1993. ACCEPTED revised manuscript January 9, 1995.

Mechanisms of Thermal and Photodegradations of Bisphenol A Polycarbonate

Arnold Factor

General Electric Research and Development Center, PO Box 8, Schenectady, NY 12301

This chapter presents an up-to-date review of the literature on the thermal and photodegradations of bisphenol A polycarbonate as well as new results from my laboratory. The areas described include hydrolysis, thermal and thermal oxidative reactions, and photodegradation. Special emphasis is given to the mechanisms of these reactions and to correlating the past work in these areas with more recent results.

BISPHENOL A POLYCARBONATE (BPA-PC; **1**) has the desirable combination of toughness, transparency, and high heat-distribution temperature. This combination makes BPA-PC an ideal material for demanding applications where resistance to hydrolytic, thermal, and photochemical degradation are important. A key part of designing more stable systems that allow the reliable use of BPA-PC at the limits of its stability range has been the application of a detailed understanding of the hydrolytic, thermal, thermal oxidative, and photochemical degradation. This chemistry will be reviewed, and an emphasis will be placed on correlating the past work in these areas with more recent results and identifying the likely species responsible for discoloration.

Hydrolysis

Perhaps the most important but most easily overlooked aspect of BPA-PC stability is its vulnerability to reaction with water (Scheme I). The reaction is important both during melt processing and in end-use applications involving

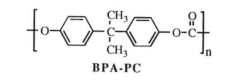

BPA-PC

1

Scheme I. Hydrolysis of BPA-PC.

exposure to water at elevated temperatures, such as sterilization by autoclaving. The keys to minimizing this reaction with water are as follows:

- removal of catalytic acidic or basic process residues from the resin during manufacture
- lowering the water content of the resin to <0.05 wt% before melt processing by carefully drying the resin (e.g., 2–4 h at 120 °C)
- use of only neutral and hydrolytically stable stabilizers, such as hindered phosphites
- addition of a buffering agent such as an appropriate epoxy compound (*1*) when the end-use application requires extra hydrolytic stability

Thermolysis

Because of the highly aromatic structure of BPA-PC, which possesses only methyl and aromatic hydrogens, it is one of the most stable commercial polymers (*2*) and shows thermogravimetric analysis weight loss (N_2, 10 °C/min) only above 550 °C. However, it is generally melt processed well below this temperature: that is, between its T_m of 220–230 °C and 340 °C. Heating BPA-PC above 340 °C leads to both molecular weight loss and gelation reactions. This gelation reaction was observed by Schnell (*3*) to occur during the base-catalyzed melt synthesis of BPA-PC from diphenyl carbonate (DPC) and BPA. The reaction was postulated to be due to the base-catalyzed formation of *o*-phenoxybenzoic acids and esters previously reported by Fosse (*4*) (*see* Scheme II).

Using 360 °C thermolysis of DPC as a model system for BPA-PC, Davis and Golden (*5*) reported the formation of significant quantities of CO_2, phenol, xanthone, diphenyl ether, and phenyl *o*-phenoxybenzoate. Similar to Schnell, they postulated that these products were formed by conversion of DPC to *o*-

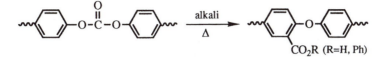

Scheme II. Base-catalyzed rearrangement of diaryl carbonates (4).

phenoxybenzoic acid, which in turn formed xanthone (by self-condensation), diphenyl ether (by decarboxylation), and phenyl *o*-phenoxybenzoate (by transesterification with DPC).

In these studies no mechanism was offered for the Fosse reaction (Scheme II). The best explanation for this reaction was given by Mercier and co-workers (6) in their study of the chemistry involved in the melt crystallization of BPA-PC induced by sodium benzoate. As summarized in Scheme III, they reported both kinetic and product evidence that *o*-phenoxybenzoates are formed by rearrangement of phenyl salicylates. These compounds in turn were derived from sodium salicylate, which formed from carboxylation of sodium phenolate by a diaryl carbonate and by CO_2, the Kolbe–Schmitt reaction.

Recently, Hoyle and co-workers reported (7) that thermal treatment of BPA-PC in either air or nitrogen at temperatures as low as 250 °C gave rise to a structured fluorescence emission mainly due to the formation of dibenzofuran and phenyl-2-phenoxybenzoate, which is the Fosse product described in Scheme II. Finally, BPA-PC pyrolysis/mass spectral studies by Wiley (8), Ballistreri and co-workers (9), Montaudo and co-workers (10), and McNeill and Rincon (11) indicated the formation of BPA-PC cyclic trimer plus dimer or tetramer via ionic intramolecular ester exchange (10) or homolytic (11) reactions. More recently, Montaudo and Puglisi (12) presented convincing arguments that the primary thermal degradation mechanism of PCs involves ionic processes rather than the radical mechanism proposed by McNeill and Rincon.

Thermal Oxidation

BPA-PC is quite stable in air having a continuous-use temperature of 100–125 °C. The major problem encountered at these elevated-use temperatures is yellowing. The extent of discoloration as measured by changes in yellowness index (ΔYI, Figure 1) is sensitive to both time and temperature (13). Early oxygen-uptake studies by Kelleher (14) and Davis (15) showed negligible O_2 absorption after more than 625 days at 100 °C, 6 mL O_2/g BPA-PC absorbed after 83 days at 140 °C, and 4 mL O_2/g BPA-PC absorbed after 100 h at 200 °C. Kovarskaya and co-workers (16, 17) reported that BPA-PC autoxidation was catalyzed by free BPA and sodium salts and that the rate was autoaccelerating above 240 °C. Kinetic and spectral studies by Gorelov and Miller (18)

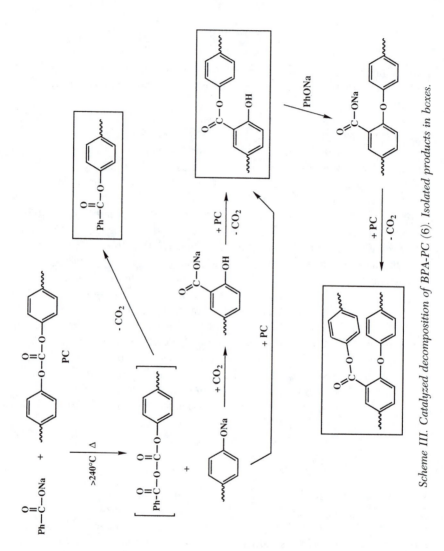

Scheme III. Catalyzed decomposition of BPA-PC (6). Isolated products in boxes.

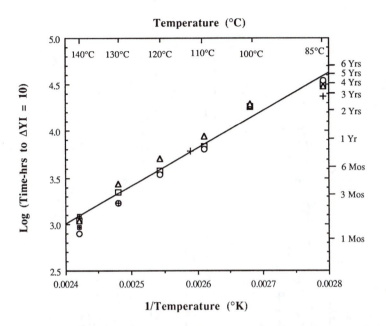

Figure 1. The effect of temperature on the thermal oxidative yellowing of BPA-PC. The different symbols represent data points obtained from four different studies.

indicated BPA-PC autoxidation produced phenols, ketones, *o*-phenoxybenzoic acid, and unsaturated compounds. In addition, based on IR evidence, they suggested simultaneous loss of both methyl groups from the BPA isopropylidene group. They suggested that the key color forming reactions involved the formation and subsequent oxidation of phenolic end groups (Scheme IV).

Evidence for oxidation of the geminal methyl groups in BPA-PC was reported by Factor and Chu (*19*), who found *p*-hydroxybenzoic acid and *p*-hydroxyhacetophenone in the base hydrolysis products derived from a sample of BPA-PC that was oven-aged for 6 months at 140 °C. Most recently, MacLaury and Stoll (*20*) analyzed the oxidation products from a sample of BPA-PC aged in an oven at 250 °C for 16 h after base hydrolysis followed by high-performance liquid chromatography (HPLC) separation (Figure 2). The individual degradation products were manually isolated by using HPLC. The

$$\text{BPA-PC} \xrightarrow{\text{O}_2} \text{H}_2\text{O} \xrightarrow{\text{BPA-PC}} {\sim}\text{PhOH} \xrightarrow{\text{O}_2} \text{yellow polyconjugated products}$$

Scheme IV. Thermal oxidative pathway for the formation of colored products (18).

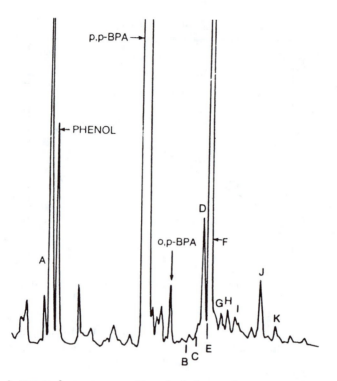

Figure 2. HPLC chromatogram of base hydrolysate of BPA-PC oxidized at 250 °C for 16 h. For key, see Chart I.

IR, NMR, UV, and mass spectrometric analyses of these materials showed them to consist entirely of BPA-derived compounds (Chart I). The mechanism proposed for the formation of these products (Scheme V, Chart I) suggests the operation of several simultaneous reactions:

- oxidation of the geminal methyl groups to form *p*-hydroxyacetophenone (A), *p*-hydroxybenzoic acid, and H_2O
- an acid-catalyzed Fries rearrangement reaction to form the salicylate products (F and E-2)
- hydrolysis of BPA-PC to form free BPA end groups
- oxidative coupling of these BPA end groups to form dimeric products (D, I, and E-1)
- cracking of the BPA end groups to form isopropenylphenol (IPP), which in turn reacts to form the well-known IPP–BPA reaction products (B, C, G, and H) (3)

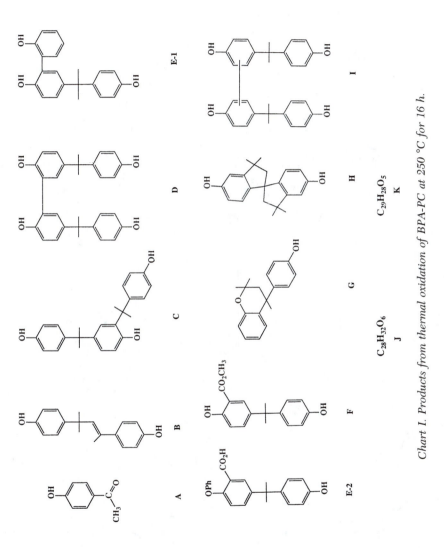

Chart I. Products from thermal oxidation of BPA-PC at 250 °C for 16 h.

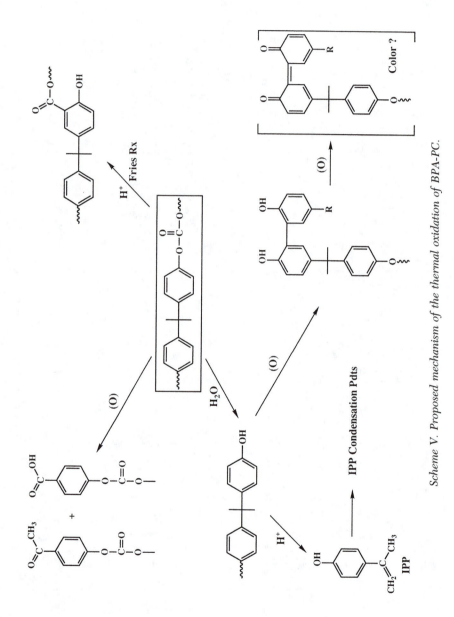

Scheme V. Proposed mechanism of the thermal oxidation of BPA-PC.

The formation of p-hydroxyacetophenone and p-hydroxybenzoic acid is envisioned as coming about via a rearrangement reaction of an initially formed methyl radical to form the more stable benzylic radical (Scheme VI), followed by reaction with O_2 to form both products. Even though p-hydroxybenzoic acid was not actually detected in this study, its absence was attributed to the inability of the HPLC to resolve it. This reaction sequence was first postulated by Lee (21) to explain the catalytic effect of O_2 on the thermal decomposition of BPA-PC. This sequence also explains the formation of p-hydroxybenzoic acid and p-hydroxyacetophenone derivatives during the photooxidation of BPA-PC (19).

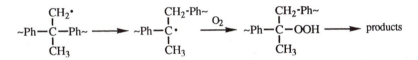

Scheme VI. Proposed pathway for the oxidation of the 2,2-diphenyl propyl unit.

Photoaging of BPA-PC

When unstabilized BPA-PC is exposed to UV light, such as encountered during outdoor exposure, the surface of the resin will become yellow and will often erode. The process is a surface phenomenon and generally extends only ~25 μm into the exposed surface (19). As indicated in Scheme VII, depending on the specific exposure conditions, the chemistry underlying these changes has been ascribed to three general processes: Fries photorearrangement and fragmentation/coupling reactions (2, 22–25), side-chain oxidation (19, 26–32), and ring oxidation (26). Recent spectral studies by Lemaire and co-workers (27, 28) and Pryde (31) clearly illustrated that Fries photoreactions were favored when light with $\lambda \leq 300$ nm was used, whereas photooxidation reactions were increasingly more important as UV light of higher wavelengths was used. Nonetheless, the relative roles of each of these reactions and the chemical nature of the compounds responsible for the observed color formation are not clear.

To better answer these questions, we recently undertook a product study of an unstabilized sample of BPA-PC that was exposed outdoors to four years of Florida weathering by reductive cleavage of the photoaged material with lithium aluminum hydride (LAH) (and lithium aluminum deuteride) followed by analysis by tandem gas chromatography (GC/GC)–high-resolution MS (30). In this way nearly 40 degradation products were characterized. The most important ones are listed in Chart II and are grouped according to the most

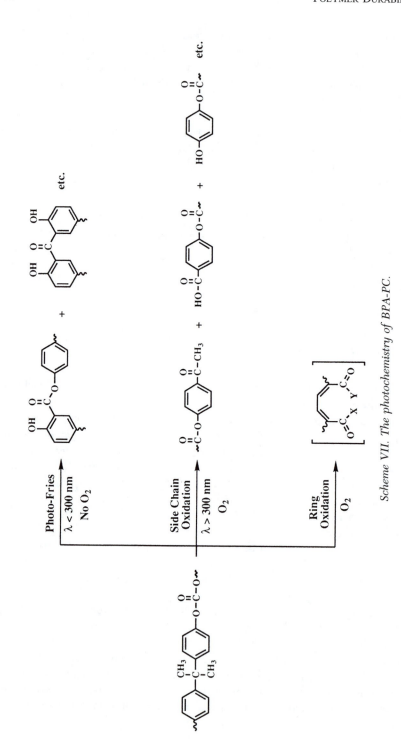

Scheme VII. The photochemistry of BPA-PC.

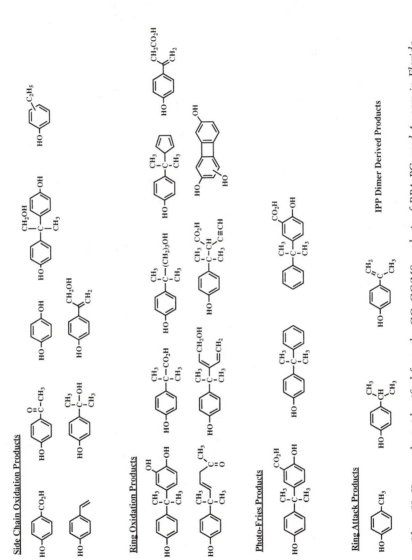

Chart II. Key products identified from the GC–GC/MS analysis of BPA-PC aged 4 years in Florida.

likely mechanism for their production. To analyze for higher molecular weight products, a direct-probe MS experiment was also carried out and revealed the presence of a number of BPA coupling products (Figure 3). The production of these coupling products can best be rationalized by the occurrence of the Fries photofragmentation/coupling reactions illustrated in Scheme VIII.

The results of these experiments indicate that the outdoor weathering of BPA-PC is quite complex and involves the operation of at least four processes: side-chain oxidation, ring oxidation, Fries photorearrangement and fragmentation/coupling reactions, and ring-attack reactions. Ring-attack reactions were not previously reported and are thought to come about by a free-radical reaction of phenolic end groups with methyl radicals (30). Most of the compounds found are not expected to be deeply colored; however, a few, such as the BPA resorcinol derivative listed in Chart II and the ortho-coupled BPA product in Scheme VIII, probably come from LAH reduction of highly colored o-quinone and o-diphenoquinone structures. Even though the products of photooxidation predominate and the Fries photoproducts constitute only a minor part of the product mixture from outdoor-weathered BPA-PC (19, 30–32), the Fries photoprocess likely plays a key role in the autocatalytic photooxidative process. During the initiation of the photodegradation process, the

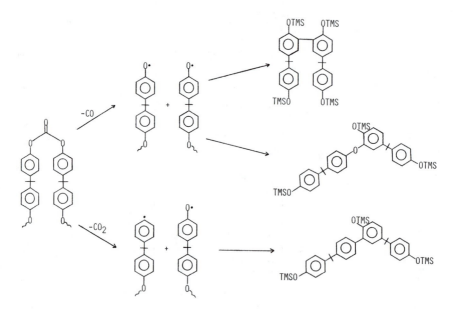

Scheme VIII. Probable mechanism for the formation of higher molecular weight photoproducts. (Reproduced with permission from reference 39. Copyright 1989 Technomic.)

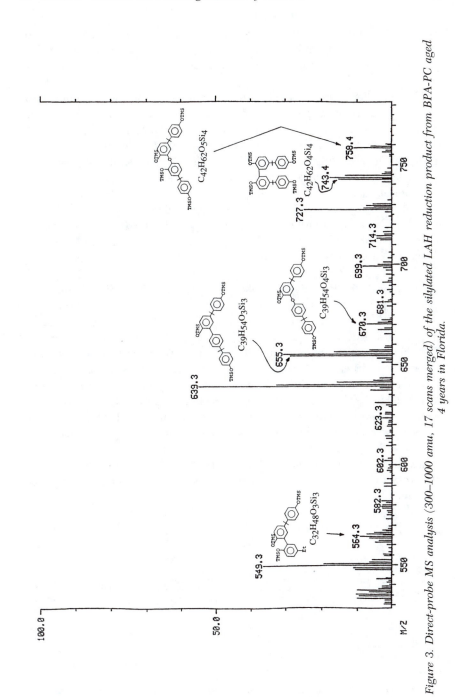

Figure 3. Direct-probe MS analysis (300–1000 amu, 17 scans merged) of the silylated LAH reduction product from BPA-PC aged 4 years in Florida.

Fries photoreactions are probably a major source of free radicals leading to photolabile oxidation products such as hydroperoxides and aromatic ketones, which in turn account for the final autocatalytic stage of BPA-PC photodegradation.

In an earlier study of BPA-PC photochemistry, Webb and Czanderna (33) found Fourier transform IR evidence for the presence of hydrogen bonding between free phenolic hydroxyl end groups and the carbonyl of the carbonate group when the phenolic end group concentration exceeded that of the water in the polymer. They proposed that upon photolysis these hydrogen-bonded moieties underwent a hydrogen atom transfer reaction giving rise to reactive free radicals that led to cross-linking (Scheme IX). However, because commercial BPA-PCs are greater than 97% capped, this pathway probably does not play an important role in the early stages of BPA-PC photogradation. These reactions could very well be important in the later stages, where, as shown by Moore (34), the concentration of phenolic end groups becomes significant.

In recent publications by Hoyle and co-workers (35, 36), fluorescence spectroscopy was used as an extremely sensitive tool for the detection of the primary products of the thermal and photodegradation of BPA-PC. For example, short exposure in air to UV light having a λ_{max} of ~300 nm produced Fries photoproducts, such as salicylic acid and biphenolic type species. Also, as described earlier (7), thermal treatment of BPA-PC in either air or nitrogen at temperatures as low as 250 °C gave rise to a structured fluorescence emission mainly due to the formation of dibenzofuran and phenyl-2-phenoxybenzoate. However, photolysis studies of representative BPA-PC thermal degradation products like dibenzofuran, BPA, and xanthone using broad spectrum UV light with a λ_{max} of 305 nm showed that the presence of these structures in BPA-PC would not greatly effect the photodegradation of BPA-PC other than to form photoproducts that subsequently underwent photobleaching.

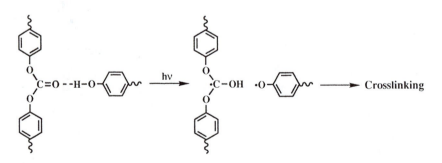

Scheme IX. Proposed photoreaction of hydrogen-bonded carbonate groups.

Finally, further proof of the occurrence of the Fries photoprocess during outdoor exposure was provided by an experiment in which an unstabilized molded sample of BPA-PC was sealed in high vacuum (<0.133 Pa) in a Pyrex tube and exposed outdoors for 2.3 years in Schenectady, NY. After this time the sample was quite yellow ($\Delta YI = 31$). Analysis by LAH reduction/GC–MS of the yellowed top surface revealed the presence of the following Fries photorearrangement and fragmentation/coupling products (**2–4**):

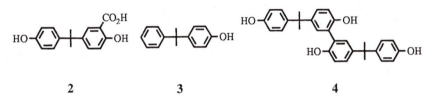

      2   **3**   **4**

Source of Photoyellowing

From the previous research cited, we know that the photodegradation of BPA-PC involves two major processes: the Fries photoreaction and the oxidative attack of the geminal dimethyl groups and the aromatic groups of BPA units. However, it is not known which of the two processes is the principal source of the yellow photoproducts found for BPA-PC.

In an effort to shed more light on this question, the side-chain free PC derived from 3,3'-dihydroxydiphenyl ether (BPO-PC; **5**) (37) and poly(oxy-carbonyloxy-1,4-phenylenedimethylsilane-1,4-phenylene) (BPSi-PC; **6**) (38), which is the polymer based on bis(p-hydroxyphenyl)dimethylsilane, were prepared and evaluated in outdoor Florida exposure. Weathering of the side-chain free BPO-PC showed that it photoyellowed about 3 times faster than BPA-PC; however, the importance of ring oxidation to photoyellowing was clouded because BPO-PC absorbed twice as much UV light as BPA-PC. In contrast, comparative weathering of BPSi-PC and BPA-PC films indicated that

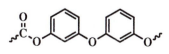
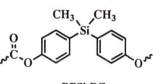

          BPO-PC   **BPSi-PC**

              5   **6**

BPSi-PC was about 5 times more resistant to weathering than BPA-PC. This result can be attributed to the lower likelihood for side-chain photooxidation, the approximately 50% lower UV absorptivity, and the higher water repellency of BPSi-PC.

Even though this result is not definitive, it does suggest that side-chain photooxidation rather than Fries photoreactions or direct-ring oxidation is the major source of BPA-PC photoyellowing. If this situation is the case, the actual compounds giving rise to color in BPA-PC are probably not the compounds produced by direct photooxidation of the geminal dimethyl group in BPA, but rather are the products resulting from further oxidation of these materials, because the products expected from the direct oxidation of these geminal dimethyl groups are not expected to be highly colored. For example, any phenolic compounds produced by side-chain photooxidation probably would lead to highly colored ring-oxidation products in secondary reactions.

Conclusions

In the absence of air, the most important thermal reactions of BPA-PC are hydrolysis and the Fosse backbone rearrangement reaction (Scheme II). Hydrolysis can be avoided by careful removal of catalytic acidic and basic residues during manufacture and by properly drying the resin before processing. Whereas the Fosse backbone rearrangement reaction is known to be base-catalyzed, recent spectroscopic evidence indicates that it proceeds to a small extent in base-free systems at temperatures as low as 250 °C. The thermal oxidation of BPA-PC involves the simultaneous operation of several processes:

- oxidation of the geminal methyl groups
- acid-catalyzed Fries rearrangement
- hydrolysis of the carbonate groups to produce BPA phenolic end groups
- oxidative coupling of these phenolic groups
- cracking of these BPA phenolic groups to form isopropenylphenol and ultimately isopropenylphenol coupling products

Photodegradation of BPA-PC involves at least four processes:

- Fries photorearrangement and fragmentation/coupling reactions
- photooxidation of the geminal methyl groups
- ring oxidation
- ring attack of BPA phenolic end groups

The most likely pathway involved in the outdoor weathering of BPA-PC is a cascade of reactions where the Fries photoreaction is responsible for the initial formation of free-radical species that then lead to the formation of photolabile oxidation products such as hydroperoxides and aromatic ketones. These products in turn are responsible for the final autocatalytic stage of BPA-PC photodegradation. Finally, even though the identities of the species responsible for the yellow color formed during BPA-PC thermal oxidation and photooxidation is as yet unknown, they probably are derived from oxidation of the phenolic and bisphenolic compounds produced by degradation of the geminal dimethyl and carbonate groups.

References

1. Bialous, C. A.; Macke, G. F. U.S. Patent 3 839 247 (1974).
2. Davis, A.; Golden, J. H. *J. Macromol. Sci., Rev. Macromol. Chem.* **1969,** *C3,* 49.
3. Schnell, H. *Chemistry and Physics of Polycarbonates;* Interscience Publications: 1964.
4. Fosse, M. R. *C. R. Acad. Sci. A* **1903,** *136,* 1074.
5. Davis, A.; Golden, J. H. *J. Chem. Soc. B* **1968,** 40.
6. Bailly, C.; Daumerie, M.; Legras, R.; Mercier, J. P. *J. Polym. Sci. Part B: Polym. Phys.* **1985,** *23,* 493.
7. Rufus, I. B.; Shah, H.; Hoyle, C. E. *J. Appl. Polym. Sci.* **1994,** *51,* 1549.
8. Wiley, R. H. *Macromolecules* **1981,** *4,* 255.
9. Ballistreri, A.; Montaudo, G.; Scamporrino, E.; Puglisi, C.; Vitalini, D.; Cucinella, S. *J. Polym. Sci. Part A: Polym. Chem.* **1988,** *26,* 2113
10. Foti, S.; Giuffrida, M.; Maravigna, P.; Montaudo, G. *J. Polym. Sci. Part A: Polym. Chem.* **1983,** *21,* 1567.
11. McNeill, I. C.; Rincon, A. *Polym. Degrad. Stab.* **1991,** *31,* 163.
12. Montaudo, G.; Puglisi, C. *Polym. Degrad. Stab.* **1992,** *37,* 91.
13. Factor, A.; Olson, D. R.; Carhart, R. R.; Webb, K. K., General Electric Corporation, unpublished data.
14. Kelleher, P. G. *J. Appl. Polym. Sci.* **1966,** *10,* 843.
15. Davis, A. *Makromol. Chem.* **1970,** *132,* 23.
16. Kovarskaya, B. M. *Sov. Plast. (Eng. Transl.)* **1962,** *10,* 12.
17. Kovarskaya, B. M.; Kolesnikov, G. S.; Levantovskaya, I. J.; Smirnova, O. V.; Dralynk, G. V.; Poletakina, L. S.; Kovovina, E. V. *Sov. Plast. (Eng. Transl.)* **1967,** *6,* 41.
18. Gorelov, P.; Miller, V. B. *Polym. Sci. USSR (Eng. Transl.)* **1979,** *20,* 2134–2141.
19. Factor, A.; Chu, M. L. *Polym. Degrad. Stab.* **1980,** *2,* 203.
20. MacLaury, M. R.; Stoll, R., General Electric Corporation, personal communication.
21. Lee, L.-H. *J. Polym. Sci. Part A: Polym. Chem.* **1964,** *2,* 2479.
22. Bellus, D.; Hrdlovic, P.; Manasek, Z. *Polym. Lett.* **1966,** *4,* 1.
23. Humphrey, J. S., Jr.; Roller, R. S. *Mol. Photochem.* **1971,** *3,* 35.
24. Humphrey, J. S., Jr.; Shultz, A. R.; Jaquiss, D. B. G. *Macromolecules* **1973,** *6,* 305.
25. Gupta, A.; Rembaum, A.; Moacanin, J. *Macromolecules* **1978,** *11,* 1285.
26. Clark, D. T.; Munro, H. S. *Polym. Degrad. Stab.* **1982,** *4,* 441; Clark, D. T.; Munro, H. S. *Polym. Degrad. Stab.* **1984,** *8,* 195.

27. Rivaton, A.; Sallet, D.; Lemaire, J. *Polym. Photochem.* **1983,** *3,* 463.
28. Rivaton, A.; Sallet, D.; Lemaire, J. *Polym. Degrad. Stab.* **1986,** *14,* 1.
29. Webb, J. D.; Czanderna, A. W. *Sol. Energy Mater.* **1987,** *15,* 1.
30. Factor, A.; Ligon, W. V.; May, R. *Macromolecules* **1987,** *20,* 2461.
31. Pryde, C. A. In: *Polymer Stabilization and Degradation;* Klemchuk, P. P., Ed.; ACS Symposium. Series 280; American Chemical Society: Washington, DC, 1985; p 329.
32. Munro, H. S.; Allaker, R. S. *Polym. Degrad. Stab.* **1985,** *11,* 349.
33. Webb, J. D.; Czanderna, A. W. *Macromolecules* **1986,** *19,* 2810.
34. Moore, J. E. In: *Photodegradation and Photostabilization of Coatings;* Pappas, S. P.; Winslow, F. H., Ed.; ACS Symposium. Series 151; American Chemical Society: Washington, DC, 1981; p 97.
35. Hoyle, C. E.; Shah, H.; Nelson, G. L. *J. Polym. Sci. Part A: Polym. Chem.* **1992,** *30,* 1526.
36. Shah, H.; Rufus, I. B.; Hoyle, C. E. *Macromolecules* **1994,** *27,* 553.
37. Factor, A.; Lynch, J. C.; Greenburg, F. H. *J Polym. Sci. Part A: Polym. Chem.* **1987,** *25,* 3413.
38. Factor, A.; Engen, P. T. *J. Polym. Sci. Part A: Polym. Chem.* **1993,** *31,* 2231.
39. Factor, A.; Ligon, W. V.; May, R. J.; Greenberg, F. H. In *International Conference on Advances in the Stabilization and Controlled Degradation of Polymers;* Patsis, A. V., Ed.; Technomic Publishing: Lancaster, PA, 1989; Vol. II, pp 45–57.

RECEIVED for review December 6, 1993. ACCEPTED revised manuscript September 26, 1994.

Photoaging of Substituted and Unsubstituted Silicones

Jacques Lacoste, Y. Israëli, and Jacques Lemaire

Laboratoire de Photochimie Moléculaire et Macromoléculaire, Unité de Recherche Associée, Centre National de la Recherche Scientifique 433, UFR Sciences et ENS Chimie de Clermont-Ferrand, Université Blaise Pascal, 63177 Aubière Cedex, France

The accelerated photoaging of solid and liquid silicones is reviewed, and particular attention is paid to the influence of the main substituents of silicone chemistry. The decomposition of functional groups as $-H$, $-CH=CH_2$, $-CH_2-CH_2-$, $-CH_2-CH_2-CH_2OH$, and $-Ph$ was followed mainly by Fourier transform IR, ^1H-, ^{13}C-, and ^{29}Si-NMR spectroscopy, and gas chromatography–mass spectroscopy. From the identification of the main oxidation products, photooxidation mechanisms are suggested.

SILICONES AND POLYDIMETHYLSILOXANES (PDMS) have very high stabilities against oxidation. Because of this property, these kinds of material are sometimes used in very severe environmental conditions. In addition, to improve some properties of PDMS such as reactivity or rigidity, methyl groups are often substituted by various substituents (e.g., $-H$, $-CH=CH_2$, $-CH_2-CH_2-$, $-CH_2-CH_2-CH_2OH$, $-OH$, and $-Ph$), with as a probable consequence the modification of the oxidation resistance. The aim of this chapter is to review the recent results obtained in the study of the photooxidation under accelerated conditions representative of the natural photoaging of solid or liquid unsubstituted silicones or of silicones modified by the substituents usually met in silicone chemistry. Some of these materials are often used for the reinforcement and the waterproofing of stones. Their resistance to long-term photoaging represents a crucial importance, particularly in the case of the restoration of ancient monuments. Another more fundamental interest is the ability to analyze the photostability of functional groups in an inert medium, because unsubstituted PDMS presents a fairly high durability under exposure in accelerated photoaging (1).

0065–2393/96/0249–0077$12.00/0

Experimental Details

Pure PDMS and modified PDMS containing functional groups such as SiH, SiCH=CH$_2$, Si–(CH$_2$)$_3$–OH, Si–Ph, Si–Ph$_2$, and SiOH were provided by Rhône-Poulenc Silicones (France); Si–CH$_2$–CH$_2$–Si groups were synthesized by the hydrosilylation reaction (2, 3) in the presence of platinium salt as catalyst. Linear PDMSs exist in the liquid or waxy state (very low glass transition temperature), and silicone oils are also obtained with monofunctional substituents (e.g., –H, –CH=CH$_2$, –Ph, and –OH). Solid silicones result from the presence of a bifunctional group such as siloxane [–O–Si(CH$_3$)$_2$– or –O–Si(CH$_3$)(Ph)–], dimethylene (–CH$_2$–CH$_2$–), methylene (–CH$_2$–), or acrylate (–CH$_2$–CH–COO—Si–O–) and ensure a cross-linked structure. The content of each substituent and the structures of most of the analyzed compounds are shown in Chart I.

Liquid silicones were photooxidized in a Pyrex reactor in which oxygen was bubbled. The temperature was maintained at 65 °C. The source was a Pyrex-filtered medium-pressure mercury lamp (MAZDA, 400 W) supplying radiations of wavelengths longer than 300 nm. Source and reactor were placed along the focal axis of an aluminium cylinder with elliptical base. Toluene solutions of silicone resins were cast on KBr plates from which solvent was carefully evaporated. Photoirradiations were performed in a SEPAP 12–24 device (4), which was the same source for the photooxidation of liquid PDMS. The temperature was maintained at 60 °C. Changes during photooxidations were analyzed by Fourier transform IR (FTIR) spectroscopy, and photoproducts were identified by using derivatization techniques (e.g., NH$_3$, SF$_4$, and HCl) (5, 6). Most of the oils were also analyzed by Raman and ^{1}H-, ^{13}C-, and ^{29}Si-NMR spectroscopies; viscosity measurements; and chemical titrations. Gas chromatography/mass spectrometry (GC/MS) was also used in the case of low-molecular weight compounds. The details of analytical conditions were reported in previous papers (3, 7–9).

Results and Discussion

Unfunctionalized PDMS.
Liquid silicones (oils) of various molecular weights as well as solid silicones (resins), where the rigidity was ensured by cross-links [tri (T)- and quadri (Q)-linkages], show a very high stability under accelerated or ultra-accelerated photoaging (1). No oxidation product was detected by FTIR even after 10,700-h exposure in the SEPAP chamber. That result means that these kinds of silicones are highly dependable for outdoor applications. If we assume that the SEPAP accelerating factor (f) for silicone is similar to the polyethylene one (f_{PE} = 12), the lifetime of silicone in outdoor exposure is supposed to be more than 30 years.

PDMS Substituted with Hydrides.
The photooxidation of silicones substituted by hydride groups with various contents and locations (the end or inside the macromolecular chain) is a much faster process when compared with photooxidation of pure PDMS. The consumption rate of SiH groups determined by FTIR in the range of 2000 cm^{-1} (Figure 1a) or by chemical titration (10, 11) increased with the SiH content; however, the rate increased much more if the SiH groups were located at the chain end (Figure

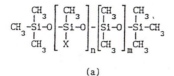

(a)

(b)

X (Inside the Chain)

X (Chain End)

X	Weight (g) of Reactive Functional Group Per Weight (100 g) of Polymer	
-H	0.12 , 0.4 , 0.5 ,1.5 (a) 0.22 , 1.49 (b)	
$-CH=CH_2$	0.4 , 3.1 , 7.1 , 12.3 (a) 2.9 , 29 (b)	
$- CH_2 - CH_2$	-H (0.22) -H (1.49) -H (0.5)	$-CH=CH_2$ (29) (b) $-CH=CH_2$ (29) (b) $-CH=CH_2$ (12.3) (a)

Cross-Link Types in Resins (Solid PDMS)

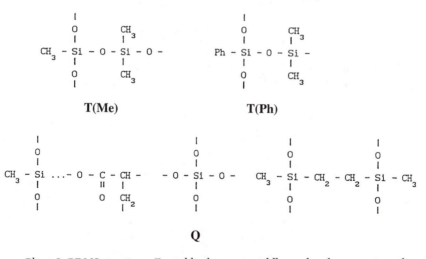

T(Me)   **T(Ph)**

Q

Chart I. PDMS structure. For table shown at middle, each value represents the functional group content in each oil studied.

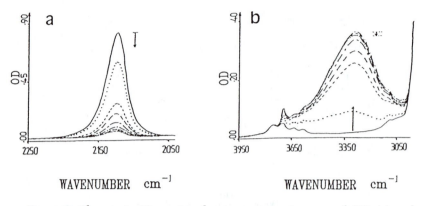

Figure 1. Changes in IR spectra showing consumption rate of SiH (a) and formation of silanol groups (b) during irradiation at λ > 300 nm and 65 °C of SiH–PDMS oil (%H = 0.22). Irradiation times: 0, 6, 21, 28, 44, 67, 90, and 118 h. (Reproduced with permission from ref. 7. Copyright 1992 Elsevier Science.)

2). FTIR analysis reveals that the photoreaction results mainly in the formation of silanol groups (Figure 1b; *see* Scheme I).

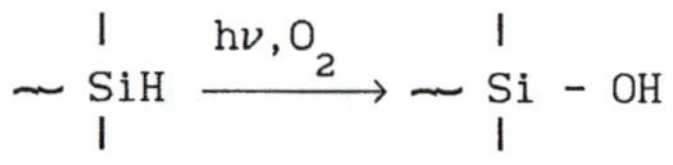

$$\sim \overset{|}{\underset{|}{Si}}H \xrightarrow{\text{h}\nu, O_2} \sim \overset{|}{\underset{|}{Si}} - OH$$

Scheme I.

A part of these silanol groups is converted into new siloxane groups by a condensation reaction (*see* Scheme II).

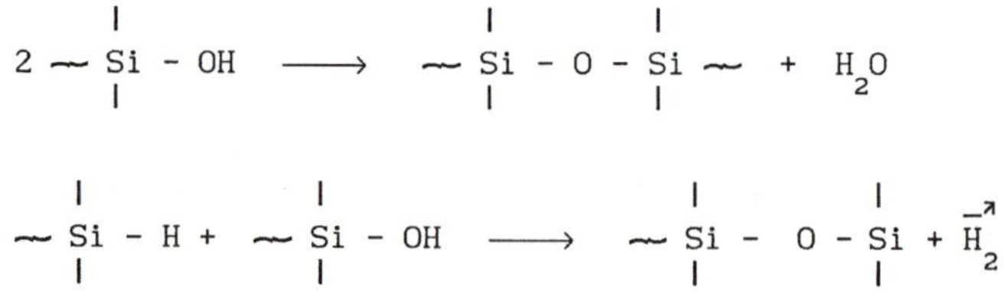

$$2 \sim \overset{|}{\underset{|}{Si}} - OH \longrightarrow \sim \overset{|}{\underset{|}{Si}} - O - \overset{|}{\underset{|}{Si}} \sim + H_2O$$

$$\sim \overset{|}{\underset{|}{Si}} - H + \sim \overset{|}{\underset{|}{Si}} - OH \longrightarrow \sim \overset{|}{\underset{|}{Si}} - O - \overset{|}{\underset{|}{Si}} + H_2 \nearrow$$

Scheme II.

This conclusion was supported by experimental evidence:

1. The SiH/oxygen stoichiometry can be demonstrated by oxygen consumption measurement, but the concentration of silanol is always lower than the concentration of SiH involved in the reaction (Figure 2).

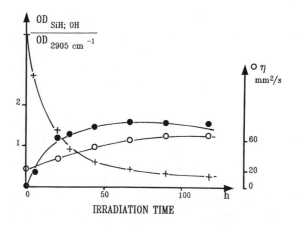

Figure 2. Kinetic changes of SiH (+), hydroxyl (●), and viscosity (○) vs. irradiation duration of SiH–PDMS oil (%H = 0.22). (Reproduced with permission from ref. 7. Copyright 1992 Elsevier Science.)

2. The viscosity increases during the photooxidation, and this increase is particularly important when the initial Si–H groups are located inside the chain, probably because of the formation of a tridimensional structure (*see* Scheme III).

3. New siloxane groups [CH_3–$Si(O\sim)_3$] were detected by ^{29}Si-NMR at –65 ppm (*11, 12*).

As in classical oxidation, hydroperoxide groups can be considered as the primary photoproducts. Their content, determined by a method based on the oxidation of Fe^{2+} in solution (*13, 14*), is very low in photooxidation. Photolysis

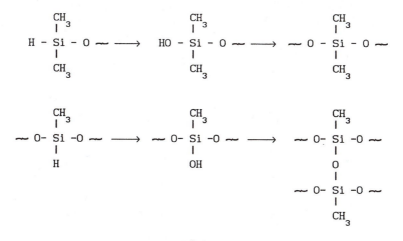

Scheme III.

under vacuum of preoxidized PDMS showed a high photoinstability of these groups and resulted in the low stationary concentration observed. Therefore, the photoaging of SiH–PDMS fits to the mechanism shown in Scheme IV and can be considered as a relatively clean conversion of silicon hydride into silanol.

Vinyl PDMS.   Vinyl groups bring a low photostability at wavelengths longer than 300 nm. The consumption rate of vinyl groups determined by FTIR spectroscopy is lower than for the SiH. The decrease of vinyl groups results in the formation of products absorbing in the hydroxyl region (mainly silanols) as well as in the carbonyl region (mainly acid and ester groups were revealed by derivatization reaction by SF_4 and NH_3) of the IR spectra.

However, the concentration of oxidation products is much lower than the concentration of converted vinyl groups. GC analysis of the gaseous phase and GC/MS analysis of a low molecular weight compound after irradiation detected a large quantity of volatile products resulting from chain-scission reactions. In fact, the scission of the Si–vinyl bond during photooxidation of vinyl–PDMS seems to be the main reaction in comparison with direct oxidation. In this case, direct oxidation represents approximately only 20%. The formation of most of the photooxidation products can be explained by the instability of the vinyl group under irradiation at wavelengths longer than 300 nm (attack by radicals, R•, resulting from the photoirradiation of extrinsic chromophores; *see* Scheme V). The presence of silanol groups and their subsequent conversion to siloxane results again in a large increase of the viscosity in the time of irradiation.

Dimethylene PDMS.   In silicones, the dimethylene group generally results from the addition reaction between a SiH–PDMS and a vinyl–PDMS. This hydrosilylation reaction (2) is of practical interest because it can be used as a polyaddition reaction when performed with bifunctional molecules or when the groups are located inside the PDMS chain.

From a fundamental point of view, this material can also be considered as a copolymer of ethylene and PDMS with isolated ethylene units. In addition, the hydrosilylation reaction results in the formation of two isomers (*15*). The second isomer is the methine form [–CH(CH$_3$)–] and represents 10–15% of the reaction. This type of group, containing a tertiary carbon, is very sensitive to photooxidation (*16*). In fact, by using ^{13}C-NMR (*3*), we proved that the photooxidation occurred mainly on the dimethylene sites (*9*). The photooxi-

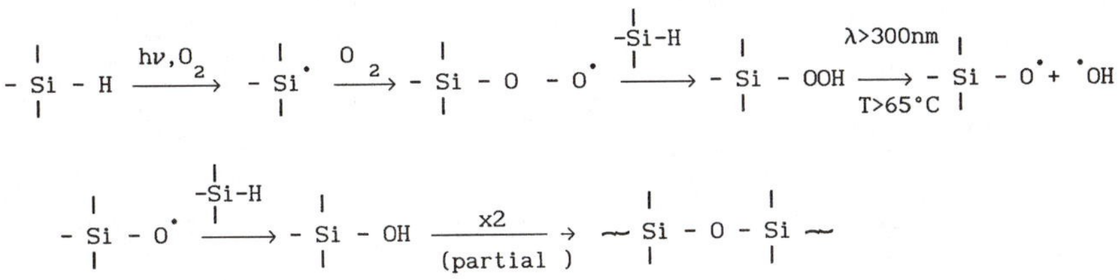

Scheme IV.

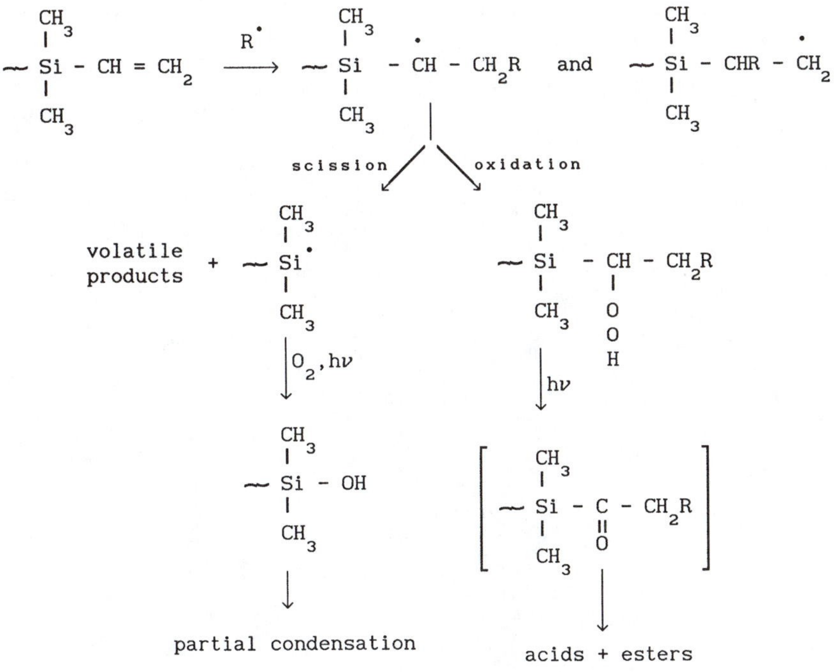

Scheme V.

dation of the dimethylene–PDMS can be described as an oxidation of the dimethylene groups that leads to the formation of *sec*-alcohols, *sec*-hydroperoxides, and chain ketones, as in the case of polyethylene (*17*) (*see* Scheme VI).

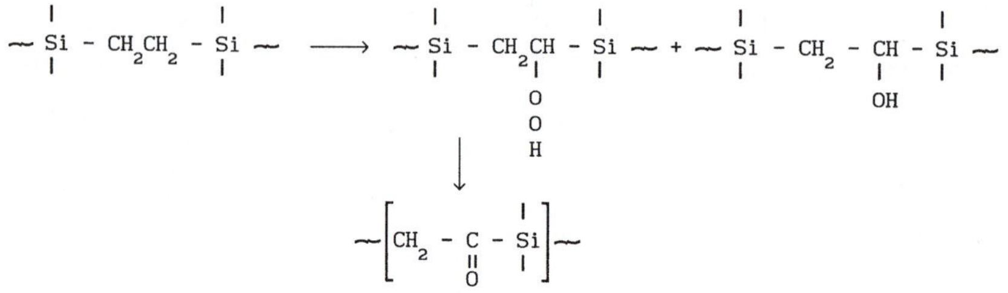

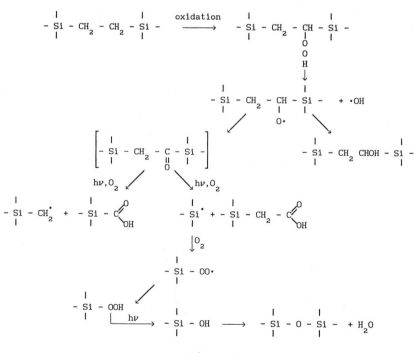

Scheme VI.

The first step is formation and decomposition of hydroperoxides (*18*). Silyl ketones are very unstable under the light used, and they are readily decomposed into scission products as carboxylic acid, silanol, and ester groups (*9*) (*see* Scheme VII). Detailed analysis of a low molecular weight compound by

Scheme VII.

^{13}C-NMR, FTIR/SF$_4$, and GC/MS proved the presence of a β-carboxylic acid (~Si–CH$_2$–COOH).

Hydroxypropyl PDMS. This monofunctional substituent results from a hydrosilylation reaction with allyl alcohol (*see* Scheme VIII). The photoreactivity of the pendant hydroxypropyl groups can be revealed by FTIR, Raman, and ^{13}C-NMR spectroscopies. The FTIR changes shown in Figure 3 are consistent with a phototransformation of the functional groups (CH bands at 2877 and 2935 cm^{-1} and OH band at 3334 cm^{-1}) into carbonyl compounds

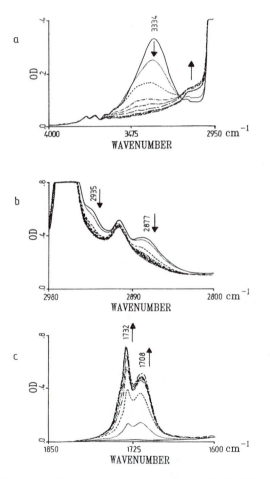

Figure 3. Changes in IR spectra during irradiation (λ > 300 nm, 65 °C) of hydroxypropyl–PDMS: a, hydroxyl vibration region; b, C–H bond-stretching vibration region; c, carbonyl vibration region. Key: — 0 h; ··· 27 h; - - - 68 h; —·— 111 h; –··– 156 h; – – – 180 h; – – – – 225 h; —···— 254 h; and –····– 265 h. For a particular wavenumber, arrows indicate an increase (↑) or decrease (↓) in spectral value as irradiation time increases.

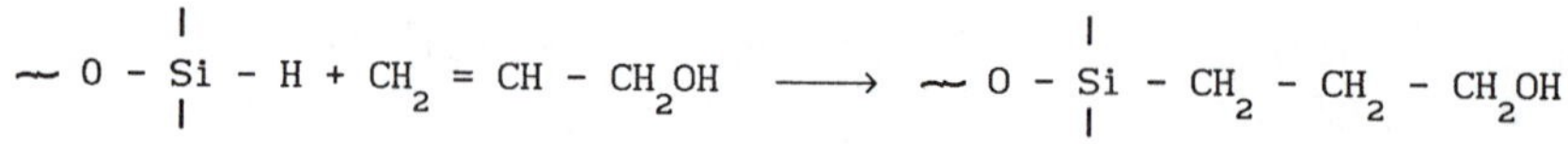

$$\sim O - \underset{|}{\overset{|}{Si}} - H + CH_2 = CH - CH_2OH \longrightarrow \sim O - \underset{|}{\overset{|}{Si}} - CH_2 - CH_2 - CH_2OH$$

Scheme VIII.

(C=O bands at 1708 and 1732 cm^{-1}). Derivatization by SF$_4$ (Figure 4) reveals that the band absorbing at 1732 cm^{-1} can be attributed to ester groups (inactive with SF$_4$), whereas the band at 1708 cm^{-1} belongs to the absorbance of two kinds of carboxylic acid groups whose acid fluorides absorb at 1830 and 1848 cm^{-1}.

^{13}C-NMR analysis (Figure 5) allowed us to identified the two carboxylic acid as Si–CH$_2$COOH (28.1 ppm) and Si–CH$_2$–CH$_2$–COOH (–*CH$_2$–CH$_2$COOH at 12.5 ppm and –*CH$_2$COOH at 28.5 ppm). The ^{13}C-NMR analysis is also consistent with a lower reactivity of the (α-carbon as revealed by the relative intensities of the three carbons before and after irradiation (C$_\alpha$, 13.4 ppm; C$_\beta$, 26.8 ppm; and C$_\gamma$, 65.4 ppm). Therefore, the main site of the photooxidation is the pendant hydroxypropyl group. Main oxidation products are β- and γ-carboxylic acids. The formation of silanol can also be suggested from the observation of both an increase of the viscosity and the formation of trifunctional siloxane (T-siloxane) structures at –66 ppm (*see* Scheme IX).

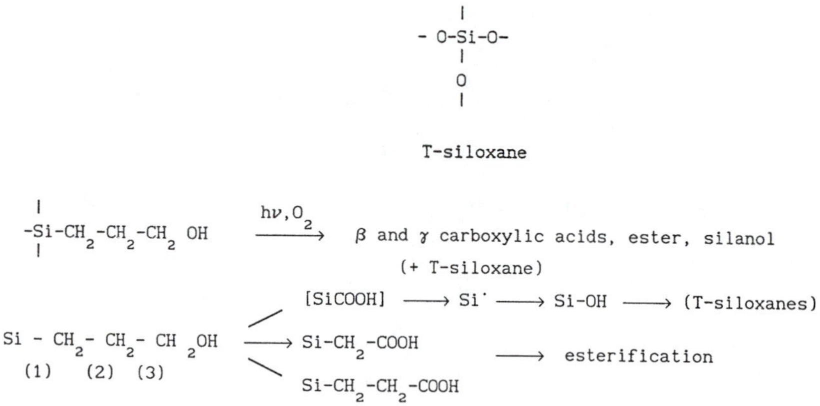

$$- O - \underset{|}{\overset{|}{Si}} - O -$$
$$\underset{|}{\overset{}{O}}$$

T-siloxane

$$-\underset{|}{\overset{|}{Si}}-CH_2-CH_2-CH_2 OH \xrightarrow{h\nu, O_2} \beta \text{ and } \gamma \text{ carboxylic acids, ester, silanol}$$

(+ T-siloxane)

$$[SiCOOH] \longrightarrow Si^\cdot \longrightarrow Si-OH \longrightarrow (T-siloxanes)$$

$$Si - \underset{(1)}{CH_2} - \underset{(2)}{CH_2} - \underset{(3)}{CH_2} OH \longrightarrow Si-CH_2-COOH \longrightarrow \text{esterification}$$

$$Si-CH_2-CH_2-COOH$$

Scheme IX.

Phenyl-PDMS. Phenyl substituents on PDMS are well known to increase both the resistance to thermal oxidation (*19, 20*) and the mechanical properties of silicones (*19*). The long-term behaviors of both liquid and solid silicones containing phenyl groups (Ph or Ph$_2$) were examined in accelerated photoaging conditions ($\lambda > 300$ nm). The photooxidation rate was comparatively lower with other substituents. The analysis of photoproducts by the

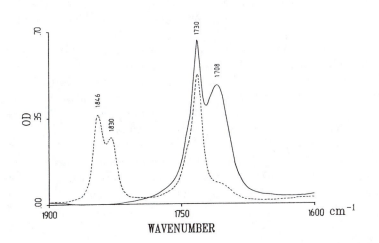

Figure 4. Changes in the carbonyl vibration region of hydroxypropyl–PDMS after SF₄ treatment. Key: (solid line) irradiated for 180 h; (broken line) treated with SF₄ for 30 min.

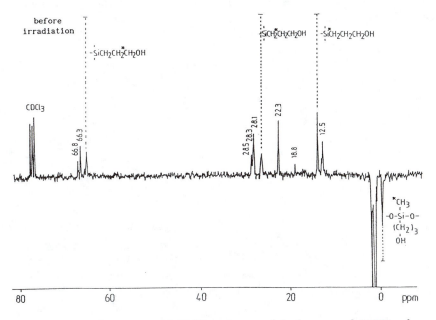

Figure 5. Changes in ¹³C-NMR spectrum of hydroxypropyl–PDMS after irradiation (λ > 300 nm, 65 °C). Key: (broken line) before irradiation; (solid line) irradiated for 265 h.

experimental techniques described clearly shows that this slow photooxidation can be characterized by chain scissions and by the oxidation of the phenyl ring.

The new IR absorption bands appeared only after long exposure times (13,200 h) of phenylated resins. These bands are consistent with the presence of phenolic groups (21) (revealed also by fluorescence spectroscopy) and of carboxylic acids and esters that probably arose from the oxidation of phenolic groups. Condensation reactions in relation to the presence of the residual silanol group have also been pointed out. This reaction was revealed in PDMS oil containing only silanol groups. Notably, the content of phenolic groups was not sufficient to start an important yellowing of phenylated silicones, but the presence of impurities or the use of low irradiation wavelength ($\lambda = 254$ nm; not representative of weathering) resulted in an increase of the yellowing.

Conclusion

Solid and liquid silicones present remarkable stability to the photoaging performed in conditions representative of natural aging. Nevertheless, any substitution of the methyl group results in a decrease of the photostability according to the following sequence:

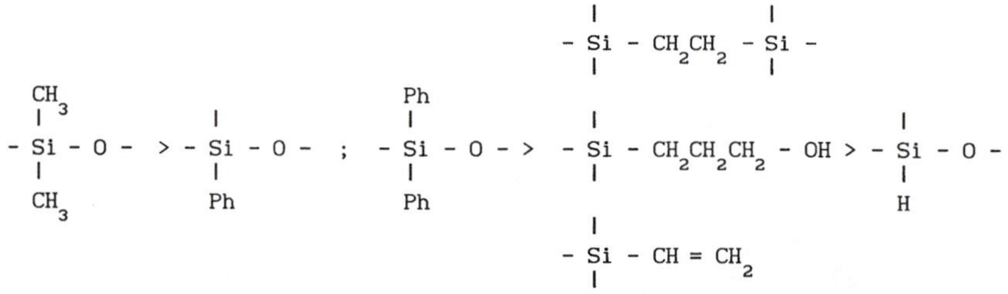

$$
\begin{array}{l}
\quad\quad\quad\quad\quad\quad\quad\quad\quad\quad\quad\quad\quad\quad |\quad\quad\quad\quad | \\
\quad\quad\quad\quad\quad\quad\quad\quad\quad\quad\quad\quad\quad - Si - CH_2CH_2 - Si - \\
\quad\quad\quad\quad\quad\quad\quad\quad\quad\quad\quad\quad\quad\quad |\quad\quad\quad\quad | \\
CH_3\quad\quad\quad\quad\quad\quad\quad Ph\quad\quad\quad\quad\quad | \\
|\quad\quad\quad\quad\quad |\quad\quad\quad\quad\quad | \quad\quad\quad\quad\quad | \\
- Si - O -\; > - Si - O - \;;\; - Si - O - > \; - Si - CH_2CH_2CH_2 - OH > - Si - O - \\
|\quad\quad\quad\quad\quad |\quad\quad\quad\quad\quad |\quad\quad\quad\quad\quad |\quad\quad\quad\quad\quad | \\
CH_3\quad\quad\quad\quad\quad Ph\quad\quad\quad\quad Ph\quad\quad\quad\quad\quad | \quad\quad\quad\quad\quad H \\
\quad\quad\quad\quad\quad\quad\quad\quad\quad\quad\quad\quad\quad\quad - Si - CH = CH_2 \\
\quad\quad\quad\quad\quad\quad\quad\quad\quad\quad\quad\quad\quad\quad |
\end{array}
$$

Silicon hydride groups are very photo-unstable. Oxidation can be described simply by the formation of a silanol group, and the reaction can be suggested tentatively as an original synthesis method for silanol groups. As was demonstrated in many cases, the formation of silanol groups is always accompanied by the correspondent condensation reaction (*see* Scheme X).

$$
\begin{array}{l}
\quad | \quad\quad\quad\quad\quad\quad\quad\quad\quad\quad | \quad\quad\quad | \\
2 - Si - OH \quad\longrightarrow\quad - Si - O - Si - \; + H_2O \\
\quad | \quad\quad\quad\quad\quad\quad\quad\quad\quad\quad | \quad\quad\quad |
\end{array}
$$

Scheme X.

Conversely, phenyl groups do not result in an important photooxidation at long wavelengths, and the oxidation products after long exposure correspond to an attack on the phenyl ring. Substituents containing aliphatic carbons as dimethylene or hydroxypropyl groups [similar results were also obtained in the case of silicone polyacrylate (22)] represent an intermediate reactivity characterized by scissions of the Si–C and C–C bonds.

References

1. Israëli, Y.; Lacoste, J.; Lemaire, J.; Cavezzan, J. *Rev. Gen. Caout. Plast.* **1993**, *724*, 71–74.
2. Pataï, S. *The Chemistry of Organo Silicon Compounds*; Wiley: New York, 1989; Vols. 1 and 2.
3. Israëli, Y.; Lacoste, J.; Cavezzan, J.; Dauphin, G. *Analusis* **1992**, *20*, 245–253.
4. Penot, G.; Arnaud, R.; Lemaire, J. *Die Angew. Makromol. Chem.* **1983**, *117*, 71.
5. Carlsson, D. J.; Brousseau, R.; Zhang, C.; Wiles, D. M. *Chemical Reactions in Polymers*; Benham, J. L.; Kinstle, J. F., Eds.; ACS Symposium Series 364; American Chemical Society: Washington, DC, 1988; p 376.
6. Allinger, N. L.; Cava, M. P.; De Jongh, D. C.; Johnson, C. R.; Lebel, N. A.; Stevens, C. L. *Chimie Organique: Réactions;* McGraw-Hill: Paris, 1983; Vol. 2, p 554.
7. Israëli, Y.; Philippart, J. L.; Cavezzan, J.; Lacoste, J.; Lemaire, J. *Polym. Degrad. Stab.* **1992**, *36*, 179.
8. Israëli, Y.; Cavezzan, J.; Lacoste, J. *Polym. Degrad. Stab.* **1992**, *37*, 201–208.
9. Israëli, Y.; Cavezzan, J.; Lacoste, J. *Polym. Degrad. Stab.* **1993**, *42*, 267–279.
10. Noll, W. *Chemistry and Technology of Silicone;* Academic: New York, 1968.
11. Pawlenko, S. *Organo-Silicon Chemistry;* de Gruyter: New York, 1986; pp 139–144.
12. Engelhart, G.; Jancke, H. *J. Organomet. Chem.* **1971**, *28*, 293–300.
13. Petruj, J.; Zchnacker, S.; Marchal, J. *XIV ème Réunion Annuelle de la Coopération Franco Tchécoslovaque pour l'Etude du Vieillissement des Polymères Talsky-Mlyn;* 1980.
14. Petruj, J.; Marchal, J. *Radiat. Phys. Chem.* **1980**, *16*, 27.
15. *Comprehensive Organometallic Chemistry;* Wilkinson, G., Ed.; Elsevier: Amsterdam, Netherlands, 1982; Vol. 2, p 117.
16. Howard, J. A.; Scaiano, J. C. In *Radical Reaction Rates in Liquid: Oxy-, Peroxy-and Related Radicals;* Fisher, H.; Hellwege, K. H., Eds.; Springer-Verlag: Berlin, 1984; Vol. 13 of Landolt-Bornstein New Series II, part D.
17. Arnaud, R.; Moisan, J. Y.; Lemaire, J. *Macromolecules* **1984**, *17(3)*, 332–336.
18. Lacoste, J.; Carlsson, D. J. *J. Polym. Sci. Polym. Chem. Ed.* **1992**, *30*, 493.
19. Guivier, H. *Silicones, Techniques de l'Ingénieur;* Debeane: Strasbourg, 1982; Vol. A9, II, p A3475.
20. Goldovskii, E. A.; Kuzminskii, A. S. *Rev. Gen. Caout. Plast.* **1968**, *45(4)*, 458.
21. Biscontin, G. *ICOM Comm. Conserv. Sydney* **1987**, *2*, 785.
22. Israëli, Y. Ph.D. Thesis, Blaise Pascal University, Clermont-Ferrand (France), 1992.

Received for review January 26, 1994. Accepted revised manuscript December 5, 1994.

Photolysis of Methylene 4,4'-Diphenyldiisocyanate-Based Polyurethane Ureas and Polyureas

C. E. Hoyle, H. Shah, and K. Moussa

Department of Polymer Science, University of Southern Mississippi, Hattiesburg, MS 39406

The photodegradation of polyurethane ureas and polyureas has been shown to depend on hydrogen bonding as well as the inherent reactivity of the N-arylurethane and N-arylurea chromophores. The molecular weight of the polyetherdiamine that constitutes the soft segment effects the lability of the photoreactive functional groups. Comparable polyureas degraded at a faster rate than polyurethane ureas exposed to the same lamp source. Laser-flash photolysis indicated that a cleavage process is responsible for the primary photochemistry of urethane and urea groups.

POLYURETHANES BASED ON AROMATIC DIISOCYANATES [methylene 4,4'-diphenyldiisocyanate (MDI) and toluenediisocyanate (TDI)] have a wide variety of commercial applications. However, aromatic polyurethanes undergo severe discoloration and loss of mechanical properties upon exposure to UV radiation (*1–6*). As a result, several reports have been published (*1–30*) dealing with the photodegradation of polyurethanes over the last three decades. Also, a report (*31*) on the photodegradation of polyurethane ureas suggests that the photostability is influenced by the molecular weight of the polyol component. The nature of the matrix was related to the photostability: a correlation to the results of polyurethane photochemistry was made. Indeed a review of the polyurethane literature indicates two major pathways for photodecomposition, photo-Fries rearrangement and photooxidation (Scheme I). Model compound studies initially conducted by Beachell and Chang (*32*) and later substantiated by others (*33–39*) clearly demonstrated the occurrence of the photo-Fries rearrangement process. Photooxidation was proposed to proceed via hydro-

0065–2393/96/0249–0091$12.25/0

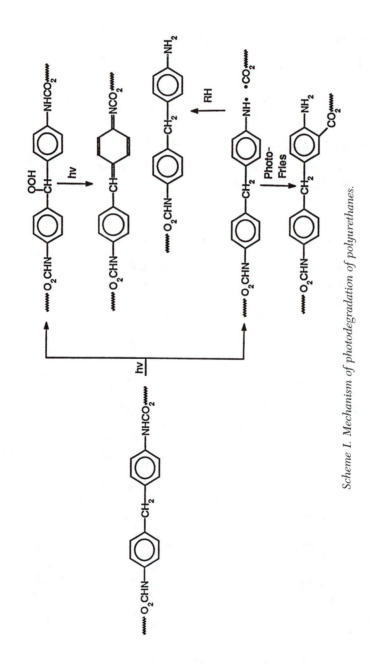

Scheme I. Mechanism of photodegradation of polyurethanes.

peroxide formation followed by a subsequent arrangement to quinoneimide-type structures. In support of this mechanism, Gardette and Lemaire (27) detected hydroperoxides during photolysis of polyurethanes. Interestingly, work by Lemaire and co-workers (28–30) also suggested that the photochemistry of polyurethanes is wavelength-dependent. They proposed that photo-Fries rearrangement is predominant at photolysis wavelengths of less than 340 nm, whereas photooxidation is predominant at irradiation wavelengths of greater than 340 nm. It is well known that the photo-Fries rearrangement products themselves absorb at longer wavelengths and form secondary photoproducts, which are highly colored and no doubt contribute to the discoloration. It has also been proposed that photooxidation processes can contribute to discoloration through formation of quinoneimide-type structures.

Polyurethanes consist of both hard and soft segments (40–42). The hard segments, comprised of the diisocyanate and a diol, are characterized by extensive hydrogen bonding. The soft segments are comprised of polyethers or polyesters with little hydrogen bonding. The morphology of polyurethanes can be varied readily by altering the ratio of hard to soft segments. It has been demonstrated (38) that, for an MDI-based polyurethane, as the flexibility of the polymer increases (with an increase in the soft segment content), photo-Fries rearrangement occurs to a greater extent. For a polyurethane elastomer based on MDI, butane diol, and polytetramethylene glycol, an increase in the amount of hard segments results in an increase in photo-Fries rearrangement and a decrease in gel formation (39). The reduced photoactivity has been attributed to a decrease in polymer flexibility and an increase in hydrogen-bonded carbonyl groups (with increasing hard-segment concentration).

In this chapter we consider the contribution of hydrogen bonding to the photodegradation process of polyurethane ureas and polyureas. As discussed previously, hydrogen bonding in polyurethanes can be varied by altering the ratio of hard to soft segments. Urea linkages exhibit substantial hydrogen bonding (43, 44). To demonstrate the effect of hydrogen bonding in urea linkages on photodegradation processes, we synthesized and characterized a series of polyurethane urea and polyurea compounds. We not only consider the effect of hydrogen bonding, but also take into account the inherent difference in the photolability of the urethane and urea linkages in the systems investigated.

Experimental Details

Materials. MDI (Miles) was purified by vacuum sublimation. The polyetherdiamines (Jeffamines ED series, $M_r = 600$ to 6000) were obtained from Texaco Chemical Company and dried at 45 °C under vacuum for 24 h before use. Ethylene glycol (99+%, anhydrous), ethylenediamine (99+%), dimethyl sulfoxide (99+%, anhydrous), 4-methyl-2-pentanone (99+%, anhydrous), phenylisocyanate

(98+%), propyl alcohol (99+%), propyl amine (99+%), and trifluoroaceticanhy-dride (99+%) were obtained from Aldrich Chemical Co. and used without further purification. 1,4-Dioxane (99%, Aldrich) was distilled and stored over molecular sieves before use. N,N-Dimethylformamide (ACS grade, Fisher) was vacuum distilled and stored over molecular sieves before use.

Synthesis of Polymers and Model Compounds. *Polymer and Model Compound Synthesis.*

Polyurethane ureas and polyureas were synthesized by a prepolymer method (45) according to the route shown in Scheme II. For polyurethane ureas, 1.25 g (5×10^{-3} mol, 5% molar excess) of MDI in 30 mL solvent (50/50 mixture of dimethyl sulfoxide and 4-methyl-2-pentanone) was reacted with 0.15 g (2.38×10^{-3} mol) of ethylene glycol at 100°(C for 2 h under flowing nitrogen. After cooling to room temperature, the prepolymer was chain-extended by dropwise addition of 2.38×10^{-3} mol of the appropriate polyetherdiamine in 40 mL solvent (50/50 mixture of dimethyl sulfoxide and 4-methyl-2-pentanone). The reaction mixture was then heated to 45 °C for 2 h. After cooling, the reaction mixture was poured into 500 mL of deionized water to precipitate the polymer. The polymer was then isolated by vacuum filtration and dried in a vacuum oven at 40 °C for one week.

For polyureas, 1.25 g (5×10^{-3} mol, 5% molar excess) of MDI in 30 mL solvent (50/50 mixture of dimethyl sulfoxide and 4-methyl-2-pentanone) was reacted with 2.38×10^{-3} mol of the appropriate polyetherdiamine at 45 °C for 2 h under a steady flowing nitrogen stream. After cooling to room temperature, the prepolymer was chain-extended by dropwise addition of 0.143 g (2.38×10^{-3} mol) of ethylene diamine in 10 mL solvent (50/50 mixture of dimethyl sulfoxide and 4-methyl-2-pentanone). The reaction mixture was then heated to 45 °C for 2 h. After cooling, the reaction mixture was poured into 500 mL of deionized water to precipitate the polymer. The polymer was then isolated by vacuum filtration and dried in a vacuum oven at 40 °C for one week.

Polymers were characterized by loss of the diisocyanate band at 2250 cm^{-1} and the appearance of urethane bands [3300 cm^{-1} (N–H stretch), 1750–1700 cm^{-1} and 1540 cm^{-1}]. Molecular weights were determined by gel permeation chromatography (GPC). For GPC analysis, the polymers were solubilized in tetrahydrofuran (THF) by addition of a drop of trifluoroaceticanhydride. On the basis of polystyrene standards, the polymer molecular weights were roughly determined to have a maximum of 5–10 polyetherdiamine units. The molecular weight of these polymer–oligomers was low because a catalyst was not used during the synthesis. For photodegradation studies, a small amount of polymer solution in dimethylformamide (DMF) (2 mg/mL) was cast on a NaCl or quartz plate and placed in

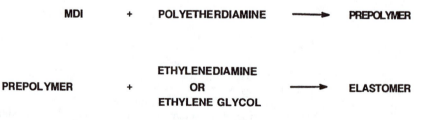

Scheme II. Synthesis of polyurethane ureas and polyureas.

vacuum oven at 50 °C for 1–2 h. A small amount of surfactant (F99, 3M Corporation) was added to the casting solutions to improve wettability.

Model Compound Synthesis. Propyl *N*-phenyl carbamate and the bis(propylurethane) of MDI were synthesized using a previously reported procedure (*46*).

Propyl *N*-phenyl urea and the bis(propyl urea) of MDI were synthesized using the following procedure. A solution of appropriate isocyanate in 5 mL 1,4-dioxane (or DMF) was added dropwise to a stirred solution of propyl amine in 5 mL of 1,4-dioxane (or DMF) in a nitrogen atmosphere. A white precipitate was formed immediately. The reaction was continued for 3 h at room temperature, and the solution was vacuum filtered to obtain a white precipitate. The precipitate was recrystallized from an appropriate solvent to yield a white powder. Elemental analysis calculated for propyl *N*-phenyl urea ($C_{10}H_{14}N_2O$: C, 67.39; H, 7.92; N, 15.72) found C, 68.33; H, 7.96; N, 15.87. Elemental analysis calculated for bis(propyl urea) of MDI ($C_{21}H_{28}N_4O_2$: C, 68.45; H, 7.66; N, 15.20) found C, 68.25; H, 7.73; N, 15.14.

Instrumentation. Fourier transform IR spectra were recorded on a Perkin-Elmer 1600 series instrument. UV-visible spectra were recorded on a Perkin-Elmer Lambda 6 UV-visible spectrophotometer.

Fluorescence lifetimes were determined on a single photon counter from Photochemical Research Associates. GPC was performed on a Waters system (solvent delivery pump, Model 6000A; refractive index detector, Model 410; and 10^5, 10^4, 500, and 100 Å columns) with THF as a mobile phase. The laser-flash photolysis system, shown in Scheme III, consisted of a Lumonics 861 excimer laser, an Applied Photophysics Xenon lamp/monochromator/photomultiplier tube/auto offset probe, a Phillips PM 3323 digitizer, an Archimedes 410/1 computer (data acquisition software from Applied Photophysics), and an Applied Photophysics

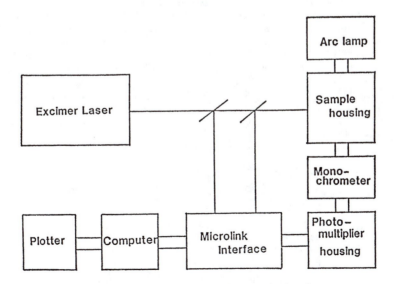

Scheme III. Schematic of a nanosecond laser flash photolysis system.

control unit. The laser, operated at 248 nm (Krypton Fluoride) in a charge-on-demand mode, had a nominal output 140 mJ/pulse.

A long-neck quartz fluorescence cuvet ($1 \times 1 \times 3$ cm) was used as a sample holder. All solutions were purged with nitrogen for 10 min before the experiment. The concentrations of all samples were adjusted to give an approximate absorbance of about 1.0 at 248 nm. Decay plots for the sample were acquired at 10-nm intervals, and the UV-vis spectrum of the transient was constructed from fitted decay plots.

Results and Discussion

The structures of the basic components used to make the polymers that are the subject of this chapter are shown in Table I. From these components, we synthesized (Scheme II) the series of polyurethane ureas and polyureas listed in Table II. The polyurethane ureas based on MDI, a polyetherdiamine, and ethylene glycol (EG) in a ratio of 2:1:1 are designated PUU600 to PUU6000; the "PUU" stands for polyurethane urea, and 600 through 6000 represent the approximate molecular weight of the polyetherdiamine. The polyureas based on MDI, a polyetherdiamine, and ethylenediamine (EDA) in the ratio of 2:1:1 are named accordingly PU600 to PU6000; "PU" stands for polyurea, and 600 through 6000 defines the approximate molecular weight of the polyetherdiamine.

Photodegradation of Polyurethane Ureas. Before proceeding directly to the results of the photodegradation study, we consider the hydrogen bonding trend in polyurethane ureas by observing the urethane carbonyl region in the IR spectrum. For all the polyurethane ureas investigated, a peak around 1705 cm^{-1} in the FTIR spectrum (Figure 1) can be attributed to hydrogen-bonded urethane carbonyl groups, whereas the peak around 1730 cm^{-1} can be attributed to nonhydrogen-bonded urethane carbonyl groups (41). Much more difficult to quantitatively resolve is the nonhydrogen-bonded (1695 cm^{-1}) (43, 44) and hydrogen-bonded urea carbonyl (1640–1660 cm^{-1}; may consist of singly and doubly hydrogen-bonded urea carbonyl groups) (43, 44) regions, especially for the PUU600, PUU900, and PUU2000 polyurethane ureas. Based on the relative intensity of the urea carbonyl peaks in the 1640–1700 cm^{-1} region, as well as an examination of the N–H stretching region at 3200–3500 cm^{-1} (not shown), we conclude that the extent of hydrogen-bonded urea carbonyl is greatly diminished for the PUU4000 and PUU6000 samples. In the ensuing paragraphs, we will correlate a qualitative trend in carbonyl hydrogen bonding with the photostability of the polyurethane ureas.

Table I. Polyether Diamine Oligomers

Name	$a + c$	b
ED-600	2.5	8.5
ED-900	2.5	15.5
ED-2001	2.5	40.5
ED-4000	2.5	86.0
ED-6000	2.5	131.5

NOTE: Values for $a + c$ and b refer to $H_2NC(CH_3)HCH_2-$
$(OC(CH_3)HCH_2)_a-(OCH_2CH_2)_b-(OCH_2C(CH_3)H)_c-NH_2$.

**Table II. List of Polyurethane Ureas
and Polyureas**

Polymer	Composition
PUU600	MDI:ED-600:EG is 2:1:1
PUU900	MDI:ED-900:EG is 2:1:1
PUU2000	MDI:ED-2001:EG is 2:1:1
PUU4000	MDI:ED-4000:EG is 2:1:1
PUU6000	MDI:ED-6000:EG is 2:1:1
PU600	MDI:ED-600:EDA is 2:1:1
PU900	MDI:ED-900:EDA is 2:1:1
PU2000	MDI:ED-2001:EDA is 2:1:1
PU4000	MDI:ED-4000:EDA is 2:1:1
PU6000	MDI:ED-6000:EDA is 2:1:1

NOTE: *See* text for structures of polyetherdiamines.

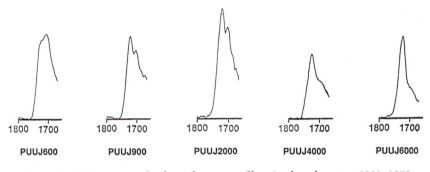

Figure 1. FTIR spectra of polyurethane urea films (carbonyl region, 1800–1650 cm^{-1}).

For photodegradation studies, polymer films, cast on NaCl quartz plates from a dilute polymer solution in DMF, were photolyzed (in the presence of air) in a Rayonet Reactor with broad band lamps having maxima at 300 nm. The progress was monitored by FTIR spectroscopy. Figures 2 and 3 show FTIR spectra of the PUU600 film photolyzed for 0, 1, and 3 h in the Rayonet Reactor. The appearance of a new band around 1780 cm^{-1} and the increase in the intensity and subsequent broadening of the band at 1700 cm^{-1} indicate that new carbonyl groups are being formed in the system. New bands are also observed around 3500 cm^{-1}, 1650 cm^{-1}, and at 1180 cm^{-1}. New bands around 1780 cm^{-1} and 1180 cm^{-1}, which are not observed when photolysis of MDI-based polyurethanes are conducted in vacuum, can be attributed to photooxidation products (*47*). The new band at 3500 cm^{-1} can be attributed to amines (*5, 18–21*), whereas the broadening of the band at 3300 cm^{-1} is indicative of the formation of hydroxyl groups (*5*) and peroxides (*5, 18–21, 27*). The overall broadening of the spectrum in the 3700–2700 cm^{-1} region is indicative of acid functionalities forming during photolysis. The decrease in intensity of the bands in the 3000–2800 cm^{-1} region with photolysis time is indicative of the loss of aliphatic CH$_2$ groups and has been observed in similar systems (*5, 18–21*).

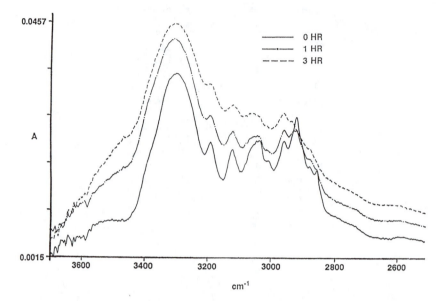

Figure 2. FTIR spectra (3700–2500 cm⁻¹) of PUU600 film after 0-, 1-, and 3-h photolysis in Rayonet Reactor with 300-nm lamps.

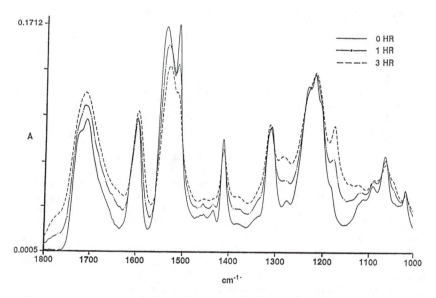

Figure 3. FTIR spectra (1800–1000 cm⁻¹) of PUU600 film after 0-, 1-, and 3-h photolysis in Rayonet Reactor with 300-nm lamps.

The loss of aromatic structure can be assessed by following either the decrease in the intensity of the absorbance in the 3000–3100 cm^{-1} region or the band around 1600 cm^{-1}. The decrease in the 3000–3100 cm^{-1} region is particularly difficult to assess quantitatively because there is a broadening that also occurs during photolysis. The loss of urethane structure can be assessed by the decrease in the intensity of bands around either 3300 cm^{-1} or 1540 cm^{-1}. We choose to monitor the N–H deformation band at 1540 cm^{-1} because the intensity of this band decreases uniformly and can be followed quantitatively as a function of photolysis time to provide an accurate assessment of the photostability of polyurethane ureas. Decker and co-workers (48) monitored the photostability of polyurethanes by following the intensity of bands at 1540 cm^{-1}, as well as at 1600 cm^{-1}. Figure 4 shows plots of the ratio I/I_o the intensity of the band at 1540 cm^{-1} before and after photolysis as a function of irradiation time for each of the polyurethane ureas. The relative loss of urethane structure (i.e., photodegradation) for PUU600, PUU900, and PUU2000 is much less than for PUU4000 and PUU6000. This result indicates a lower photostability for PUU4000 and PUU6000 and might be expected based on a lower degree of hydrogen-bonded carbonyls in PUU4000 and PUU6000. We cannot account at this time for the exact ordering of the photolability of PUU600, PUU900, and PUU2000 exhibited in Figure 4. Several morphological factors may be involved in addition to both doubly and singly hydrogen-bonded carbonyls.

Photodegradation of Polyureas. Polyurethane urea photostability is quite sensitive to the extent, and probably type, of hydrogen bonding. In a previous

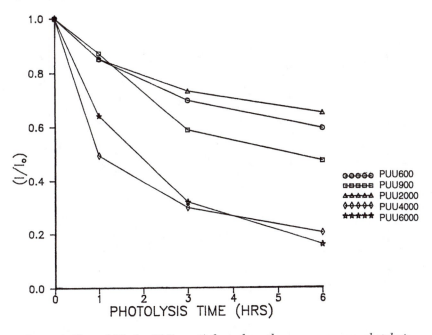

Figure 4. Plots of I/I_o (at 1540 cm^{-1}) for polyurethane ureas versus photolysis time in Rayonet Reactor with 300-nm lamps.

report (39), a direct relationship was observed between hydrogen bonding in polyurethanes and their photostability. It is reasonable to expect that pure polyureas
would also display a correlation between hydrogen bonding and the extent of
photodegradation. Before proceeding to a consideration of the effect of hydrogen
bonding in polyureas, we first concentrate on photolysis of the polyurea PU600
made from the lowest molecular weight polyetherdiamine. Figures 5 and 6 show
FTIR spectra of a PU600 film photolyzed for 0, 1, and 3 h with the 300-nm (broad
band) lamps in the Rayonet Reactor. Comparison of the results in Figures 2 and
3 with the results in Figures 5 and 6 reveals that FTIR spectra of unphotolyzed
and photolyzed PUU600 and PU600 are similar except in the carbonyl region,
1800–1640 cm^{-1}. In the case of the polyurea, PU600, only the nonhydrogen-
bonded urea carbonyl ($\sim$1695 cm^{-1}) and the hydrogen-bonded carbonyl(s)
($\sim$1640–1660 cm^{-1}) are present: There are no urethane carbonyls, of course, in
the pure polyureas. With an increase in photolysis time, the carbonyl region
around 1700 cm^{-1} increases in intensity and undergoes broadening and subsequently develops a discernible peak around 1730 cm^{-1} that is indicative of the
formation of new carbonyl groups. In addition, a new peak is also observed at
1190 cm^{-1}, which, as already discussed in our analysis of the polyurethane urea
photolysis in the previous section, is assigned to photooxidation product(s). The
new band at 3500 cm^{-1} is most likely due to amines (5, 18–21).

On the basis of a comparison with the polyurethane ureas, the broadening of
the band around 3300 cm^{-1} can be assigned to hydroxyl groups and peroxides,
and the overall broadening of the spectrum in the 3700–2700 cm^{-1} region is
tentatively assigned to acid groups formed during photolysis of the primary photoproducts. The loss of aliphatic methylene groups can be inferred from the decrease in the absorption intensity in the 3000–2800 cm^{-1} region. With increasing

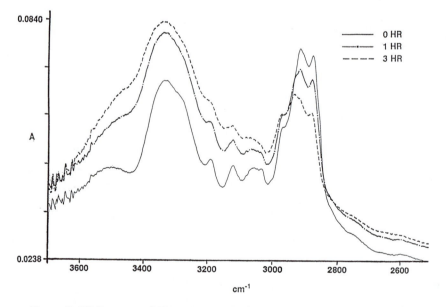

*Figure 5. FTIR spectra (3700–2500 cm^{-1}) of PU600 film after 0-, 1-, and 3-h
photolysis in Rayonet Reactor with 300-nm lamps.*

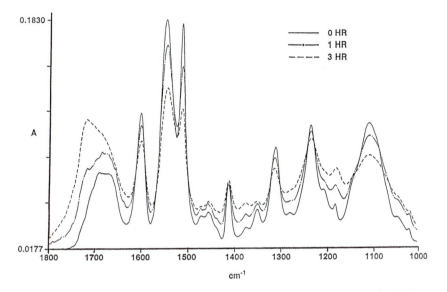

Figure 6. FTIR spectra (1800–1000 cm^{-1}) of PU600 film after 0-, 1-, and 3-h photolysis in Rayonet Reactor with 300-nm lamps.

photolysis time, the band intensities at 3100–3000 cm^{-1} and 1600 cm^{-1} decrease and indicate the loss of aromatic structure during photolysis. In addition, the decrease in intensity of the bands around 3300 cm^{-1} and 1540 cm^{-1} indicates that the urea structure is being destroyed during photolysis. As for the polyurethane ureas, the loss in the band at 1540 cm^{-1} , due to an N–H deformation, is a direct measure of the loss of urea functionality.

As in the case of the polyurethane ureas with polyetherdiamines of progressively higher molecular weight, we also monitored the loss in the N–H deformation band as a function of photolysis time for the series of polyureas made from polyetherdiamines ranging in molecular weight from 600 to 6000. As shown in Figure 7, the photostability of the polyurea decreases dramatically and systematically with increasing polyetherdiamine molecular weight. Does the trend in photolability directly correlate with a decrease in the extent of hydrogen bonding in the polyureas examined? Figure 8 shows representative IR spectra for PU600 and PU2000 for direct comparison in the carbonyl-stretching region. It is quite obvious that the intensity of the hydrogen-bonded carbonyl peaks (1640–1660 cm^{-1}) decreases dramatically in the PU2000 polyurea with respect to the nonhydrogen-bonded (free) carbonyl. This trend, an increase in total nonbonded carbonyl content with increasing polyetherdiamine molecular weight, follows the order: PU600 < PU900 < PU2000 < PU4000 < PU6000. Hence, we conclude that for polyureas, hydrogen bonding plays an important role in the photostability.

Comparison of Photostability of Polyurethane Ureas and Polyureas. Two polymers composed of the polyetherdiamine with molecular weight of about 2000 (PUU2000 and PU2000) were selected as representative examples for a direct comparison of the photostability of polyurethane urea and polyureas. Figures 9

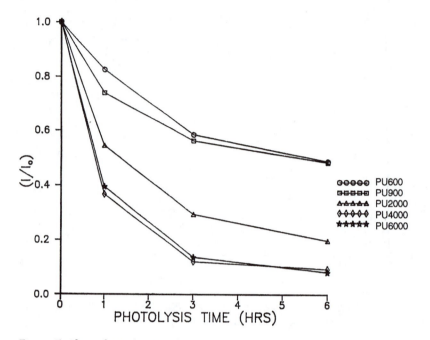

Figure 7. Plots of I/I_o (at 1540 cm^{-1}) for polyureas versus photolysis time in Rayonet Reactor with 300-nm lamps.

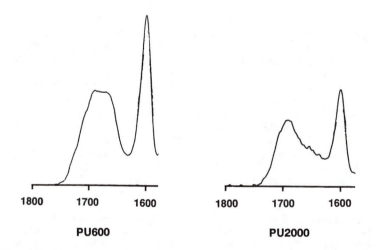

Figure 8. FTIR spectra of polyurea films (carbonyl region, 1800–1630 cm^{-1}).

and 10 show FTIR spectra of PUU2000 and PU2000 films, respectively (adjusted so that their absorbances were approximately equal in the wavelength range of the photolysis source), photolyzed for 0, 1, and 3 h with the 300-nm lamps in the Rayonet Reactor. Initially, the spectra of the polyurethane urea and polyurea are very similar except for the carbonyl region. Comparison of the photolyzed spectra reveals that the polyurea degrades much faster than the corresponding polyurethane urea. The loss of intensity in the N–H deformation region (1540 cm⁻¹) and the aromatic structure (1600 cm⁻¹), accompanied by buildup of new peaks in the carbonyl region (1800–1700 cm⁻¹), occur very rapidly for PU2000 compared with PUU2000. This result indicates, at least in this case, the enhanced photolability of polyureas.

Interestingly, a simple visual observation of both PU2000 and PUU2000 films during photolysis indicated that yellowing occurs to a greater extent in the polyurethane ureas. To confirm this observation, PUU2000 and PU2000 films photolyzed for 0, 1, and 3 h at 300 nm (Rayonet Reactor) were monitored by UV-vis spectroscopy. Figures 11 and 12 show UV-visible absorption spectra of PUU2000 and PU2000, respectively, as a function of photolysis time. From the results in Figures 11 and 12, we conclude that a higher degree of yellowing is indeed observed for the polyurethane urea (PUU2000) compared with the pure polyurea (PU2000). However, these results may be somewhat misleading because the rate of disappearance of the primary aryl absorption bands between 250 and 300 nm is markedly greater for PU2000 than PUU2000; PU2000 degrades faster than PUU2000 upon photolysis. This loss in UV absorbance is in agreement with FTIR results from Figures 9 and 10. Exactly the same trend (i.e., the polyurea is more photolabile than the corresponding polyurethane urea) was found for each of the member pairs in the series when directly compared (PUU600 is more stable than

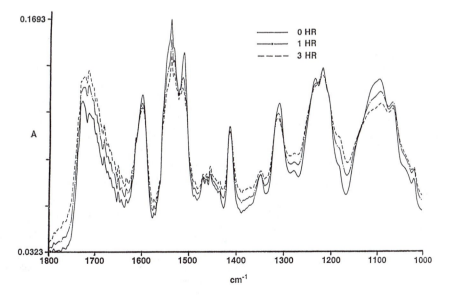

Figure 9. FTIR spectra of PUU2000 film after 0-, 1-, and 3-h of photolysis in Rayonet Reactor with 300-nm lamps.

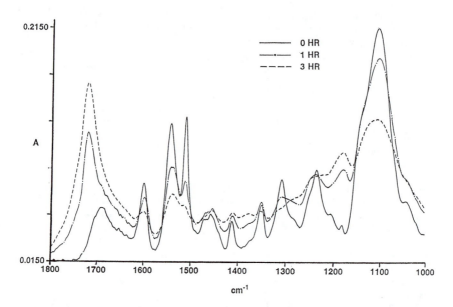

Figure 10. FTIR spectra of PU2000 film after 0-, 1-, and 3-h of photolysis in Rayonet Reactor with 300-nm lamps.

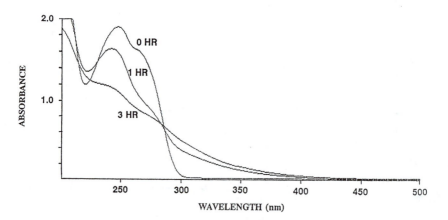

Figure 11. UV-vis spectra of PUU2000 film after 0-, 1-, and 3-h photolysis in Rayonet Reactor with 300-nm lamps.

PUU600; PUU900 is more stable that PU900; PUU4000 is more stable than PU4000; PUU6000 is more stable that PU6000). The results in Figures 9–12 raise important questions. Is the basic urea linkage more photolabile than the urethane linkage? Are the primary modes for photolytic decomposition of the urethane and urea linkages equivalent? To answer these questions, we conducted model compound studies.

Model Compound Studies. A major obstacle in photodegradation studies of polymers in general is the inability to conveniently measure and identify primary photochemical events. Accurate measurement of quantum efficiencies of photoprocesses occurring in polymer films is also extremely difficult. One way to circumvent these obstacles, at least to a limited extent, is to study appropriate, small-molecule model compounds for which one can easily separate and identify products as well as measure quantum efficiencies for various photolysis pathways. Bearing this in mind, we consider the photolysis of the propyl N-phenyl carbamate (**1**) and propyl N-phenyl urea (**2**) model compounds.

Both model compounds, adjusted to have an equal absorbance at 254 nm in dichloromethane, were photolyzed at 254 nm (medium-pressure mercury lamp, 254-nm bandpass filter, and 10-nm bandpass) for 15 min. The percent disappearance of the urea model compound, monitored by gas chromatography, was three times greater than for the urethane compound, indicating that the basic urea linkage is more photolabile than the urethane linkage. A similar conclusion was

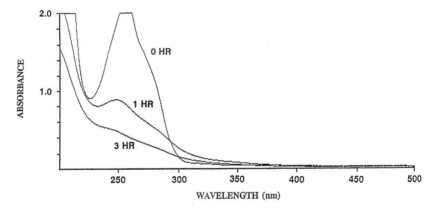

Figure 12. UV-vis spectra of PU2000 film after 0-, 1-, and 3-h photolysis in Rayonet Reactor with 300-nm lamps.

reached for other urea systems (33, 49). From the reported quantum yield of disappearance for the urethane model compound (Φ_{Loss} = 0.03, ref. 46), the quantum yield of disappearance for the urea model compound is estimated to be 0.09.

Because reactions occur from the singlet excited state of the molecule, the fluorescent lifetimes of the model compounds or polymers are important. If the singlet lifetime is long, one could envision using appropriate energy quenchers to inhibit the resulting photochemistry. Singlet lifetimes (Table III) of the simple urethane and urea models measured in our laboratory, as well as the bis(propyl urethane) (3) and bis(propyl urea) (4) model compounds, are all reasonably close to the reported value of 1.0 ns for ethyl N-phenyl carbamate (50). The singlet lifetimes in Table III are all very short and hence preclude the use of traditional quenchers at reasonable concentrations. Although not shown herein, polyurethane ureas and polyureas appear to have even shorter singlet lifetimes (perhaps due to self quenching) than those for the model compounds in Table III.

Having shown that the urea linkage is more photolabile than the urethane linkage, we proceed to answer the question: Does urea photochemistry occur by a different mechanism than for urethanes? To answer this question, we identified the transient intermediates produced during the photolysis of both models. Figures 13 and 14 show the transient absorption spectra of propyl N-phenyl carbamate and propyl N-phenyl urea, respectively [2.0 μs delay after pulsing with excimer laser (krypton fluoride at λ_{ex} = 248 nm)], in nitrogen saturated solutions. Both spectra are very similar and have a strong, fairly sharp peak around 300 nm and a very weak absorption between 400 and 440 nm. Land and Porter (51) have shown that a strong peak at 310 nm and a weak, broad absorbance centered around 400 nm is characteristic of an anilinyl or N-substituted anilinyl radical.

Table III. Singlet Lifetime of Model Compounds

Model Compound	Lifetime (ns)
Propyl N-phenyl carbamate	1.2
Bis(propyl carbamate) of MDI	1.6[a]
Propyl N-phenyl urea	1.8
Bis(propyl urea) of MDI	2.5–3.0

[a]Data reproduced from reference 49.

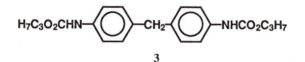

3

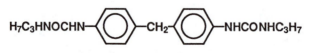

4

Osawa and co-workers (*12, 17*) reported the presence of an anilinyl radical during photolysis of MDI-based polyurethane by using electron spin resonance. On the basis of Land and Porter's (*51*) investigation of anilinyl and N-substituted anilinyl transients and the expectation from Osawa work (*12, 17*) that such radicals may be present, the relatively strong absorption at 300 nm and the weak absorbance in the 390–410-nm region in Figures 13 and 14 can be tentatively assigned, at least in part, to anilinyl and N-propylcarboxy phenyl radicals following previous assignments from our lab (*52, 53*). Incidentally, for completeness sake, we note that there may be some minor contribution from a radical cation (*54*) (formed via electron ejection) to the weak absorbance between 400 and 440 nm in Figures 13 and 14.

Figure 15 shows the transient absorption spectrum of the bis(propyl urea) of MDI in dichloromethane at a 2.0-(μs delay after the laser pulse in a nitrogen saturated solution. On the basis of comparison with transients in Figures 13 and 14, the absorbance around 300 nm can be assigned to anilinyl- and N-propylcarboxy-anilinyl-type radicals, and the absorbance above 400 nm can be assigned to a combination of an anilinyl radical, an N-propylcarboxy phenyl radical, and a radical cation. In past publications (*52, 53*), we identified the transient peak at 370 nm for the bis(propyl carbamate) of MDI as a methylene radical produced by cleavage of the C–H bond in the bridging methylene group. This cleavage is a very facile process due to the inherent stability of the diphenylmethyl radical: The weak peak at 370 nm for the bis(propyl urea) model may indicate the presence of the methylene radical. Finally, we conclude that, except for the question of the relative amounts of transients formed during photolysis of both the urethane and urea model compounds, the photodecomposition pathways appear to be the same. On the basis of photoproduct analysis for the simple urea and urethane models, we conclude that the N–C bond cleavage is especially efficient for the urea.

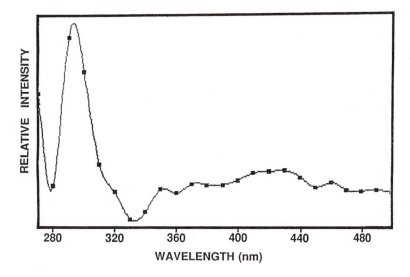

Figure 13. Transient absorption spectrum (2 μs) of 5.0 × 10⁻⁴ M propyl N-phenyl carbamate in nitrogen saturated dichloromethane (10 μs/division, λ_{ex} = 248 nm).

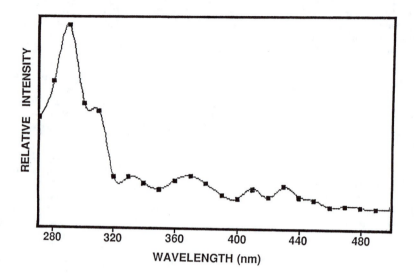

Figure 14. Transient absorption spectrum (2 μs) of 1.5 × 10⁻⁴ M propyl N-phenyl urea in nitrogen saturated dichloromethane (10 μs/division, λ_{ex} = 248 nm).

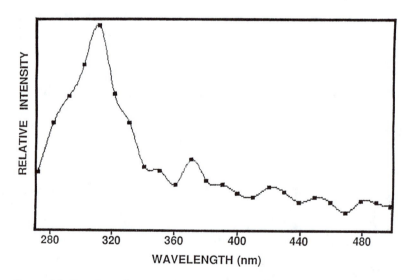

Figure 15. Transient absorption spectrum (2 μs) of (optical density = 0.83 at 248 nm) bis(propyl urea) of MDI in nitrogen saturated dichloromethane (10 μs/division, λ_{ex} = 248 nm).

Conclusions

Results for the effect of hydrogen bonding on the photodegradation of polyurethane ureas and polyureas are presented. The primary photochemical processes for urethane and urea model compounds are compared and quantum efficiencies for product formation determined. Specific conclusions are as follows:

1. The nature of the hydrogen bonding in polyurethane ureas and polyureas depends on the molecular weight of polyetherdiamine-based soft segment.
2. Polyurethane ureas exhibit enhanced photostability compared with comparable polyureas. However, a higher degree of yellowness was observed for photolyzed polyurethane ureas.
3. Model compound studies indicate that the basic cleavage mechanisms involved in the primary photolysis process of urethane and urea compounds are similar. However, the quantum yield for the disappearance of a urea model compound is three times greater than for a structurally similar urethane compound and may be responsible for the relatively greater photolability of the polyureas.

In general, we demonstrated the effect of hydrogen bonding on the photodegradation of polyurethane urea and polyureas. The effect of hydrogen bonding on the photodegradation of polyurethane and ureas remains to be demonstrated by varying the hard segment content of the polymer and measuring the effect on the photodegradation process. Other morphological effects should also be considered in order to develop a complete description of the photodegradation process of polyurethane ureas and polyureas.

Acknowledgment

We acknowledge the financial support of the Office of Naval Research, the NSF-EPSCoR Program, and the State of Mississippi.

References

1. Schollenberger, C. S.; Scott, H.; Moore, G. R. *Rubber World* **1958**, *4*, 549.
2. Schollenberger, C. S.; Pappas, L. G.; Park, J. C.; Vickroy, V. V. *Rubber World* **1960**, *139*, 81.
3. Schollenberger, C. S.; Dinsberg, K. *SPE Trans.* **1961**, *1*, 31.
4. Schollenberger, C. S.; Steward, J. D. *J. Elastoplast.* **1972**, *4*, 294.
5. Schollenberger, C. S.; Stewart, F. D. In *Advances in Urethane Science and Technology;* Frisch, K. D.; Reegan, S. L., Eds.; Technomic: Westport, CT, 1973; Vol. II, p 71.

6. Schollenberger, C. S.; Stewart, F. D. *In Advances in Urethane Science and Technology;* Frisch, K. D.; Reegan, S. L., Eds.; Technomic: Westport, CT, 1975; Vol. IV, p 68.
7. Nevskii, L. V.; Tarakanov, O. G.; Belyakov, V. *Soviet Plastics* **1966,** *7,* 45.
8. Nevskii, L. V.; Tarakanov, O. G. *Soviet Plastics* **1967,** *9,* 47.
9. Nevskii, L. V.; Tarakanov, O. G.; Belyakov, V. *Soviet Plastics* **1967,** *10,* 23.
10. Nevskii, L. V.; Tarakanov, O. G.; Belyakov, V. *J. Polym. Sci.* **1968,** *23,* 193.
11. Nevskii, L. V.; Tarakanov, O. G.; Kafengauz, A. P. *Soviet Plastics* **1972,** *1,* 46.
12. Osawa, Z.; Cheu, E. L.; Ogiwara, Y. *J. Polym. Sci. Polym. Lett Ed.* **1975,** *13,* 535.
13. Cheu, E. L.; Osawa, Z. *J. Appl. Polym. Sci.* **1975,** *19,* 2947.
14. Osawa, Z.; Cheu, E. L.; Nagashima, K. *J. Polym. Sci. Polym. Chem. Ed.* **1977,** *15,* 445.
15. Osawa, Z.; Nagashima, K. *Polym. Degrad. Stab.* **1979,** *1,* 311.
16. Osawa, Z.; Nagashima, K.; Ohshima, H.; Cheu, E. L. *J. Polym. Sci. Polym. Lett. Ed.* **1979,** *17,* 409.
17. Osawa, Z. In *Developments in Polymer Photochemistry;* Allen, N. S., Ed.; Applied Science: London, England, 1982; Vol. 3, p 209.
18. Rek, V.; Bravar, M. *J. Elast. Plast.* **1983,** *15,* 33.
19. Rek, V.; Bravar, M.; Jocic, T. *J. Elast. Plast.* **1984,** *16,* 256.
20. Rek, V.; Mencer, H. J.; Bravar, M. *Polym. Photochem.* **1986,** *7,* 273.
21. Rek, V.; Bravar, M.; Jocic, T.; Govorcin, E. *Die. Ang. Makr. Chem.* **1988,** *158/159,* 247.
22. Barbalata, A.; Caraculacu, A. A.; Lurea, V. *Eur. Polym. J.* **1978,** *14,* 427.
23. Abu-Zeid, M. E.; Nofal, E. F.; Tahseen, L. A.; Abdul-Rasoul, F. A.; Ledwith, A. *J. Appl. Polym. Sci.* **1984,** *29,* 2443.
24. Allen, N. S.; McKeller, J. F. *J. Appl. Polym. Sci.,* **1976,** *20,* 1441.
25. Noack, R.; Schwetlick, K. *Plaste Kautschuk* **1978,** *25,* 259.
26. Stumpe, J.; Schwetlick, K. *Polym. Deg. Stab.* **1987,** *17,* 103.
27. Gardette, J. L.; Lemaire, J. *Makromol. Chem.* **1981,** *182,* 2723.
28. Gardette, J. L.; Lemaire, J. *Makromol. Chem.* **1982,** *183,* 2415.
29. Gardette, J. L.; Lemaire, J. *Polym. Degrad. Stab.* **1984,** *6,* 135.
30. Lemaire, J.; Gardette, J.; Rivaton, A.; Roger, A. *Polym. Degrad. Stab.* **1986,** *15,* 1.
31. Thapliyal, B. P.; Chandra, R. *Polym. Int.* **1991,** *24,* 7.
32. Beachell, H. C.; Chang, I. L. *J. Polym. Sci. Part A-1,* **1972,** *10,* 503.
33. Schultz, H. *Makromol. Chem.* **1973,** *172,* 57.
34. Schultz, H. Z. *Naturforsch.* **1973,** *28,* 339.
35. Schwetlick, K.; Stumpe, J.; Noack, R. *Tetrahedron* **1979,** *35,* 63.
36. Hoyle, C. E.; Kim, K. J. *J. Polym. Sci. Part A: Polym. Chem.* **1986,** *24,* 1879.
37. Hoyle, C. E.; Chawla, C. P.; Kim, K. J. *J. Polym. Sci. Part A: Polym. Chem.* **1988,** *26,* 1295.
38. Hoyle, C. E.; Kim, K. J. *J. Polym. Sci. Part A: Polym. Chem.* **1987,** *25,* 2631.
39. Hoyle, C. E.; Kim, K. J.; No, Y. G.; Nelson, G. L. *J. Appl Polym. Sci.* **1987,** *34,* 763.
40. Cooper, S. L.; Tobolsky, A. V. *J. Appl. Polym. Sci.* **1967,** *11,* 1361.
41. Seymour, R. W.; Estes, G. M.; Cooper, S. L. *Macromolecules* **1970,** *3(5),* 579.
42. Seymour, R. W.; Cooper, S. L. *Macromolecules* **1973,** *6,* 48.
43. Ishihara, J.; Kimura, I.; Saito, K.; Ono, H. *J. Macromol. Sci. Phys.* **1974,** *10,* 591.
44. Paik Sung, C. S.; Smith, T. W.; Sung, N. H. *Macromolecules* **1980,** *13,* 117.
45. Lyman, D. J.; Hill, D. W.; Stirk, R. K.; Adamson, C.; Mooney, B. R. *Trans. Am. Soc. Artif. Intern. Organs* **1972,** *18,* 19.
46. Herweh, J. E.; Hoyle, C. E. *J. Org. Chem.* **1980,** *45,* 2195.

47. No, Y. G. Ph.D. Dissertation, University of Southern Mississippi, 1988.
48. Decker, C.; Moussa, K.; Bendaikha, T. *J. Polym. Sci., Polym. Chem. Ed.* **1991,** *29,* 739.
49. Schwetlick, K.; Noack, R.; Schmieder, G. *Z. Chem.* **1972,** *12,* 109.
50. Kim, K. J., Ph.D. Dissertation, University of Southern Mississippi, 1987.
51. Land, E. J.; Porter, G. *Trans. Faraday Soc.* **1963,** *59,* 2027.
52. Hoyle, C. E.; No, Y. G.; Malone, K. G.; Thames, S. F.; Creed, D. *Macromolecules* **1988,** *21,* 2727.
53. Hoyle, C. E.; Ezzell, K. S.; No, Y. G.; Malone, K. ; Thames, S. F. *Polym. Degrad. Stab.* **1989,** *25,* 325.
54. Lewis, G. N.; Lipkin, D. *J. Am. Chem. Soc.* **1942,** *64,* 2801.

RECEIVED for review December 27, 1993. ACCEPTED revised manuscript February 6, 1995.

Formation and Removal of Singlet (a¹Δ$_g$) Oxygen in Bulk Polymers: Events That May Influence Photodegradation

P. R. Ogilby[1], M. Kristiansen[1], D. O. Mártire[1], R. D. Scurlock[1],
V. L. Taylor[2], and Roger L. Clough[2]

[1]Department of Chemistry, University of New Mexico, Albuquerque, NM
87131
[2]Sandia National Laboratories, Albuquerque, NM 87185

*The lowest excited state of molecular oxygen, singlet oxygen [$O_2(a^1 \Delta_g)$],
is an intermediate in many photooxygenation reactions. In polymeric
materials, these reactions may play a role in the photodegradation of
the macromolecule or of additives (such as dyes) dissolved in the matrix. We used time-resolved spectroscopy to directly monitor the behaviors of $O_2(a^1\Delta_g)$ generated in bulk polymer matrices. This chapter
describes some of our recent results concerning (a) mechanisms and
quantum yields for the photo-induced formation of $O_2(a^1\Delta_g)$ in polymers, (b) $O_2(a^1\Delta_g)$ lifetimes in different types of polymers, (c) comparison of rate constants for the quenching of $O_2(a^1\Delta_g)$ by additives in
amorphous polymer glasses and in liquid solutions, and (d) characteristics of chemical reactions between $O_2(a^1\Delta_g)$ and solute molecules dissolved in a polymer glass.*

THE GROUND STATE of molecular oxygen is a triplet-spin state [$O_2(X^3\Sigma_g^-)$].
The lowest excited electronic state of oxygen is a singlet, $O_2(a^1\Delta_g)$, which lies
94.3 kJ/mol (22.5 kcal/mol) above $O_2(X^3\Sigma_g^-)$. The $O_2(a^1\Delta_g)$–$O_2(X^3\Sigma_g^-)$ energy
gap corresponds to a phosphorescent transition in the near IR region (1270
nm), which provides a method by which $O_2(a^1\Delta_g)$ can be directly monitored
(1). This spectroscopic probe can be used in several ways: The kinetics of
$O_2(a^1\Delta_g)$ formation and removal can be quantified in time-resolved measure-

0065–2393/96/0249–0113$12.00/0

ments, and the intensity of the phosphorescence can be used to determine $O_2(a^1\Delta_g)$ yields. Data from $O_2(a^1\Delta_g)$ phosphorescence measurements can be supplemented with luminescence and flash-absorption data from organic molecules to provide a unique and informative perspective of processes that occur in the oxygen–organic molecule photosystem. We have applied these techniques in studies of a variety of organic polymers both above and below the glass transition temperature (T_g) (2–6).

$O_2(a^1\Delta_g)$ is an intermediate in many photooxygenation reactions (7). In some cases, these reactions may be important in the degradation of organic materials (7–12). For example, the production of an allylic hydroperoxide via the "ene" reaction may precede a variety of radical reactions that result from cleavage of the labile O–O bond (see Scheme I).

$O_2(a^1\Delta_g)$ may also be important in the photobleaching of dyes or other additives dissolved in the polymer. Polymeric materials susceptible to these reactions are used in a wide range of applications, including new electrooptic systems (e.g., nonlinear optical devices and organic light emitting diodes). However, even if $O_2(a^1\Delta_g)$ is not an intermediate in reactions that result in the degradation of an organic polymer, experiments in which $O_2(a^1\Delta_g)$ is monitored can be useful probes of other photoinduced processes which, in turn, may be directly related to the degradation of the material. In this chapter, we discuss the formation and removal of $O_2(a^1\Delta_g)$ in bulk organic polymers from the perspective of events that may influence the photodegradation of polymeric materials.

$O_2(a^1\Delta_g)$ Formation

$O_2(a^1\Delta_g)$ can be formed in bulk polymers by several photoinduced processes. It has been shown by several investigators, including ourselves, that $O_2(a^1\Delta_g)$ can be produced in polymers by energy transfer from an excited-state organic molecule to $O_2(X^3\Sigma_g^-)$. We recently demonstrated that $O_2(a^1\Delta_g)$ can also be formed upon irradiation into the oxygen–organic molecule charge-transfer (CT) band. This CT absorption band derives from the interaction between

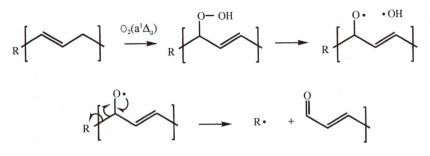

Scheme I.

oxygen and the polymer in which it is dissolved or between oxygen and a solute dissolved in the polymer.

Photosensitized Production of $O_2(a^1\Delta_g)$.

$O_2(a^1\Delta_g)$ can be formed by energy transfer from a molecule dissolved as a solute in the polymer matrix, or from a chromophore that is an integral part of the polymer itself, such as the carbonyl group in polycarbonate (2, 3) (Scheme II).

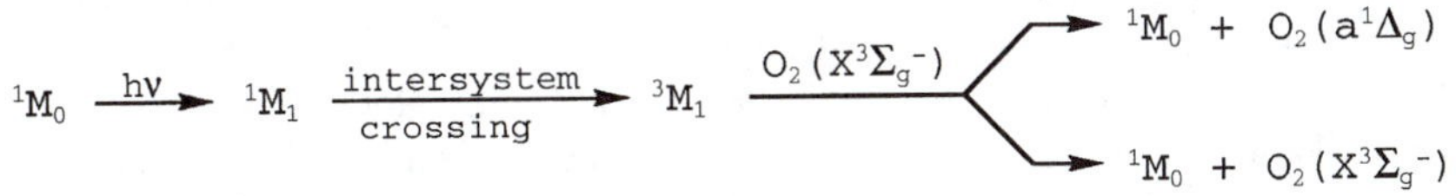

$$^1M_0 \xrightarrow{h\nu} {}^1M_1 \xrightarrow[\text{crossing}]{\text{intersystem}} {}^3M_1 \xrightarrow{O_2(X^3\Sigma_g^-)} \begin{cases} {}^1M_0 + O_2(a^1\Delta_g) \\ {}^1M_0 + O_2(X^3\Sigma_g^-) \end{cases}$$

Scheme II.

The singlet state ($^1M_1$) lifetime of organic-molecule photosensitizers is typically less than 20 ns. [In designating organic molecule (M) electronic states, numerical superscripts identify the spin state, and subscripts identify either the ground (0) or first excited (1) state.] In liquid solutions where solute diffusion coefficients are on the order of $1–5 \times 10^{-5}$ cm²/s, it is possible for $O_2(X^3\Sigma_g^-)$ to encounter $^1M_1$ within the $^1M_1$ lifetime. Indeed, the oxygen-induced deactivation of $^1M_1$ is usually limited by diffusion, and it is possible, in certain cases, to produce $O_2(a^1\Delta_g)$ by energy transfer from $^1M_1$ (13–17). In amorphous polymer glasses, however, the oxygen diffusion coefficient is typically less than $\sim 1 \times 10^{-7}$ cm²/s, and unimolecular decay channels of $^1M_1$ are more efficient than bimolecular quenching by $O_2(X^3\Sigma_g^-)$. Thus, in rigid media, $^1M_1$ should rarely, if ever, be a precursor to $O_2(a^1\Delta_g)$.

Organic molecule triplet state ($^3M_1$) lifetimes, however, are sufficiently long that quenching by $O_2(X^3\Sigma_g^-)$ is an efficient process, even in amorphous polymers below T_g. Nevertheless, because solute diffusion coefficients in the glass are small, $^3M_1$ deactivation by $O_2(X^3\Sigma_g^-)$, and the corresponding rate of $O_2(a^1\Delta_g)$ production, is significantly slower than in a liquid solvent (2, 3). In aerated or oxygenated polymer glasses, rates of $O_2(a^1\Delta_g)$ decay often exceed the rate of $O_2(a^1\Delta_g)$ formation in a photosensitized process (2, 3). Thus, in analyzing time-resolved $O_2(a^1\Delta_g)$ phosphorescence data from such polymers, we must deconvolute the decay function of the $O_2(a^1\Delta_g)$ precursor to obtain the intrinsic $O_2(a^1\Delta_g)$ decay function (2, 3). The $O_2(a^1\Delta_g)$ precursor decay function can be obtained in a flash absorption experiment in which $^3M_1$ is monitored.

The Quantum Yield of $O_2(a^1\Delta_g)$ in a Bulk Polymer.

In liquid solvents, the quantum yield, ϕ_Δ, of $O_2(a^1\Delta_g)$ in a photosensitized reaction depends on the sensitizer, M, and on the solvent, among other variables (18).

Under conditions where $^3M_1$ is the sole precursor to $O_2(a^1\Delta_g)$, the product $\phi_T f_\Delta^T$ defines ϕ_Δ. The parameter ϕ_T is the $^3M_1$ quantum yield, and f_Δ^T is the fraction of $^3M_1$ states that yield $O_2(a^1\Delta_g)$ upon interaction with $O_2(X^3\Sigma_g^-)$.

In independent experiments (6), we measured the quantum yield of acridine-sensitized $O_2(a^1\Delta_g)$ production in an amorphous polystyrene glass by using a reaction in which $O_2(a^1\Delta_g)$ was chemically trapped and by comparison of the $O_2(a^1\Delta_g)$ phosphorescence intensity to a known standard. Upon 355-nm pulsed-laser irradiation of acridine at ~1.3 mJ/pulse, the $O_2(a^1\Delta_g)$ quantum yield in polystyrene [ϕ_Δ^{PS}(acridine) = 0.56 ± 0.05] is somewhat smaller than that obtained in the liquid solvent analog toluene [ϕ_Δ^{TOL}(acridine) = 0.83 ± 0.06]. The ϕ_Δ^{PS} value also depends inversely on the incident laser power. This latter phenomenon is not uncommon when pulsed lasers are used as the photolysis source, where nonlinear effects can contribute at high incident powers (18).

Several additional experiments were performed to help in the interpretation of these data. Because solute diffusion coefficients are smaller in the glass, it was important to ascertain whether or not the polymer data reflected a less-efficient scavenging of the acridine triplet state by oxygen. Changes in the triplet state lifetime on the admission of oxygen into the system indicated that in both the glass and liquid, 99.9% of the acridine triplets formed are quenched by $O_2(X^3\Sigma_g^-)$. [The lifetime of ^{1}acridine in both media is short enough to preclude quenching by oxygen (13–15)]. It is also possible that the inequality ϕ_Δ^{PS}(acridine) < ϕ_Δ^{TOL}(acridine) may reflect a larger component of $O_2(a^1\Delta_g)$ quenching by the sensitizer itself in the solvent cage of the more rigid polymer. If such quenching was more pronounced in the polymer, one might expect the ratio $\phi_\Delta^{PS}/\phi_\Delta^{TOL}$ to be smaller for sensitizers that can more-efficiently quench $O_2(a^1\Delta_g)$. Results obtained by using different sensitizers, however, do not appear to support this hypothesis. At higher incident-laser powers, despite efficient quenching of ^{3}acridine by $O_2(X^3\Sigma_g^-)$, the lower $O_2(a^1\Delta_g)$ yields may result (1) from an "inner-filter" effect in which absorption by ^{3}acridine effectively shields ground-state acridine from the incident light, and/or (2) from depletion of the initial ^{3}acridine population by multiphoton absorption to form species that are not $O_2(a^1\Delta_g)$ precursors.

One objective of this ϕ_Δ study was to provide more insight into the potential contribution of $O_2(a^1\Delta_g)$ in photooxygenation reactions in a solid matrix. Specifically, if the $O_2(a^1\Delta_g)$ yield was substantially less in a solid polymer than in a liquid, it would then appear that $O_2(a^1\Delta_g)$ was not a likely intermediate in photoinduced reactions that degrade either the macromolecule itself or low molecular weight additives within the solid matrix. Whatever the reasons for the slightly lower ϕ_Δ value obtained in the macromolecular matrix, it is clear that a photosensitized process in a solid, air-saturated polymer can indeed produce a substantial amount of $O_2(a^1\Delta_g)$.

Production of $O_2(a^1\Delta_g)$ upon Charge-Transfer Band Irradiation. $O_2(a^1\Delta_g)$ can be produced in polymers that are free of solutes or chromophores, which can otherwise act as photosensitizers, by irradiation into the oxygen–polymer CT absorption band (2, 4). This feature in the absorption spectrum usually appears as a red shift in the absorption onset subsequent to dissolution of oxygen in the material, and it is attributed to a transition from a ground-state oxygen–organic molecule complex to a CT state $(M^+O_2^{-\bullet})$ (19, 20). The CT state is an $O_2(a^1\Delta_g)$ precursor (16, 20).

In a triplet-state photosensitized process, photon absorption precedes the rate-limiting encounter between $^3M_1$ and $O_2(X^3\Sigma_g^-)$. Consequently, the rate of signal appearance in a pulsed-laser, time-resolved $O_2(a^1\Delta_g)$ phosphorescence experiment can be comparatively slow in an amorphous polymer glass due to the small diffusion coefficients. On the other hand, the encounter between oxygen and the organic component necessarily precedes the absorption of a photon in the transition that produces the CT state. Thus, the rate of $O_2(a^1\Delta_g)$ signal appearance upon pulsed irradiation into the CT band is much faster than that observed in a photosensitized reaction in polymer glasses. This difference in the rates of $O_2(a^1\Delta_g)$ signal appearance in time-resolved measurements has two important ramifications:

1. Upon CT band irradiation, it is not necessary to deconvolute the decay kinetics of the $O_2(a^1\Delta_g)$ precursor from the observed $O_2(a^1\Delta_g)$ phosphorescence signal to obtain the intrinsic rate of $O_2(a^1\Delta_g)$ decay.
2. $O_2(a^1\Delta_g)$ phosphorescence measurements can be used to monitor the production of $O_2(a^1\Delta_g)$ sensitizers that result from the thermal or photochemical degradation of the medium.

An example of how $O_2(a^1\Delta_g)$ data can be used to follow events associated with the photodegradation of a polymer is shown in Figure 1. Upon pulsed irradiation into the oxygen–polymer CT band of a freshly prepared polystyrene sample, the rate of $O_2(a^1\Delta_g)$ phosphorescence signal appearance is rapid. After prolonged irradiation into the CT band, however, the *intensity* of the $O_2(a^1\Delta_g)$ phosphorescence signal progressively increases, and the *rate* of $O_2(a^1\Delta_g)$ signal appearance becomes slower. These results indicate that, subsequent to prolonged irradiation, $O_2(a^1\Delta_g)$ is also being produced by energy transfer from a photosensitizer. These sensitizers arise by the photooxidation of the polymer. If sites of unsaturation are already present in the polymer (as chain defects, for example), the $O_2(a^1\Delta_g)$ produced upon CT band irradiation may, via the "ene" reaction, form sensitizing chromophores (e.g., carbonyls) with a concomitant cleavage of the macromolecular chain (*vide supra*). Alternatively, chromophores capable of producing $O_2(a^1\Delta_g)$ may arise at a saturated center through the intermediacy of the CT state. A saturated hydroperoxide, for example, could be formed by the sequence of events in Scheme III (21).

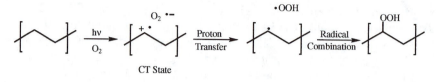

Scheme III.

Alkyl and aryl hydroperoxides serve as both photochemical and thermal precursors of products such as ketones, aldehydes, and alkenes that characterize a "degraded" polymer (*4*). These products may not only sensitize the production of $O_2(a^1\Delta_g)$ during the absorption of subsequent photons, but some may also provide a center suitable for reaction with $O_2(a^1\Delta_g)$ or for reaction via other mechanisms.

$O_2(a^1\Delta_g)$ *Removal*

$O_2(a^1\Delta_g)$ can be removed from a particular system by physical or reactive quenching channels. A physical quencher deactivates $O_2(a^1\Delta_g)$ to $O_2(X^3\Sigma_g^-)$.

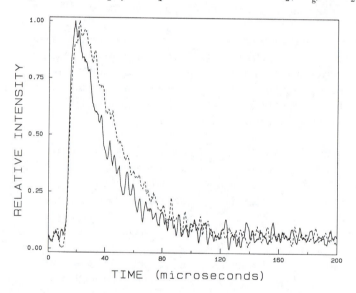

Figure 1. Time-resolved $O_2(a^1 \Delta_g)$ phosphorescence recorded from a polystyrene glass subsequent to pulsed-laser irradiation at 341 nm. Data recorded from a freshly prepared sample (——) that had not previously been irradiated had faster rates of both appearance and disappearance than data recorded from the same sample subsequent to three separate 30-min periods of photolysis at 341 nm (---). Between photolysis sessions, the sample was allowed to stand at 25 °C under 84 kPa of oxygen for 4 days. The data have been scaled to the same intensity to better show rate differences. (Reproduced from reference 4. Copyright 1990 American Chemical Society.)

Even in the absence of an added quencher, the host medium or solvent will itself deactivate $O_2(a^1\Delta_g)$ to $O_2(X^3\Sigma_g^-)$ (*1, 22*). However, the addition of specific solutes can often provide a more efficient process by which $O_2(a^1\Delta_g)$ may be removed. Many amines, for example, are particularly effective at inducing $O_2(a^1\Delta_g)$ deactivation (*7*). $O_2(a^1\Delta_g)$ can also be removed by chemical reaction to form oxygenated products (*vide supra*).

Intrinsic Lifetime of $O_2(a^1\Delta_g)$ in Bulk Polymers.

The lifetime of $O_2(a^1\Delta_g)$ is strongly influenced by the surrounding medium (*1, 7*). The probability of $O_2(a^1\Delta_g)$ participating in a chemical reaction depends, in part, on its intrinsic lifetime in that medium. For example, the medium-induced deactivation of $O_2(a^1\Delta_g)$ may be rapid compared with the rate at which $O_2(a^1\Delta_g)$ can encounter a reaction partner by diffusion. Measuring the lifetime of $O_2(a^1\Delta_g)$ in bulk polymers is thus important in evaluating the potential role of $O_2(a^1\Delta_g)$ in the photochemistry of polymeric materials.

We have determined the lifetime of $O_2(a^1\Delta_g)$ in a number of common polymer matrices at 25 °C (Table I). The lifetimes are similar to those obtained in liquid solvents of analogous molecular composition. The data are consistent with a model in which the electronic excitation energy of $O_2(a^1\Delta_g)$ is deposited in vibrational modes of the host medium, particularly in the C–H and O–H stretching modes. Thus, the $O_2(a^1\Delta_g)$ lifetime decreases with increasing concentration of C–H and O–H bonds in the polymer. There is no indication from our results that matrix rigidity has a significant influence on $O_2(a^1\Delta_g)$ lifetimes. For example, the $O_2(a^1\Delta_g)$ lifetime in a poly(methyl methacrylate) (PMMA) glass is ~22 μs. In poly(ethyl acrylate), a rubbery material (i.e., T_g < 25 °C) that has a molecular composition closely related to that of PMMA, $O_2(a^1\Delta_g)$ has a lifetime of ~31 μs.

Rate Constants for $O_2(a^1\Delta_g)$ Removal in Bulk Polymers.

The removal of $O_2(a^1\Delta_g)$ by physical and reactive quenching channels can be described, respectively, by the kinetic reactions in Scheme IV

$$O_2(a^1\Delta_g) + Q \underset{k_{-diff}}{\overset{k_{diff}}{\rightleftarrows}} {}^1[\,O_2(a^1\Delta_g)\cdots Q\,] \overset{k_{isc}}{\longrightarrow} {}^3[\,O_2(X^3\Sigma_g^-)\cdots Q\,] \longrightarrow O_2(X^3\Sigma_g^-) + Q$$

$$O_2(a^1\Delta_g) + Q \underset{k_{-diff}}{\overset{k_{diff}}{\rightleftarrows}} {}^1[\,O_2(a^1\Delta_g)\cdots Q\,] \overset{k_{rxn}}{\longrightarrow} Q\text{-}O_2$$

Scheme IV.

where Q is the $O_2(a^1\Delta_g)$ quencher, k_{diff} is the bimolecular rate constant for the diffusion-dependent encounter of two solutes, and k_{-diff} is the unimolecular

Table I. Singlet Oxygen Lifetimes in Bulk Polymer Matrices

Polymer	Structure	Intrinsic Lifetime (μs)
Poly(4-methyl-1-pentene)		18 ± 2
Polystyrene		19 ± 2
Poly(methyl methacrylate)		22 ± 3
Polycarbonate		29 ± 2
Poly(ethyl acrylate)		31 ± 1
Poly(dimethyl siloxane)		46 ± 1
Perdeuteriopolystyrene		250 ± 15
DuPont Teflon AF		1700 ± 100

rate constant for diffusion-dependent dissociation of an encounter pair. The overall physical quenching rate constant, k_q, can be expressed in terms of k_{diff}, k_{-diff}, and k_{isc} by invoking the steady-state approximation for the Q–O$_2$ encounter pair. The rate constant for quencher-induced intersystem crossing in oxygen is denoted by k_{isc}. A similar treatment for the reactive scheme yields the overall rate constant for $O_2(a^1\Delta_g)$ removal by reaction, k_r, in terms of k_{diff}, k_{-diff}, and k_{rxn}. The rate constant for reaction from the Q–O$_2$ encounter pair is denoted by k_{rxn}.

$$k_q = k_{diff} \, k_{isc}/(k_{-diff} + k_{isc})$$

$$k_r = k_{diff} \, k_{rxn}/(k_{-diff} + k_{rxn})$$

The rate constants k_q and k_r can be measured by quantifying the rate constant k_Δ for $O_2(a^1\Delta_g)$ removal as a function of the quencher concentration [k_0 is the reciprocal of the intrinsic $O_2(a^1\Delta_g)$ lifetime] (5).

$$k_\Delta = k_0 + k_q[Q] \qquad \text{or} \qquad k_\Delta = k_0 + k_r[Q]$$

For very efficient quenchers, k_r or k_q approaches k_{diff}, and we say that the process is limited by diffusion (with the comparatively small molecule oxygen, $k_{diff} \approx 3 \times 10^{10} \text{ s}^{-1} \text{ M}^{-1}$ in liquids). At this limit, a change from a liquid solvent to a polymer glass, with the concomitant reduction in diffusion coefficients, should result in a *decrease* in k_r or k_q. The data are indeed consistent with this expectation (Table II).

For poor quenchers, a change from a liquid solvent to a polymer glass results in an *increase* in the quenching rate constant (Table II). This phenomenon is understood by recognizing that in a solid matrix, where comparatively small diffusion coefficients yield correspondingly small values of k_{-diff}, $O_2(a^1\Delta_g)$, and Q will undergo more collisions within the solvent cage and thus increase the chance for quenching before dissociation.

The data in Table II indicate that rate constants for $O_2(a^1\Delta_g)$ removal in a polymer glass are not directly proportional to those recorded in a liquid solution, and in fact the data span a much smaller range compared with liquid solvents. Thus, in attempts to ascertain whether or not a polymer additive acts as a stabilizer by removing $O_2(a^1\Delta_g)$, k(polymer) rather than k(liquid) should be used in interpreting longevity data from quencher-doped polymers.

Removal of $O_2(a^1\Delta_g)$ by Reaction: Unique Features of a Solid Matrix.

Removal of $O_2(a^1\Delta_g)$ by *chemical* reaction provides a model for investigating photooxygenations in solid materials. A variety of polycyclic aromatic hydrocarbons, such as 9,10-diphenylanthracene (DPA), react with $O_2(a^1\Delta_g)$ generated either chemically (7), by UV/vis irradiation (7), or by ion-

Table II. Rate Constants for $O_2(a^1\Delta_g)$ Removal at 25 °C

Quencher	Toluene	Polystyrene	Poly(methyl methacrylate)
[nickel complex structure: S,P=Ni,S,P]	$(6.5\pm0.1)\times10^9$	$(2.0\pm0.1)\times10^8$	$(1.4\pm0.2)\times10^7$
[amine structure]	$(2.4\pm0.1)\times10^8$	$(9\pm1)\times10^7$	$(3.5\pm0.4)\times10^6$
[anthracene structure: Ph, OCH$_3$]	$(1.3\pm0.1)\times10^8$	$(2.4\pm0.3)\times10^7$	
[naphthacene structure: Ph, Ph, Ph, Ph]	$(4.5\pm0.3)\times10^7$	$(1.6\pm0.15)\times10^7$	
[anthracene structure: Ph]	$(7.8\pm0.3)\times10^5$	$(9.8\pm0.7)\times10^5$	
$(Ph)_3N$	$(1.6\pm0.1)\times10^5$	$(3.0\pm0.4)\times10^5$	$(6.2\pm0.5)\times10^5$

NOTE: Representative values for the oxygen diffusion coefficient (cm^2 s^{-1}) are as follows: 5×10^{-5} for liquid toluene, 2×10^{-7} for polystyrene glass, and 1×10^{-8} for poly(methyl methacrylate) glass. Rate constant units are in M^{-1} s^{-1}. The nickel complex and the two amines are physical quenchers. The three aromatic hydrocabons remove $O_2(a^1\Delta_g)$ by reaction. Errors are two standard deviations in the slope of k_Δ vs. [Q].

izing radiation (23) via a $_\pi2 + _\pi4$ cycloaddition to yield an endoperoxide (see Scheme V).

Data from photolysis experiments involving amorphous polymer glasses doped with an aromatic hydrocarbon are consistent with a process that removes $O_2(a^1\Delta_g)$ by chemical reaction (24). In polystyrene samples containing rubrene (5,6,11,12-tetraphenylnaphthacene) or DPA

1. UV/vis absorption measurements indicate that the aromatic hydrocarbon is being consumed during the experiment.
2. Upon dissolution of the irradiated polymer sample in benzene, the aromatic endoperoxide can be recovered.

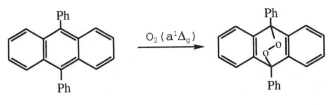

Scheme V.

3. $O_2(a^1\Delta_g)$ lifetime measurements indicate that $O_2(a^1\Delta_g)$ is quenched by the aromatic hydrocarbons.

Furthermore, with each successive photolysis pulse, the intensity of the $O_2(a^1\Delta_g)$ phosphorescence signal and the decay rate of the $O_2(a^1\Delta_g)$ precursor both decrease, which is consistent with consumption of the oxygen dissolved in the polymer matrix. The $O_2(a^1\Delta_g)$ phosphorescence intensity and precursor decay rate recover during re-equilibration of the sample in air.

In the self-sensitized photooxygenation of rubrene, the quantum yield of the $O_2(a^1\Delta_g)$ reaction product (rubrene endoperoxide) is greater in the liquid than in the polymer glass, particularly at high rubrene concentrations. This result derives principally from differences between liquid- and solid-phase solute diffusion coefficients. Thus, in the liquid, where oxygen and rubrene are more mobile, the $O_2(a^1\Delta_g)$ generated by one excited-state rubrene molecule can more readily interact with other rubrene molecules in the system. Also in the liquid, $O_2(X^3\Sigma_g^-)$ can more readily induce deactivation of ^{1}rubrene ($\tau_s \approx 16$ ns) to produce both $O_2(a^1\Delta_g)$ and ^{3}rubrene, which is itself a $O_2(a^1\Delta_g)$ precursor. [Because ϕ_T for rubrene is inherently small in the absence of oxygen, the deactivation of ^{1}rubrene by $O_2(X^3\Sigma_g^-)$ ultimately results in a higher yield of $O_2(a^1\Delta_g)$]. In the polymer glass, however, where the mobility of oxygen is reduced and rubrene is essentially immobilized, the reaction of $O_2(a^1\Delta_g)$ with rubrene molecules other than the "parent" photosensitizing molecule is restricted on the time scale defined by the $O_2(a^1\Delta_g)$ lifetime, and $O_2(X^3\Sigma_g^-)$-induced deactivation of ^{1}rubrene is precluded.

Useful information can also be obtained from experiments in which two different quenchers compete for the $O_2(a^1\Delta_g)$ produced. In a sample containing both rubrene and the physical quencher 1,4-diazabicyclo[2.2.2]octane (DABCO), the $O_2(a^1\Delta_g)$ produced during rubrene irradiation can either be deactivated to $O_2(X^3\Sigma_g^-)$ by DABCO or react with rubrene to yield the endoperoxide. In such a system, where rubrene sensitizes its own $O_2(a^1\Delta_g)$-mediated photooxygenation, it can be expected that at the limit of high DABCO concentration, all the $O_2(a^1\Delta_g)$ would be quenched except that reacting with the "parent" sensitizer molecule within a solvent cage (*see* Scheme VI).

Data obtained by using polystyrene samples are consistent with this expectation (Figure 2); an increase in the DABCO concentration blocks all but a fixed amount of rubrene endoperoxide formation. Furthermore, and as expected, the quantum yield of rubrene endoperoxide at high DABCO concentration ($\phi = 0.021$) is independent of the rubrene concentration.

When these same competitive quenching experiments are performed in toluene, the quantum yield of rubrene endoperoxide at high DABCO concentration ($\phi = 0.01$) is a factor of 2 smaller than that observed in polystyrene. This difference between the solid and liquid phase data is even more pronounced for DPA, which has a quantum yield of unquenchable endoperoxide in polystyrene ($\phi = 0.0022$) that is 22 times larger than that in toluene ($\phi = 0.0001$). These results are consistent with a "cage escape" channel (*vide supra*)

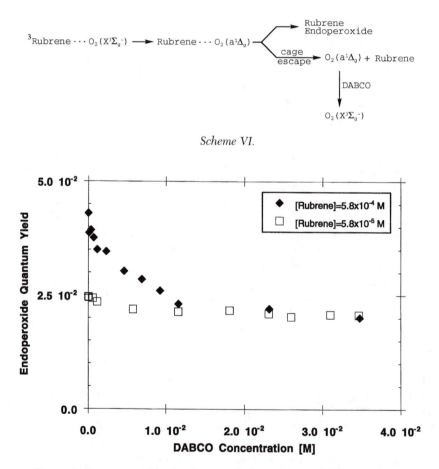

Scheme VI.

Figure 2. Quantum yield of rubrene endoperoxide as a function of DABCO concentration in an amorphous polystyrene glass.

that is more pronounced in the liquid where diffusion coefficients are much larger. In the polymer, the rigid matrix inhibits the escape of $O_2(a^1\Delta_g)$ from its parent sensitizer. Thus, when a molecule capable of acting both as a sensitizer and a reactant is present in a solid polymer at the limit of *low* concentration, where an out-of-cage reaction will be insignificant, the quantum yield of reaction with $O_2(a^1\Delta_g)$ can be *greater* than that observed in an analogous liquid-phase system. These experiments also indicate that in solid polymer matrices that contain low concentrations of a species that can act both as $O_2(a^1\Delta_g)$ sensitizer and reactant, the addition of a $O_2(a^1\Delta_g)$ quencher (i.e., stabilizer) may not inhibit the oxygenation reaction. Thus, cage effects can play an important role in reactions that may influence the degradation of additives, or of the macromolecule itself.

Summary and Conclusions

The formation and removal of $O_2(a^1\Delta_g)$ in bulk polymers can be monitored using time-resolved spectroscopy. Such experiments indicate that $O_2(a^1\Delta_g)$ can arise within solid materials by two photo-induced mechanisms: a photosensitized process involving either dissolved dye molecules or chromophores on the macromolecule, and subsequent to irradiation into the CT band that arises upon oxygen dissolution in the polymer. The quantum yield of $O_2(a^1\Delta_g)$ in a photosensitized process was determined in one polymer glass (polystyrene) and found to be substantial ($\sim$0.6). Lifetimes of $O_2(a^1\Delta_g)$ in common polymers are similar to those in liquid-solvent analogs, and these lifetimes vary inversely with the concentration of C–H bonds in the macromolecule. Rate constants for quenching of $O_2(a^1\Delta_g)$ by additives can differ substantially in polymers versus liquid solvents. Efficient quenchers exhibit greatly reduced rate constants in polymer glasses, whereas poor quenchers exhibit somewhat larger rate constants. The quantum yield of $O_2(a^1\Delta_g)$-derived reaction products, arising from low concentrations of molecules able to serve as both sensitizer and reaction partner, can be significantly higher in a polymer glass than in a liquid solution. Differences in the diffusion coefficient for oxygen, along with attendant changes in the dynamics of the solvent cage, play an important role in the photophysics and photochemistry of bulk solids as compared with liquids.

Acknowledgments

The assistance of M. Malone (Sandia) is gratefully acknowledged. This work was supported by the Department of Energy under contract numbers DE-AC04-4AL85000 and DE-AC04-76DP00789 and by a grant from the Army Research Office.

References

1. Gorman, A. A.; Rodgers, M. A. J. In *CRC Handbook of Organic Photochemistry*; Scaiano, J. C., Ed.; CRC: Boca Raton, FL, 1989; Vol. 2, pp 229–247 and references cited therein.
2. Ogilby, P. R.; Dillon, M. P.; Gao, Y.; Iu, K.-K.; Kristiansen, M.; Taylor, V. L.; Clough, R. L. In *Structure–Property Relations in Polymers;* Urban, M. W.; Craver, C. D., Eds.; Advances in Chemistry Series 236; American Chemical Society: Washington, DC, 1993; pp 573–598.
3. Clough, R. L.; Dillon, M. P.; Iu, K.-K.; Ogilby, P. R. *Macromolecules* **1989**, *22*, 3620–3628.
4. Ogilby, P. R.; Kristiansen, M.; Clough, R. L. *Macromolecules* **1990**, *23*, 2698–2704.
5. Ogilby, P. R.; Dillon, M. P.; Kristiansen, M.; Clough, R. L. *Macromolecules* **1992**, *25*, 3399–3405.
6. Scurlock, R. D.; Mártire, D. O; Ogilby, P. R.; Taylor, V. L.; Clough, R. L. *Macromolecules* **1994**, *27*, 4787–4794.
7. *Singlet Oxygen;* Frimer, A. A., Ed.; CRC: Boca Raton, FL, 1985.
8. Ranby, B.; Rabek, J. F. *Photodegradation, Photooxidation, and Photostabilization of Polymers*; Wiley: New York, 1975.
9. Allen, N. S.; McKellar, J. F. *Macromol. Rev.* **1978**, *13*, 241–281.
10. *Photochemistry of Dyed and Pigmented Polymers;* Allen, N. S.; McKeller, J. F., Eds.; Applied Science: London, 1980.
11. Carlsson, D. J.; Wiles, D. M. *J. Polym. Sci. Polym. Phys.* **1973**, *11*, 759–765.
12. Carlsson, D. J.; Wiles, D. M. *Rubber Chem. Technol.* **1974**, *47*, 991–1004.
13. Scurlock, R. D.; Ogilby, P. R. *J. Photochem. Photobiol. A* **1993**, *72*, 1–7.
14. Iu, K.-K.; Ogilby, P. R. *J. Phys. Chem.* **1987**, *91*, 1611–1617. (Erratum: *J. Phys. Chem.* **1988**, *92*, 5854.)
15. Iu, K.-K.; Ogilby, P. R. *J. Phys. Chem.* **1988**, *92*, 4662–4666.
16. Kristiansen, M.; Scurlock, R. D.; Iu, K.-K.; Ogilby, P. R. *J. Phys. Chem.* **1991**, *95*, 5190–5197.
17. Saltiel, J.; Atwater, B. W. *Adv. Photochem.* **1988**, *14*, 1–90.
18. Wilkinson, F.; Helman, W. P.; Ross, A. B. *J. Phys. Chem. Ref. Data* **1993**, *22*, 113–262.
19. Tsubomura, H.; Mulliken, R. S. *J. Am. Chem. Soc.* **1960**, *82*, 5966–5974.
20. Scurlock, R. D.; Ogilby, P. R. *J. Phys. Chem.* **1989**, *93*, 5493–5500.
21. Onodera, K; Furusawa, G.-I.; Kojima, M.; Tsuchiya, M.; Aihara, S.; Akaba, R.; Sakuragi, H.; Tokumara, K. *Tetrahedron* **1985**, *41*, 2215–2220.
22. Scurlock, R. D.; Ogilby, P. R. *J. Phys. Chem.* **1987**, *91*, 4599–4602.
23. Clough, R. L. *J. Am. Chem. Soc.* **1980**, *102*, 5242–5245.
24. Clough, R. L.; Taylor, V. L.; Kristiansen, M.; Scurlock, R. D.; Ogilby, P. R., submitted for publication.

RECEIVED for review January 24, 1994. ACCEPTED revised manuscript January 17, 1995.

Photodegradation Mechanisms of Poly(p-phenylene sulfide) and Its Model Compounds

Z. Osawa, S. Kuroda, S. Kobayashi, and F. Tanabe

Department of Chemistry, Faculty of Engineering, Gunma University, Kiryu, Gunma 376, Japan

Photodegradation of poly(p-phenylene sulfide) (PPS) and its model compounds diphenyl sulfide, 1,4-bis(phenylthio)benzene, and bis(4-phenylthiophenyl)sulfide was performed. Products were analyzed by using various techniques. During photoirradiation of PPS, yellowing and an increase in absorbance in the UV-vis spectrum were observed. The insoluble portion in hot 1-chloronaphthalene was formed, and the molecular weights of the samples estimated by differential scanning calorimetry increased with irradiation time. These results suggest that cross-linking took place. IR spectra indicated the formation of sulfoxide, 1,2,4-substituted benzene, and other products. Photoirradiation of the model compounds resulted in the formation of sulfoxide and biphenyl structures. On the basis of the results obtained, a photodegradation mechanism of PPS was proposed.

Engineering plastics usually contain aromatic structures that are very susceptible to UV light. The increasing use of these plastics under UV light has prompted photodegradation studies. Significant studies on the photodegradation of engineering plastics such as polycarbonate, poly(phenylene oxide), polyoxymethylene, polyesters, and aliphatic polyamide have been conducted to date. However, little work has been done on the photodegradation of poly(p-phenylene sulfide) (PPS) (1). Yellowing of PPS during photoirradiation prevents its extensive commercial application when exposed to light. In this chapter, a photodegradation mechanism of PPS is proposed on the basis of results obtained in the photodegradation of PPS and its model compounds.

0065–2393/96/0249–0127$12.00/0

Experimental Details

Materials. Polystyrene film (thickness, 38 μm) was supplied by Toray Co., Ltd., Japan. Model compounds were the following oligomers of *p*-phenylene sulfide: diphenyl sulfide (M-2), 1,4-bis(phenylthio)benzene (M-3), and bis(4-phenyl-thiophenyl)sulfide (M-4). They were taken as the dimer, trimer, and tetramer models of PPS, respectively. M-2 was a commercial reagent and was purified in the usual manner. M-3 and M-4 were synthesized according to Koch and Heitz (2). The following compounds were also synthesized as models of photoreaction products: 4-biphenylyl phenyl sulfide (4-BPS), 2-biphenylyl phenyl sulfide (2-BPS), and 4,4'-bis(phenylthio)biphenyl (4,4'-BPTB).

Photoirradiation. Photoirradiation of PPS was carried out in air with a 300-W, high-pressure Hg lamp (ORC, HML-300/A-OM) through a soft glass filter that cut off light of <300 nm. The PPS films were placed 20 cm from the light source and were kept at ca. 45 °C during irradiation. Another irradiation was performed to monitor changes in IR spectra. The combination of a Xe lamp (SUGA, XBF-2500W/1) and a TOSHIBA L-42 glass filter made the wavelength of the incident light longer than 400 nm and reduced the secondary reactions. The temperature of the films placed 30 cm away from the light source was maintained at 25 °C. Model compounds placed in petri dishes or test tubes were irradiated with a high-pressure Hg lamp through a Pyrex glass filter in an air or nitrogen atmosphere.

Measurements. Characterization of the photoirradiated PPS was carried out with UV-vis and IR spectrometries and gel permeation chromatography (GPC) and differential scanning calorimetry (DSC). Photodegradation products of the model oligomers were analyzed by gas chromatography (GC), GC–mass spectrometry (MS), and UV-vis and IR spectrometries. UV-vis spectra were recorded on a HITACHI spectrophotometer (model 228). IR spectra were recorded on a JASCO Fourier transform IR (FTIR) 8000 spectrophotometer. GPC measurements were carried out with 1-chloronaphthalene as the eluent by using two Shodex A-80M/S columns at 210 °C (3). Thermal analyses of PPS films were performed on a RIGAKU DSC 8230B combined with a TAS 100. Gas chromatograms were obtained by using a Silicone SE-30 (5%) column and heating from 70 to 300 °C. GC–MS measurements were performed with a SHIMADZU GC–MS-QP1000 at electron impact of 70 eV and heating of the column from 50 to 300 °C.

Results and Discussion

Photocoloration of PPS Film. During photoirradiation of PPS films, a yellowing was observed. Figure 1 shows the changes in UV-vis spectra during photoirradiation in air. The increase in absorbance is pronounced in the absorption tail (i.e., the 400–500-nm region).

The increase in absorbance at 400 nm is plotted in Figure 2. The absorbance increases continuously during the photoirradiation, but the rate is slowed in the later stage. Considering the incident-light wavelength and the absorptivity of PPS, we assumed that photoreactions proceeded near the film surface. Actually, decreasing the film thickness from 38 to 5 μm had little

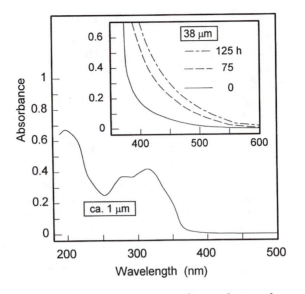

Figure 1. Changes in UV-vis spectra during photoirradiation.

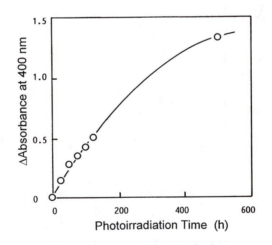

Figure 2. Changes in absorbance at 400 nm during photoirradiation.

effect on the changes in the UV-vis spectra. This superficial degradation seemed to become pronounced as coloration proceeded.

Cross-Linking Reaction.   Photoirradiation of PPS also led to cross-linking. The amount of the insoluble portion in hot 1-chloronaphthalene, which had a dark-brownish color, increased with irradiation time and became ca. 4 wt% of the original polymer after 125-h exposure.

The molecular weights of the irradiated PPS films were determined for their 1-chloronaphthalene soluble parts by GPC using the polystyrene samples with narrow molecular weight distributions as standards. The changes in the GPC traces are shown in Figure 3, and the evaluated molecular weights are summarized in Table I.

The fraction with high molecular weight, around 10^6, increased with irradiation time in Figure 3. In addition, the value of average weight-average molecular weight divided by average number-average molecular weight $(\overline{M}_w/\overline{M}_n)$, a measure of molecular weight distribution, increased slightly. These facts support the mechanism that cross-linking occurred during the photoirradiation of PPS.

On the other hand, M_n decreased appreciably, whereas M_w changed little. Two interpretations are possible for these results. Simultaneous chain scission and cross-linking may have proceeded as in the case of photodegradation of polysulfones (4, 5). Also, the average molecular weights may have been un-

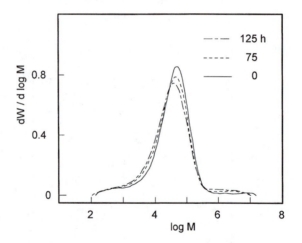

Figure 3. Changes in molecular weight distribution during photoirradiation in air (1-chloronaphthalene soluble portion).

**Table I. Changes in Molecular Weight of 0.5 wt%
1-Chloronaphthalene Soluble Parts Determined
by GPC at 210 °C**

IRT (h)	$\overline{M}_n \times 10^{-3}$	$\overline{M}_w \times 10^{-5}$	$\overline{M}_n/\overline{M}_w$
0	6.69	1.93	28.9
25	6.17	1.77	28.7
50	6.18	1.99	32.2
75	6.07	1.99	32.8
100	5.81	1.82	31.4
125	5.69	1.85	32.5

NOTE: IRT is irradiation time.

derestimated because the insoluble portions were not real gels but were materials with low solubility due to the very high molecular weight (6). In either case, an alternative method was required to estimate the number of cross-links formed during the photodegradation.

The alternative method for monitoring the changes in molecular weight was carried out to measure the heat of crystallization (ΔH_c) (7, 8). To prevent the cross-linking reaction during a DSC measurement, ΔH_c was recorded on cooling from 305 °C in a N_2 atmosphere immediately after the sample reached 305 °C.

Changes in the DSC curve during photoirradiation are shown in Figure 4. Photoexposure broadens the exothermic peak as a result of crystallization. The peak area decreased with irradiation time. This decrease is ascribed to the formation of cross-linked structures that restrict the mobility of PPS molecules, widen the crystallizing temperature range, and decrease the ΔH_c value.

Port and Still (7) reported that M_n of PPS is proportional to $\Delta H_c^{-5.4}$. Suwa et al. (8) concluded that the increasing entanglements with larger molecular weight decrease ΔH_c. The molecular weights of the cross-linked polymers, therefore, might be underestimated because the cross-linking points act as the permanent entanglements. However, Port's equation can still be applied to the present system for two reasons. First, the equation was obtained for the nonlinear PPS synthesized by the Macallum polymerization, which is known to form the branched structures (9, 10). Second, the PPS irradiated in the present study contained only a small amount of insoluble portions, which means the number of cross-links were small enough even if the insoluble portions were real gels.

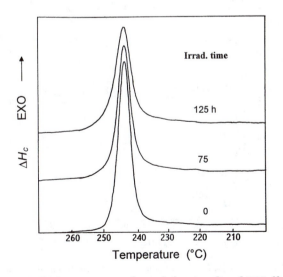

Figure 4. DSC curves on cooling of photoirradiated PPS films.

Consequently, the molecular weights of the irradiated PPS were estimated as:

$$\overline{M}_{\Delta H_c} = 3.38 \times 10^9 \, \Delta H_c^{-5.4} \tag{1}$$

where $\overline{M}_{\Delta H_c}$ is the average molecular weight estimated by ΔH_c (cal/g) measurements. The subscript, ΔH_c, distinguishes the molecular weight from M_n. The front factor in eq 1 was estimated by using M_n obtained by GPC and ΔH_c for the undegraded polymer. The evaluated molecular weights are plotted in Figure 5.

Theory of cross-linking predicts the number-average molecular weights of degraded polymers as follows (11):

$$\overline{M}_n = \frac{\overline{M}_{n,0}}{1 - u_n x}$$

where $\overline{M}_{n,0}$ and u_n are the number-average molecular weight and the number-average degree of polymerization of the initial undegraded polymer, respectively, and x is the number of cross-links per structural unit. By using $\overline{M}_{\Delta H_c}$ instead of $\overline{M}_n$, the apparent number of cross-links per initial molecule, $\zeta_{x,app}$, was successively calculated by the following equation:

$$\zeta_{x,app} = 1 - \frac{\overline{M}_{n,0}}{\overline{M}_{\Delta H_c}}$$

$\zeta_{x,app}$ increased during photoirradiation in a manner similar to the increase in the absorbance at 400 nm. The relation between $\zeta_{x,app}$ and change in absorb-

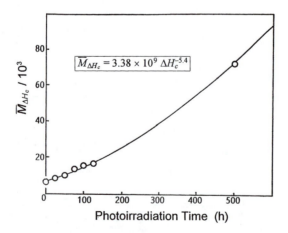

Figure 5. Changes in molecular weight during photoirradiation estimated from the heat of crystallization.

ance at 400 nm is shown in Figure 6. The first-order correlation between the two parameters is in good agreement, which suggests that both the cross-linking and the coloration took part in the same photoreaction pathway of PPS.

Chemical Changes during Photoirradiation. Changes in FTIR spectra were dependent on the incident-light wavelength. When irradiation was performed with a high-pressure Hg lamp through a soft glass filter in air, the phenolic O–H stretching bands appeared around 3400 and 3200 cm^{-1} and broad absorption bands appeared around 1700 and 1200 cm^{-1}. On the other hand, when irradiation was carried out with a Xe lamp through a L-42 glass filter in air, significant changes were observed. The differential FTIR spectra of PPS film irradiated for 100 h are illustrated in Figure 7. In addition to the formation of phenolic moieties (ν_{O-H} at 3466 and 3225 cm^{-1}), the formation of sulfoxide ($\nu_{S=O}$, 1049 cm^{-1}) and the transformation of 1,4-disubstituted benzene into 1,2,4-trisubstituted benzene (δ_{C-H} out of plane, 1718 cm^{-1}; δ_{C-H} in plane, 1161 cm^{-1}) were noticed. The results are summarized in Table II. Among these changes, the formation of the 1,2,4-trisubstituted benzene is definitely for the cross-linking of PPS.

Photodegradation of Model Compounds. To clarify the photo-degradation mechanism of PPS, photoreaction products of the model oligo-mers were analyzed. The simplest model oligomer was the dimer model, M-2. M-2 changed into a brownish color during photoirradiation in air as well as in N$_2$. The colored fraction of the irradiated M-2 was difficult to dissolve in

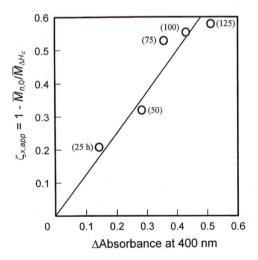

Figure 6. Relation between $\zeta_{x,app}$ and change in absorbance at 400 nm. (Numbers in parentheses indicate irradiation time.)

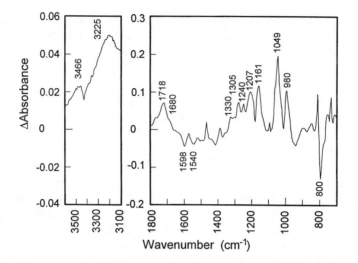

Figure 7. Differential FTIR (attenuated total reflection) spectra of PPS film irradiated for 100 h by using a Xe lamp and an L-42 glass filter in air.

Table II. Changes in IR Bands of PPS during Photoirradiation

Substances	Vibrational Mode	Wavenumber (cm^{-1})	Change
1-Substituted benzene	Skeletal	1598	−
1,4-Disubstituted	Skeletal	1540	−
benzene	δ C–H out of plane	800	−
1,2,4-(1,3-)Substituted	δ C–H out of plane	1718	+
benzene	δ C–H in plane	1161, 980	+
Sulfoxide	ν S=O	1049	+
Sulfone	ν_s SO$_2$	1330, 1305	+
	ν_{as} SO$_2$	1161	+
Phenol	ν O–H	3466, 3225	+
	ν C–O	1240	+
Aryl ether	ν C–O	1240	+
Cyclohexadienone	ν C=O	1680	+

NOTE: Substances and vibrational modes are tentative assignments; − means decrease; + means increase.

methanol and was easily condensed. Gas chromatograms of the photoirradiated M-2 and the colored fraction shown in Figure 8 indicate that the photoirradiation of M-2 in bulk forms diphenyl sulfoxide (P9), dibenzothiophene (P6), thianthrene (P10), thiols (P5 and P8), and compounds containing disulfide (–SS–) or biphenyl structures. These compounds were confirmed by GC–MS. No oxidation products such as phenol or benzenesulfenic acid were detected in addition to the diphenyl sulfoxide. Also, under an N$_2$ atmosphere

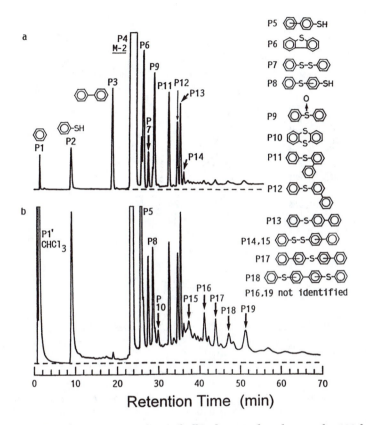

Figure 8. Gas chromatogram of M-2 (bulk) photoirradiated in air for 130 h in a Pyrex tube (a) and the M-2 colored fraction (b).

no formation of the sulfoxide was observed. Sulfides with the biphenyl structure (P17 and P18) are pronounced in the colored fraction.

The trimer (M-3) and the tetramer (M-4) models of PPS were also photoirradiated and analyzed. Sulfoxide, M-4, and compounds containing biphenyl structure were identified in the photoreaction products of M-3 irradiated in bulk. Irradiation of M-4 in benzene resulted in the formation of 4-BPS and 4,4'-BPTB as predominant products. Formation of these compounds apparently indicates that oxidation of sulfur and C–S bond cleavage followed by phenylation and arylation take place during the photoirradiation of the model oligomers. Study of photodegradation of the model oligomers will be reported in more detail elsewhere.

UV-vis spectra of the colored fraction of the photoreaction products of M-2 and the model compounds are shown in Figure 9. Three absorption maxima [232 (I), 251 (II), and 275 (III) nm] are observed for M-2. These peaks and the absorption edge shift toward the longer wavelength region with

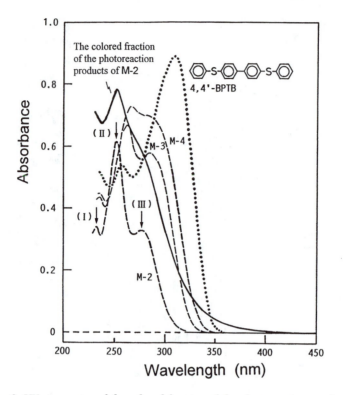

Figure 9. UV-vis spectra of the colored fraction of the photoreaction products of M-2 and model compounds in tetrahydrofuran.

an increase in molecular weight of the model compounds (*see* M-3 and M-4 spectra). 4,4'-BPTB, which is one of the photodegradation products of the model compounds and has a lower molecular weight than M-4, has a maximum peak at 310 nm and absorption edge in the longer wavelength region. Therefore, the biphenyl structure introduced into PPS molecule is probably one of the origins of the photocoloration of PPS.

Photodegradation Mechanism of PPS. On the basis of the results mentioned in the previous section, Scheme I is proposed for the photodegradation mechanisms of PPS. The first stage in the main reactions is the cleavage of a sulfur–phenyl bond in the main chain. The two kinds of aryl radicals formed, phenyl-type radical and thiophenoxy-type radical, initiate radical reactions.

Substitution of the phenyl-type radical to the other aromatic ring and recombination of the radicals induce biphenyl structures in the main chain of PPS. Recombination of the thiophenoxy-type radicals result in a disulfide moiety. These structures were confirmed by GC–MS of the photoreaction prod-

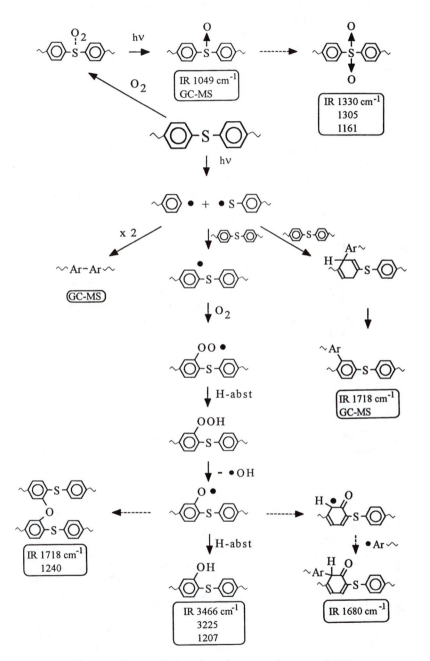

Scheme I. Proposed photodegradation mechanisms of PPS.

ucts of the model compounds. Substitution of the thiophenoxy-type radical, however, remains ambiguous. The thiophenoxy-type radical may prefer hydrogen abstraction and form thiol rather than arylation. The thiol formed could be oxidized into disulfide or hydrogen-abstracted, regenerating thiophenoxy radical.

On the other hand, hydrogen abstraction from the main chain leads to the on-chain radical. It is possible for this radical to be attacked by molecular oxygen followed by the formation of phenolether, and ketone structures as detected by IR spectroscopy of the photoirradiated PPS.

Photoirradiation of PPS in air also leads to the formation of sulfoxide and sulfone moieties. In these reactions, PPS is thought to be directly photooxidized via the excited complex between sulfide and oxygen as reported for M-2 (12, 13). However, the oxidation of the primary radicals obtained by scission of PPS seems negligible, because neither phenol nor benzenesulfenic acid was detected in the photodegradation products of M-2.

Conclusion

We concluded that the cross-linking and the formation of sulfoxide take place during the photodegradation of PPS in air. One of the possible reasons for the coloration of PPS seems to be the formation of biphenyl structures in the chain through recombination and substitution of the phenyl-type radicals formed by C–S bond cleavage.

Acknowledgment

Support by Toray Co., Ltd., Japan, is gratefully acknowledged. Thanks are due to A. Kinugawa, Toray Research Center, for performing GPC measurements.

References

1. Osawa, Z. In *Handbook of Polymer Degradation;* Hamid, S. H., Ed.; Dekker: New York, 1992; Chapter 6.
2. Koch, W.; Heitz, W. *Makromol. Chem.* **1983,** *184,* 779–792.
3. Kinugawa, A. *Kobunshi Ronbunshu* **1987,** *44,* 139–141.
4. Kuroda, S.; Nagura, A.; Horie, K.; Mita, I. *Eur. Polym. J.* **1989,** *25,* 621–627.
5. Kuroda, S.; Mita, I. *Polym. Degrad. Stab.* **1990,** *27,* 257–270.
6. Kuroda, S.; Terauchi, K.; Nogami, K.; Mita, I. *Eur. Polym. J.* **1989,** *25,* 1–7.
7. Port, A. B.; Still, R. H. *J. Appl. Polym. Sci.* **1979,** *24,* 1145–1164.
8. Suwa, T.; Takehisa, M.; Machi, S. *J. Appl. Polym. Sci.* **1973,** *17,* 3253–3257.
9. Lenz, R. W.; Carrington, W. K. *J. Polym. Sci.* **1959,** *41,* 333–358.
10. Lenz, R. W.; Handlovits, C. E. *J. Polym. Sci.* **1960,** *43,* 167–181.
11. David, C.; Baeyens-Volant, D. *Eur. Polym. J.* **1978,** *14,* 29–38.
12. Tezuka, T.; Miyazaki, H.; Suzuki, H. *Tetrahedron Lett.* **1978,** *22,* 959–1960.
13. Akasaka, T.; Yabe, A.; Ando, W. *J. Am. Chem. Soc.* **1987,** *109,* 8085–8087.

RECEIVED for review January 26, 1994. ACCEPTED revised manuscript February 14, 1995.

Far-Ultraviolet Degradation of Selected Polymers

Mark R. Adams[1] and Andrew Garton

Polymer Program and Chemistry Department, Box U-136,
University of Connecticut, Storrs, CT 06269-3136

The far-UV (100–250 nm) photodegradations in vacuum of bisphenol A polycarbonate (PC), poly(tetrafluoroethylene-co-hexafluoro-propylene) (FEP), and a polyimide-block-polysiloxane (PISX) are reviewed. The results of controlled far-UV irradiation in vacuum are compared with the results of exposure in a National Aeronautics and Space Administration (NASA) low Earth orbit (LEO) simulation facility (a source of both far-UV radiation and high-kinetic-energy atomic oxygen), and exposure on the NASA Long Duration Exposure Facility (LDEF) mission. Far-UV at >190 nm rapidly decarboxylated the PC surface with quantum efficiency of about 0.07 and lead to extensive mass loss and the production of a cross-linked skin. The chemical functionality in the surface of PISX was also rapidly changed. FEP was little affected at >190 nm because of its weak absorption at these wavelengths. However, radiation from an argon plasma (100–150 nm) defluorinated the FEP surface. The effects of atomic oxygen overwhelmed the effects of far-UV in the LEO simulation facility. The surface chemistry of FEP facing the ram direction in the LDEF mission was also dominated by the effects of atomic oxygen, but specimens in the wake direction showed effects consistent with extensive photodegradation.

Aⅼᴍᴏsᴛ ᴀʟʟ sᴛᴜᴅʏ ᴏf ᴘᴏʟʏᴍᴇʀ ᴘʜᴏᴛᴏᴅᴇɢʀᴀᴅᴀᴛɪᴏɴ has been limited to wavelengths >250 nm (*1*), probably because far-UV radiation forms a negligible component of terrestrial sunlight and so there is little practical concern. However, spacecraft experience the full solar spectrum, unshielded by the ozone layer. The continuum of solar radiation has significant intensity even below 200 nm (*2*), and plasma lines occur at lower wavelengths, notably the

[1]Current address: Franklin International, 2020 Bruck Street, Columbus, OH 43207

emission from atomic hydrogen at 121 nm. Examples where far-UV degradation have been considered include studies of far-UV lithography (*3, 4*), and the response of materials that may be relevant to spacecraft applications, such as fluoropolymers (*5, 6*). However, important considerations such as the effect of wavelength on the quantum yields and reaction pathways are essentially unknown.

We described previously (*7, 8*) a systematic study of far-UV (180–250 nm) irradiation of bisphenol A polycarbonate (PC), both in vacuum and in air; we demonstrated that decarboxylation is an efficient process (quantum yields about 0.07), but chain scission processes and mass loss occur in preference to the photo-Fries processes that are observed at mid-UV wavelengths. Far-UV processes are also highly surface specific, and shielding by a layer of cross-linked material reduces quantum yields at longer irradiation times in vacuum. The purpose of this chapter is to review our experience of far-UV degradation of three very different polymers (PC, poly(tetrafluoroethylene-*co*-hexafluoropropylene) [FEP], and a polyimide-*block*-polysiloxane [PISX]) and to relate this experience to the effect of the spacecraft environment, real or simulated, on these polymers.

Experimental Details

The PC was a commercial product (Lexan, General Electric) with a number-average molecular weight (M_n) of 15,300 g/mol and a weight-average molecular weight of 28,700 g/mol. Specimens were examined after the commercial stabilizer package was removed by reprecipitation of the polymer. However, the presence of stabilizer did not affect the far-UV processes (*7*). Films of PC were cast from dichloromethane solution. The PISX (*see* structure in Chart I) was a block copolymer of poly(dimethyl siloxane) (PDMS, M_n = 3000 g/mol) and a polyimide that was supplied by Jeffrey W. Gilman of Phillips Laboratories, Edwards Air Force Base, CA. The PISX was supplied as a powder in fully imidized form and was

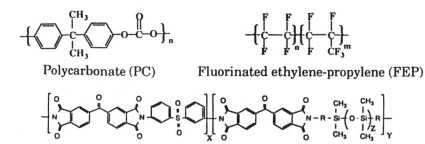

Polycarbonate (PC)   Fluorinated ethylene-propylene (FEP)

R = —CH₂-CH₂-CH₂—
Polymer endcapped with phthalimide

Polyimide-siloxane (PISX)

Chart I. Polymer structures.

spin-cast at 2000 rpm from a 10% solution in dimethylacetamide on to silicon wafers in a Class 100 clean room. The PISX was annealed (to bring the siloxane component to the surface) by heating to 230 °C for 24 h. The final film thicknesses were 150–1200 nm, as measured by IR spectroscopy, and were chosen depending on the anticipated level of chemical modification and the analytical technique to be employed. The FEP was a thermal control blanket (Sheldahl Inc.) consisting of a 25-μm FEP film backed with thin vacuum-deposited layers of inconel and silver. The FEP was used as received except for cases where transmission spectroscopic measurements (IR and UV) were desired, when the metallic layers were removed by mild abrasion and washing with acetone. The chemical structures of the polymers are shown in Chart I.

Controlled irradiation at 180–250 nm was carried out in a stainless steel vacuum chamber with a Suprasil grade quartz window. A vacuum of <0.1 Pa was maintained, and the small IR component of the radiation (about 1 mW/cm²) produced no significant heating of the specimen. The light source was a deuterium lamp (Oriel 63162, 30 W). The spectral distribution is shown in Figure 1, and the intensity axis was calculated by using a thermopile detector (7). The 180-nm cutoff was the result of the quartz windows on the lamp and cell. The deuterium lamp had a relatively low output at >250 nm (i.e., in the mid-UV and beyond). The typical solar irradiance (2) is also shown in Figure 1 for comparison, plotted on the same intensity axis as that of the deuterium lamp. The integrated far-UV intensity of the deuterium lamp for the chosen optical configuration was therefore approximately two "far-UV suns".

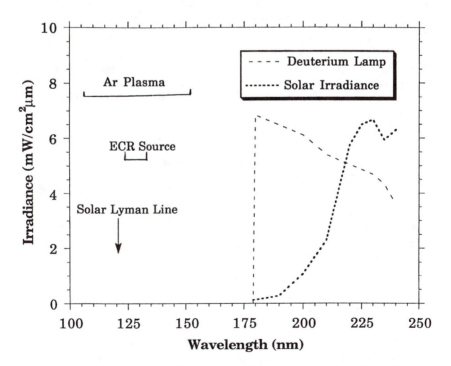

Figure 1. Far-UV spectral distributions of light sources used in this study.

Far-UV irradiation at shorter wavelengths (predominantly 150 nm) was carried out by using a low-pressure argon plasma. The specimen, in a Petri dish covered with a calcium fluoride window, was placed just below the glow region of 50 W radio-frequency-generated argon plasma. Such an arrangement exposes the specimen to the far-UV component of the plasma without bombardment by the high energy ions in the plasma. No information is available on the radiation intensity.

Specimens of PC, PISX, and FEP were exposed in the National Aeronautics and Space Administration (NASA) low Earth orbit (LEO) simulation facility at NASA Lewis, courtesy of Bruce Banks and Sharon Rutledge. This facility consisted of an electron cyclotron resonance (ECR) source containing low pressure oxygen. The ECR source produced atomic oxygen and far-UV radiation at about 130 nm. The chamber pressure was 0.027 Pa and the atomic flux was 2×10^{15} atoms/ ($cm^2 \cdot s$). No information is available on the intensity of the far-UV radiation.

Specimens of FEP were also obtained from the NASA Long Duration Exposure Facility (LDEF) space experiment, courtesy of Philip Young of NASA Langley. The FEP specimens were located such that they experienced the extremes of possible exposure environments. Two specimens were taken from opposite sides of the spacecraft, but the sides had experienced nearly identical conditions (about 90° from the ram direction, 3×10^{21} atoms/cm² of atomic oxygen, 7000 UV h). Two other specimens were taken from ram and wake directions. They had experienced 11,200 h and 9×10^{21} atoms/cm² (ram) and 10,500 h and 9×10^4 atoms/cm² (wake).

Because far-UV degradation is predominately a surface phenomenon (see discussion in subsequent section), greater emphasis was placed on careful surface analysis or the examination of ultrathin film specimens. Transmission IR spectra were obtained on thin (<1 μm) films, either supported on a metal grid (PC) or cast on a silicon wafer (PISX), by using a Mattson Polaris FTIR system equipped with a mercury–cadmium–telluride (MCT) detector and operated at 4 cm^{-1} resolution. Surface IR spectra of FEP were obtained by the internal reflection spectroscopy (IRS) technique using either a germanium of KRS-5 IRS element at 45° or 60° incidence (9). The sampling depth of the IRS technique was about 1 μm for the KRS-5 element and about 0.3 μm for the germanium, both at mid-IR wavelengths (the sampling depth in IRS is proportional to the wavelength).

Surface IR information on the PC and PISX was obtained by reflection absorption IR (RA-IR) analysis of thin films deposited on silver mirrors (9). More surface specific information (<10 nm sampling depth) was obtained by the X-ray photoelectron (XPS) technique using a Perkin Elmer PHI 5300 spectrometer with monochromatized Alk$_\alpha$ source (characteristic X-ray from aluminum source; kinetic energy = 1486.6 eV), a hemispherical analyzer, and a position sensitive detector. Spectra were obtained at $<7 \times 10^{-7}$ Pa by using a low energy electron neutralizer gun to compensate for charging. Mass changes were measured by using a microbalance, and molecular weight changes were measured by size exclusion chromatography (SEC, Waters 150-C, tetrahydrofuran solvent, 10–10,000 pore size microstyragel columns).

Results and Discussion

Far-UV Irradiation Alone.   The UV spectra of PC, PISX, and FEP are shown in Figure 2. The spectrum of FEP below 180 nm is assumed to be similar to that of polytetrafluoroethylene (PTFE), which absorbs strongly

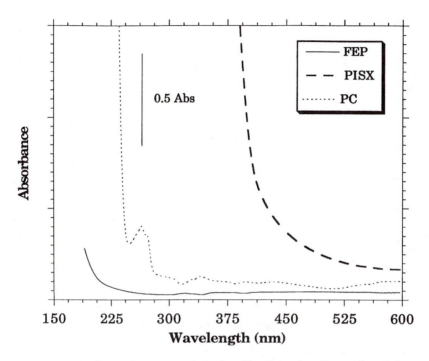

Figure 2. UV absorption spectra of PC thin film, PISX thin film, and FEP thin film.

at 160 nm and below. Energy absorption, and therefore possible photochemistry, will depend on the overlap of the emission spectrum of the source (Figure 1) and the absorption spectrum of the specimen. The PC will absorb all the far-UV photons yet be virtually transparent to the small mid-UV component of the deuterium lamp. PISX will absorb all far-UV and mid-UV photons. FEP will absorb little of the energy from the deuterium lamp but will absorb energy from the lower wavelength sources. The very high absorptivity of the PC and PISX in the far-UV ($>10^5$ cm^{-1}) means that energy absorption will also be highly surface specific (75% in less than the first 100 nm).

Transmission IR spectra of thin PC films as a function of irradiation time show a decrease in the characteristic absorptions of carbonate functionality (1780 cm^{-1}) and phenylene functionality (1510 cm^{-1}). The fractional loss of these absorptions with time is shown in Figure 3. Because the initial film thickness, film density, and polymer chemical structure were known, the rate of chemical reaction in equivalents/(cm^2 • s) could then be calculated. When this result is compared with the far-UV flux from the deuterium lamp [typically 3.4×10^{14} photons/(cm^2 • s) or 5.7×10^{-10} einstein/(cm^2 • s)], an initial quantum yield for carbonate loss of about 0.07 can be calculated (7). It is obvious

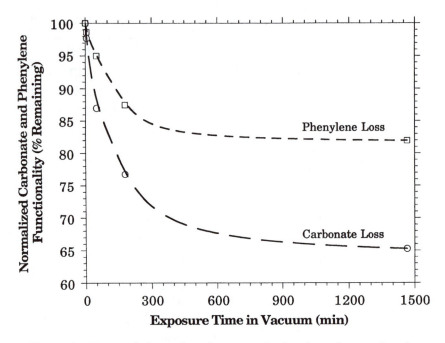

Figure 3. Fractional loss of carbonate and phenylene functionality by transmission IR analysis of PC thin film irradiated with the deuterium source.

from Figure 3 that the quantum yield for loss of carbonate functionality decreases by an order of magnitude with increasing irradiation time. We attribute this result to the production of a cross-linked polymer skin. Other measurements of interest include a marked decrease in molecular weight (M_n dropped by half after 3 h irradiation), an increase in water contact angle from 80° to 87°, considerable mass loss, the production of volatiles such as benzene dicarboxylic acids, and the formation of a skin that was insoluble in dichloromethane (7).

The XPS spectrum (Figure 4) shows that almost complete destruction of the carbonate functionality in the top 10 nm occurred in the first few hours of irradiation. Deconvolution of the carbon 1s spectra showed that only 20% of the carbonate remained after 50 min irradiation from the deuterium lamp. The use of a variable take-off angle for the XPS measurement showed that chemical change was concentrated in the first few nanometers of the surface.

The lack of doubly bonded oxygen in the XPS spectra of irradiated specimens, the low yield of substituted benzophenones and polymer phenyl salicylates as detected by IR, and the occurrence of mass loss by volatilization of the PC was not by the photo-Fries process that has been well studied at mid-UV wavelengths (1, 10, 11). Possibly, the higher energy of far-UV photons

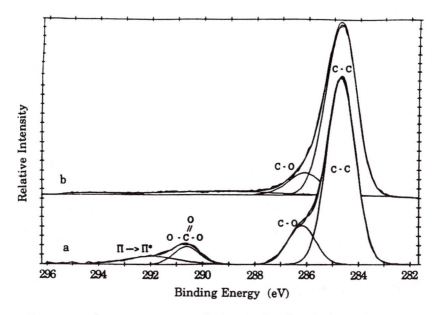

Figure 4. Carbon 1s XPS spectra of PC, 62° take-off angle, for (a) unexposed and (b) after 10 h far-UV exposure in vacuum.

allows escape of the radical pair from the reaction cage before reorganization and recombination can take place to yield the photo-Fries products. An alternative explanation is that the highly surface-specific nature of the far-UV process leads to a high local concentration of photo-Fries products that are then photolyzed further.

The PISX specimens were also appreciably modified by far-UV radiation from the deuterium lamp source (mostly 180–250 nm). The RA-IR spectra of thin films of PISX on a silver mirror (Figure 5) show that the methyl functionality (2961 cm^{-1}, originating from the polyimide block) was largely destroyed in a few hours of irradiation. The characteristic absorptions of the imide functionality (1726 and 1780 cm^{-1}), sulfone functionality (1159 cm^{-1}), and phenylene functionality (1480 cm^{-1}) were reduced by about 25% after 40 min irradiation, a result that implies that considerable photoablation was taking place. Mass spectrometry of volatile fragments (*12*) was consistent with these observations.

XPS analysis of the PISX (Figure 6 and Table I) confirmed that the PDMS block was appreciably degraded by far-UV exposure. The carbon 1s spectra (Figure 6) show that the C–Si (284.5 eV) was greatly reduced in proportion to the aromatic carbon (285.6 eV) along with a smaller decrease in the carbons of the imide functionality (288.9 eV) and adjacent to the sulfone functionality. After 24-h irradiation, the imide and C–Si functionalities were essentially de-

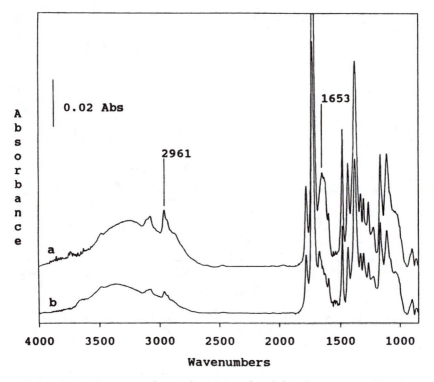

Figure 5. RA-IR spectra of PISX for (a) initial and (b) after 40 min of far-UV irradiation from the deuterium source (>180 nm).

stroyed. The atomic composition data (Table I) confirm a loss of silicon (from the PDMS block), nitrogen, and sulfur (both from the polyimide block). The variable-angle XPS data show that, as expected (13), the silicone block was preferentially located in the first few nanometers of the surface. A 25° take-off angle gives a sampling depth of about 3 nm, whereas a 62° take-off angle gives a sampling depth of about 7 nm. The fractional reduction of S, N, and Si species after irradiation was also greater in the first few nanometers, and this result is consistent with the highly surface-specific nature of the energy absorption (Figures 1 and 2). This response to far-UV alone is very different from the combined far-UV/atomic oxygen response, which is also shown in Table I but will be discussed in a subsequent section.

Far-UV radiation from the deuterium source had little effect on FEP, as might be expected from a comparison of the source output and the FEP absorption spectrum. Ten hours of irradiation at two far-UV suns produced no detectable change in the IR spectrum (both transmission and IRS) and only minor changes in the XPS spectrum [very broad, weak signals, possibly originating from defluorinated or oxygenated material (12)].

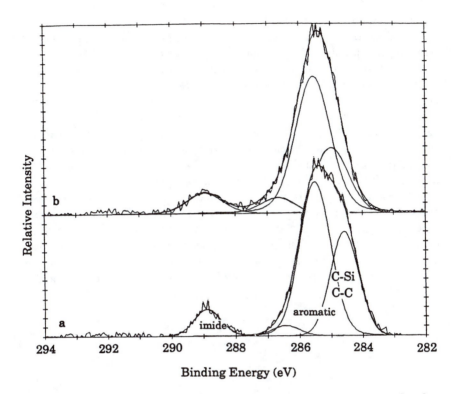

Figure 6. Carbon 1s XPS spectra of PISX, 62° take-off angle, for (a) initial and (b) after 3 h irradiation from deuterium source (>180 nm).

Table I. XPS Analysis of PISX

Specimen	C (%)	O (%)	N (%)	Si (%)	S (%)
Deuterium source (>180 nm)					
Unexposed, 62° take-off angle	63.4	21.2	2.9	10.9	1.6
10 h UV, 62° take-off angle	64.3	24.2	2.4	8.4	0.8
24 h UV, 62° take-off angle	66.8	23.5	1.7	7.0	1.0
Unexposed, 25° take-off angle	57.8	23.2	2.0	16.0	1.0
10 h UV, 25° take-off angle	60.9	26.0	1.3	11.1	0.9
24 h UV, 25° take-off angle	68.1	22.4	1.1	7.9	0.5
ECR source (far-UV + atomic oxygen)					
Unexposed, 62° take-off angle	63.2	21.2	2.9	11.3	1.4
39 min ECR, 62° take-off angle	12.9	54.3	1.1	31.2	0.5
Unexposed, 25° take-off angle	57.4	23.4	1.6	17.0	0.6
39 min ECR, 25° take-off angle	9.0	59.1	0.6	30.1	1.2

The response to far-UV of about 150 nm wavelength from the argon plasma was appreciably greater, although we have no quantitative information on radiation intensity and so cannot calculate quantum yields. Figure 7 shows the IR-IRS spectra of FEP before and after 6.8 h exposure to the 150-nm source. We assign the absorptions at 1717–1790 cm^{-1} to vinylic unsaturation and ketonic carbonyls and the absorption at 1883 cm^{-1} to acid fluoride formation (14). These spectra were obtained with a 45° KRS-5 IRS element, and so the sampling depth was about 1 μm in this wavenumber region of the spectrum. The XPS spectra (Figure 8) are much more surface specific (<10 nm sampling depth) and showed a greater chemical change with exposure to the argon plasma radiation (about 150 nm wavelength). As expected, the unexposed FEP had mostly CF_2 functionality and a small amount of CF_3 (294 eV) and CF (290 eV) from the perfluoropropylene comonomer. The 4-h exposed specimen (argon plasma radiation) showed a marked increase in the CF_3 functionality (presumably as chain ends, and hence an indication of chain scission), and a range of deflourinated and oxygenated functionality was introduced when the irradiated surface was exposed to air. These data are presented in

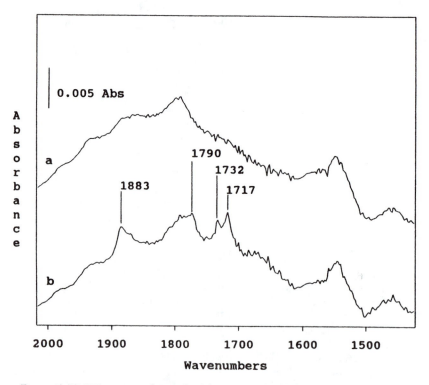

Figure 7. IR-IRS spectra of FEP for (a) unexposed and (b) after 6.8 h exposure to far-UV from argon plasma (150 nm).

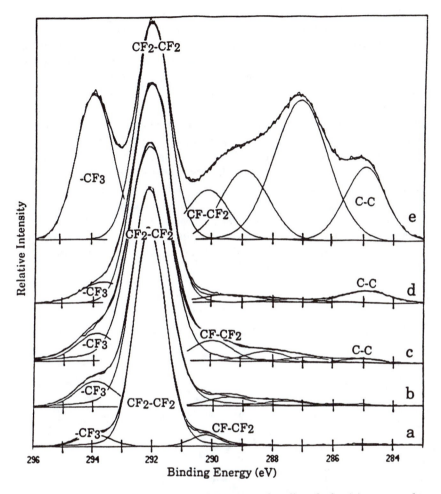

Figure 8. Carbon 1s XPS spectra of FEP 62° take-off angle for (a) unexposed;
(b) 4 h far-UV (argon plasma); (c) 2.4 h far-UV/atomic oxygen (ECR source);
(d) LDEF experiment, ram direction; and (e) LDEF experiment, wake direction.

Table II, along with data from FEP under other exposure conditions that will
be discussed. Again, the variable angle studies indicate that chemical change
is concentrated in the first few nanometers of the surface.

The scission and deflourination of FEP is consistent with the far-UV studies of Kim and Liang (15) and George et al. (6) who showed by electron
paramagnetic resonance (EPR) spectroscopy that a range of fluorinated radicals are produced when FEP is exposed to far-UV radiation. We also note
that unsaturated products will shift the FEP energy absorption to longer wavelengths, and so the photodegradation may be autocatalytic in nature.

Far-UV/Atomic Oxygen (LEO Simulation, ECR Source). The transmission IR spectra of PC before and after exposure to the far-UV/atomic oxygen environment of the NASA LEO simulation facility (Figure 9) indicate the dominant mechanism was erosion by atomic oxygen, and little change occurred to the remaining polymer. The 2.4-h exposed specimen lost 2,200 nm in thickness but showed little evidence of chemical change; the IR spectrum is unchanged but reduced in intensity. This result should be contrasted with the change in carbonyl:phenylene ratio produced by far-UV radiation alone (Figure 3; more detail in ref. 7). The XPS spectra showed that oxygenation of the surface occurred in the ECR source rather than the deoxygenation of far-UV exposure alone. After 2.4-h exposure, the carbon content of the surface 7 nm (62° take-off angle) decreased from 84 to 75%, and the oxygen content increased from 16 to 24%. The carbon 1s spectra (Figure 10) showed that a range of singly and doubly bonded oxygenated species was produced. A comparison of the 62° and 25° take-off angle data (Figure 10) indicates that oxygenation was concentrated in the first few nanometers of the surface. In

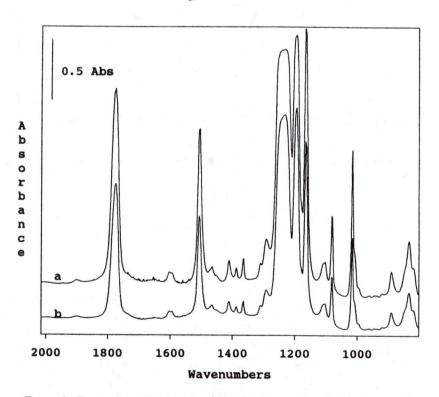

Figure 9. Transmission IR spectra of PC for (a) initial and (b) after 2.4 h exposure to far-UV/atomic oxygen (ECR source).

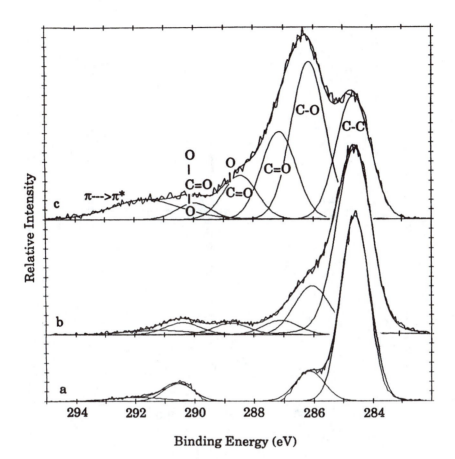

Figure 10. Carbon 1s XPS spectra of PC for (a) initial, (b) after 2.4 h exposure to far-UV/atomic oxygen (ECR source) with 62° take-off angle, and (c) after 2.4 h exposure to far-UV/atomic oxygen (ECR source) with 25° take-off angle.

the case of PC, therefore, we conclude that the effect of atomic oxygen erosion overwhelmed the far-UV effects.

The behavior of the PISX specimen also appeared to be dominated by the effects of atomic oxygen. After 39-min exposure, the carbon content of the top 7 nm was reduced dramatically (Table I and Figure 11), and the surface composition resembled that of silica. Similar behavior of polyimide–polysiloxane block copolymers was reported in atomic oxygen alone (*16*). There was appreciable thickness loss (350 nm after 60 min), but the protective layer of silica that was produced slowed that erosion process such that it was appreciably less than the equivalent thickness loss for PC. Scanning electron microscopy (SEM) of the exposed PISX (Figure 12) shows that the silica-like

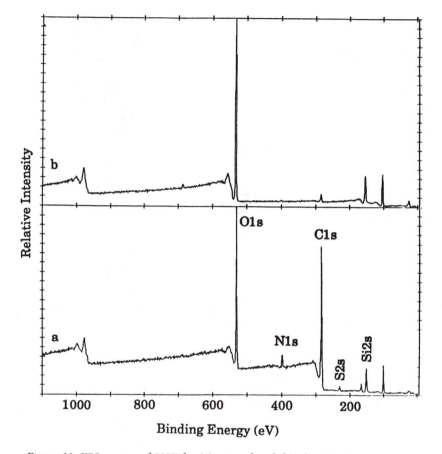

Figure 11. XPS spectra of PISX for (a) original and (b) after 39 min exposure to far-UV/atomic oxygen (ECR source).

surface was not continuous; considerable shrinkage and surface cracking had taken place.

When FEP was exposed to the far-UV/atomic oxygen environment of the ECR source, there was evidence of chain scission (increase in CF_3, Figure 8c) and a higher level of oxygenation (about 288 eV, Figure 8c, and atomic compositions in Table II) than observed with far-UV alone. IR-IRS measurements (Figure 13) showed that acid fluoride formation (1880 cm^{-1}) occurred in the FEP surface, but the amount of oxidation did not increase with exposure time. This result implies that any mass loss was by ablation rather than by modification of the FEP to any significant depth. The thickness loss could not be measured with these specimens. The surface chemistry resembled that of fluoropolymers exposed to atomic oxygen alone (downstream from an oxygen

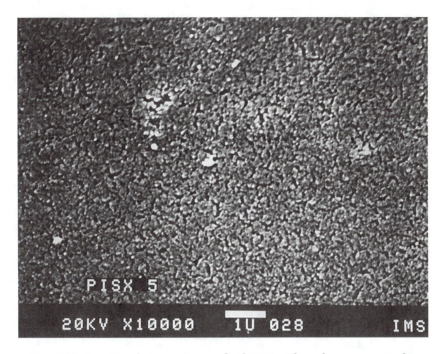

Figure 12. Scanning electron micrograph of PISX surface after exposure to far-UV/atomic oxygen (39 min, ECR source).

Table II. XPS Analysis of FEP

Specimen	C(%)	F (%)	O (%)
Control, 62° take-off angle	39.5	60.5	0
Control, 25° take-off angle	38.5	61.5	0
6.8 h/far-UV (argon plasma), 62°	40.5	58.4	1.1
2 min far-UV/AO (ECR source), 62°	40.8	57.0	2.2
2 min far-UV/AO (ECR source), 25°	36.4	61.8	1.8
2.4 h far-UV/AO (ECR source), 62°	41.9	56.4	1.7
2.4 h far-UV/AO (ECR source), 25°	39.4	58.9	1.7

Note: AO is atomic oxygen.

plasma, ref. 17), and so we again conclude that the polymer response was dominated by atomic oxygen effects in the LEO simulation facility.

LDEF Exposure. Of the series of polymers described here, only FEP specimens were exposed on the LDEF mission. The configuration of the specimens is shown in Figure 14. The specimens from rows 1 and 7 were at an oblique angle to the flight direction ("ram" direction, and hence direction of atomic oxygen impact). The specimen in row 10 was close to the ram

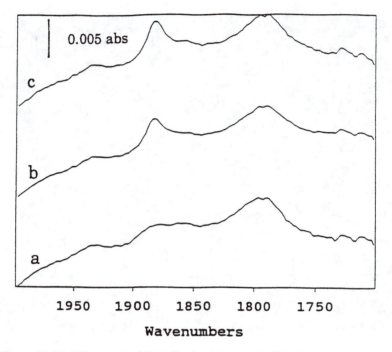

1950	1900	1850	1800	1750

Wavenumbers

Figure 13. IR-IRS spectra of FEP film for (a) original, (b) after 2 min exposure to far-UV/atomic oxygen (ECR source), and (c) after 2.4 h exposure.

direction, whereas the specimen in row 4 was close to the wake direction. The amount of UV exposure and flux of atomic oxygen are given for each specimen in the experimental section.

The specimen from the ram direction showed a surface chemistry by XPS similar to specimens exposed in the LEO simulation facility (Figure 8 and Table III). Even though there was an appreciable mass loss from these specimens (18), the change in surface chemistry was relatively small. There was some chain scission (CF_3 at 294 eV), defluorination, and oxygenation (about 288 eV). The peak at 285 eV may be from carbonaceous contamination, as has been observed with other LDEF specimens (6). The specimens with oblique orientation were little different in surface chemistry from the ram-direction specimen. However, the specimen taken from the wake direction showed a very different surface chemistry consistent with a large amount of photodegradation rather than atomic oxygen erosion. Obviously the XPS spectrum in Figure 8e shows that a wide range of chemical functionalities are present, and a complete chemical assignment would be unwise without confirmation by derivatization reactions.

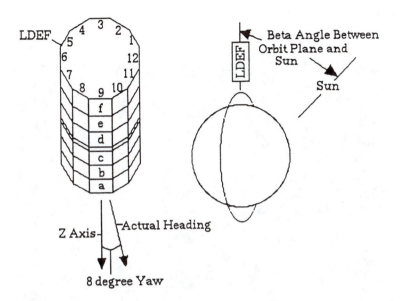

Figure 14. Specimen location on LDEF experiment and orientation of spacecraft during 69-month mission.

Table III. XPS Analysis of FEP from LDEF Mission

Specimen	Takeoff Angle	C (%)	F (%)	O (%)
Control	62°	39.5	60.5	0
Control	25°	38.5	61.5	0
Row 10, tray E	62°	40.2	58.0	1.8
(Ram direction)	25°	40.2	55.1	4.7
Row 1, tray D	62°	43.1	53.0	3.9
(Oblique orientation)	25°	41.4	54.5	4.1
Row 7, tray D	62°	39.5	59.1	1.4
(Oblique orientation)	25°	40.6	55.5	3.9
Row 4, tray F	62°	50.7	43.2	6.1
(Wake direction)	25°	49.7	44.0	6.3

However, the CF_3 peak at 294 eV is assignable with some confidence, and the only reasonable mechanism for CF_3 production would be through chain scission. This observation of a transition from atomic-oxygen-dominated behavior to photodegradation-dominated behavior as the specimen location changes from ram to wake is consistent with the conclusions of others (5, 6, 18). The possibility of synergism between far-UV and atomic oxygen has been raised (5). The data presented here offer no evidence for or against this hypothesis, and we are presently exploring the subject further (12). A crucial

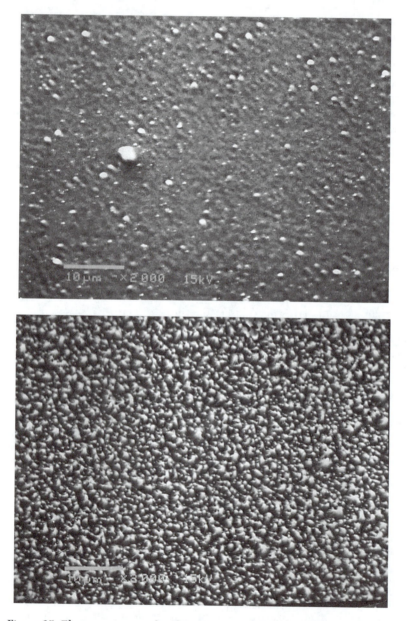

Figure 15. Electron micrographs of LDEF-exposed FEP surfaces: ram direction
(bottom panel) and wake direction (top panel).

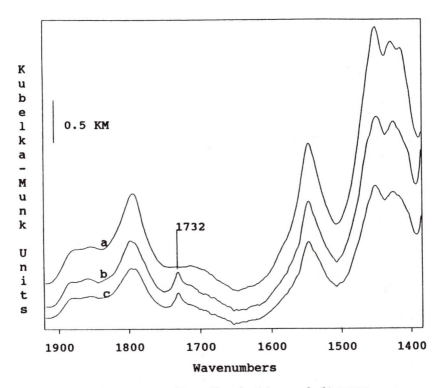

Figure 16. DRIFT spectra of FEP films for (a) control, (b) LDEF specimen exposed in wake direction, and (c) LDEF specimen exposed in ram direction.

experimental requirement is accurate intensity information on the far-UV radiation so that quantum yields can be calculated with precision.

Figure 15 shows the physical state of the FEP surfaces of the ram and wake specimens. There was considerable roughening of specimen facing the ram direction, which showed the characteristic carpet-like morphology often reported for atomic oxygen eroded materials (19) due to the directionality of the atomic oxygen flux. The roughness of the surfaces precluded the use of IR-IRS analysis [good optical contact is needed between the specimen and the IRS element (9)], but the roughness proved advantageous in obtaining diffuse reflectance (DRIFT) IR spectra. The most noticeable feature in Figure 16 is an absorption at 1732 cm^{-1} that occurred in both the ram and wake specimens. A similar absorption was also reported by Young et al. (18) following transmission IR analysis of LDEF-exposed FEP specimens. It is reasonable to assign this to carbonyl species produced by oxidation of the FEP, but the possibility of contamination during such a prolonged experiment cannot be ignored.

Acknowledgment

We wish to thank the Connecticut NASA Space Grant Consortium for financial support.

References

1. Rabek, J. F. *Mechanisms of Photophysical Processes and Photochemical Reactions in Polymers*; Wiley: New York, 1987.
2. Heath, D.; Thekaekara, M. *The Solar Output and IB Variation*; Colorado Associated University Press: Boulder, CO, 1977.
3. Srinivasan, R. *Polymer* **1982,** *23,* 1863.
4. Srinivasan, R.; Lazare, L. *Polymer* **1985,** *26,* 1297.
5. Stiegman, A. E.; Brinza, D. E.; Laue, E. G.; Anderson, M. S.; Liang, R. H. *J. Spacecr. Rockets* **1992,** *29,* 150.
6. George, G. A.; Hill, D. J. T.; O'Donnell, J. H.; Pomery, P. J.; Rasoul, F. A. In *LDEF—69 Months in Space*; NASA Conference Publication 3194; National Aeronautics and Space Administration: Washington, DC, 1993; pp 867–876.
7. Adams, M. R.; Garton, A. *Polym. Degrad. Stab.* **1993,** *41,* 265.
8. Adams, M. R.; Garton, A. *Polym. Degrad. Stab.* **1993,** *42,* 145.
9. Garton, A. *Infrared Spectroscopy of Polymer Blends, Composites and Surfaces*; Hanser: New York, 1992.
10. Guillet, J. E. *Polymer Photophysics and Photochemistry*; Cambridge University: Cambridge, England, 1985.
11. Lemaire, J.; Gardette, J.; Rivaton, A.; Roger, A. *Polym. Degrad. Stab.* **1986,** *15,* 1.
12. Adams, M. R.; Garton, A., Ph.D. dissertation, University of Connecticut, 1993.
13. Gilman, J. W.; Schlitzer, D. S. *Proceedings of the 6th Annual Conference on Surface Reactions in the Space Environment*; Northwestern University: Chicago, IL, 1991, p 46.
14. Ha, K.; McLain, S.; Suib, S. L.; Garton, A. *J. Adhes.* **1991,** *33,* 169.
15. Kim, S. S.; Liang, R. H. *Polym. Prep. Am. Chem. Soc. Div. Polym. Chem.* **1990,** *31,* 389.
16. Arnold, D. H.; Chen, R. O.; Waldbauer, M. E.; Rogers, M. E.; McGrath, J. E. *High Perform. Polym.* **1990,** *2,* 83.
17. Golub, M. A.; Wydeven, T.; Cormia, R. D. *Polymer* **1989,** *30,* 1571.
18. Young, P. R.; Slemp, W. S.; Chang, A. C. *LDEF—69 Months in Space*; Levine, A. S., Ed.; NASA Conference Publication 3194; National Aeronautics and Space Administration: Washington, DC, 1993; pp 827–835.
19. Garton, A.; McLean, P. D.; Wiebe, W.; Densley, R. J. *J. Appl. Polym. Sci.* **1986,** *32,* 3941.

RECEIVED for review January 31, 1994. ACCEPTED revised manuscript February 8, 1995.

Physical Spreading and Heterogeneity in Oxidation of Polypropylene

M. Celina[1,4], G. A. George[2,4] and N. C. Billingham[3]

[1] Department of Chemistry, The University of Queensland, 4072 Brisbane, Australia
[2] School of Chemistry, Queensland University of Technology, Brisbane, Australia
[3] School of Chemistry and Molecular Sciences, University of Sussex, Brighton BN1 9QJ, United Kingdom

The oxidation of solid polypropylene measured by chemiluminescence (CL) has been interpreted as involving heterogeneous initiation that leads to high oxidation rates in localized zones and is followed by the physical spreading of oxidation. Evidence of the high activity of oxidizing centers to promote further oxidation of even physically separated PP powder particles has been obtained. In such a model of highly reactive centers existing from the earliest onset of oxidation, an induction period was related to the physical characteristics of the material rather than the chemical interpretation in liquid-state kinetic models. Therefore, failure of the material may be related to the spreading of a few centers of initial oxidation of the polymer.

THE THERMAL AND PHOTOCHEMICAL OXIDATION of solid polyolefins has traditionally been studied within a kinetic framework developed for the autooxidation of liquid hydrocarbons. The chemical analysis of polymer films during oxidation by spectroscopic methods, particularly transmission and attenuation total reflectance–IR spectrophotometry, has produced concentration profiles of oxidation products such as ketones, aldehydes, acids and alcohols as a function of time. These profiles show an *induction period* before a rapid increase in concentration to a steady increase with time, which has been interpreted as the limiting oxidation rate of the polymer (1).

[4] Current Address: G.A. George and M. Celina, School of Chemistry, QUT GPO Box 2434, Brisbane 4001, Australia.

0065–2393/96/0249–0159$12.00/0

Oxygen uptake experiments confirmed this general oxidation profile for polyethylene and polypropylene (PP) powders and film. The similarity of these curves to those from liquid hydrocarbons during autooxidation supported the application of homogeneous free-radical kinetics in which the degenerate branching agent is the polymer hydroperoxide and the limiting oxidation rate is controlled by the rate of the propagation reaction (2). The steady-state approximation may then be applied to determine the kinetic parameters and thus predict the extent of oxidation at any time. Such an approach, if applied to the solid oxidizing polymer, would theoretically enable the ultimate service life of the material to be determined from the kinetic curve provided a failure criterion is established and the appropriate rate coefficients can be determined accurately enough. The attractiveness of such an approach had led to the adoption of very sensitive analytical methods such as XPS (X-ray photoelectron spectroscopy) (3), microoxygen uptake using sensitive pressure transducers (4), and chemiluminescence (CL) (5) to determine the rate at the earliest stages of oxidation.

Measurement of the ultra-weak CL that accompanies oxidation has been of interest because of the high sensitivity with which weak emitted light can be measured when compared with the difference between two intense trans-mitted beams as in absorption spectrophotometry. However, the process is intrinsically inefficient and has quantum yields for the overall production of light from the polymer oxidation around 10^{-9}. Controversy has surrounded the nature of the light-emitting reaction in the free-radical oxidation sequence, but it has generally been regarded as the exoenergetic termination reaction of peroxy radicals by the Russell mechanism (6). This mechanism requires one of the terminating radicals to be either primary or secondary so that a six-membered transition state can be formed by the two peroxy radicals, which would lead to an alcohol, singlet oxygen, and a triplet excited carbonyl chro-mophore. The emission from this excited state is the measured CL.

Even though it is easy to envisage such a mechanism for polymers such as polyethylene and polyamides, in the case of PP the chain-carrying radical is tertiary and the usual termination reaction of the peroxy radicals involves an intermediate tetroxide that cannot lead to an emissive carbonyl. This re-striction has led to the proposal of a variety of light-emitting reactions (7). Similarly, PP hydroperoxide prepared by controlled oxidation will produce CL when heated under nitrogen. Several reaction mechanisms have been sug-gested to account for this phenomenon, and they do not involve peroxy radical intermediates. By heating a sample to around 180 °C, all of the hydroperoxides may be decomposed. The integral of the glow curve was shown (8) to be a measure of the concentration of peroxides, although the absolute correlation has been questioned at all concentrations (9).

Although measurements in inert atmosphere provide one method of measuring oxidation product concentration in the early stages of thermal (10) or photooxidation (8), the most common CL experiment involves the continuous measurement of the CL intensity during isothermal oxidation. A family of such CL curves for the thermal oxidation of PP powder is shown in Figure 1. They show features of instantaneous oxidation rate curves generated by a method such as oxygen uptake. A low but nonzero rate is maintained for a time, and the rate decreases with temperature before it increases exponentially to a limiting value.

The integral of these curves is analogous to the concentration–time profile that is obtained by IR analysis of carbonyl group concentration. The integral also shows the features expected of a homogeneous, branching chain reaction, such as an apparent induction period. This induction period is frequently used as a measure of the stability of the polymer, because it increases when stabilizers such as free-radical scavenging antioxidants are included in the solid polymer. In the liquid state the induction time is taken as that time for the total consumption of antioxidants, after which the oxidation proceeds at the uninhibited rate. The linear part of the concentration–time curve is considered to be a measure of the steady oxidation rate.

The application of this approach to the integral CL curve for PP powder during oxidation at 150 °C immediately reveals a difficulty in the definition of both the induction period and the steady rate of oxidation. Figure 2 shows that as the sensitivity of the CL analysis is increased, the apparent induction

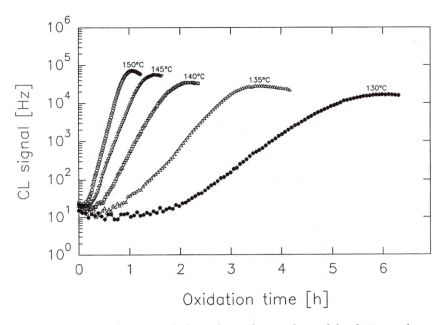

Figure 1. Typical CL signals from the oxidation of unstabilized PP powder samples at different temperatures under oxygen. (Reproduced with permission from reference 13. Copyright 1993 Elsevier Science Ltd.)

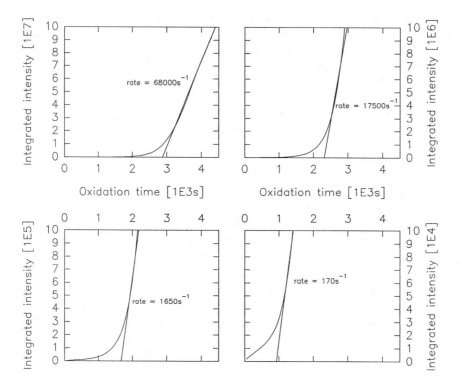

Figure 2. Integrated CL signals from the oxidation of an unstabilized PP powder sample at 150 °C presented in relation to different instrumental sensitivities.

time and the limiting rate decrease due to the continuing expansion of what is actually an exponential growth curve. The true nature of the induction period is revealed in the highest sensitivity curve as a short region of linear increase in integrated CL intensity with time before the exponential growth. This increase corresponds to the region of low but steady emission intensity seen in the CL curves of Figure 1.

Consequently, the magnitude of the induction period in any oxidation experiment depends on the sensitivity of the method used to measure it. The sensitivity of CL measurements has enabled the early stages of both thermal and photooxidation to be studied in detail. In this chapter we wish to present some of our recent results on the oxidation of PP powder and film that bring into question the often-used interpretation of the oxidation curves in Figure 2 in terms of homogeneous free-radical chain reactions. These results support a view that the oxidation of even single particles of powder is highly hetero-geneous and requires a new interpretation of the kinetic data.

Evidence for Heterogeneous Oxidation from CL Studies of Photooxidation

CL has been applied as a technique to measure the hydroperoxide formation during the photooxidation of PP-film samples (8). The integrated CL emission from a ramped temperature experiment is related to the concentration of hydroperoxides present in the sample. PP-film samples, when subjected to UV irradiation at 340 nm, immediately formed hydroperoxides at a concentration that could not be detected by ATR-IR, XPS, or a related general increase in the carbonyl index. A detailed analysis of hydroperoxide concentration during these very early stages of photooxidation indicated a possible kinetic scheme of a consecutive reaction involving a rapid formation of hydroperoxides followed by their photolysis to secondary oxidation products. Figure 3 shows this rapid initial hydroperoxide buildup in relation to the overall oxidation of the sample as measured by the carbonyl index.

Interestingly, further studies of stabilized samples revealed that this peak was not affected or inhibited by the presence of many different stabilizers. A

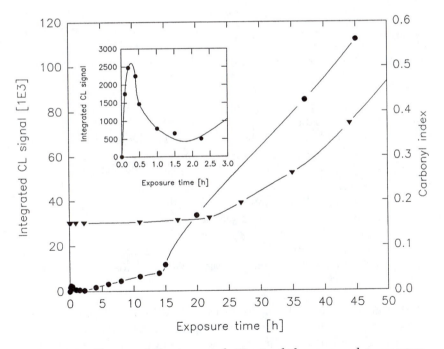

Figure 3. Changes of the integrated CL signal from ramped temperature experiments (●) and carbonyl index (▼) of unstabilized PP film samples upon UVA photooxidation. Inset shows an expansion of the early stage of photooxidation. (Reproduced with permission from reference 8. Copyright 1991 Elsevier Science Ltd.)

phenolic antioxidant or hindered amine light stabilizer (HALS) was unable to suppress the initial reactivity of the sample during irradiation (11). Therefore, we concluded that the hydroperoxides were formed in localized centers of high reactivity. Limited concentrations and mobility of the additives were seen as the reasons for uninhibited reactions taking place in very small zones of the material. Such heterogeneous behavior during photooxidation had been discussed before (12).

Thermal Oxidation

The CL emission from the isothermal oxidation of PP is fundamentally related to the formation of an oxidation product (excited carbonyl) and therefore similar to other techniques of oxidation measurements such as carbonyl index or oxygen uptake. The main advantage of using CL has to be seen as the larger dynamic sensitivity range enabling a continuous monitoring of the oxidation from the very weak early stages to the main oxidation of the material (see Figures 1 and 2).

Isothermal CL curves were obtained from the oxidation of PP powder and film samples in the solid state at various temperatures between 90 °C and 150 °C. Some of the curves are presented in Figure 1. The analysis of such curves over a wide temperature range resulted in the following conclusions (13):

1. The maximum intensity (I_{max}) appeared at a nearly constant total emission or extent of oxidation.
2. In the very early stages of the oxidation and at the highest instrumental sensitivity, it was possible to measure an induction period during which the signal was on a constant level and significantly above the baseline. This finding indicated oxidation and the immediate formation of secondary oxidation products commencing from the earliest time possible.
3. Arrhenius plots of parameters such as induction period (t_{ind}), maximum signal time (t_{max}), initial intensity during induction period (I_{ini}), and maximum signal intensity (I_{max}) were possible. Activation energies for I_{ini} and I_{max} were similar (113 kJ/mol for PP powder), whereas I_{ind} was 149 kJ/mol for PP powder.

Similar results were obtained for unstabilized PP powder and film samples, and we concluded that the induction period is a separate process from the remaining oxidation. In particular, the different activation energies of initial CL emission intensity (I_{ini}) and induction period (t_{ind}) make it impossible to predict the oxidative stability of the material from a simple measurement of I_{ini} without knowledge of the temperature behavior of the actual induction period. The product of I_{ini} and t_{ind} at different temperatures is not a constant, which means that the end of the induction period corresponds to different extents of oxidation of the polymer.

Once the induction period is regarded as a separate process and removed from the CL curves, it is possible to plot the remaining oxidation curves in reduced coordinates. A presentation of I as a fraction of I_{max} versus t as a fraction of t_{max} (after subtraction of t_{ind}) results in a master curve for all oxidation curves over a wide temperature range (Figure 4). From the appearance of a universal sigmoidal curve shape, we suggest that all oxidation curves following the induction period are governed by a common fundamental process.

Mathematical analysis of the master curve showed that statistical functions such as parts of Gaussian or Weibull distributions could be easily fitted to the sigmoidal increase of oxidation intensity. Therefore, we concluded that the

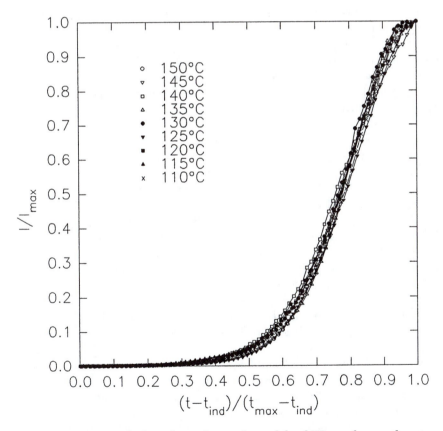

Figure 4. CL signals from the oxidation of unstabilized PP powder samples at different temperatures presented in relative units after subtraction of the induction period and resulting in a master curve. (Reproduced with permission from reference 13. Copyright 1993 Elsevier Science Ltd.)

main oxidation may in fact be closely related to a statistical progress or spreading of the oxidation. Such a process would lead to a continuously increasing fraction of oxidizing material until a maximum was reached that originated from a depletion of readily oxidizable material. This process is then followed by a final decaying signal to reach complete oxidation of the sample. The observed master curve may also be closely fitted with mathematical models of fractal growth or percolation through the solid polymer. In such a model of heterogeneous oxidation based on statistical progress and increase of oxidizing sites, the induction period has been proposed as the time required before the oxidation starts to spread from initially localized centers (13).

The occurrence of the Russell mechanism of bimolecular termination of alkyl peroxy radicals requires the participation of at least one primary or secondary alkyl peroxy radical in the light-producing reaction (14). The observation of CL emission in itself during the induction period requires, therefore, a degenerately branched chain reaction leading to either primary or secondary alkylperoxy radicals. The usual picture of PP oxidation is that tertiary alkylperoxy radicals are the chain carrier in the early stages of oxidation. If this were the case, CL would not be observed (5). Thus, higher extents of oxidation must be occurring that involve these other radical species. Heavy oxidation during the induction period may quickly lead to secondary oxidation products, such as water or CO_2. Such oxidation products were observed (15) from the earliest onset of oxidation. However, this high extent of oxidation must be confined to only a small fraction of the polymer powder or film for the duration of the apparent induction period. After this induction period, the oxidation spreads and leads to the characteristic sigmoidal growth curve (Figure 4) that represents an increasing fraction of the polymer that is oxidizing.

Heavily oxidized centers that suddenly start spreading will also lead to rapid crack formation soon after the end of the induction period. Mechanical failure due to microcrack formation is closely related to the actual induction period (13). In a heterogeneous model, heavy oxidation confined only to certain locations can easily explain the observation of secondary oxidation products such as water and CO_2 and crack formation commencing from the very early stages of the main oxidation of the material that would be inexplicable by a classical homogeneous kinetic model. In the classical homogeneous model, such as that applied to liquid-state oxidation kinetics, the induction period is only seen as the time necessary to produce a critical concentration of a certain species before chain branching and autoacceleration commences.

Homogeneous interpretation of the integrated CL curve as arising from an autoaccelerating chain reaction may also result in misleading conclusions about an average oxidation rate as the slope of parts of the curve. Tiny centers may in fact oxidize much more rapidly and cause immediate failure. Different materials may only be compared by measuring true induction periods and initial oxidation rates rather than comparing average rates, for example, determined by carbonyl index or oxygen uptake.

Physical Spreading

Photo- and thermal oxidation studies as discussed in the preceding sections led to conclusions about a highly heterogeneous oxidation in which the physical spreading from an initial center may play an important role. This finding is in agreement with many other studies indicating heterogeneous behavior (*16*).

Unlike other methods, CL is sensitive enough to allow study of very small (<20 μg) samples of polymer. Therefore, we were interested in further identifying evidence for the actual spreading of the oxidation by studying the CL from individual polymer particles. It has long been recognized that catalyst remnants from modern polymer manufacturing processes remain in the polymer and can interfere with the thermooxidative behavior of the material (*16, 17*). It has been difficult to define catalyst residue composition and their reactivities, but complex activities and effects of their presence were reported (*17, 18*). Individual powder particles were, therefore, expected to display heterogeneous behavior in such a way that some particles were more sensitive to oxidation than others. Higher rates of initiation due to catalytic influences or impurities should considerably shorten the lifetime of some particles.

Isothermal CL was applied to investigate the thermooxidative behavior of unstabilized PP powder particles (*19*). Oxidation of many single powder particles (10–500 μm) did reveal individual lifetimes over a large range regardless of the actual particle size (Figure 5). A signal was obtained from the simultaneous oxidation of many individual and separated particles (25 as shown in Figure 5), and this finding was interpreted as a simple sum of the behavior of each particle. In contrast, the oxidation of a group of particles in loose physical contact (again 25 as shown in Figure 5) resulted in a standard sigmoidal oxidation curve similar to the data obtained from other larger samples (*13*).

Groups of particles either as individuals or in physical contact were oxidized in a fundamentally different way. We concluded that when in contact with other particles, the particle with the shortest induction period was able to infect its neighbors to cause rapid oxidation of the complete sample. Thus, oxidation spread effectively from particle to particle. A similar result is presented in Figure 6, which shows the oxidation of only five particles at a lower temperature. Individual behavior of separated particles is much more pronounced. A tiny kink in the signal from the combined sample may indicate where the first particle to oxidize was able to infect the rest of the group and the oxidation started to spread. Physical spreading of the oxidation starting

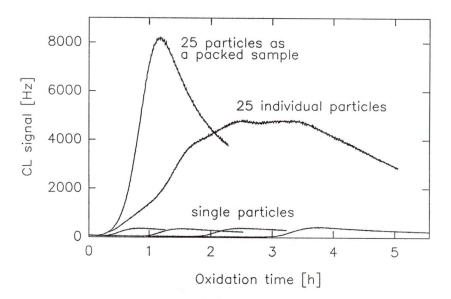

Figure 5. CL signals from the oxidation of different PP powder samples at 150 °C in oxygen that show the change on proceeding from isolated particles to particles in physical contact. (Reproduced with permission from reference 19. Copyright 1993 Elsevier Science Ltd.)

from an initial center and the following statistical progress can, therefore, be seen as the fundamental mechanism behind the observed sigmoidal shape of the oxidation curve.

Chemiluminescence Imaging

Another way in which the heterogeneity and physical spreading of the oxidation may be researched is by direct imaging of the CL from the oxidizing polymer by using either a sensitive charge-coupled device (CCD) or resistive anode encoder-based cameras (20–22). Cameras capable of imaging at the low levels of light associated with polymer CL have only appeared in the last few years. The low quantum yield of CL (10^{-9}) and the small fraction of the polymer initially oxidizing still require relatively long integration times at low temperatures at the pixel resolution necessary to resolve the oxidizing centers.

A commercially available CCD system (22) was used to obtain some preliminary images from the oxidation of PP powder particles. Figure 7 shows some of the images that were obtained during the oxidation of a PP powder sample at 150 °C in oxygen. The first image (Figure 7a) was obtained by artificially illuminating the sample and shows the original location of groups and individual particles. After 60 min of oxidation (Figure 7b), the image

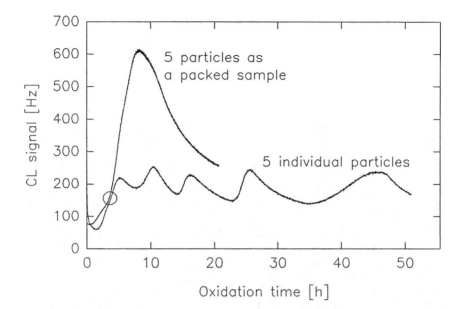

Figure 6. CL signals from the simultaneous oxidation of five individual and packed PP powder particles at 125 °C in oxygen. (Reproduced with permission from reference 19. Copyright 1993 Elsevier Science Ltd.)

clearly shows that only a fraction of the particles is oxidizing. The formation of oxidizing clusters of particles is clearly shown after 90 min of oxidation (Figure 7c). Figure 7d, taken after 120-min oxidation, shows isolated particles that resisted oxidation for a long time. This finding is consistent with our model of heterogeneous initiation of oxidation followed by the physical spreading and formation of oxidizing groups of particles.

One of the limitations of CL imaging seems to be the increased intensity (aura) around oxidizing centers. This aura is probably due to light scattering and reflection effects reducing the overall resolution of the image. However, both resistive anode encoder (20, 21) and CCD (22) cameras were used to image the CL emission from the oxidation of polymers.

Efficiency of Spreading

Identifying the efficiency and activity of the actual spreading was of further interest. In a packed sample some particles may be joined by surface contact, whereas others may be separated by tiny gaps. In a preliminary experiment, we tried to gain some information about the lateral activity of an oxidizing center or particle. Groups of five particles were arranged in straight lines with different surface-to-surface separations and were oxidized (19). Some of the

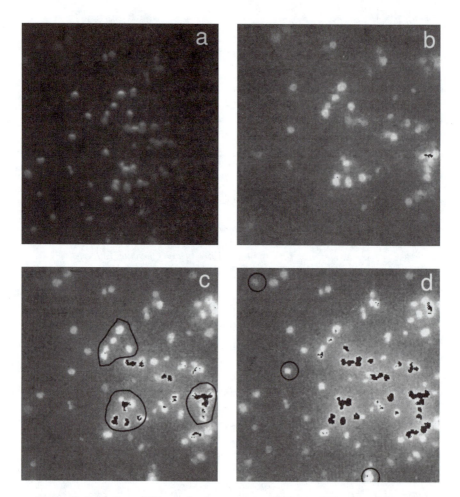

Figure 7. Oxidation of PP powder in oxygen at 150 °C: a is the original sample illuminated by visible light; b–d are CL images showing the initial oxidation of isolated particles (b) after 60 min, the formation of oxidizing clusters (outlined) after 90 min (c), and some particles (circled) remaining inactive for long times (d, after 120 min). The full width of each image corresponds to 1 cm at the sample. (The software used for producing these images truncates the original 12-bit data to 256 gray levels; highly emitting particles, therefore, appear as black.) (Reproduced with permission from reference 22. Copyright 1995 Elsevier Science Ltd.)

oxidation curves are presented in Figure 8. A group of particles separated from each other by approximately 100 μm gave a signal identical to a packed sample. Increasing the separation to 400 or 800 μm showed more pronounced individual behavior of the particles.

The exact nature of the initiating species still needs study. It appears that the physical spreading of the oxidation is, however, not limited by precise surface-to-surface contact of the particles but can jump via the gas phase. The production of volatile hydroperoxide fragments is one possible mechanism for gas-phase transport of oxidation, although the lifetimes of free-hydroxyl and hydroperoxy radicals may be long enough to allow for participation in such an initiation process.

Physical Spreading in a Stabilized Sample

Phenolic antioxidants incorporated into PP to improve the thermooxidative stability are heterogeneously distributed (16) mainly due to a limited solubility (23). During the extended induction period a nonhomogeneous consumption of stabilizers may also occur. This consumption may easily lead to some centers of the polymer losing their effective stabilization earlier than the rest of the material. The oxidation may start to spread from such centers of early instability. This spreading would be followed by a retarded increase in the

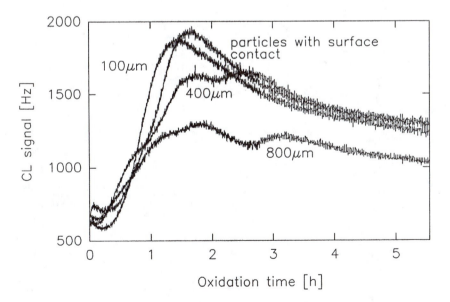

Figure 8. CL signals from the oxidation of five particles linearly arranged with varying surface-to-surface separations at 150 °C under oxygen. (Reproduced with permission from reference 19. Copyright 1993 Elsevier Science Ltd.)

oxidizing fraction of the material as the remaining stabilized centers in the polymer resist the oxidation for some time. Therefore, we attempted to compare the oxidation of unstabilized and stabilized PP-film samples after subtraction of the induction period.

The resulting signals are presented in Figure 9. The main oxidation of the unstabilized film proceeds rapidly, whereas the oxidation of the stabilized film requires considerable time to increase significantly. We concluded that for the stabilized film the remaining stabilized centers interfere with the physical spreading and result in an overall retarded oxidation of the sample.

Conclusion

The analysis of the CL data from the thermal and photooxidation of PP indicates that the oxidation of the solid polymer can be interpreted within a heterogeneous model with the following features:

1. Initiation occurs at high rates in localized zones, possibly associated with catalyst residues or other defects in the polymer.
2. Stabilizers are unable to inhibit this process, but they limit the spreading of the oxidation.

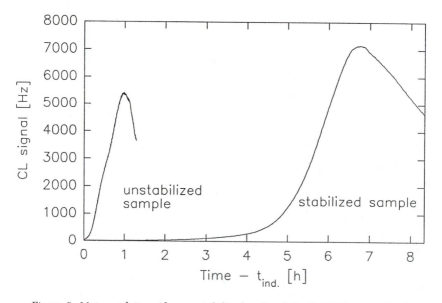

Figure 9. Main oxidation of an unstabilized and stabilized PP film sample after subtraction of the induction period. (Reproduced with permission from reference 19. Copyright 1993 Elsevier Science Ltd.)

3. The induction period represents a period of apparent constant rate of oxidation for both unstabilized and stabilized polymers. It measures the time taken for the oxidation to spread from the reactive centers. In these reactive centers, a branched chain reaction may be occurring that produces a wide range of oxidation products including water and carbon dioxide.

4. An induction period in chemical kinetic terms does not exist in the solid-state oxidation. The observed induction period is controlled by the physical oxidation behavior of the polymer.

5. The observed sigmoidal oxidation profile represents the statistical increase of the oxidizing fraction and not a kinetic curve corresponding to a homogeneous free-radical chain reaction.

6. From the concerted oxidation of closely spaced particles it appears that the spreading may occur through mobile, long-lived species that may percolate through the polymer and initiate further oxidation unless scavenged by a stabilizer molecule or otherwise deactivated.

Such a model is able to reconcile the following conflicting data in the oxidation of polymers: the evolution of water and carbon dioxide in the early stages of oxidation, the rapid crack growth that occurs soon after the end of the induction period, the lack of correlation between the useful lifetime of a polymer and that predicted from a homogeneous kinetic analysis, and the inability of many stabilizers to totally inhibit oxidation.

References

1. Vink, P. *J. Appl. Polym. Sci. Appl. Polym. Symp.* **1979**, *35*, 265.
2. Zlatkevich, L. *J. Polym. Sci. Polym. Phys. Ed.* **1985**, *23*, 1691.
3. Briggs, D. *Polymer* **1984**, *25*, 1379.
4. Grattan, D. W.; Carlsson, D. J.; Wiles, D. M. *Chem. Ind.* **1978**, *April*, 228.
5. George, G. A. *Dev. Polym. Deg.* **1981**, *3*, 173.
6. George, G. A. In *Luminescence Techniques in Solid State Polymer Research*; Zlatkevich, L., Ed.; Dekker: New York, 1989; p 93.
7. Matisova-Rychla, L.; Rychly, J.; Vavrekova, M. *Eur. Polym. J.* **1978**, *14*, 1033.
8. George, G. A.; Ghaemy, M. *Polym. Degrad. Stab.* **1991**, *33*, 411.
9. Billingham, N. C.; Then, E. T. H.; Gijsman, P. J. *Polym. Degrad. Stab.* **1991**, *34*, 263.
10. George, G. A.; Willis, H. A. *High Perform. Polym.* **1989**, *1*, 335.
11. George, G. A.; Ghaemy, M. *Polym. Degrad. Stab.* **1991**, *34*, 37.
12. Carlsson, D. J.; Wiles, D. M. *J. Macromol. Sci. Rev. Macromol. Chem.* **1976**, *C14*, 65.
13. Celina, M.; George, G. A. *Polym. Degrad. Stab.* **1993**, *40*, 323.
14. Vassil'ev, R. F. *Prog. React. Kinet.* **1976**, *4*, 305.
15. Iring, M.; Laszlo-Hedvig, S.; Barabas, K.; Kelen, T.; Tudos, F. *Eur. Polym. J.* **1978**, *14*, 439.
16. Billingham, N. C. *Makromol. Chem., Macromol. Symp.* **1989**, *28*, 145.
17. Gijsman, P.; Hennekens, J.; Vincent, J. *Polym. Degrad. Stab.* **1993**, *39*, 271.
18. Scheirs, J.; Bigger, S. W.; Billingham, N. C. *Polym. Degrad. Stab.* **1992**, *38*, 139.

19. Celina, M.; George, G. A.; Billingham, N. C. *Polym. Degrad. Stab.* **1993,** *42,* 335.
20. Fleming, R. H.; Craig, A. Y. *Polym. Degrad. Stab.* **1992,** *37,* 173.
21. Mattson, B.; Kron, A.; Reitberger, T.; Craig, A.; Fleming, R. *Polym. Test.* **1992,** *11,* 357.
22. Celina, M.; George, G. A.; Lacey, D. J.; Billingham, N. C. *Polym. Degrad. Stab.* **1995,** *47,* 311–317.
23. Billingham, N. C.; Calvert, P. D.; Manke, A. S. *J. Appl. Polym. Sci.* **1981,** *26,* 3543.

RECEIVED for review January 26, 1994. ACCEPTED revised manuscript March 13, 1995.

Inherent Relations of Chemiluminescence and Thermooxidation of Polymers

L. Matisova-Rychla and J. Rychly

Polymer Institute, Slovak Academy of Sciences, 842 36 Bratislava, Slovak Republic

Deterioration of properties of polymer products by oxidation generally is accompanied by weak light emission (chemiluminescence). The most important factors affecting the chemical properties of the polymer and the experimental conditions that play a role in the mechanism of chemiluminescence emission from polymers are examined. The variety of possible chemical and physicochemical pathways leading to the appearance of the light from thermooxidized polymer and the relation between the kinetics of chemiluminescence and the kinetics of polymer oxidation are discussed.

$\mathbf{B}$Y THE TIME Ashby (*1*) and later Schard and Russell (*2, 3*) published their papers on chemiluminescence from thermooxidation of polymers, Vassil'ev and co-workers (*4*) in Moscow had already done the detailed study of the oxidation of model low-molecular hydrocarbons. At that time, the group in Moscow deliberately stopped the work on chemiluminescence from thermooxidized polymers because they believed that the ultimate kinetic information could not be obtained from chemiluminescence measurements and that the unambiguous source of light emission could not be determined unequivocally. The title of Billingham's (*5*) presentation at a 1993 ACS meeting, "Chemiluminescence from Polymers—Powerful Tool or Illusion?", demonstrates that the reservations on the output of the method remain even after more than 30 years of investigation of the phenomenon.

Because we have witnessed the development of chemiluminescence investigations of polymers from the experimental and theoretical viewpoints in laboratories throughout the world, we will present an overview on the current

0065–2393/96/0249–0175$12.00/0

state of understanding of polymer oxidation as seen by the "eyes" of the chemiluminescence method. Our aim is to underline what the thermooxidation of polymers has in common with chemiluminescence signal output, which may be the only practical use. Our attention will not be focused on the variety of instruments and equipment that were designed for the study of chemiluminescence of polymers; these descriptions can be found elsewhere (6).

One fact should be kept in mind: light emission from the polymer thermooxidation does not necessarily originate from one kind of initiation event. This caveat is particularly true when experiments performed at low (20–100 °C), medium (100–150 °C) and high (150–250 °C) temperature regions of polymer oxidation are compared. The registration of several different procedures that differ in intensity very easily can mislead the interpretation.

Light Generation Mechanism

McCapra and Perring (7) expressed a general rule about searching for the possible source of chemiluminescence: chemiluminescence reactions involve rapid exothermic steps such as fragmentations, rearrangements, and electron transfers. Several elementary reactions can be considered as possible candidates providing chemiluminescence from the thermooxidized polymer. Most researchers (8) believe that the following reaction of self-recombination of secondary peroxy radicals provides the chemiluminescence:

$$R–CXH–OO^{\bullet} + ROO^{\bullet} \rightarrow R–C(X)=O^* + ROH + O_2^* \qquad (1)$$

where X is alkyl or H.

The energy gain of this reaction is more than 400 kJ/mol. Ketone is formed in an excited triplet state. Because of the spin conservation rule, oxygen should be formed in excited singlet state. Both particles can thus emit light when passing to their ground state. The possibility that reaction 1 is the light-emitting process is supported by experiments where the switching from oxygen into nitrogen atmosphere of thermooxidized polymer, or vice versa (9), leads to the immediate fall or increase, respectively, of light emission. This finding is in accordance with the fast replacement of peroxy to alkyl radicals or alkyl to peroxy radicals.

Obviously, other elementary reactions involving different routes of hydroperoxide decomposition, which are sometimes considered as the light-emitting processes, cannot easily be differentiated from reaction 1 (10). The decomposition of hydroperoxides is the initiation, whereas self-recombination of peroxy radicals is the termination reaction step of the one chain-reaction scheme. The rate of this reaction is governed by the rate of the slowest process: that is, the rate of initiation.

As proposed by Quinga and Mendenhall (*11*) for oxidized solutions of hydrocarbons, chemiluminescence may appear in disproportionation of alkoxy radicals. This reaction is energetically favorable for the excitation of carbonyl groups.

The objections against the self-recombination of secondary peroxy radicals as a source of chemiluminescence in thermooxidized polymer are based mostly on the difference of chemiluminescence between polypropylene (PP) and polyethylene (PE). PP gives higher light intensity when thermally oxidized than does PE. However, peroxy radicals on tertiary carbon that are the main chain carriers in oxidation of PP recombine to provide dialkyl peroxide and oxygen:

$$R-C(X_1X_2)-OO^\bullet + {}^\bullet OO-C(X_1X_2)-R \rightarrow R-C(X_1X_2)-OO-C(X_1X_2)-R + O_2 \quad (2)$$

The energy gain of reaction 2 is only about 300 kJ/mol, which is not enough when compared with the exothermicity of reaction 1.

The following reaction

$$-CH_2-C(O^\bullet)(CH_3)-CH_2- \rightarrow -CH_2-C^\circ O-CH_3 + {}^\bullet CH_2- \quad (3)$$

of β-scission of alkoxy radicals was proposed by Audouin-Jirackova and Verdu (*12*) for interpretation of chemiluminescence from PP and PE. This reaction does not require the presence of oxygen, and it explains the discrepancy in the higher value of chemiluminescence intensity from PP than from PE. Even though this reaction is not exothermic enough to provide the energy for the excitation of carbonyl groups, the structure $-C^\bullet-O^\bullet$ formed at the first step of the C–C cleavage of alkoxy radical is a biradical that is identical with excited carbonyls in photochemical processes of Norrish type II. If so, the light might be emitted when the biradical decays and the structure passes into the ground state.

Recently Verdu (*13*) suggested that the β-scission of C–C bonds is also involved in self-recombination of peroxy radicals (Russel's scheme) and in disproportionation of alkoxy radicals (Mendenhall's scheme). These schemes occur via tetraoxides or dialkyl peroxides in a concerted manner and result in a simultaneous formation of new bonds.

In an oxidized polymer, the situation is complicated by a nonspecified extent of quenching of triplet carbonyls by a ground state of oxygen that can change the resulting chemiluminescence–time pattern considerably. Notably, traces of oxygen are difficult to reliably remove from any polymer, and sometimes chemiluminescence observed in the inert atmosphere is simply the chemiluminescence in the presence of low quantities of oxygen. The whole spectrum of different peroxy radicals of low-molecular and macromolecular nature that are present during the oxidation process under such conditions and the alkyl radicals complicate the interpretation considerably.

The peaks with maximum around 500 nm for nylon 6,6 (14), 475 nm for PP (15), and 500 nm for epoxides (16) in chemiluminescence spectra of oxidized polymers were attributed either to polymeric C=C conjugated carbonyl groups or single carbonyl groups.

The fact that singlet oxygen also contributes to the overall light emission cannot be excluded. The excitation of oxygen is energetically less demanding [the $^1\delta$ state has absorption maximum at 762 nm, which corresponds to 156 kJ/mole (14)] and singlet oxygen is less quenched. No systematic study of this phenomenon has been undertaken to date. This omission may be due to broad spectral bands observed in chemiluminescence spectra and thus the low intensity of chemiluminescence emission. Another inhibiting aspect may be the low sensitivity of photomultipliers toward wavelengths above 600 nm in equipment usually used for investigation of polymer oxidation. In our opinion, the light emission from singlet oxygen should be considered at least as important as that from carbonyl groups, especially in the presence of end amino and hydroxyl groups, and from basic impurities in polyamides, polyethers, epoxides, or other heteropolymers. It may acquire a particular importance when hydrogen peroxide is among the primary products of polymer oxidation. This scenario may occur in oxidation of structural moieties carrying hydroxyl groups:

$$-C(OH)H- + O_2 \rightarrow HOOH + -CO- \tag{4}$$

$$2\ HOOH \rightarrow 2\ H_2O + {}^1\!/_2\ O_2{}^* \tag{5}$$

Osawa and Kuroda (17) suggested that the luminescence particles in commercial isotactic PP are foreign impurities such as polynuclear aromatics from the atmosphere. Together with co-workers, they (18, 19) demonstrated this identification by the disappearance of chemiluminescence when the polymer was extracted by hexane. After the exposure of extracted PP film to the ambient atmosphere, the fluorescence and phosphorescence emission reappeared. This interesting observation can change the approach to the interpretation of chemiluminescence from polymers but further investigation is required.

Decomposition of Low-Molecular Initiators of Free Radical Reactions

During the 1970s, our interests were focused mainly on the use of the chemiluminescence method for investigation of the decomposition of low-molecular initiators of free radical reactions, such as benzoyl peroxide in a polymer matrix. Our aim was to ascertain how the polymer medium affects the decomposition when compared with low-molecular solvents. Low-molecular initiators give rise to chemiluminescence in polymer medium even in inert atmosphere.

This phenomenon may be exemplified by the case of benzoyl peroxide used as an initiator of bulk polymerization of poly(alkyl methacrylates) (*20, 21*). At low conversions of monomer to polymer the chemiluminescence intensity is very low; however, at solidification of the polymer after the gel-point the intensity increases to a sufficiently higher level that depends on the character of the alkyl group. The highest intensity is obtained from methyl, and the lowest is obtained from *n*-butyl in the series from methyl methacrylate to *n*-butyl methacrylate.

We suggested (*22*) that benzoyl peroxide forms microheterogeneous domains in the polymer bulk as a consequence of some reprecipitation due to the fast polymerization at the gel-point. In these domains the highly exothermic chain reaction of peroxide decomposition may occur and give rise to light emission. However, the source of the light emission was not specified, and the problem is far from being solved. The question is how the light emission is influenced by the aggregation state of the initiator in a given medium. The light may originate from the local overheating of the reacting microdomain resulting from worsened transport of the heat and subsequent electronic excitation of some reaction product. The effect should be pronounced by much weaker quenching of excited states in the solid state than in the liquid (*22*).

Not every peroxide can produce light when decomposed thermally under inert atmosphere. Billingham et al. (*23*), for example, did not observe any significant increase of chemiluminescence intensity in nitrogen when *t*-butyl hydroperoxide was introduced into PP. *t*-Butyl pivalate, a model perester yielding tertiary alkoxy radicals, also did not give luminescence in PP under nitrogen (*23*). On the other hand, methyl linoleate hydroperoxide adsorbed on neutral alumina yielded rather strong chemiluminescence that decayed with second-order kinetics (*24*). Benzoyl peroxide adsorbed from benzene solution on activated synthetic faujasite NaX produced intense bursts of light emission that decayed within 20 min at 81 °C (*25*).

These results support the idea that microheterogeneous domains of predominantly peroxidic reactants in the solid state are much more beneficial to the generation of light emission than homogeneously dispersed domains in the solid system. George and co-workers (*26, 27*) stated that isothermal chemiluminescence curves for oxidation of PP may be interpreted by the accumulation of centers of high local oxidation that prevail over the homogeneous oxidation. By using the staining technique applied on oxidized PP and its observation by UV microscopy, Billingham (*28*) confirmed a high degree of oxidation nonhomogeneity at the micron level that appeared to be associated with catalyst residues. George et al. (*26*) showed that all chemiluminescence–time curves can be reduced to a single master curve by plotting the data in reduced coordinates over a wide range of temperature. The resulting curve fits Gaussian statistics and apparently represents the spreading of the oxidation from the initial reactive zones.

Other Potential Sources of Light Emission

A release of volatiles that are either inherently present in the polymer or are formed during the oxidation process may be another prerequisite for light emission from polymer samples. Provided that low-molecular products of oxidation are reactive enough to undergo the further process with oxygen, light emission may appear at the instant when these products are released from the polymer phase.

On the other hand, the microbubbles of a gas captured in a solid polymer can become the source of rather high strain. For instance, 10 molecules, each with a volume of 1×10^{-24} dm^3, kept in a 10×10^{-24} dm^3 space exert about 50 atm at 373 K. This pressure can initiate the mechanochemical cleavage of bonds in a polymer if efficiently focused. Obviously, this effect is pronounced with the presence of very small molecules and higher temperatures.

Low-molecular products kept in a polymer bulk have a tendency to escape and thus to generate the rupture of bonds. This effect may be particularly important at the glass transition temperature (T_g). The mechanical scission of C–C bonds combined with counter diffusion of oxygen may produce peroxy radicals and give rise to the chemiluminescence.

Indeed, at the programmed heating (2 °C/min), the poly(methyl methacrylate) (PMMA) and polystyrene (PS) foils without free radical initiators yield the maximum of a very weak chemiluminescence situated at T_g. Signal output for PMMA is more distinct than that for PS (Figure 1). Also for poly(vinylpyrrolidons), the temperature at which maximum chemiluminescence emission occurs correlates with T_g (29).

An observation by Billingham et al. (23) supports the idea of chemiluminescence resulting from the escape of some low-molecular products from the polymer. They found that the rate of chemiluminescence decay of preoxidized PP in nitrogen was considerably higher than that of peroxide decomposition. They suggested that chemiluminescence is associated with decomposition of a small fraction of the total peroxide content, which is the most reactive.

Another explanation of a faster decay of chemiluminescence that corresponds to the reduction of the total peroxide content requires that the chemiluminescence be associated with the recombination of peroxy radicals. A so-called "oxygen drop" (30), represented by a sudden fall of chemiluminescence intensity, occurs at the initiated oxidation of liquid hydrocarbons performed in a closed system. The oxygen dissolved in hydrocarbon is consumed gradually because of the initiation reaction of peroxide decomposition. A certain critical level of oxygen in the system is then reached, below which not all alkyl radicals are converted into peroxy radicals. The kinetics of the fast decay of the light intensity after oxygen consumption has, however, nothing to do with the kinetics of peroxide decomposition. Such oxygen drop is considerably

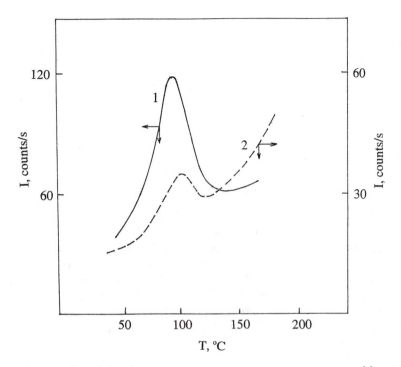

Figure 1. Plot of chemiluminescence vs. temperature at a programmed heating (2 °C/min) for PMMA (1, solid line) and PS (2, broken line) foils in air. Foils were cast from chloroform solution.

faster than any chemiluminescence decay connected with preoxidized polymer.

Schard and Russell (2, 3) proposed that the chemiluminescence accompanying the thermooxidation of polymers could be used as a tool for determination of structural chain branching. They observed that among the polymers having a carbon–carbon backbone, the maximum intensity of light emission increased in parallel to the number of tertiary carbon atoms. They demonstrated this occurrence on a series of high-density PE, low-density PE, and three types of PP in which the maximum chemiluminescence intensity (150 °C, anode of photomultiplier at 1000 V) was expressed in analogous signal (in amperes) increased as follows: 1.2×10^{-12} for HDPE, 6.7×10^{-10} for LDPE, and 1.9–2.7×10^{-8} for PP. Polystyrene exhibits very low luminosity at the same conditions (6.9×10^{-11} A).

The degree of polymer branching has, however, another interesting aspect that could explain the origin of chemiluminescence from polymers. For example, PP has a higher concentration of tertiary carbon atoms in the polymer chain and a higher concentration of terminal C=C bonds than does PE. If the light emission arises predominantly from the oxidation of terminal C=C bonds in polymers one could easily explain the variability in chemilumines-cence–time patterns of different series of the same polymer and of one pol-ymer compared with another polymer. In the case of PMMA, new terminal double bonds are gradually formed during the polymer thermooxidation by depolymerization. Chemiluminescence from PMMA increased steadily over time (2, 3). PP gives a typical autoaccelerated increase to a maximum chem-iluminescence and then a subsequent decay or steady course. However, pol-ymers such as polyethers [poly-(2,6-dimethylparaphenylene oxide) (31)], poly(ethyleneterephthalate) (4), and cured epoxies (32, 33) have a steadily decreasing chemiluminescence intensity from some initial value. Provided that unsaturations or other reactive structures are converted quickly to hydrope-roxides and their initial concentration is above the equilibrium level of hydro-peroxides corresponding to the given conditions of polymer oxidation, the decay of chemiluminescence from the beginning of the experiment may well be understandable. Moreover, hydroperoxides formed in initial oxidation of unsaturations and defect structures may be those of higher reactivity as re-quired by Billingham et al. (23). For example, two adjacent hydroperoxy groups or peroxy radicals might be formed from terminal unsaturation as shown by structure 1. Peroxidic structures such as **1** decompose with consid-erably more ease than isolated backbone hydroperoxide groups.

$$
\begin{array}{ccc}
\overset{\overset{\displaystyle O\cdot}{\underset{\displaystyle O}{|}}}{-\,CH}\;-\;\overset{\overset{\displaystyle O\cdot}{\underset{\displaystyle O}{|}}}{CH_2} & \text{or} & -\;CH_2-\;\overset{\overset{\displaystyle H}{|}}{\underset{\overset{\displaystyle O}{\underset{\displaystyle O-H}{|}}}{C}}\;-\;\overset{\overset{\displaystyle O-H}{\underset{\displaystyle O}{|}}}{CH_2}
\end{array} \qquad (1)
$$

Recently, we have found that formic acid or H_2O_2 induces a fast, exo-thermic release of bromine from 1,3-dibromo-5,5-dimethylhydantoin (DBH). The overall reaction heat was proportional to the concentration of both the formic acid and hydrogen peroxide, whereas the concentration of DBH in water was constant. We performed the oxidation of 100 mg of PP powder in a closed ampoule, filled the system with water after cooling it to ambient temperature, and filtered the polymer powder to receive the filtrate with wa-ter-soluble products of polymer oxidation.

To our surprise these products reacted rapidly with DPH and released the bromine. Simultaneously, an appreciable amount of heat was detected by solution calorimetry. Accumulation of such products calibrated to formic acid is shown in the Figure 2. The reaction of DBH occurred only with products of oxidation that were collected in the upper, colder parts of the reaction ampoule; oxidized polymer did not react.

At the same time, the chemiluminescence–time pattern of the same pol-ymer (Figure 2) matches quite well with the accumulation of these reactive products at 140 °C and suggests that the chemiluminescence can appear at

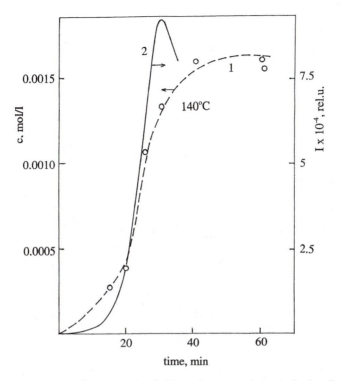

Figure 2. Accumulation of water-soluble oxidation products (1, broken line) and chemiluminescence curve (2, solid line) for oxidation of 100 mg of isotactic PP at 140 °C in oxygen atmosphere.

the oxidation process ending at formic acid or hydrogen peroxide. The maximum amount of these products corresponds to the oxidation of one group in polymer per average molecular mass of about 17,000 g/mol. (The average molecular mass of polymer was 220,000 g/mol.)

Nonisothermal Chemiluminescence Experiments

One main chemiluminescence peak, distorted by some shoulders or other smaller peaks, can be observed when the polymer sample is heated in oxygen with a linear increase of temperature from 100 to 250 °C. The maximum of this main peak lies above 200 °C for almost all polymers examined (*34, 35*). A mathematical approach to modeling and determination of kinetic parameters from nonisothermal chemiluminescence glow curves were shown elsewhere (*33*).

Two first-order reactions can be responsible for an apparently single-peak glow curve for PE (Figure 3). The process of very low activation energy (41

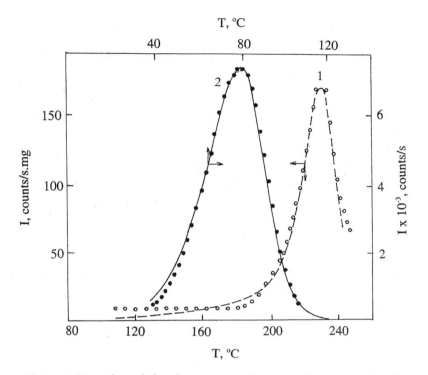

Figure 3. Nonisothermal chemiluminescence glow curves for 3 mg powdered PE Marlex 55180 (1, broken line) in oxygen and for a foil of PMMA containing 1 wt% of tetramethyldioxetane (2, solid line) in air and heated at 2 °C/min. Points denote the experimental curve; lines represent theoretical curves corresponding to respective fits.

kJ/mol) predominates at low temperatures, and that of high activation energy (245 kJ/mol) predominates at higher temperatures. The high activation energy value is an indication that the C–C bond cleavage and subsequent reactions of peroxy radicals can contribute to the light emission, particularly at higher temperatures. For polymers such as PE or poly(2,6-dimethylparaphenylene), the process is accompanied by ignition or incandescence.

The lower temperature region and its relation to isothermal experiments is not, for our purposes, easily interpreted. The problem arises from the role of various defects in the polymer structure that can acquire an importance in initiation of polymer oxidation. A two-peak process was also assumed for nonisothermal chemiluminescence of epoxies cured by amine with or without the presence of monofunctional phenylglycidyl ether (33). The values of apparent activation energies are 120 and 80 kJ/mol, respectively. The activation energies of decomposition of low-molecular initiators obtained from chemiluminescence–temperature experiments appear reasonable. The nonisothermal chemiluminescence record of tetramethyldioxetane decomposition in a PMMA matrix that has the maximum of emission at considerably lower temperature yields an activation energy of about 90 kJ/mol (Figure 3).

Chemiluminescence from Stabilized Polymer Systems

The most practical output of the chemiluminescence technique occurs because the interruption of oxidation chains by inhibitors leads to the considerable decrease of light intensity. If substance X depressed the stationary concentration of peroxy radicals or hydroperoxides, then the chemiluminescence intensity of polymers in the presence of X is reduced proportionally to the reactivity and concentration of X. We proved (36) that the identical sequence of reactivity of substituted 4-alkyl-2,6-di(*tert*-butyl)phenols in PP (0.02 mol/kg of polymer; temperature, 168 °C) was obtained with both the chemiluminescence and oxygen-absorption methods from induction periods of oxidation.

This same correspondence was demonstrated for stabilizing mixtures of different inhibitors such as Hostanox OSP1 (2-(1,1-dimethylethyl)-4-[[3-(1,1-dimethylethyl)-4-hydroxy-5-methylphenyl]thio]-5-methylphenol), Irganox 1010 (3,5-bis(1,1-dimethylethyl)-4-hydroxybenzenepropanoic acid) and distearyl-thiodipropionate (190 °C, PP) (37).

The antioxidation efficiency of newly synthesized stabilizers based on N-phenyl-N-pyrazolamines, N-phenyl-N-benzimidazolamines, and N-(arylamine)-imides in PP was verified. Relatively good coincidence of the results was obtained from differential scanning calorimetry and thermogravimetry measurements (38–40).

The different inhibitors and inhibiting systems can thus be attempted, and coincidence with other tests should be verified. Unfortunately, papers devoted to this problem are not very numerous (41–43).

Inhibition period on chemiluminescence–time curves can be observed only for polymers that oxidize with a sufficiently long length of kinetic chains. Polymers such as epoxides and other polyethers such as poly(2,6-dimethyl-paraphenylene), however, do not give autocatalytic chemiluminescence curves with a distinct induction period. Instead, these compounds show a more or less fast decrease of chemiluminescence from some initial value. In such a case, no inhibition period can be obtained with common phenolic or aminic antioxidants; the antioxidant efficiency is displayed only in the lowering of the overall chemiluminescence signal. The application of the method to the study of stabilization of such polymers has not been fully explored.

Significant suppression of chemiluminescence from PP was also achieved when PbO was mixed mechanically with the polymer (0.5–3 wt%, 195 °C) (44). This mixing can stimulate a contemplation over the possible mechanism of the light emission from volatile products. As early as 1965 and 1966, Niclause, Fusy, and co-workers (45, 46) indicated that the surface of PbO activated by oxygen is an efficient scavenger of methyl radicals. These radicals are destroyed fast enough so that they cannot propagate to isobutane, which was used as substrate. Because the methyl radicals can be assumed to be possible intermediates of β-cleavage of *tert*-alkoxy radicals from a PP chain, the reaction sequence ending at formaldehyde and formic acid is probably as follows:

$$CH_3^{\cdot} + O_2 \rightarrow CH_3OO^{\cdot}$$

$$CH_3OO^{\cdot} + RH \rightarrow CH_3OOH + R^{\cdot}$$

$$CH_3OOH \rightarrow CH_3O^{\cdot} + {}^{\cdot}OH \rightarrow CH_2O^* + H_2O$$

$$2CH_3OO^{\cdot} \rightarrow CH_2O^* + CH_3OH + O_2$$

These reactions also can be presumed as the possible sources of light emission, but the chemiluminescence process should be localized rather to the gaseous side of the polymer–oxygen interface, which was not proved.

Modeling of Oxidation and Chemiluminescence–Time Curves

No numerical computations that verify a reaction pathway can be found in the literature on different reaction schemes related to the observation of chemiluminescence. Until now, analytical solutions involving numerous simplifications were preferred (47, 48).

We have undertaken the approach of numerical computation on a model consisting of 14 elementary reaction steps, including mono- and bimolecular decomposition of hydroperoxides, three termination reactions, and reactions with inhibitors.

Although the numerical solution is not as illustrative as the analytical one, some interesting points arise. The numerical solutions enable wide variations in elementary reactions involved, which is not possible in analytical solutions without "a priori" simplifications.

The model of a polymer (RH) oxidation involved the following elementary steps:

$$ROOH \rightarrow R^{\cdot} + \text{products} \qquad k_1$$

$$2ROOH \rightarrow R^{\cdot} + RO_2^{\cdot} + \text{products} \qquad k_2$$

$$R^{\cdot} + O_2 \rightarrow RO^{\cdot} \qquad k_3$$

$$RO_2^{\cdot} + RH \rightarrow ROOH + R^{\cdot} \qquad k_4$$

$$RH \rightarrow R^{\cdot} + \text{products} \qquad k_5$$

$$RO_2^{\cdot} + RO_2^{\cdot} \rightarrow O_2 + \text{products} + h\nu \qquad k_6$$

$$RO_2^{\cdot} + R^{\cdot} \rightarrow \text{products} \qquad k_7$$

$$R^{\cdot} + R^{\cdot} \rightarrow \text{products} \qquad k_8$$

$$RO_2^{\cdot} + InH \rightarrow ROOH + In^{\cdot} \qquad k_{10}$$

$$R^{\cdot} + InH \rightarrow RH + In^{\cdot} \qquad k_{11}$$

$$RO_2^{\cdot} + In^{\cdot} \rightarrow \text{products} \qquad k_{12}$$

$$R^{\cdot} + In^{\cdot} \rightarrow \text{products} \qquad k_{13}$$

$$RH + O_2 \rightarrow R^{\cdot} + \text{products} \qquad k_{14}$$

$$In^{\cdot} + In^{\cdot} \rightarrow \text{products} \qquad k_{15}$$

where In is the inhibitor. Oxygen is supplied by the rate $k_9(O_{2_0} - O_2)$, where k_9 is proportionality constant and $O_2 0$ is the initial concentration of oxygen in the surrounding atmosphere.

In addition to the mono- and bimolecular decomposition of hydroperoxides, we discuss two other independent initiation steps: those characterized by the reaction with k_5, which corresponds to cleavage of the weak bond in the polymer; and reaction with k_{14}, which represents a direct attack of oxygen on polymer. The reactions are represented schematically without the intermediate formation of alkoxy radicals from hydroperoxides. Alkoxy radicals are assumed to react rapidly with a polymer chain and to yield polymeric alkyl radicals. We include neither the reaction of initiation due to inhibitor-radical transfer to polymer nor the formation of volatile products, which accompanies each thermooxidation process. The parameters for which the model computations were performed are given in Figures 4–7. The chemiluminescence intensity, I, is defined by the relation:

$$I = \mu k_6 [RO_2^{\cdot}]^2$$

where μ is a proportionality constant.

For a given set of parameters (Figure 4), the chemiluminescence decay in the inert atmosphere is about 4 times faster than the decay of hydroperoxides; both curves were evaluated by the first-order scheme. This discrepancy follows from the properties of the oxidized system itself, including three termination steps.

In Figure 4 we see how chemiluminescence–time curves look when the highly efficient inhibitor (high value of k_{10} constant of the reaction of peroxy radicals with the inhibitor) is present in an oxidized polymer. Induction period of oxidation is very distinct and increases with the concentration of inhibitor; the plot is linear for lower concentrations of inhibitors.

The value of k_{10} determines the slope in the point of inflexion. As follows from Figure 5, the lower the slope, the lower the value of k_{10}. For inhibited

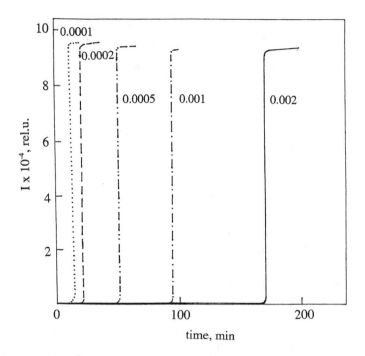

Figure 4. Chemiluminescence–time curves for a model of polymer oxidation at different initial concentrations of inhibitor $[InH]_o$ (indicated at a respective line). The values of parameters of the kinetic model used in this chapter are as follows: $k_1 = 0.001$, $k_2 = 0.02$, $k_3 = 1 \times 10^7$, $k_4 = 2$, $k_5 = 1 \times 10^{-7}$, $k_6 = 1 \times 10^5$, $k_7 = 1 \times 10^6$, $k_8 = 1 \times 10^7$, $k_9 = 1$, $k_{10} = 3 \times 10^6$, $k_{11} = 5 \times 10^6$, $k_{12} = 1 \times 10^7$, $k_{13} = 2 \times 10^7$, $k_{14} = 5 \times 10^{-6}$, $k_{15} = 100$, $[RH]_o = 1$, $[ROOH]_o = 0.01$, $\mu = 1 \times 10^{10}$, $[O_2]_o = 1$. Except for oxygen, where $[O_2]_o = 1$ represents the pure gas, concentrations are in moles per kilogram and time is in minutes.

oxidation of low-molecular hydrocarbons, the theoretical relation was derived (30):

$$(dI/dt)_{max} = \alpha k_{10}(w_i/k_6)^{1/2}$$

where t is time, w_i is the constant rate of initiation and α is the proportionality constant (I is related to 1). The relationship between $(dI/dt)_{max}$ and k_{10} is linear even though the rate of initiation of a polymer changes with changes of concentration of hydroperoxides.

An increase of initial concentration of hydroperoxides leads to an increase also of the stationary value of chemiluminescence and reduces the induction periods (Figure 6). Figure 6 is an important indication of how the efficiency rating of an inhibitor is dependent on the history of the polymer expressed by the degree of previous oxidation. Certain types of inhibitor at a given

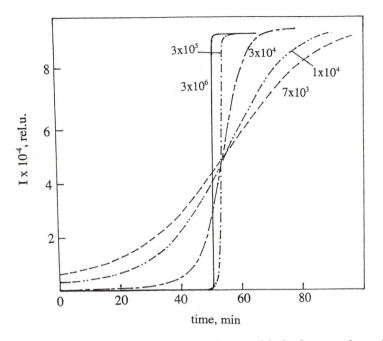

Figure 5. Chemiluminescence–time curves for a model of polymer oxidation for different values of transfer rate constant k_{10} *(indicated at a respective line) of peroxy radicals to inhibitor, InH. Parameters:* $[InH]_o = 0.0005$*; other parameters are the same as in Figure 4.*

concentration may thus yield rather large scatter of results when used in a polymer of the same kind but from different sources.

The rate constant k_1 (or k_2) of hydroperoxide decomposition, which is predominately the function of the temperature, can modify the observed chemiluminescence–time curves so that a maximum replaces the steady level. The higher concentration of inhibitor shifts the curve to longer times (Figure 7). In such a way we can examine the effect of each parameter of the model on resulting chemiluminescence–time curves as well as its effect on the character of plots of integral intensity versus oxygen consumed or hydroperoxide decomposed.

Further Perspectives of Chemiluminescence Method in Oxidation of Polymers

As stated in the previous section, the factors affecting the chemiluminescence of polymers are so numerous that without a systematic approach, we cannot thoroughly understand the phenomenon. For such a purpose, model computations involving the large variations of elementary reactions taking place in

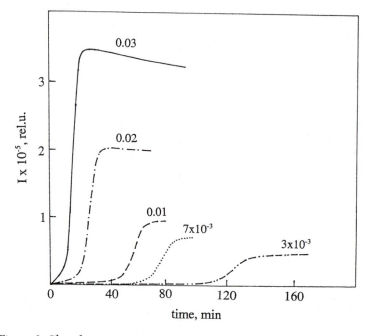

Figure 6. Chemiluminescence–time curves for a model of polymer oxidation at different initial values of hydroperoxide concentration (the number at a respective line). Parameters: $k_{10} = 3 \times 10^4$, $[InH]_o = 0.0005$; other parameters are the same as in Figure 4.

homogeneous or heterogeneous conditions with different rate constants can be good tools for analysis of respective experimental results.

Any optical hindrance can modify the absolute value of chemiluminescence intensity considerably. In this chapter, we enumerated the reduction of intensity due to the volatile products released from the polymer that absorb the light or quench the excited states and the absorption of light by colored admixtures in a polymer.

We have shown (33) that the volatiles that are formed during epoxide decomposition suppress chemiluminescence provided that they are resistant toward further oxidation, but the opposite may be true for other polymers.

Because of the extreme sensitivity of the chemiluminescence technique, attention should be paid to better characterization of polymer samples. Characterization is concerned with anomalous structures, end groups, the residual amount of catalysts, and other molecular compounds. Such studies would throw some light on the nature of the process in a respective polymer under respective conditions. Also, the effect of sample coloration and quenching of chemiluminescence from polymers by oxygen should be examined more systematically.

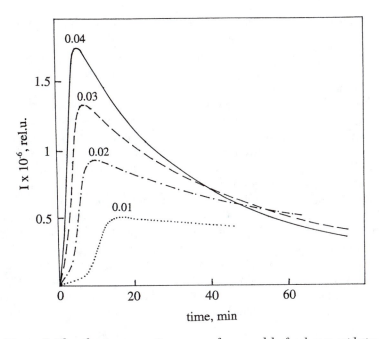

Figure 7. Chemiluminescence–time curves for a model of polymer oxidation at different initial values of the first-order rate constant k_1 of hydroperoxide decomposition (the number at a respective line). Parameters: $k_{10} = 3 \times 10^4$, $[InH]_o = 0.0005$; others parameters are the same as in Figure 4.

The comparison of decay of polymer free radicals in oxygen with the decay of chemiluminescence should also be of interest. Polymer radicals can be generated by γ-irradiation or by mechanical destruction. Cross-linked polymers such as epoxides can provide free radicals by a simple grinding in air atmosphere. The chemiluminescence decay for powder obtained by grinding for 35 s of hardened epoxide containing 30 wt% phenyl glycidyl ether was composed of two parts. The first part was very fast and the second was slower: First-order rate constants were 0.02 s^{-1} and $1–3 \times 10^{-4}$ s^{-1}, respectively, at 20 °C. The ESR spectrum of the powder is a singlet of width 1.4 mT and rather low intensity; the rate constant of peak-to-peak decay of this spectrum was about 2×10^{-4} s^{-1} at 20 °C.

The most promising use of the chemiluminescence method is in tests of antioxidant efficiency of various stabilizing systems. Quite reliable orders of efficiency in different stabilizing series for polymers that oxidize with sufficiently large kinetic chains can be obtained. The chemiluminescence imaging technique enabling the examination of selected parts of a polymer sample during its oxidation appears to be a powerful tool in industrial tests of thermooxidation stability (49).

Interpretation of the chemiluminescence–time curves from oxidation of hardened epoxies, polyethers, and even from polyamides remains an unsolved problem, especially when the common stabilizing systems based on phenols or amines are considerably less efficient. The study of the effect of end groups appears to be of the highest importance.

However, the chemiluminescence method has large potential application in other areas of analytical chemistry because of its high sensitivity. In comparison with other methods of detection of free radicals such as ESR, chemiluminescence is more sensitive by at least two orders of concentration of free radicals. This advantage may be exemplified by the experiments where the oxygen containing gas was passed over a polymer sample and bubbled through a chemiluminescent probe solution of a hydrocarbon such as ethylbenzene (50). The efficiency of free radical acceptors being released from the polymer and captured in a hydrocarbon was determined from the changes of the hydrocarbon oxidation curves monitored by chemiluminescence method.

The combination of chemiluminescence with photodetectors can be used for characterization of permeation through polymer film of liquids such as H_2O_2. If alkaline solution of luminol is present at the other side of the film, the instant of permeation as well as the film homogeneity can well be imaged (51).

References

1. Ashby, G. E. *J. Polym. Sci.* **1961,** *50,* 99.
2. Schard, M. P.; Russell, G. A. *J. Appl. Polym. Sci.* **1964,** *8,* 985.
3. Schard, M. P.; Russell, G. A. *J. Appl. Polym. Sci.* **1964,** *8,* 997.
4. Vassil'ev, R. F.; Karpukhin, O. N.; Shlyapintokh, V.Ya. *DAN SSSR.* **1959,** *125,* 106 (in Russian).
5. Billingham, N. C. *Proceedings of the ACS Symposium on Lifetime, Degradation, and Stabilisation of Macromolecular Materials;* Chicago, IL, 1993.
6. Zlatkevich, L. *Luminescence Techniques in Solid State Polymer Research;* Dekker: New York, 1989.
7. McCapra, F.; Perring, K. D. *Chemi and Bioluminescence;* Burr, J. G., Ed.; Dekker: New York, 1985.
8. George, G. H.; Egglestone, G. T.; Riddell, S. Z. *Polym. Chem. Eng. Sci.* **1983,** *23,* 412.
9. George, G. A., *Dev. Polym. Degrad.* **1983,** *3,* 173.
10. Matisova-Rychla, L.; Fodor, Z.; Rychly, J.; Iring, M. *Polym. Degrad. Stab.* **1981,** *3,* 371.
11. Quinga, E. M. Y.; Mendenhall, G. D. *J. Am. Chem. Soc.* **1983,** *105,* 6520.
12. Audouin-Jirackova, L.; Verdu, J. *J. Polym. Sci.* **1987,** *25,* 1205.
13. Verdu, J., private communications.
14. George, G. A. *Polym. Degrad. Stab.* **1979,** *1,* 217.
15. Matisova-Rychla, L.; Rychly, J.; Vavrekova, M. *Eur. Polym. J.* **1978,** *14,* 1033.
16. George, G. A.; Sweinsberg, P. D. *J. Appl. Polym. Sci.* **1987,** *33,* 2281.
17. Osawa, Z.; Kuroda, H. *J. Polym. Sci. Polym. Lett. Ed.* **1982,** *20,* 577.
18. Osawa, Z.; Kuroda, H.; Kobayashi, Y. *J. Appl. Polym. Sci.* **1984,** *29,* 2843.
19. Kuroda, H.; Osawa, Z. *Makromol. Chem. Macromol. Symp.* **1989,** *27,* 97.
20. Matisova, L.; Pavlinec, J.; Lazar, M. *Eur. Polym. J.* **1970,** *6,* 785.

21. Matisova-Rychla, L.; Rychly, J.; Lazar, M. *Eur. Polym. J.* **1972,** *8,* 655.
22. Matisova-Rychla, L.; Rychly, J.; Lazar, M. *Makromol. Chem.* **1975,** *176,* 2701.
23. Billingham, N. C.; Then, E. T. H.; Gijsman, P. J. *Polym. Degrad. Stab.* **1991,** *34,* 263.
24. Cash, G. A.; George, G. A.; Bartley, J. P. *Chem. Phys. Lipids* **1987,** *43,* 265.
25. Rychly, J.; Matisova-Rychla, L.; Lazar, M. *J. Lumin.* **1971,** *4,* 161.
26. George, G. A.; Celina, M.; Ghaemy, M. *Proceedings of the 14th Annual Conference on Advances in the Stabilisation and Degradation of Polymers;* Lucerne, Switzerland, May 1992, p 75.
27. Celina, M.; George, G. A. *Polym. Degrad. Stab.* **1993,** *40,* 323.
28. Billingham, N. C. *Makromol. Chem. Macromol. Symp.* **1989,** *28,* 145.
29. Scheirs, J.; Bigger, S. W.; Then, E. T. H.; Billingham, N. C. *J. Polym. Sci. Part B: Polym. Phys.* **1993,** *31,* 287.
30. Shlyapintokh, V.Ya.; Karpukhin, O. N.; Postnikov, L. M.; Zakharov, I. V.; Vichutinskii, A. A.; Tsepalov, V. F. *Chemiluminescence Methods of Investigation of Slow Chemical Processes* (in Russian); Publishing House Nauka: Moscow, 1966.
31. Matisova-Rychla, L.; Chodak, I.; Rychly, J.; Bussink, J. *J. Appl. Polym. Sci.* **1993,** *49,* 1887.
32. Tcharkhtchi, A.; Audouin, L.; Verdu, J. *J. Polym. Sci.* **1993,** *31,* 683.
33. Matisova-Rychla, L.; Rychly, J.; Verdu, J.; Audouin, L.; Tcharkhtchi, A.; Janigova, I. *J. Appl. Polym. Sci.* **1994,** *53,* 1375.
34. Wendlandt, W. W. *Thermochim. Acta* **1984,** *72,* 363.
35. Wendlandt, W. W. *Thermochim. Acta* **1987,** *71,* 129.
36. Matisova-Rychla, L.; Ambrovic, P.; Kulickova, M.; Rychly, J.; Holcik, J. *J. Polym. Sci. Polym. Symp.* **1976,** *57,* 181.
37. Rychla, L.; Rychly, J.; Krivosik, I. *Polym. Degrad. Stab.* **1988,** *20,* 325.
38. Matisova-Rychla, L.; Rychly, J.; Meske, M.; Schulz, M. *Polym. Degrad. Stab.* **1988,** *21,* 323.
39. Matisova-Rychla, L.; Rychly, J.; Csomorova, K.; Velikov, A. A.; Kluge, R.; Schulz, M. *Angew. Makromol. Chem.* **1990,** *176/177,* 231.
40. Rychla, L.; Rychly, J.; Ambrovic, P.; Csomorova, K.; Mogel, L.; Schulz, M. *Polym. Degrad. Stab.* **1986,** *14,* 147.
41. Forstrom, D.; Kron, A.; Mattson, B.; Reittberger, T.; Steinbergrg, B.; Terselius, B. *Rubber Chem. Technol.* **1992,** *65,* 736.
42. Osawa, Z.; Tsurumi, K. *Polym. Degrad. Stab.* **1989,** *26,* 151.
43. Jipa, S.; Setuescu, R.; Setuescu, T.; Cazac, C.; Budrugeac, P.; Mihalcea, I. *Polym. Degrad. Stab.* **1993,** *40,* 101.
44. Matisova-Rychla, L.; Rychly, J.; Vavrekova, M. *Polym. Degrad. Stab.* **1980,** *2,* 187.
45. Niclause, M.; Martin, R.; Combes, A.; Dzierzynski, M. *Can. J. Chem.* **1965,** *43,* 1120.
46. Fusy, J.; Martin, R.; Dzierzynski, M.; Niclause, M. *Bull. Soc. Chim. Fr.* **1966,** 3783.
47. Goldberg, V. M.; Vidovskaya, L. A.; Zaikov, G. E. *Polym. Degrad. Stab.* **1988,** *20,* 93.
48. Iring, M.; Tudos, F. *Prog. Polym. Sci.* **1990,** *15,* 217.
49. Lacey, D. J; Dudler, V.; Krohnke, Ch. *Proceedings of the 5th European Polymer Federation Symposium;* Basel, Switzerland, 1994; p 244.
50. Belyakov, V. A.; Fedorova, G. F.; Vassil'ev, R. F. *J. Photochem. Photobiol. A* **1993,** *72,* 73.
51. Hiramatsu, M.; Muzaki, H.; Ito, T. *J. Polym. Sci. Part C: Polym. Lett.* **1990,** *28,* 133.

RECEIVED for review January 26, 1994. ACCEPTED revised manuscript February 6, 1995.

Chemiluminescence of Polymer Materials during Thermal Oxidation and Stress

Satoru Hosoda, Hayato Kihara, and Yoshinori Seki

Sumitomo Chemical Co., Ltd., Petrochemicals Research Laboratory, Anesakikaigan 5 - 1, Ichihara - city, Chiba, 299-01, Japan

The chemiluminescence (CL) time courses of polyolefins during heating and UV irradiation in air and inert gas were investigated kinetically. CL parameters such as peak-top intensity, integrated intensity during measurement, steady-state intensity, and CL decay rate correlated strongly with the stability of the polymer materials that underwent durability tests. Luminescence induced by mechanical stress was observed for various polymers. Stress-induced CL for nylon 6 and the blend of rubber-modified polystyrene with polyphenylene oxide was related to chain scission during the uniaxial extension. Imaging of CL was successfully carried out during heating or stress loading by using two-dimensional photon detection. The difference in the oxidation reaction rate and in the degree of the stress concentration was easily observed.

THE DEGRADATION OF POLYMER MATERIALS is caused by exposure to various factors such as heat, UV light, irradiation, ozone, mechanical stress, and microbes. Degradation is promoted by oxygen, humidity, and strain and results in such flaws as brittleness, cracking, and discoloration. Durability of polymers has been evaluated by exposure to such conditions as thermal heating, outdoor weathering, UV irradiation, and repeated stress. However, resistance tests often require 100–1000 h, and outdoor exposure tests often require a few months.

On the other hand, it is well known that degradation is accompanied by very weak emission of chemiluminescence (CL). Since publication of the precedent-setting studies of CL in polymers by Ashby (1) and Schard and Russell (2), the high sensitivity of CL has been capitalized on for the evaluation of

0065–2393/96/0249–0195$12.00/0

the stability of polymer materials exposed to heat (*3*, *4*), UV light (*5*, *6*), electron beams (*7*), and γ-irradiation (*8*).

We investigated the CL phenomena induced by heat and UV irradiation for polyolefins, and this phenomenon is discussed in connection with the results of actual durability tests of the same materials. Stress-induced CL, which may be related to chain scission during deformation, was also investigated for polymers such as nylon 6 and high-impact polystyrene to study the mechanism of fracture. Imaging of CL is a promising technique for visualizing and predicting the sites of failure induced by aging and mechanical stress in polymer materials.

In this review, we summarize our studies on CL induced by various factors. We also discuss imaging of CL under heat and mechanical stress.

Oxidative Degradation and Chemiluminescence

Kinetic Approach to Thermal Oxidation. Radical species ($R^{\cdot}$) produced by main-chain scission or proton elimination caused by heat, UV light, and mechanical stress partly react with O_2 and readily produce peroxy radicals ($ROO^{\cdot}$) (activation energy, ΔE, is 0 kJ/mol) according to Reich and Stivala (*9*). A simplified scheme of the oxidation process for polymers is shown in Scheme I (*10*). Among the various reactions, such as proton elimination from polymer main chain by $ROO^{\cdot}$, decomposition of ROOH, and recombination of $R^{\cdot}$, the bimolecular termination reaction of $ROO^{\cdot}$ has a smaller activation energy ($\Delta E = 12$ kJ/mol; reference 9) and is largely exothermic (change in enthalpy, ΔH, is 462 kJ/mol; reference 9) compared to the other reactions. Therefore, only this bimolecular reaction of $ROO^{\cdot}$ is able to excite a product ketone to its first triplet state ($RR'C=O^{*}$) (*11*). Visible light, or CL, is emitted as the ketone returns to its ground state. The intensity of light emitted, I_{CL}, will be proportional to the rate of the bimolecular reaction and overall efficiency, f (apparatus constants, yield of formation, and emission of excited state):

$$I_{CL} = fk_b \,[ROO^{\cdot}]^2 \qquad (1)$$

where k_b is the rate constant of the bimolecular reaction of $ROO^{\cdot}$. In the absence of inhibitor, the following equation is valid:

$$d[ROO^{\cdot}]/dt = R_i - k_b[ROO^{\cdot}]^2 \qquad (2)$$

where t is time, and R_i is the rate of generation of $R^{\cdot}$ including the rates of generation of $R^{\cdot}$ by heat, the return from $R^{\cdot}$ to R, and the disappearance by recombination of $R^{\cdot}$. Scheme I suggests that the measurement of CL intensity

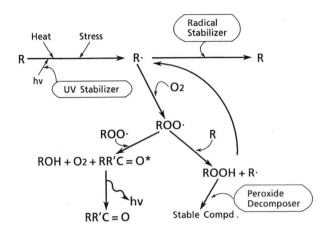

Scheme I. Representation of oxidation and luminescence processes for polymers.

gives information about the rate of degradation and the efficiency of additives because stabilizers eventually reduce the concentration of excited ketone.

A typical CL time course of stabilized polypropylene (PP) heated in air is shown in Figure 1. A dotted line shows the time course in which the sample is heated in an inert atmosphere like argon or nitrogen. One of the purposes of our study was to investigate how to use the parameters obtained from the time course curve, such as the peak-top intensity (I_p), the area intensity of the peak (S_p), the steady-state intensity (I_s), and the time from which the intensity rises up abruptly (t_i). Even though t_i has been used occasionally to evaluate the stability of polymers (*12*), the use of this parameter is not convenient because it becomes rather long with the increasing efficiency of stabilizers added in the sample.

A CL time course was simulated for heating in an inert gas and heating in air (*13, 14*). As a result, I_p and S_p of the CL time course in an inert gas corresponded to the initial concentration of ROO⁺. This result suggests that I_p and S_p indicate the history of oxidation of the sample. The experimental results for the preheated polyethylenes were consistent with the result of the simulation. The same kind of conclusion concerning the peak intensity was reported recently by Billingham et al. (*15*) for the isothermal CL curve and temperature-ramp CL curve measured in nitrogen for PP.

The CL time course measured in air during heating was simulated by using eq 2, and the CL intensity was shown to converged to some finite value during the time course. In this steady-state condition, $d[ROO⁺]^2/dt$ in eq 2 is zero, and CL intensity in this condition (I_s) can be expressed as follows:

$$I_s = fk_b[ROO^·]^2 = fR_i \tag{3}$$

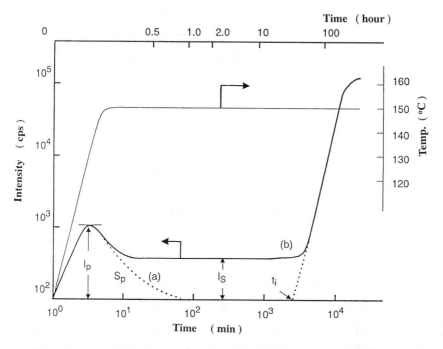

Figure 1. A typical CL time course of stabilized PP measured at 150 °C in air (solid lines) and in argon gas (dotted line).

Equation 3 indicates that I_s is proportional to the rate of generation of ROO$^{\bullet}$ (R_i). Thus, the ease of oxidation of polymers and the ability of inhibitors to act as radical captors can be evaluated from I_s. Figure 2 shows the dependence of I_s on the concentrations of the hindered phenols for PP heated in air at 150 °C; I_s decreased with increasing concentration of the phenols. The hindered phenol PH-2 had a stronger ability to catch radicals in PP than PH-1. A reciprocal Gear–Oven life (150 °C), which is one of the important parameters of durability, of the same samples in air had a good linear relationship with I_s (Figure 3). This result suggests that the CL intensity at the steady state measured in air corresponds to the rate of oxidative degradation.

Table I shows another example of the relationship between I_s and Gear–Oven life for PP sheets containing various phenols (*14*). The results indicate that 2,4,6-trimethylphenol is the most effective phenol shown in Table I for stabilizing PP.

If $R_i = 0$, the differential of eq 2 gives the following equation:

$$1/[\text{ROO}^{\bullet}] = -k_b t + 1/C_0 \qquad (4)$$

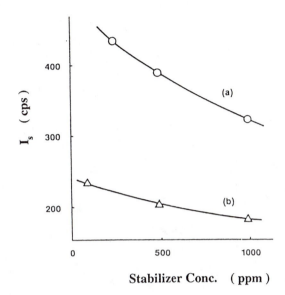

Figure 2. Concentration dependence of I_s on hindered phenols PH-1 (a) and PH-2 (b) for PP measured at 150 °C in air. (Reproduced with permission from reference 14. Copyright 1990 Society of Polymer Science, Japan.)

where C_0 is the initial concentration of ROO˙. Substituting eq 4 into eq 1 gives eq 5:

$$1/\sqrt{I_{CL}} = t\sqrt{k_b/f} + 1/(C_0\sqrt{fk_b}) \qquad (5)$$

This equation reveals that the reciprocal of the square root of the CL intensity is proportional to time. Simulated time course of $1/\sqrt{I_{CL}}$ (Figure 4) gives good linearity when R_i is zero, whereas the $1/\sqrt{I_{CL}}$ value saturates with increasing time when R_i is not zero. The experimental results for PP heated in argon gas and in air support this simulated result (13, 14). From eq 5, the slope of the plot corresponds to the square root of the bimolecular reaction rate of ROO˙. This slope is called the CL decay rate. Because the CL decay rate is dependent on the molecular mobility, we can use this rate to evaluate the degree of aging that is accompanied with the chemical cross-linking, as in the case of polyethylene (PE) (14).

If the peroxide decomposer is contained in a sample, eq 2 is replaced by eq 6:

$$d[ROO˙]/dt = R_i - k_b [ROO˙]^2 - k_d [ROO˙] \qquad (6)$$

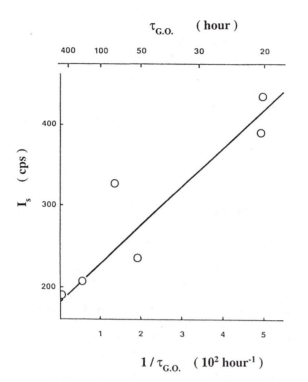

Figure 3. Relationship between reciprocal Gear–Oven life and I_s of PP measured at 150 °C in air. (Reproduced with permission from reference 14. Copyright 1990 Society of Polymer Science, Japan.)

Table I. Steady-State CL Intensity of PP Sheets Containing Various Phenolic Stabilizers

Specimen	Stabilizer[a]	GO Life[b] (h)	I_s^c
1	A	530	3.8
2	B	1010	1.8
3	C	1320	1.7
4	D	2130	1.4

[a] A is 3,4-dimethylphenol; B is 4-methyl-2,6-di-*tert*-butylphenol; C is 2-*tert*-butyl-4-methylphenol; and D is 2,4,6-trimethylphenol.
[b] Gear–Oven life measured at 150 °C (following Japan Industrial Standard K7212).
[c] Steady-state CL intensity measured at 150 °C; values are counts $\times 10^{-3}$ per 30 s.
SOURCE: Reproduced with permission from ref. 14. Copyright 1990 Society of Polymer Science, Japan.

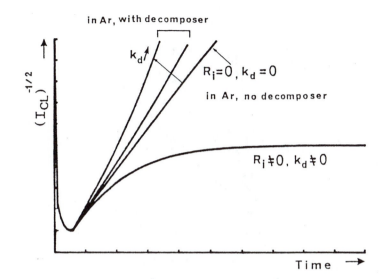

Figure 4. Simulated time course of the reciprocal of square root of CL intensity:
$R_i = 0$ *and* $R_i \neq 0$ *correspond to conditions under inert gas and in air, respectively.*

where k_d is the overall rate constant of the reaction from ROO˙ to a stable compound under the action of a peroxide decomposer. From the simulation (*13*) of eq 6 when $R_i = 0$, the reciprocal of the square root of the CL intensity is proportional to t and the slope of the plot increases with increasing k_d (Figure 4). This relationship suggests that this type of plot yields the efficiency of the peroxide decomposers. Figure 5 shows the plot of $1/\sqrt{I_{CL}}$ vs. t (measured in argon at 150 °C) for PP containing phosphorous and sulfuric peroxide decomposers. Figure 5 shows that PP stabilization increases with increasing concentration of the sulfuric peroxide decomposer (DS-1). Also, the phosphorus peroxide decomposer (DP-1) is less effective than DS-1.

Chemiluminescence during Photooxidation.   The degradation or aging of polymers caused by UV irradiation is another serious problem in polymer technologies. Especially for outdoor use, it is very important to stabilize polymer materials against UV light. A long period of time is required to evaluate the durability of polymers and the efficiency of stabilizers by use of an outdoor exposure test or a sunshine weatherometer test. Therefore, a more efficient and quicker method was needed to judge durability against UV irradiation. Some studies on CL induced by UV light were reported (*5, 6*), but CL was not necessarily correlated with the durability of the sample polymers. We investigated (*16*) UV-induced CL of PE and tried to obtain, from

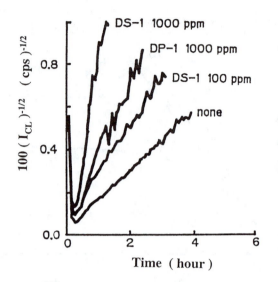

Figure 5. Plot of $1/\sqrt{I_{CL}}$ vs. time for PP containing sulfuric peroxide (DS-1) and phosphorus peroxide (DP-1) decomposers measured at 150 °C in argon gas.

a kinetic analysis of the CL time course, a parameter that evaluates durability against UV light.

The rate of generation of ROO· by UV irradiation in the presence of oxygen is expressed as follows, in the same manner as eq 2:

$$d[\text{ROO·}]/dt = R_{UV} - k_b\,[\text{ROO·}]^2 \qquad (7)$$

where R_{UV} is the rate of the initiation reaction generating peroxy radical during UV light exposure. At the steady state the following expression holds:

$$I_0 = f(R_{UV}) \qquad (8)$$

where I_0 is the saturated CL intensity at the steady state. In the treatment above, the thermal oxidation reaction is ignored because the measurement is carried out at room temperature.

Even though the method for the measurement of UV-induced CL was developed by using a mechanically chopped, pulsive UV irradiation as a light source (17), very weak CL is difficult to detect exclusively during UV irradiation. We also considered the CL time course after stopping UV irradiation. Fluorescence and phosphorescence from impurities and PE induced by UV light might be annihilated within 10 s. Therefore, the observed luminescence probably is due to the oxidation reaction induced by UV irradiation. Under this condition, eq 9, which shows that the reciprocal of the square root of CL intensity after stopping UV light is proportional to t, finally can be obtained:

$$1/\sqrt{I_{CL}} = t\sqrt{k_b/f} + 1/\sqrt{I_0} \tag{9}$$

The intercept at $t = 0$ gives $1/\sqrt{I_0}$ in the plot of $1/\sqrt{I_{CL}}$ vs. t. The procedure to obtain I_0 is as follows (Figure 6):

1. Sample specimen was irradiated 60 s in air at 25 °C.
2. UV irradiation was stopped and sample was kept in the dark during decay of fluorescence and phosphorescence.
3. Measurement of CL was started after 30 s from the stop of UV irradiation.
4. The CL parameter, I_0, was obtained by using the least-squares method from the above plot.

The plot of $1/\sqrt{I_0}$ vs. t for the CL time course for PEs that were UV-stabilized and unstabilized showed a good linear relationship and indicated the validity of eq 8. The I_0 parameter of UV-stabilized PE always showed a smaller value than that of the unstabilized sample (*16*). Further, I_0 for some linear low-density PE-blown films for an agricultural use (before use) was measured and compared with the result of the outdoor exposure test, which is represented as the half-life of tensile elongation. As shown in Figure 7, the film having a smaller value of I_0 showed a longer life and good durability. Interestingly, the CL parameter for the photooxidation reaction obtained from

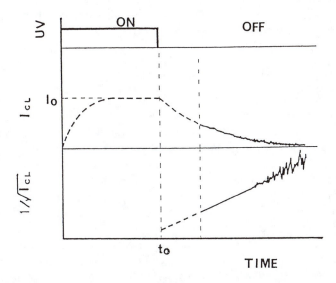

Figure 6. Schematic UV irradiation and CL time course for UV-induced CL.

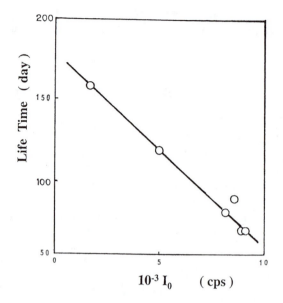

Figure 7. Relationship between I_0 and outdoor life for various linear low-density PE films used in agriculture. (Reproduced with permission from reference 16. Copyright 1992 Springer Verlag.)

the measurement in short time shows a good relationship with the result of the outdoor exposure test after a few months. By obtaining much more data on the durability and elucidating the mechanism of photodegradation for each kind of polymer material, we believe that the CL parameter becomes useful for evaluating the stability of polymer materials against UV irradiation.

Chemiluminescence during Stress

Friction among molecular chains and structural changes such as expansion of the bond angle, molecular orientation, and bond scission occur when polymer material is deformed by mechanical stress. Some of these stresses generate weak light emission (18), and the luminescence is attributed to several processes including CL from the bimolecular termination reaction of peroxy radicals. George et al. (19) reported detailed results for stress-induced chemiluminescence (SICL) during the deformation of nylon 6,6 fibers and concluded that the main part of the CL resulted from progressive scission of taut tie molecules in the amorphous phase. Fanter and Levy (20) and Monaco et al. (21) manufactured the drawing machine for SICL measurement and studied the luminescence phenomenon by evaluating durability during mechanical stress.

We also set up a system for SICL measurement and investigated SICL for various polymer materials (22–25). For instance, a strong SICL was observed for a slowly cooled nylon 6 specimen when it was uniaxially drawn at room temperature, whereas a quenched specimen showed very weak SICL (about 17% of that of the slowly cooled one compared at the same drawing rate). The SICL gradually increased with increasing drawing rate. The dots in Figure 8 show the experimental data of the SICL as a function of the applied stress during extension (drawing rate, 2.5 mm/min) of a slowly cooled nylon 6 specimen.

Free-radical formation during the deformation of PE and nylons was revealed by electron spin resonance measurement (26). The rate of free-radical formation due to the chain scission, R, on an applied stress, σ, at a temperature, T, is given by the Zhurkov equation (27):

$$R = n_c \, \omega_0 \, \exp \left\{ -(U_0 - \beta\sigma)/RT \right\} \qquad (10)$$

where n_c is the number of chains to be cut off, ω_0 is the fundamental bond vibration frequency, U_0 is the main-chain bond strength, β is the activation volume for bond scission, and R is the gas constant. If the applied stress on the sample specimen is assumed to be the average stress on the chains, and the SICL observed is proportional to the rate of peroxy radical formation, then the intensity of the SICL can be expressed as follows:

$$I_{CL} = I_k \, \exp \, (\beta\sigma - U_0) + I' \qquad (11)$$

where I_k is a constant including the overall efficiency of emission, and I' is the intensity of emission other than SICL. As shown in Figure 8 (24), the theoretical relationship described by eq 11 exhibits good agreement with the experimental results and suggests the validity of the mechanism estimated previously: that is, tie-chain scission by the applied stress, alkyl radical formation, the oxidation reaction, and then CL.

Blends of high-impact polystyrene (HIPS) and 2,4-dimethylphenylene oxide exhibited three-step SICL against the applied stress (23) as shown in Figure 9. For the first step, a slow increase of SICL intensity (strain ε = 0–10%) might be attributable to the thermal oxidation process induced by the internal friction. At the second step (ε = 10–20%), the intensity increased steeply with the stress whitening of the specimen, and the micro crazes increased in number with drawing and were observed around the rubber particles in HIPS by the transmission electron microscope (TEM). At the final step (ε = 20–30%), the elongation of the craze length rather than the increase in number of the crazes was observed. Results of the TEM observation suggest that the SICL of this blend system might be due to the radical formation caused by the chain scission in the micro crazes.

The SICL of this type, in which the intensity increased with increasing stress, was observed for other polymer materials such as acrylonitrile butadiene styrene resin and polyetheretherketone. On the other hand, a pulsed

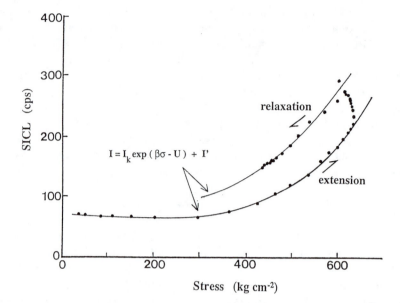

Figure 8. Plot of SICL as a function of the applied stress for a slowly cooled nylon 6 specimen. Closed circles are experimental results and lines express theoretical relationship obtained from eq 11. Sample specimen was drawn uniaxially at room temperature at the rate of 2.5 mm/min. (Reproduced with permission from reference 24. Copyright 1993 Elsevier Science Ltd.)

emission was observed at room temperature during drawing and at break for unmodified polystyrene, bisphenol-A polycarbonate (Figure 10), poly(methyl methacrylate), and poly(butylene terephthalate). This pulsed luminescence is a kind of triboluminescence induced by a new surface formation with the progress of the crack.

In addition to luminescence induced by mechanical stress, luminescence induced by electric stress was also reported by Bamji et al. (28, 29). Emission of UV-vis light was observed by charging with an alternating current voltage larger than some threshold value of a low-density PE specimen in which a needle tip electrode was buried. The luminescence is probably due to the recombination of injected electrons with the holes in PE and is called electroluminescence (EL). Because the continuous charge of the electric stress causes the degradation of the PE specimen by the generation of an electric tree, EL is considered to be a premonitory phenomenon of the degradation and could be used in an evaluation method for the durability of insulation cable materials against electric stress.

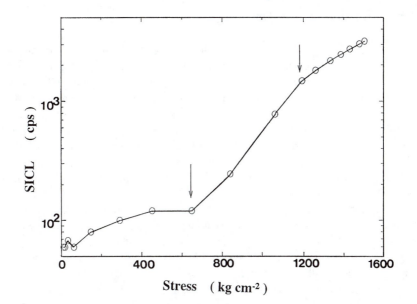

Figure 9. Relationship between SICL intensity and applied stress for blend of 2,6-dimethylphenylene oxide and rubber-modified polystyrene: average size of rubber particle, 4.9 μm; drawing rate, 2.5 mm/min at room temperature.

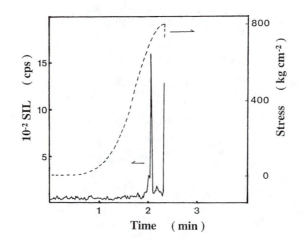

Figure 10. Time course of stress-induced luminescence (SIL) and applied stress during drawing at 2.5 mm/min of bisphenol-A polycarbonate in air at room temperature.

Chemiluminescence Imaging

A photomultiplier is used as a photon detector for CL detection, and the total number of photons emitted from the sample surface is counted in these conventional instruments. Recently, a new detector system equipped with a series of multichannel plates used as an intensifier has made the photon-counting measurement extremely sensitive. By combining this system with a two-dimensional detector (a vidicon or a position-sensitive detecting system), CL imaging has become possible. Hiramatsu et al. (*30*) reported real time imaging of the permient distribution on a polymer film surface. Fleming and Craig (*31*) also reported CL imaging for the thermal oxidation of hydroxy-terminated polybutadiene gumstock.

We investigated (*24*) CL imaging induced by heat and mechanical stress by using this detection system for polymer materials such as nylon 6, high-pressure low-density PE, polystyrene, and styrene–butadiene triblock copolymer (SBS). The press-molded sheets of nylon 6 and polystyrene exhibited strong CL, especially at the trimmed edge of the specimen and the cracked part produced by mechanical trimming when the sheets were heated in air at an elevated temperature.

An example of a PE–SBS multilayered specimen is shown in Figure 11. The multilayered sheet was obtained by compression molding at 150 °C by

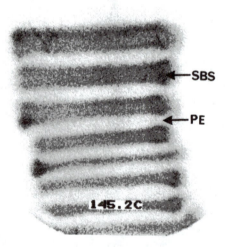

Figure 11. The CL image obtained at 145 °C for PE–SBS triblock copolymer multilayered sheet. Each individual sheet was originally 0.1-mm thick. (Reproduced with permission from reference 24. Copyright 1993 Elsevier Science Ltd.)

using eight SBS sheets and seven PE sheets that were layered alternately. The multilayered sheet about 1.5-mm thick was cut with a razor blade, and the section was used for the imaging experiment. A clear image was obtained above 120 °C. The image shown in Figure 11 was obtained at 145 °C and indicates that the SBS layers exhibit stronger CL (about 50 cps/pixel) than the PE layers (5–10 cps/pixel) during heating in air. Although the layers might expand to some extent at this temperature, each PE and SBS layer, which was originally 0.1-mm thick at room temperature, is quite distinguishable. Considering the clearness of the image obtained, the lateral special resolution with the lens system would be 30–50 μm. From these results it would be possible to visualize the difference in the rate of oxidation reactions among component polymers of systems such as multilayer extruded laminates.

Imaging of SICL at room temperature was tried for a nylon 6 sample that possessed a hole at the center of the specimen. By using this hole, good reproducibility for the stress concentration in drawing the sample could be attained. The intensity of the visual SICL became stronger with an increasing extension ratio, and SICL was strong especially around the hole in a transverse direction relative to the machine direction Figure 12. This figure indicates the SICL real-time image of the area around the hole of the sample drawn for 45 s (ε is estimated to be 4.7%). The explanatory diagram is superposed so that the dimension of the specimen can be understood easily, and the arrows

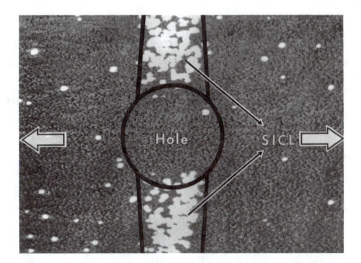

Figure 12. Real-time image of region around hole of nylon 6 specimen drawn at the rate of 2.5 mm/min at room temperature (ε = 4.7%). Arrows show the machine direction. White area near the hole in the transverse direction shows strong luminescence. (Reproduced with permission from reference 24. Copyright 1993 Elsevier Science Ltd.)

show the machine direction. The hole at the center could not be distinguished in the SICL image because the luminescence intensity was very weak at the less-stressed region and at the hole space.

To understand the regional peculiarity for SICL, simulation by a finite element method was carried out for the stress distribution when drawing a nylon 6 specimen at $\varepsilon = 5\%$. The simulation indicates that the stress is concentrated around the hole in the transverse direction, which is coincident with the region of observed SICL (24).

The crystalline phase of crystalline polymers is connected by a number of tie molecules in the amorphous phase. Chain scission of the tie molecules occurs predominantly at the most-strained molecules. The radicals thus formed could easily react with oxygen in air to produce peroxy radicals, which are assumed to exhibit CL through a bimolecular termination reaction. Hence, it is quite reasonable that the deformation of crystalline polymers like nylon 6 produces strong SICL at the area of stress concentration as a result of tie-chain scission.

Heat and mechanical deformation were identified as primary causes of an increase in CL, but other external causes such as electric stress (25, 32), UV light, and γ-ray irradiation can be used. Accordingly, CL imaging could be developed in various fields to investigate the effects of these causes on the aging of polymer materials.

References

1. Ashby, G. E. *J. Polym. Sci.* **1961,** *L,* 99.
2. Schard, M. P.; Russell, C. A. *J. Appl. Polym. Sci.* **1964,** 8, 985.
3. George, G. A. In *Development in Polymer Degradation 3;* Grassie, N., Ed.; Applied Science Publishers: London, 1981; p 173 and references in this review.
4. Stivala, S. S.; Kimura, J.; Reich, L. In *Degradation and Stabilization of Polymers;* Jellinek, H. G., Ed.; Elsevier: Amsterdam, 1982 ; Vol. 1, p 1 and references in this review.
5. Osawa, Z.; Konoma, F.; Wu, S.; Cen, J. *Polym. Photochem.* **1986,** 7, 337.
6. Cen, J.; Konoma, F.; Osawa, Z. *Polym. Photochem.* **1986,** 7, 467.
7. Yoshii, F.; Sasaki, T.; Makuuchi, K; Tamura, N. *Ikigaku* **1985,** 55, 252.
8. Yoshii, F.; Sasaki, T.; Makuuchi, K; Tamura, N. *J. Appl. Polym. Sci.* **1986,** 31, 1343.
9. Reich, L.; Stivala, S. S. In *Autooxidation of Hydrocarbons and Polyolefins;* Dekker: New York, 1969.
10. George, G. A. *Polym. Degrad. Stab.* **1979,** 1, 217.
11. Brauman, S. K.; Pronkp, J. G. *J. Polym. Sci. Part B: Polym. Phys.* **1988,** 26, 1205.
12. Zlatkevich, L. *Polym. Degrad. Stab.* **1987,** 19, 51.
13. Hosoda, S.; Kihara, H. *ANTEC* **1988,** 88, 941.
14. Kihara, H.; Hosoda, S. *Polym. J.* **1990,** 22, 763.
15. Billingham, N. C.; Then, E. T.; Gijsman, P. *J. Polym. Degrad. Stab.* **1991,** 34, 263.
16. Kihara, H; Yabe, T.; Hosoda, S. *Polym. Bull.* **1992,** 29, 369.
17. Gromek, M.; Derrick, R. *Polym. Degrad. Stab.* **1993,** 39, 261.

18. Streletskii, A.; Butyagin, P. *Polym. Sci. USSR (Engl. Transl.)* **1973,** *15,* 739.
19. George, G.; Egglestone, G.; Riddell, S. *J. Appl. Polym. Sci.* **1982,** *27,* 3999.
20. Fanter, D.; Levy, R. *CHEMTECH.* **1979,** 682.
21. Monaco, S.; Richardson, J.; Breshears, J.; Lanning, S.; Bowman, J.; Walkup, C. *Ind. Eng. Chem. Prod. Res. Dev.* **1982,** *21,* 546.
22. Kihara, H.; Hosoda, S. *Polym. Prepr. (Japan)* **1989,** *38,* 1195.
23. Kihara, H.; Seki, Y.; Hosoda, S. *Annual International Symposium Polymer '91 (Melborne),* **1991,** 329.
24. Hosoda, S.; Kihara, H.; Seki, Y. *Polymer* **1993,** *34,* 4602.
25. Seki, Y.; Kihara, H.; Hosoda, S. *Encyclopedia of Polymer Materials*; Salamone, J., Ed.; CRC Press: Florida, in press.
26. Kausch, H. In *Polymer Fracture*; Springer–Verlag: Heidelberg, Germany, 1978.
27. Lloyd, A.; De Vries, L.; Williams, L. *J. Polym. Sci., Part A–2: Polym. Phys. Ed.* **1972,** *10,* 1415.
28. Bamji, S.; Bulinski, A.; Densley, J *J. Appl. Phys.* **1986,** *61,* 694.
29. Bamji, S.; Bulinski, A.; Densley, J. *Makromol. Chem. Macromol. Symp.* **1989,** *25,* 271.
30. Hiramatsu, M.; Muraki, H.; Itoh, T. *J. Polym. Sci. Polym. Lett. Ed.* **1990,** *28,* 133.
31. Fleming, R.; Craig, A. *Polym. Degrad. Stab.* **1992,** *37,* 173.
32. Hosoda, S.; Kihara, H.; Seki, Y. *Sumitomo Kagaku (Osaka)* **1993,** *1993-II,* 86.

RECEIVED for review January 26, 1994. ACCEPTED revised manuscript March 23, 1995.

Luminescence from Nile Red in Poly(*n*-butyl methacrylate)

G. D. Mendenhall

Department of Chemistry, Michigan Technological University, Houghton, MI 49931

Films of poly(n-butyl methacrylate) containing the fluorescent dye Nile Red displayed both spontaneous and stimulated luminescence emission. Brief exposure of the dyed film to fluorescent light resulted in a delayed luminescence that decayed in a time scale of minutes and whose intensity was reduced to negligible levels when the film was first cooled to 0 °C. The conventional fluorescence lifetime of the dye in the polymer matrix, 5 ns, differed little from the value of 4 ns in ethyl acetate solution. The temperature-dependence (4–35 °C) of the lifetime in both media was small. The dyed film also showed spontaneous luminescence that was ascribed to reactions of benzoyl peroxide present as residual initiator and was greatly reduced by prior purification of the polymer. When Nile Red was added to unpurified solvents, the solutions displayed spontaneous luminescence intensities that roughly correlated with the tendency of the solvent to autoxidize.

W̲E̲ ̲H̲A̲V̲E̲ ̲D̲E̲S̲C̲R̲I̲B̲E̲D̲ ̲A̲ ̲D̲E̲L̲A̲Y̲E̲D̲ ̲L̲U̲M̲I̲N̲E̲S̲C̲E̲N̲C̲E̲ that appears on a time scale of minutes to days after brief exposure of a large (20 cm²) surface area of both polymeric and nonpolymeric solids to visible (*1, 2*), UV (*2*), or ionizing radiation (*3*). This light emission constitutes the slowest of all observable, stimulated luminescent processes in the material. In nearly every instance in which we have been able to observe this delayed luminescence, it decayed with a hyperbolic rather than exponential law, that is $I(t) \approx at^{-n}$, over a given decade. In this equation, I is the rate of luminescence emission, a is a constant, and n is usually around 1.0. This rate law is characteristic of a charge-recombination processes (*4*), although much more complicated decay forms have been derived (*5*).

0065–2393/96/0249–0213$12.00/0

Billingham et al. (6) and Mendenhall and Guo (7) described large changes in delayed luminescence from two-component epoxy systems undergoing cure. In an attempt to examine the relation between the luminescence and physical changes in a polymeric system, we have investigated the effect of temperature on the recombination luminescence from a dye, Nile Red, dispersed in a polymer with a glass transition temperature (T_g) close to ambient temperature. This dye is highly fluorescent, dissolves in a variety of solvents, and displays the additional useful feature that the position and intensity of the fluorescence maximum are highly solvent-dependent (8). If a diffusive recombination of charges were involved in the delayed emission from such a system, we hoped that any changes in the rate of diffusion above and below T_g would be reflected in large changes in the rate of luminescence decay.

Experimental

Poly(n-butyl methacrylate) was obtained from DuPont as Elvacite grade 2044 and had a stated weight-average molecular weight of 337K, a number-average molecular weight of 95.9K, and a PD 3.51. The polymer, which smelled strongly of n-butanol, was purified by precipitation from reagent acetone (5.6 mL/g polymer) into reagent methanol (26 mL/g polymer). The spongy mass of polymer was collected and pressed between paper towels to remove as much liquid as possible. The precipitation was repeated twice more, and the final odorless product was dried overnight on a vacuum line at 1.3 Pa. The precipitations would presumably deplete lower molecular weight fractions of the polymer in addition to small molecule impurities, but this point was not examined. The Nile Red (1; Polysciences, Inc.) was a crystalline material that was used as received.

Both purified and unpurified polymer were made into films by dissolving 0.12 g of polymer in 4.0 mL of 7.5×10^{-5} M ethyl acetate in Nile Red in a glass scintillation vial with diameter 2.5 cm, followed by slow evaporation while covered with a beaker at room temperature in a hood. The film from the purified polymer was then covered with a 0.5-cm layer of mercury to act as a thermal ballast and to increase the efficiency of the stimulation and detection. The film was calculated to have an average thickness of 0.25 mm and a dye concentration of 0.002 m.

Steady-state luminescence from the films was measured with a thermostatted Turner Designs, Inc., Luminometer (20-s delay, 60-s measurement). Decay curves were measured in the Luminometer after stimulating the films for 10 s by resting them on a light box consisting of white fluorescent lamps and a plastic diffusion

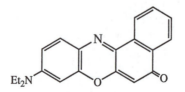

1

plate. The luminescence up to 4 min after the end of the stimulation period was monitored periodically (5-s delay, 10-s measurement), although in most cases only the initial data points were sufficiently intense. The data were fitted to the empirical equation

$$\ln(I - \text{bkg}) = b - n\ln(t)$$

where the background (bkg) is the luminescence from the film before it was irradiated, and t is in minutes from the end of the irradiation period to the midpoint of the luminescence measurement. Values of chemiluminescence in the tables are unfiltered readings directly from the digital display of the instrument.

Fluorescence emission spectra were determined from ethyl acetate solutions, 1.3×10^{-6} M in Nile Red, with a Spex Fluorolog, and fluorescence lifetimes were measured with the same solutions in an apparatus described elsewhere (9). The fluorescence measurements of the polymer containing Nile Red were conducted with a portion of the same film (from purified polymer) from which the delayed emission was measured. The dynamic probe measurements were made on a Perkin-Elmer Series 7 Thermal Analysis System.

Results and Discussion

Delayed Luminescence from Polymer–Dye Mixtures. In qualitative experiments we observed no luminescence from solid, unpurified poly(n-butyl methacrylate), but a relatively strong spontaneous luminescence was observed when the polymer was dissolved in ethyl acetate containing Nile Red. The luminescence from the film obtained from the Nile Red solution upon evaporation of the ethyl acetate was much stronger than from the original solution and was further increased by a factor of 4 after exposure of the film to fluorescent light. However, the intensity of the spontaneous luminescence in the film declined within a few days at room temperature, and the color of the film changed from red to yellow during this time.

When the experiments were repeated with purified polymer, the background luminescence of a film containing Nile Red was very small, and stimulation of the film with fluorescent light gave very weak decay curves whose initial values were fitted individually to the empirical equation

$$\ln(I - \text{bkg}) = A + n\ln[\text{t}(min)]$$

and $r = -0.985$ to -0.999. Plots of n showed considerable scatter and revealed no significant change over the temperature range of the study (Figure 1). On the other hand, the calculated intensities at 1.0 min ($= A$) and the background emission were temperature-dependent, and Figure 1 reveals a slight discontinuity at 24 ± 2 °C.

Physical Property Measurements. The discontinuity in the value of A (Figure 1) with temperature seemed to correspond to a small inflection

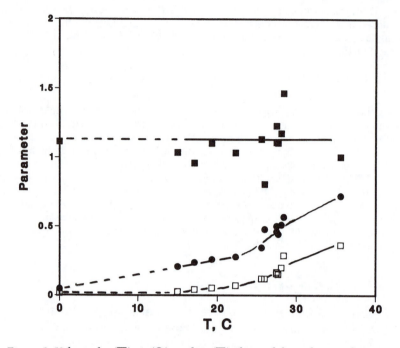

Figure 1. Values of n (■), A (●), *and* I_{ss} (□) *obtained from the initial portion of delayed emission from 0.002 m Nile Red in poly(n-butyl methacrylate). Values correspond to the equation* $I(t) = At^{-n} + I_{ss}$.

in the differential scanning calorimetry curve at 23 ± 2 °C that we observed in the same film. However, measurement of the loss tangent with a dynamic stress technique gave a more clearly defined T_g of 32 ± 2 °C (Figure 2). Values of T_g are recognized to be dependent on the method of measurement (10); therefore, the differing numerical values are not necessarily inconsistent.

Photophysical Measurements. The fluorescence characteristics of Nile Red in ethyl acetate solution and dispersed in the solid polymer film were determined by conventional measurements with steady-state and pulsed methods. The summed fluorescence decays were fitted to single-exponential decay over several half-lives. The results are summarized in Table I.

The fluorescence lifetime was about 20% longer and the emission maximum about 10-nm shorter in the polymer than in ethyl acetate. There appeared to be a slight shift in the emission to longer wavelengths in ethyl acetate as the temperature was lowered. However, no distinct differences are associated with temperatures above and below T_g of the polymeric medium. Therefore, the much stronger temperature dependence of the delayed emis-

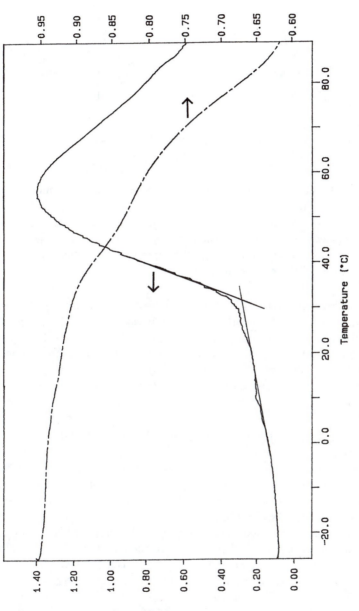

*Figure 2. Plot of loss tangent and probe position from dynamic mechanical scan of poly(n-butyl methacrylate)/
Nile Red: Dynamic stress, 90.0 mN; frequency, 1.00 Hz; scan rate, 2.0 °C/min.*

Table I. Fluorescence of Nile Red

Medium	Temperature (°C)	Lifetime (ns)[a]	Relative Intensity[b]	λ_{max} (nm)[b]
EtAc	0	4.4		
	4	—	1.8	594
	6.5	—	1.9	594
	20	4.0	—	586, 589
	24	—	1.8	590
Polymer	0	5.3		
	4	—	1.0	582
	10	—	1.0	582
	20	5.0	1.1	580
	40.2	4.8	—	—

NOTE: — is not determined.
[a] Fluorescence decay measurements with λ_{exc} = 400 or 480 nm. Estimated error is ± 0.1 ns.
[b] Steady-state experiments.

sion from Nile Red in poly(n-butyl methacrylate) are not due to unusual properties of the fluorescent state of the dye.

Attempts to measure the activation energy of the spontaneous chemiluminescence were discontinued when it was found that when the film was cooled and then warmed to the original temperature, the chemiluminescence level measured at the original temperature was not restored.

Experiments with Nile Red and Benzoyl Peroxide. Steenken (11) and others (12, 13) reported chemiluminescent reactions between benzoyl and related peroxides and aromatic amines. Because Nile Red has two amine functionalities, it was not too surprising that a solution 0.084 mM in Nile Red in freshly distilled ethyl acetate luminesced strongly when benzoyl peroxide was added to give an initial concentration of 3.2 mM. However, the time dependence of the luminescence was not simple (Table II).

Experiments with Nile Red and Solvents. Additional control experiments revealed that reagent-grade ethyl acetate gave easily measured, spontaneous chemiluminescence with dissolved Nile Red, even though the solvent did not give a positive test for peroxides with KI–starch paper. However, solvent freshly distilled from NaBH$_4$ was minimally chemiluminescent when Nile Red was dissolved in it.

Because peroxidic impurities were presumably responsible for the spontaneous luminescence of the dye both in polymers and in solution, could the dye, added as a solution to organic media, serve as a sensitive peroxide indicator? To address this question, 2-mL portions (by pipet) of a stock solution of Nile Red (0.29 mM) in benzene were treated with an equal volume of various organic liquids, and the resulting chemiluminescence was recorded for three successive 20-s readings. Except for technical-grade methyl ethyl ketone,

Table II. Luminescence from Nile Red and Benzoyl Peroxide

Time (h)	Relative CL
0	0.586
3	2.39
8	1.22
19	1.11
28	0.90
45	1.06

NOTE: Ambient temperature and concentrations described in text. Luminescence reading before addition of peroxide was 0.004.

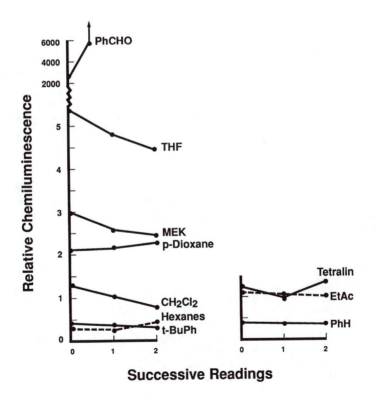

Figure 3. Initial chemiluminescence readings from mixtures of 2.0 mL of 0.29 mM Nile Red in benzene with 2.0 mL of various liquids at ambient temperature.

all of the samples were reagent grade but were not purified. The results are shown in Figure 3. Except for dichloromethane, the relative values of the luminescence correspond generally to the propensity of the solvents to undergo autoxidation. The luminescence from the solution of Nile Red and the sample of benzaldehyde, taken from an old bottle that had been stored at room temperature in air, was so intense that it tripped the safety shutter of the instrument during the third reading.

Although the experiments with solvents were convenient to carry out, it may be possible to characterize the state of oxidation of polymers by addition of Nile Red dissolved in a compatible solvent. Under these conditions the intensity of spontaneous chemiluminescence may reflect peroxide content, whereas shifts in the fluorescent wavelengths (8) may reflect the accumulation of oxidation products whose polarity differs from that of the starting polymer.

Conclusion

The prompt emission (fluorescence) from Nile Red in poly(n-butyl methacrylate) changes little on lowering the temperature through T_g. The intensity of the delayed emission from the same system, whose mechanism was not established, is temperature dependent, but the changes near T_g are so small that this approach *for this polymer* does less well than conventional methods.

The spontaneous luminescence from the dye–polymer combination is inferred to arise largely from impurities, including benzoyl peroxide present as residual polymerization initiator. The luminescence induced by Nile Red may be useful as a qualitative means to characterize the degree of oxidation of organic media.

Acknowledgment

The fluorescence lifetimes were measured in the laboratory of T. W. Wilson, Harvard University. I also thank George Chemparathy for carrying out the dynamic probe experiment.

References

1. Mendenhall, G. D.; Agarwal, H. K. *J. Appl. Polym. Sci.* **1987,** *33,* 1259–1274.
2. Guo, X.; Mendenhall, G. D. *Chem. Phys. Lett.* **1988,** *152,* 146–150.
3. Hu, X.; Mendenhall, G. D. In *Radiation Effects on Polymers;* Clough, R. L.; Shalaby, S. W., Eds.; ACS Symposium Series No. 475; American Chemical Society: Washington, DC, 1990; pp 534–553.
4. Debye, P.; Edwards, J. O. *J. Chem. Phys.* **1952,** *20,* 236–239.
5. Tachiya, M. *Int. J. Radiat. Appl. Instrument. C Radiat. Phys. Chem.* **1987,** *30,* 75–81.
6. Billingham, N. C.; Burdon, J. W.; Kozielski, K. A.; George, G. A. *Makromol. Chem.* **1989,** *190,* 3285.

7. Mendenhall, G. D. *Angew. Chem. Int. Ed. Eng.* **1990,** *29*, 362–373 (Figure 8).
8. Greenspan, P.; Mayer, E. P.; Fowler, S. D. *J. Cell. Biol.* **1985,** *100*, 965–973.
9. Wilson, T.; Frye, S. L.; Halpern, A. M. *J. Am. Chem. Soc.* **1984,** *106*, 3600–3606.
10. Seymour, R. B.; Carraher, C. E., Jr. *Polymer Chemistry*, 2nd ed.; Dekker: New York, 1988; p 32.
11. Steenken, S. *Photochem. Photobiol.* **1970,** *11*, 279–283.
12. Matisova-Rychla, L.; Rychly, J.; Lazar, M. *J. Luminesc.* **1972,** *5*, 269–276.
13. Schuster, G. B. *Acc. Chem. Res.* **1979,** *12*, 366–373.

RECEIVED for review January 26, 1994. ACCEPTED revised manuscript January 10, 1995.

15

Oxyluminescence of Cross-Linked Amine Epoxies: Diglycidylether of Bisphenol A–Diaminodiphenyl Sulfone System

L. Audouin, V. Bellenger*, A. Tcharkhtchi, and J. Verdu

Ecole Nationale Supérieure d'Arts et Métiers, 151 Bd de l'Hôpital, 75013 Paris, France

The oxidation of a stoichiometric diglycidylether of bisphenol A–diaminodiphenyl sulfone network was studied from 200–240 °C by IR spectrophotometry and chemiluminescence. Chemiluminescence reveals the existence of two kinetic stages. The first one is very brief and corresponds to the radicals resulting from the polymer thermolytic degradation under nitrogen during the preheating stage. This process leads to a sharp emissive peak whose intensity is an increasing function of the temperature and duration under nitrogen. The second stage corresponds to the oxidation propagation by hydrogen abstraction responsible for carbonyl and amide formation. An interesting peculiarity of this mechanism is that it is practically independent of the conversion ratio of the first stage process. Some mechanistic implications of these results are discussed.

INTEREST IN CHEMILUMINESCENCE AS A TOOL for the study of polymer oxidation mechanisms and kinetics is increasing (*1, 2*). Linear hydrocarbon polymers such as polypropylene have been widely studied. However, the corresponding mechanisms are still the topic of much discussion and research. The research field remains largely open to other polymers, especially tridimensional, heteroatoms containing polymers for which there is only a scarce literature. Investigations were reported, however, on epoxy networks for which chemiluminescence was tentatively used as a probe to monitor the cross-linking process (*3*). We recently reported (*4*) a comparative study concerning

* Corresponding author

0065–2393/96/0249–0223$12.00/0

oxidation mechanisms of anhydride-cured epoxy systems revealing the strong influence of hardener structure on chemiluminescence and the changes induced by the cross-linking reaction. Because data obtained by classical (especially IR spectrophotometric) analytical methods on oxidation mechanisms and kinetics of cross-linked amine diglycidylether of bisphenol A (DGEBA) systems are available (5), we decided to study the same systems by chemiluminescence with the hope of reaching a better understanding of this process and to appreciate the potential of chemiluminescence as a routine tool in this field. The present chapter is devoted to the study of a stoichiometric DGEBA–diaminodiphenyl sulfone (DDS) system.

Experimental

Materials. The network under study was based on DGEBA having an epoxide index of 5.8 mol kg^{-1} and cross-linked by DDS in stoichiometric amount. The compounds were weighed, mixed with a stirrer in an oil bath at 130 °C for 15 min, outgassed for 30 min, and then cured for 1 h at 180 °C and postcured for 3 h at 250 °C in a cylindrical mold. The final glass transition temperature (T_g), determined by differential scanning calorimetry at 20 K min^{-1}, is 225 °C and corresponds to the full conversion of the cure reaction. For spectrophotometric and chemiluminescence measurements, microtomic slides of 25-μm thickness, 20-mm diameter, were cut from the molded cylinder.

Infrared Spectrophotometry. IR spectra were recorded on a Perkin-Elmer Fourier transform IR (FTIR) 1710 spectrophotometer between 400 and 4000 cm^{-1}. The study was focused on peaks of carbonyl groups ($v_{C=O}$ = 1720 cm^{-1}, ε = 200 kg mol^{-1} cm^{-1}), amide groups (v_{CON} = 1680 cm^{-1}, (= 470 kg mol^{-1} cm^{-1}; phenylethers v_{C-O} = 1107 cm^{-1}), and alkylethers (v_{C-O} = 1035 cm^{-1}). We verified first that no oxidation occurred during processing, then we measured the carbonyl and amide build–up during the chemiluminescence experiments.

Chemiluminescence. A laboratory made, previously described apparatus (6) was used. The test chamber was swept by preheated gas at a flow rate of 50 L min^{-1}. The sample was heated under nitrogen until thermal equilibrium (for at least 0.5 min and generally 3 min). Air was then admitted. Two kinds of experiments were carried out:

Steady-State Experiments. All the exposure conditions (temperature, oxygen pressure) were constant. The chemiluminescence intensity expressed in arbitrary units was recorded versus time. Experiments were made at 200, 210, 220, 230, and 240 °C.

Break-Off Experiments. The oxidation was interrupted by switching the gas supply from air to nitrogen. The chemiluminescence decay was observed. After a given period of interruption, air was readmitted and the corresponding change of chemiluminescence intensity was recorded. Pressure measurements showed that the order of magnitude of the time constant of partial pressure change was about 5 s.

Results

Steady-State Chemiluminescence. Kinetic curves of chemiluminescence are presented in Figures 1 and 2. In Figure 1, the time of preheating
in nitrogen was constant at 3 min, and temperatures varied from 200 to 240 °C.
In Figure 2, the three curves were obtained at 240 °C, but they differ by the
preheating times of 3 min, 15 min, and 1 h.

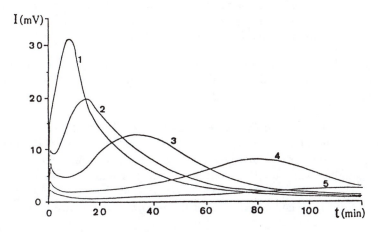

Figure 1. Kinetic curves of chemiluminescence experiments at various
temperatures: 1, 240 °C; 2, 230 °C; 3, 220 °C; 4, 210 °C; 5, 200 °C.

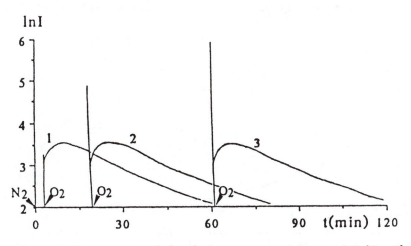

Figure 2. Kinetic curves of chemiluminescence experiments at 240 °C with
various preheating times: 1, 3 min; 2, 20 min; 3, 60 min.

On these curves, it is possible to distinguish two successive stages:

- **Stage 1.** Immediately after admission of oxygen into the test chamber, the intensity of emission increased abruptly and reached a maximum after a period of about 1 to 2 min, and this maximum was nearly independent of temperature. Then, the intensity decreased rapidly.
- **Stage 2.** This stage is characterized by time constants that are at least one order of magnitude higher than those of the stage 1. The peak width and time (t_m) corresponding to the maximum intensity were decreasing functions of the temperature. For instance, t_m was about 10 min at 240 °C and 100 min at 200 °C. Ninety percent of photons were emitted during this stage.

A gravimetric study at 240 °C (Figure 3) revealed a weight loss process. This loss was about 10% after 40 min. Consequently, it appears that the final decrease of intensity was not linked to the complete volatilization of the material.

The spectrophotometric study of samples revealed the build up of carbonyls (1720 cm^{-1}) and amides (1680 cm^{-1}) (Figure 4). Phenyl ether (1107 cm^{-1}) and aliphatic ether groups disappeared. In Figure 5, the integrated intensity of chemiluminescence at t is plotted versus the concentration of amides determined concurrently by IR for two temperatures. All the experimental points are close to a single curve that calls for the following comments:

- The two previously described stages are also present and they can be distinguished by the yield of amides, which is higher in the second stage than in the first one.
- The apparent activation energy, E_a, was determined for maximal in-

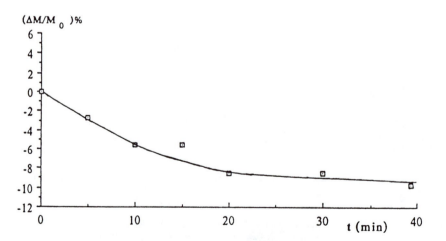

Figure 3. Weight loss of sample during chemiluminescence experiment at 240 °C.

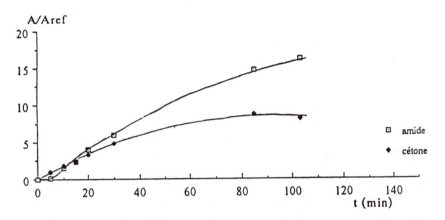

Figures 4. *Carbonyl and amide build–up during chemiluminescence experiments at 240 °C:* ◆, *carbonyl;* □, *amide.*

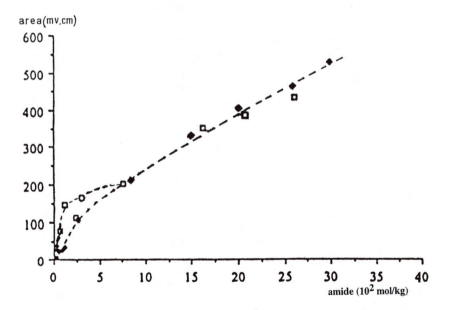

Figure 5. *Integrated luminescence intensity versus amide concentration measured concurrently during experiments at 220 °C (◆) and 240 °C (□).*

tensity of the second peak and for kinetic rate constants of formation of carbonyls and amides. For temperatures below T_g, E_a for luminescence was 130 kJ/mole. For temperatures $>T_g$, the E_a values for luminescence, carbonyls, and amides were 90, 110, and 130 kJ/mole, respectively. No simple relationship among the different values exists. In the case of second stage of chemiluminescence (second maximum), the Arrhenius plot (Figure 6) displays a discontinuity at T_g.

Break-Off Chemiluminescence. Chemiluminescence behavior of the sample during a single interruption of variable duration after 6 min of oxidation is presented in Figure 7 and during repeated interruptions of 5 min in Figure 8. To make the discussion easier, we shall call I_{sn} the stationary intensity just before the nth interruption (time t_n) under nitrogen, Δt the duration of interruption, I_{mn} the intensity of the peak immediately following the readmission of air at time $t_n + \Delta t$, and $I_{sn'}$ the stationary-state intensity immediately after the peak. These results can be summarized as follows:

1. Each change of atmosphere involves a sharp variation of luminescence intensity. It is not possible to say if time constants (maximum few seconds) are linked to the disappearance kinetics of the radicals that are responsible for luminescence, to the oxygen diffusion in the sample, or to the change in oxygen partial pressure.
2. The oxygen readmission at the end of interruption period, $t_n + \Delta t$, results in a luminescence peak for which $I_{mn} > I_{sn}$. But the intensity decreases rapidly to reach a stationary value of $I_{sn'}$. Variations of the ratio $I_{sn'}/I_{sn}$ follow the same curve as during the nonperturbed, continuous exposure. The steady-state regime represented by an envelope of

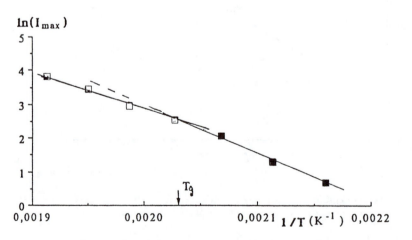

Figure 6. Arrhenius plot of the maximum luminescence intensity. ■, *below* T_g; □, *above* T_g.

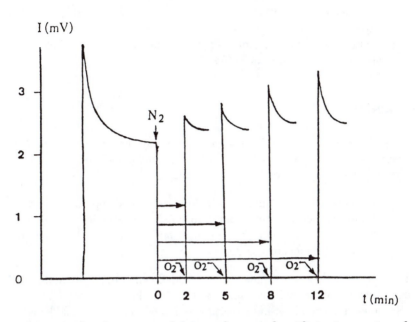

Figure 7. Chemiluminescence behavior after a single oxidation interruption of various duration.

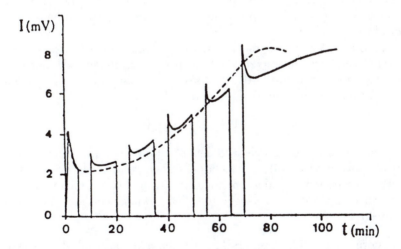

Figure 8. Chemiluminescence behavior during repeated interruptions of 5 min.

points (I_{sn}, t_n) is apparently not affected by interruptions, even when Δt (duration of interruption) is not negligible compared with the period of the cycle $(t_n + 1 - t_n)$.

3. The intensity of the peak following the readmission of air (I_{mn}) increases with the duration of interruption Δt.

Discussion

At least three types of chemiluminescence mechanisms are recognized that are more or less compatible with experimental kinetic data according to the nature of polymer and experimental conditions:

1. Luminescence resulting from hydroperoxide decomposition (7, 8)
2. Luminescence resulting from peroxy radical recombination (9)
3. Luminescence resulting from non identified processes in highly oxidized microdomains (10–13)

In the absence of data concerning eventual oxidation heterogeneity in the case under study, we will temporarily separate case 3 and consider cases 1 and 2, which can be discussed by using classical schemes of mechanisms and kinetics.

Let us consider first the hypothesis of luminescence resulting from hydroperoxide decomposition, 1. Earlier papers (14) suggested that the amide formation results from the bimolecular decomposition of a aminohydroperoxides. From this point of view, the eventual existence of a relationship between amide concentration and integrated intensity (Figure 5) seems to be in favor of this hypothesis.

Let us consider now the rate of intensity increase during air readmission: If it is induced by the hydroperoxide decomposition, hydroperoxide formation would have to be effective in the time scale of intensity variations (typically ≤ 1 s). The propagation rate would be extremely high:

$$PO_2^{\bullet} + PH \rightarrow PO_2H + P^{\bullet}$$

This condition does not comply with data of a previous study about oxidizability of epoxy networks (5) and kinetics of carbonyl and amide formation.

The most current hypothesis of luminescence originated from terminating combination of peroxy radicals seems to be *a priori* more compatible with our experimental results:

1. Formation of $POO^{\bullet}$ is very fast. Oxygen addition to alkyl radicals is diffusion controlled.
2. Rate constant of $POO^{\bullet} + POO^{\bullet}$ termination can be very high (15). Consequently, we observed very short transitions in both cases: when N_2 is admitted (time t_n) and also when oxygen is readmitted (time $t_n + \Delta t$).

Thermolysis of polymers produces alkyl radicals where the concentration just before O_2 admission is $[R^{\cdot}]_0$. The luminescence peak linked to this perturbation is undoubtedly induced by the oxidation of those radicals. According to the standard scheme, we can write:

$$\left.\begin{array}{l} \text{Polymer (RH)} \xrightarrow{\Delta T} R^{\cdot} \\ \quad\quad ROOH \xrightarrow{\Delta T} RO^{\cdot} \longrightarrow R^{\cdot} \end{array}\right\} \quad\quad r_i$$

$$R^{\cdot} + O_2 \longrightarrow ROO^{\cdot} \quad\quad\quad k_2$$

$$ROO^{\cdot} + RH \longrightarrow ROOH + R^{\cdot} \quad\quad k_p$$

$$ROO^{\cdot} + ROO^{\cdot} \longrightarrow \text{products} + h\nu \quad\quad k_t$$

where h is Planck constant and ν is frequency. The stationary concentration of aklyl radicals is

$$[R^{\cdot}]_{\infty} = \frac{r_i}{k_2 \, [O_2]} \, (1 + \lambda)$$

where

$$\lambda = \frac{k_p[RH]}{(2r_ik_t)^{1/2}}$$

is the kinetic chain length of the reaction.

We will observe a chemiluminescence sharp peak if the radical concentration $[R^{\cdot}]_0$ at the onset of oxygen admission is $[R^{\cdot}]_0 >> [R^{\cdot}]_{\infty}$. In this case, the system is initially far from equilibrium and tends rapidly toward $[R^{\cdot}]_{\infty}$ (corresponding to the steady state when the radical concentration is $[R^{\cdot}]_{\infty}$).

This condition ($[R^{\cdot}]_0 >> [R^{\cdot}]_{\infty}$) is realized only when alkyl radicals accumulate during heating under N_2; that is, when the thermolysis of the polymer has not reached a stationary state during the experiment duration. This result was justified because the intensity of the first peak increased with the time of exposition under nitrogen (Figure 2). A simplified kinetic scheme was proposed to explain this behavior (4).

Information about the nature of $R^{\cdot}$ radicals is scarce, but we can reasonably suppose that they originate principally from network scission (*see* Mechanisms I and II). Mechanism I can be justified by disappearance of ethers observed by IR spectroscopy. Formation of methylamines via Mechanism II was noted by Paterson-Jones (16), but others mechanisms are not excluded.

Indeed, it is easy to imagine that the above kinetic scheme simplifies the phenomenon too much because we can suppose that alkyl radicals resulting from propagation ($P^{\cdot}$) are probably different from radicals $R^{\cdot}$ resulting from polymer thermolysis. We should then consider a new simplified scheme:

$$\text{Polymer (RH)} \rightarrow R^{\cdot} \quad\quad r_i \, \text{(thermolysis)} \quad (1)$$

$$R^{\cdot} + O_2 \rightarrow ROO^{\cdot} \quad\quad\quad\quad\quad\quad k_{21} \quad (2)$$

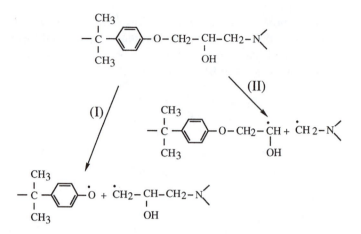

Mechanisms I and II. Origination of R· radicals via network scission.

$$ROO· + PH \rightarrow ROOH + P· \qquad k_{pl} \quad (3)$$
$$P· + O_2 \rightarrow POO· \qquad k_{22} \quad (4)$$
$$POO· + PH \rightarrow POOH + P· \qquad k_{p2} \quad (5)$$
$$ROO· + ROO· \rightarrow products \qquad k_{t1} \quad (6)$$
$$ROO· + POO· \rightarrow products \qquad k_{t2} \quad (7)$$
$$POO· + POO· \rightarrow products \qquad k_{t3} \quad (8)$$

This scheme could probably explain the existence of two distinct kinetic stages observed in experimental curves $I = f(t)$ (Figure 1).

Indeed, we can suppose that:

1. Termination 6 predominates just after O_2 admission and is responsible for the corresponding luminescence peak.
2. Termination 8 predominates long term (in the stationary state).
3. Termination 7 predominates in an intermediary term.

The experimental results, particularly the existence of an intensity minimum after the first luminescence peak, can be explained because termination 6 could be responsible for short-term luminescence, and termination 8 could be responsible of long-term luminescence. The fact that reaction 7 is progressively induced in competition with reaction 6 could be the second reason for a fast intensity decrease after the first maximum. The necessary condition of that phenomenon is a nonemissive cross termination, 7.

Species created during the second stage influence the development of the first stage. The first peak just after preheating under nitrogen is always more intensive than the nth peak whenever n is superior to 1. Here again, a nonemissive cross termination, 7, could explain the phenomenon.

More surprising is the following result (Figures 2 and 7): The first stage has no influence on the second stage. Starting from the second kinetic scheme, we can suggest that $RO_2^{\cdot}$ radicals do not propagate the oxidation reaction (k_p is very low) and that they participate only in terminations. In the opposite case, it would be difficult to imagine that the hydroperoxides ROOH created during the first stage are not involved in the initiation of oxidation chains during the second stage. It is difficult to determine the nature of $R^{\cdot}$ and $P^{\cdot}$ radicals in the absence of complementary analytical data. It seems logical to suppose that the $R^{\cdot}$ radicals are primary radicals resulting from skeleton breaking reactions, and $P^{\cdot}$ radicals are either secondary when hydrogen splitting takes place on methylene or tertiary when it takes place on secondary alcohol:

$$\overset{\backslash}{\underset{\sim}{}}N-\overset{\downarrow}{CH_2}-\underset{\underset{OH}{|}}{\overset{\downarrow}{CH}}-\overset{\downarrow}{CH_2}-O-\text{\textlangle}\bigcirc\text{\textrangle}-$$

Our interpretation is valid only in the case when long-time anaerobic degradation (a few dozens of minutes under nitrogen) does not result in noticeably decreasing concentrations of groups involved in oxidation propagation.

A thermogravimetric study under N_2 revealed that degradation was weak for $T < 350$ °C, and the weight loss observed by dynamic thermogravimetric analysis begins at 396 °C (*17*). This fact seems to be compatible (but should be verified) with the hypothesis of a relative polymer stability during exposure of a few hours at 240 °C or lower temperatures.

The oxidative degradation mechanism of DGEBA–DDS network is relatively complex and supposes the coexistence of two reaction mechanisms practically independent of one another and the participation of two kinds of alkyl radicals. Intensity variations during nonstationary experiments were extremely fast, and these results provided us with information about luminescence mechanism but did not allow us any quantitative approach to rate constants of termination or hydroperoxide decomposition. The limited sensitivity of our apparatus did not allow us to make the experiments at temperatures noticeably lower than 200 °C, where these values could probably be measurable.

The luminescence study of the DGEBA–DDS network revealed relatively little about the mechanism of oxidation in comparison with spectrophotometric or physical methods (*14*). On the contrary, our results suggest that the chemiluminescence can be a valuable tool to study the radical production induced by a thermolytic degradation under an inert atmosphere. The intensity of the peak accompanying the oxygen admission is directly linked to the concentration of radicals $R^{\cdot}$. The luminescence gives us the ability to measure, with a good sensitivity, parameters linked to the degradation rate when other methods such as thermogravimetry or IR spectrophotometry are not able to detect any changes.

References

1. Zlatkevich, L. In *Luminescence Technique In Solid State Polymer Research;* Zlatkevich, L., Ed.; Dekker: New York, 1989; pp 135–197.
2. George, G. A.; Egglestone, G. T.; Riddell, S. Z. *Polym. Eng. Sci.* **1983,** 7, 412–418.
3. George, G. A.; Schweinsberg, D. P. *J. Appl. Polym. Sci.* **1987,** 2281–2292.
4. Tcharkhtchi, A.; Audouin, L.; Verdu, J. *J. Polym. Sci.* **1993,** 31, 682.
5. Bellenger, V.; Verdu, J. *J. Appl. Polym. Sci.* **1985,** 30, 363–374.
6. Audouin, L.; Verdu, J. *J. Polym. Sci.* **1987,** A25, 1205–1217.
7. Loyd, R. A. *Trans. Farad. Soc.* **1965,** 61, 2173 and 2182.
8. Reich, L.; Stivala, S. *J. Polym. Sci.* **1965,** A3, 4299.
9. Russel, J. A. *J. Am. Chem. Soc.* **1965,** 78, 1047.
10. George, G. A.; Ghaemi, M. *Polym. Degrad. Stab.* **1991,** 34, 37–53.
11. Billingham, N. C. *Makromol. Chem. Symp.* **1989,** 28, 145.
12. Knight, I. B.; Calvert, P. D.; Billingham, N. C. *Polymer* **1985,** 26, 1713.
13. Geuskens, G. *Polym. Photochem.* **1984,** 5, 313–331.
14. Bellenger, V.; Verdu, J. *J. Appl. Polym. Sci.* **1985,** 30, 363–374.
15. Reich, L.; Stivala, S. In *Autooxidation of Hydrocarbons and Polyolefins;* Dekker: New York, 1969; p 375.
16. Paterson-Jones, J. C. *J. Appl. Polym. Sci.* **1975,** 19, 1539.
17. Bellenger, V.; Fontaine, E.; Fleishmann, A.; Saporito, J.; Verdu, J. *Polym. Degrad. Stab.* **1984,** 9, 195–208.

RECEIVED for review January 26, 1994. ACCEPTED revised manuscript January 23, 1995.

16

Chemiluminescence Analysis and Computed X-ray Tomography of Oxidative Degradation in Polymers

Bengt Mattson[1] and Bengt Stenberg

Department of Polymer Technology, Royal Institute of Technology, S-100 44, Stockholm, Sweden

Chemiluminescence analysis (CL), which measures the emission of photons during oxidative aging, and computed X-ray tomography (CT), which estimates densities, have been used to follow the oxidative aging behavior of rubber materials. Both techniques are shown to be fast and sensitive and are applicable on carbon-black-filled materials as well as on unfilled polymers. However, caution is warranted for the isolated use of either technique, because nonoxidative processes may greatly influence the measurements. By combining the techniques, a clearer picture emerges of the complex aging phenomena.

THERMOOXIDATIVE DEGRADATION OFTEN DETERMINES the lifetime of polymer products. Therefore, the understanding of oxidation phenomena is of great importance. One of the most important features is the diffusion-controlled nature of oxidation. Diffusion-limited oxidation can occur whenever the dissolved oxygen in a polymer is used up faster than it can be replenished by diffusion from the oxygen-containing atmosphere surrounding the material. This inequality will lead to a lowering in the oxygen concentration with depth into the sample and will result in maximum oxidation levels at the oxygen-exposed sample surface and reduced or nonexistent oxidation in the interior. The importance of this effect, therefore, depends on three factors (1–5):

- the geometry of the material
- the oxygen consumption rate
- the oxygen permeation rate

[1] Current address: Pharmacia Hospital Care, Franzéngatan 9, S-112 87, Stockholm, Sweden.

0065–2393/96/0249–0235$12.00/0

For typically used polymer thicknesses of about 1 mm, these effects are commonly observed in UV and thermal environments.

The dominating thermooxidative degradation products are hydroperoxides. These hydroperoxides decompose during prolonged aging, and different carbonylic compounds such as ketones, esters, and acids are the final degradation products. During the oxidation, an oxidized layer may be formed at the exposed surface. This oxidized layer has a low permeability to oxygen (1, 5) and therefore protects the interior of the material from further oxidative aging (6). This statement is valid as long as no cracks are formed in the surface of the sample. The reduced permeation rate of oxygen has been explained (5, 7) as being mainly due to the presence of carboxylic acid groups in the oxidized surface.

Several techniques have been used throughout the years to follow the oxidative aging of polymeric materials, such as measurements of the carbonyl index by using IR techniques and different mechanical techniques. However, these techniques are not sensitive enough in certain applications. This shortcoming is especially true when the degradation is in its initial phases and just minor changes in the mechanical properties have occurred, although severe degradation of the polymer structure may have taken place. The limitations of IR are most obvious in carbon-black-filled polymer systems, where the large absorbance from the carbon black obstructs the characterization.

Analyses using chemiluminescence (CL) and computed X-ray tomography (CT) have been presented, and both techniques seem promising for rapid and sensitive measurements of oxidative degradation of both carbon-black-filled and unfilled polymers. Results from these techniques, and advantages as well as drawbacks, are reviewed in this chapter.

Chemiluminescence Analysis

In the 1980s, CL was the subject of increasing interest (8, 9) due to the development of single-photon-counting equipment. CL measures the low level of light emission that accompanies the oxidation of organic materials. The occurrence of CL during the oxidation of polymeric materials has been known for many years, since the pioneering work by Ashby (10) in the early 1960s. However, during the last few years the technique has become particularly interesting.

A few reactions are capable of light emission, which requires that the reaction is sufficiently exoenergetic to produce a product in an excited state and that the excited product has a finite probability of photon emission rather than radiationless deactivation. It is generally accepted that the emitter during polymer oxidation is a triplet excited carbonyl, but there is discussion about its origin. The Russel mechanism shown in Scheme I is assumed to be the predominating mechanism responsible for CL (11). The light is emitted from

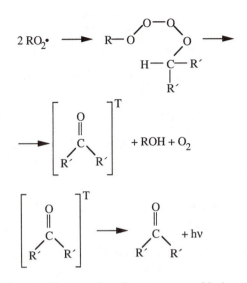

Scheme I. The Russel mechanism responsible for CL.

a termination step in the autoxidation of polymers when two peroxy radicals combine to produce a carbonyl, an alcohol, and oxygen.

A fraction of the carbonyls, usually ketones, are found in an excited, triplet state; when they relax back to the ground state, light is emitted. If the CL experiment is run in oxygen-containing atmospheres such as air or oxygen gas, oxidation processes are fed with fresh oxygen during the test. Therefore, the light intensity recorded may be regarded as being a measure of the oxidation rate. On the other hand, if the test is performed in an inert atmosphere such as nitrogen gas, the emitted light will be proportional to the hydroperoxide concentration (*12*); therefore, the recorded intensity may be regarded as a measure of the degree of oxidation.

The great advantage of the CL technique is high sensitivity permitting analyses to be made at low temperatures or during short times. An obvious drawback is that the CL phenomenon is still not completely understood and sometimes the interpretations are troublesome. Also, processes such as light absorbance in heavily oxidized layers and CL behavior of the additives themselves interfere with the measurements.

Computed X-ray Tomography

CT has recently started to be used in studies of aging behavior (*13*). Because organic materials have densities that do not differ greatly from that of human tissue, tomodensiometry (*14*), the nondestructive surveying of densities through a sample, can be performed with normal medical X-ray scanners. The

Houndsfield CT number, H, is proportional to the X-ray attenuation through an object and is represented by the gray scale of an X-ray image. For most organic and polymeric materials, a reasonable linear relationship exists between H and the density of the material.

Figure 1 shows the correlation between H and density as determined during aging of a carbon-black-filled ethylene–polypropylene–diene monomer (EPDM) rubber (4). Several samples were taken from different depths in the materials and from materials with a different degree of aging. Except for a single data point that represents the surface result for a very rapidly degraded material, there is good correlation between CT results and density. The probable reason for the lack of correlation of the single surface point is an extremely rapid drop, approximately 0.05 g/cm^3 in 0.5 mm, in the density of this sample near the surface. In this case the resolution of the CT technique is not good enough. Although the resolution of the CT technique could be improved by reducing the thickness of a CT slice or by using smaller pixel sizes (with concomitant reductions in sensitivity), such improvements are really not necessary because all the salient features of degradation are readily observable under standard conditions.

The advantages of the CT technique are that it is nondestructive, quite sensitive in detecting small variations in density, and only requires approximately 20 s for each measurement. The drawbacks are the cost of an instrument and that several processes besides oxidation affect density, such as changes in cross-link density and migration of additives.

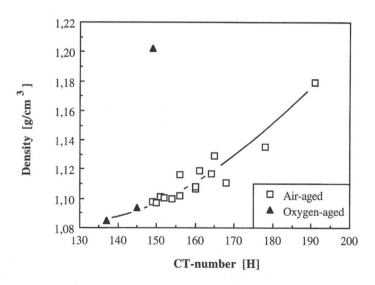

Figure 1. Correlation between density and H found by using aged EPDM materials.

Experimental

Experiments Performed using CL.   Most researchers have studied the CL light emitted from polymers within the time frame of minutes to hours. The obvious drawback of an experimental design of this type is that information present during the initial time of heating is lost. The CL experiments presented in this review were all performed on a CL instrument especially suited for studies of the initial burst of CL. The instrument was designed by modification of a commercial TLD-instrument (Thermoluminescence Dosimeter) from Alnor Instruments AB in Studsvik, Sweden (*15, 16*). A very good example of the rapid analysis possible with this instrument is shown in Figure 2. CL plots, measured in oxygen at 150 °C, are presented for stabilized and unstabilized polybutadiene (*15*). Already after a few seconds it is possible to see that the yield of luminescence from the material containing antioxidant is much less than that from the reference material. This difference is ascribed to a slower oxidative degradation.

Experiments Performed using CT.   All CT experiments presented in this review were performed by a CT instrument at the hospital in Skellefteå, Sweden, by using a General Electric 9800 Quick CT. The pixel unit in this study was 0.19 mm, and the remaining tomography parameters are reported in Table I.

Results and Discussion

The high sensitivity of the CL technique allows recording of aging at temperatures near room temperature. Figures 3a–c present Arrhenius plots for three different types of paper coatings: A, B, and C, respectively (*15*). The logarithm of the photon counts during 20 s is plotted versus the reciprocal temperature given in Kelvin. Materials A and B are styrene–butadiene rubber

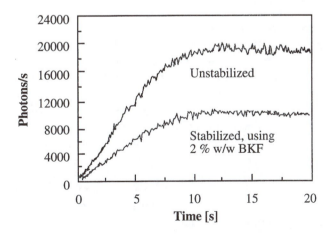

Figure 2. CL emission from stabilized and unstabilized polybutadiene when heated in oxygen at 150 °C: BKF = 2,2'-methylene-bis(4-methyl-6-tert-butylphenol).

Table I. CT Parameters

Parameter	Setting
Voltage of accelerator	140 kV
Anode current	170 mA
Scan time	4 s + overscan
Number of scans	5 per sample
Cut thickness	10 mm
Display field of view	96 mm
Filter	Smooth
Scan field of view	Large

(SBR)-latex coatings, with only minor dissimilarities in the ratio of styrene to butadiene. Material C is an acrylate coating. The materials were examined in oxygen. These plots clearly reveal the importance of measuring close to the working temperature. At 110 °C (2.6/K), all three materials were degraded at approximately the same rate, but with decreasing temperature the sensitivity to oxidation dropped more rapidly for material C than for materials A or B. If one material was to be selected from these three materials for use at 60 °C (3.0/K), material C would be the rational choice. A proper selection of materials requires monitoring of the oxidation process at a temperature close to the temperature of use. This monitoring can be done conveniently and rapidly with CL analysis.

Figure 4 shows an Arrhenius plot for natural rubber unfilled and filled with 50 phr (parts per hundred rubber) carbon black and heated in oxygen (15). There is a large difference in light emission due to absorbance by the carbon-black particles. However, the plot shows that the CL technique should be suitable for use on carbon-black-filled materials as well. In Figure 5a–c, CL plots are reported for stabilized, carbon-black-filled EPDM-rubbers measured in nitrogen at 200 °C (15). The material shown in Figure 5a is unaged, and the signal is then equivalent to the background emission. However, when the materials are preaged in air at 150 °C, a signal easily distinguishable from the background level is obtained. The material shown in Figure 5b was aged for 14 days, and the one shown in Figure 5c was aged for 45 days. The higher intensities of the CL emission for the preaged materials are ascribed to a buildup of hydroperoxides during aging. When the samples are subjected to the high temperature in the CL instrument, hydroperoxides are cleaved and the oxidation processes responsible for CL emission take place.

In a report examining the aging behavior of natural rubber vulcanizates free of carbon black (3), we found a good correlation between H_s and carbonylic content determined by attenuation total reflection–Fourier transform IR. The results are shown in Figures 6 and 7. Figure 6a shows H recorded at the surface as a function of aging time, and Figure 6b shows H as a function of depth into the materials aged for 45 days. Corresponding data showing the

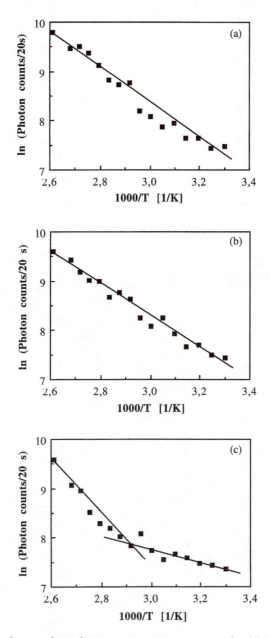

Figure 3. Arrhenius plots of CL measurements in oxygen for (a) an SBR-latex; (b) an SBR-latex; and (c) an acrylate paper coating.

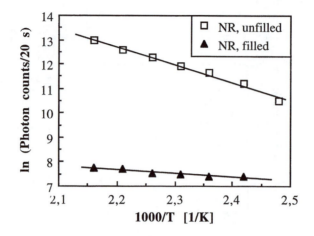

Figure 4. Arrhenius plots of CL measurements in oxygen for a carbon-black-filled and an unfilled natural rubber.

IR intensity of the ester and the ketone peaks (in arbitrary unit, I) are displayed in Figure 7a for the surfaces and in Figure 7b for values versus depth into the materials. The material denoted A is a conventionally (by using sulfur and N-cyclohexyl-2-benzothiazole-2-sulfenamide) vulcanized material. Material B is vulcanized using TMTD (tetramethylthiuram disulfide).

Because the dithiocarbamates formed during TMTD vulcanization possess antioxidant activity (*17*), material B is regarded as being stabilized. The conclusion was that the CT changes primarily were due to the incorporation of the relatively heavy oxygen atom in the material during aging. Oxygen may be present both as parts of degradation products, mainly peroxides and carbonyl compounds, and as molecular oxygen.

The effects from the diffusion-controlled oxidation in Figures 6 and 7 are important. Figure 7b shows that the depth of oxidative aging, that is, the depth into the sample where carbonyl absorbance is at its original value (I = 10), decreases when the temperature is raised from 100 °C to 150 °C for material B. This decrease is due to a higher rate of consumption of oxygen at 150 °C, and oxidation is therefore more concentrated at the surface. When the curves for material A and material B aged at 100 °C are compared, material A is obviously more highly degraded at the surface, but the oxidative penetration depth of A is much less than for material B (3 mm and 6 mm, respectively). This difference is probably due to the rapid formation of an oxidized layer that has a lower permeability toward oxygen and protects the bulk of material A from oxidative aging.

The formation of a protective oxidized layer is very slow for material B at 100 °C due to the antioxidant effect of the dithiocarbamates formed during vulcanization. The effects on depth of oxidative aging from diffusion-limited

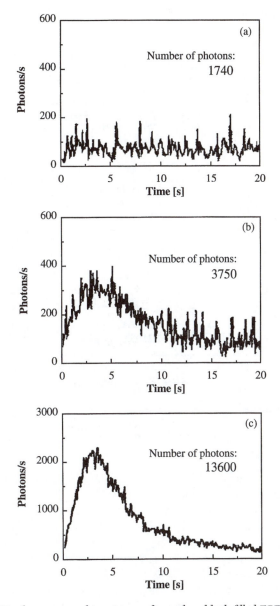

Figure 5. CL plots measured in nitrogen for carbon-black-filled EPDM, preaged in 150 °C for (a) 0, (b) 14, and (c) for 45 days.

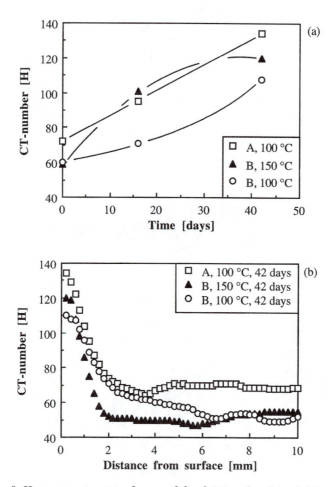

Figure 6. H *versus aging time for unstabilized (A) and stabilized (B) natural rubber (a), and* H *versus depth into the material for unstabilized (A) and stabilized (B) natural rubber (b).*

oxidation is also detectable by using CT (Figure 6b). To be able to interpret Figure 6b correctly, it is important to know that material A originally has a higher H in the bulk than material B (3) due to the composition of the vulcanizing systems. At a depth of approximately 3 mm, H for material A aged at 100 °C is back to the original, unaged level of $H = 68$, whereas a depth of approximately 6 mm is required for material B aged at 100 °C before the value of unaged material ($H = 55$) is reached.

The carbon-black-filled EPDM materials subjected to CL analysis in Figure 5 were also investigated using CT (4). Figure 8a shows a comparison

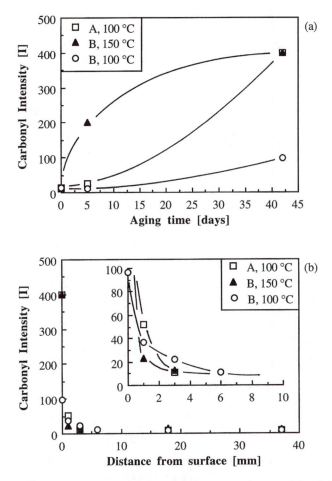

Figure 7. The carbonyl IR-peak [I] versus aging time for unstabilized (A) and stabilized (B) natural rubber (a), and carbonyl IR-peak [I] versus depth into the material for unstabilized (A) and stabilized (B) natural rubber (b). The area of greatest interest is enlarged.

between CL data and CT data for values recorded at the surface of the samples. Corresponding data from the bulk, that is, data from a depth of 3–4 mm, are reported in Figure 8b. The changes in CL intensity are small in the bulk compared with the surface values, a result indicating a much lower degree of oxidation in the bulk. This finding is in good agreement with expectations for diffusion-controlled oxidation. However, an increase in H is still evident during aging in the bulk. This increase is explained as being primarily due to non-oxidative processes such as migration of oils, because the CL changes, which

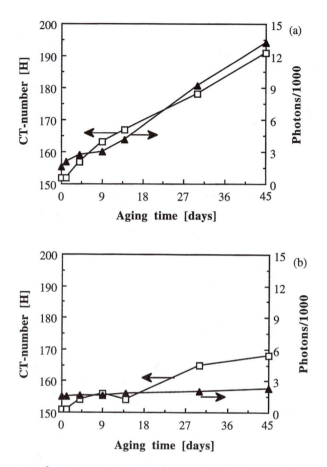

Figure 8. H *and CL emission (expressed as number of photons divided by 1000 emitted during 20 seconds) versus aging time for (a) the surface and (b) the bulk of carbon-black-filled EPDM-materials. Symbols:* □ *is CT number;* ▲ *is CL emission.*

are directly connected to oxidation, are small. A distinction between oxidation and nonoxidative processes can only be made when complementary techniques like CL and CT are used in parallel.

Table II summarizes the values from CT and CL measurements on irradiated EPDM-materials (*4*). No changes relative to unaged materials are detected by CT. On the other hand, CL measured in nitrogen at 200 °C shows large changes both at the surface and in the bulk relative to unaged material. Bulk values were recorded at a depth of 3–4 mm. However, the increase in luminescence has nothing to do with oxidation. The photon emission is due to charge recombination luminescence (*18, 19*), which is a physical phenom-

**Table II. Comparison Between CT and CL for
Irradiated Samples**

Materials	H		CL Value (Photons/20 s)	
	Surface	*Bulk*	*Surface*	*Bulk*
EPDM, unaged	152	150	1,700	1,600
EPDM, aged by electron irradiation				
8 Mrad	153	151	12,500	13,900
16 Mrad	152	149	14,100	16,400

enon involving the formation of an electronically excited chromophore that can eject an electron to give a cation–electron pair. Thermal motions in the solid result in a diffusive hopping of the charge through the material, where it resides in potential wells of various depths. Charge recombination occurs with a finite probability and is followed by radiationless decay or emission of light. The motion of charges initially is rapid but becomes progressively slower as the charges encounter deeper potential wells. The heat supplied during the CL analysis releases the charges from their wells, and recombination may occur. The emissions from these recombination processes are not dependent on thermooxidative degradation processes. Again, necessity of using complementary techniques to avoid incorrect conclusions is proved.

Conclusions

This review shows that both CL and CT can be useful for following the thermooxidative degradation of polymers even if the polymers are filled with carbon black. Both techniques appear to have great potential as rapid and sensitive evaluation methods. Caution is advised, however, for the isolated use of either technique, because nonoxidative processes may greatly influence the measurements. By combining the information obtained from the different techniques, a clearer picture emerges of the complex aging phenomena that commonly occur in polymeric materials during thermooxidative degradation.

A prominent feature with CL analysis, as compared with CT, is that CL can be used to investigate a material during its thermooxidative aging. CL and CT can determine the degree of aging for a material after it has been subjected to preaging.

Acknowledgments

Financial support from NUTEK (Swedish National Board for Industrial and Technical Development) is gratefully acknowledged. We are grateful to Wenner-Gren Center Foundation, NorFA (Nordic Academy for Advanced Study) and to SINTEF Materials Technology in Norway for support in the preparation of this paper. Thanks are due to Torbjörn Reitberger, Björn Terselius, and Håkan Lavebratt at the Royal Institute of Technology, to Erik Östman at Skega AB, and to Kenneth Gillen and Roger Clough at the Sandia National Laboratories for experimental assistance and valuable discussions.

References

1. Stenberg, B.; Shur, Y. J.; Jansson J. F. *J. Appl. Polym. Sci.* **1979,** *35,* 511.
2. Clough, R. L.; Gillen, K. T. *Polym. Degrad. Stab.* **1992,** *38,* 47.
3. Mattson, B.; Östman, E.; Persson, S.; Stenberg, B. *Rubber Chem. Technol.* **1990,** *63,* 23.
4. Mattson, B.; Gillen, K. T.; Clough, R. L.; Östman, E.; Stenberg, B. *Polym. Degrad. Stab.* **1993,** *41,* 211.
5. Mattson, B.; Stenberg, B. *J. Appl. Polym. Sci.* **1993,** *50,* 1247.
6. Björk, F.; Stenberg, B. *Polym. Test.* **1985,** *5,* 245.
7. Mattson, B. Ph.D. Thesis, Royal Institute of Technology, 1993.
8. George, G. A., *Dev. Polym. Degrad.* **1981,** *3,* 173.
9. *Luminescence Techniques in Solid-State Polymer Research;* Zlatkevich, L., Ed.; Marcel Dekker: New York, 1989.
10. Ashby, G. E. *J. Polym. Sci.* **1961,** *50,* 99.
11. Billingham, N. C.; O'Keefe, E. S.; Then, E. T. H. *Proc. 12th Ann. Conf. Adv. Stab. Cont. Polym. Degrad.;* Lucerne, Switzerland, 1990.
12. Billingham, N. C.; Burdon, J. W.; Kaluska, I. W.; O'Keefe, E. S.; Then, E. T. H. *Proc. Int. Symp.;* Lucerne, Switzerland, **1988,** *2,* 11.
13. Östman, E.; Persson, S.; Lavebratt, H. *Non-Destr. Test. Int.* **1989,** *22,* 6.
14. Persson, S. *Polymer* **1988,** *29,* 802.
15. Mattson, B.; Reitberger, T.; Stenberg, B.; Terselius, B. *Polym. Test.* **1991,** *10,* 399.
16. Forsström, D.; Kron, A.; Mattson, B.; Reitberger, T.; Stenberg, B.; Terselius, B.; *Rubber Chem. Technol.* **1992,** *65,* 736.
17. Dunn, J. R.; Scanlon, J. In *The Chemistry and Physics of Rubberlike Substances;* Bateman, L., Ed.; Maclaren & Sons: New York, 1963; Chapter 18.
18. Mendenhall, G. D. *Angew. Chem., Int. Ed. Engl.* **1990,** *29,* 362.
19. Mendenhall, G. D.; Agarwal, H. K. *J. Appl. Polym. Sci.* **1987,** *33,* 1259.

RECEIVED for review December 6, 1993. ACCEPTED revised manuscript September 15, 1994.

Odor Characterization of Low-Density Polyethylene Used for Food-Contact Applications

S. W. Bigger[1], M. J. O'Connor[1], J. Scheirs[2], J. L. G. M. Janssens[3], J. P. H. Linssen[3], and A. Legger-Huysman[3]

[1]Department of Environmental Management, Victoria University of Technology, St. Albans Campus, PO Box 14428, Melbourne Mail Centre, Melbourne 3000, Australia
[2]ExcelPlas Australia Ltd., PO Box 102, Moorabbin 3189, Australia
[3]Department of Food Science, Wageningen Agricultural University, PO Box 8129, 6700 EV Wageningen, The Netherlands

Three novel techniques were explored for the analysis of odorous oxidation products originating from low-density polyethylene (LDPE) used for food-contact applications. The technique of combined computerized gas chromatography and sniffing port analysis showed that hexanal and octanal are among the odorous compounds that can be isolated from mineral water packed in polyethylene-lined aluminium–cardboard containers. The performance of a sintered, high-temperature SnO_2 semiconductor device, a component in a commercially available odor meter, was also investigated and shown to respond linearly to the level of odor in film-grade LDPE when calibrated against a sensory evaluation panel. The technique involving the dynamic headspace trapping of volatiles in LDPE at ambient temperatures and subsequent gas chromatographic–mass spectrometric analysis, identified the presence of unsaturated C_6 aldehydes such as 2-hexenal and 2-methyl-2-pentenal. We propose that these odor-active compounds are produced in the polymer as a result of undesirable degradation reactions involving the polymer, the slip agent erucamide, or the diatomaceous silica antiblock agent.

LOW-DENSITY POLYETHYLENE (LDPE) is widely used for food packaging because of its many desirable properties. Compared with other materials, it is cheap, light-weight, tough, and flexible. It also offers excellent moisture and chemical resistance. About 65% of all LDPE is used for film production (1), much of which is used for the packaging of food. Although PE is generally recognized as a safe packaging material by the U.S. Food and Drug Administration and it is one of the most inert polymers (1), the flavor properties that it sometimes imparts to packaged food products must be taken into consideration. Odors and off-flavors can be caused by the migration into the foodstuff of minor compounds such as residual monomers, additives, or solvents from printing inks. Low-molecular-weight compounds formed during the polymerization process or compounds formed by the degradation of the polymer during processing may also cause odor problems (2, 3).

Possibly the earliest literature dealing with the problem of undesirable odors in PE used for food packaging dates back to the late 1950s (4) and coincides with the increased use of PE as a food-packaging material (5). Since the early 1960s, the characterization of volatile degradation products isolated from PE has been investigated many times (2, 3, 5–24). Of these studies, about half contain specific reference to odorous products (2, 3, 10, 13–16, 18, 19, 21, 23, 24), and most of the work on odor has been carried out since the mid-1980s.

Peled and Mannheim (2), Kiermeier and Stroh (7), Berg (25), and Bojkow (26) reported that products such as milk, water, and fruit juices can show an unfavorable off-taste after contact with PE or PE-coated cardboard. Kozinowski and Piringer (14) described an off-odor in water packed in LDPE-lined cardboard as being like candle-grease or musty, rancid, or soapy. Furthermore, a study of water packed in freshly made PE-lined aluminum test pouches showed that the water developed an off-flavor, which could be described by sensory profiling as musty, sickly, astringent, synthetic, metallic, and dry (20).

Experimental Techniques

The experimental techniques for characterizing odor-active compounds in PE traditionally involve the use of sensory evaluation panels (SEPs) or gas chromatography (GC) (2, 3, 9, 15–17, 19, 27, 28). These techniques can be performed either separately (2, 3, 13, 15, 18, 19) or in combination with each other (16, 21, 23, 24). Techniques based on GC are by far the most widely used methods for the analysis of the volatiles from PE. Of these, GC–mass spectrometry (MS) is the one most commonly used (3, 13, 15, 16, 21, 24); GC–flame ionization detection (FID) is used to a lesser extent (2, 3, 16). More recently, vacuum distillation (15), selective volatile extraction (3, 21), supercritical fluid extraction (29), GC–olfactory detection (21), and dynamic headspace GC analysis (13, 20) were used in the analysis of odorous compounds in PE resins.

Sensory Evaluation Panels

The human nose is undoubtedly a key asset in the area of organoleptic analysis of odorous volatiles and is widely used for this purpose (2, 13, 15, 18, 19).

Sensory evaluation of odor is usually achieved by carefully selecting a panel of assessors whose task is to describe the odorous properties of a given sample. It is claimed that sensory methods can sometimes be used to make measurements that cannot be made directly by conventional physical or chemical techniques (*30*).

Many polymer manufacturers engage the services of sensory evaluation panels (SEPs) to monitor organoleptic quality. However, routine analyses by SEPs are onerous, time-consuming tasks; and the results obtained from such panels can often be inherently unreliable, irreproducible, and subjective. The recent introduction of new, portable instruments that use a solid-state, sintered SnO_2 electrode to monitor the intensity of odor-active compounds (*31*) has brought with it the attractive possibility of replacing SEPs with such devices.

Hot Jar Methods and Dynamic Headspace Sampling.   Possibly the least complicated method of chemical analysis of odorous volatiles in PE involves heating the polymer in a closed chamber and directly injecting a sample of the headspace gas onto a chromatographic column (*2, 17, 32, 33*). This technique is known as the "hot jar" method and was originally developed by Wilks and Gilbert (*34*) for the analysis of residual monomer in poly(vinyl chloride). A variation of this method was suggested by Culter (*35*) in which a mixture of polymer and water is placed in a vial and heated for a period of time. The water facilitates the removal of volatile compounds by a steam-stripping mechanism.

The hot jar methods offer a quick analysis time and a simple means of introducing the sample onto the chromatographic column. However, the method suffers from the fact that odorous volatiles may be present in such small amounts that a concentration step may be required before detection is possible.

Many investigations are now performed using a method based on the dynamic headspace sampling of packaging material contained in a closed jar or vessel that is heated to a particular temperature. Temperatures ranging from room temperature to 350 °C were used for this type of analysis (*3, 11, 12, 21, 25, 36*). A clean, inert gas is used to continuously purge the headspace vapor above the sample. The volatile compounds are trapped and concentrated on an inert sorbent medium such as carbon (*16*) or a synthetic material such as Tenax-GC (2,6-diphenyl-*p*-phenylene oxide polymer). These compounds are then thermally desorbed onto a chromatographic column (*13, 20, 37*).

Compounds Isolated from PE.   Ketones, aldehydes, and low-molecular weight alkanes and alkenes are among the most commonly identified compounds isolated during the analysis of PE volatiles. For example, hexanone, butanal, butane, pentanone, heptane, and ethylene were frequently reported in the literature (*3, 5, 8–12, 16, 20*). Mechanisms for the production of some of these compounds were also proposed (*3, 38*).

Hoff and Jacobsson (11) found 44 compounds after the thermal oxidation of PE at 296 °C. These compounds were mostly hydrocarbons, alcohols, aldehydes, ketones, acids, cyclic ethers, cyclic esters, and hydroxycarboxylic acids, of which fatty acids and aldehydes constituted the majority. The reaction mechanism for the formation of these degradation products is based on the formation of radicals at high temperatures.

Bravo and Hotchkiss (3) studied the volatile compounds produced during the heating of PE in the presence of excess oxygen at 150–350 °C for 5–15 min. The sample was dispersed on glass wool to increase the surface area available for oxidation and then heated in a glass vessel. The volatiles were purged with purified air into a cryogenic glass trap cooled with liquid nitrogen, extracted from the trap with ultrapure Freon-113 (1,1,2-trichloro-1,2,2-trifluoroethane), and then concentrated for chromatographic analysis. A total of 84 compounds were identified, and these compounds had chain lengths in the range of C_5–C_{23}. The compounds were mainly aliphatic hydrocarbons, aldehydes, ketones, and olefins. Temperature and heating time determined the amount and type of products that were obtained.

In another study Bravo et al. (21) found 14 odor-active compounds in PE using GC–olfactometry analysis with one trained observer. The overall odor was described as wax-like, and the separate compounds were assigned descriptors such as fruity, herbaceous, rancid, metallic, waxy, pungent, or orange. Eight compounds could be identified as C_6–C_9 saturated or unsaturated aldehydes and ketones.

Sniffing Port Analysis. Human olfactory analysis of the effluent from a GC column is a very useful technique for determining which of the components of a complex mixture of volatiles possess odor. The technique of "sniffing port analysis" has been widely used in PE odor identification (15, 16, 23, 24). The method usually involves splitting the effluent from a chromatographic column into two streams. One stream is passed through a humidifying device and into an effluent port (or mask) to obtain a human olfactory response. The other stream is analyzed either by MS or FID. The technique thus involves the coupling of a human olfactory response with an electronic signal and enables the two signals to be compared (39). Although the technique enables odor-active compounds to be identified, the odor descriptors of the separately eluted compounds are often quite different than the descriptor for the volatile mixture (21). This difference makes it difficult to correlate a given SEP evaluation with the corresponding instrumental analysis.

Berg (25) identified the following four classes of compounds isolated from headspace samples of both PE-coated bleached paper board and PE granules sealed in closed vessels in nitrogen atmospheres:

- saturated aliphatic hydrocarbons
- unsaturated aliphatic hydrocarbons

- aromatic hydrocarbons
- aromatic hydrocarbons with unsaturated side chains

Combined GC and sniffing port analysis showed that the aromatic hydrocarbons with unsaturated side chains impart very strong odor impressions. Therefore, the converting and extrusion stages of polymer processing only slightly influence the amount of odorous compounds in the PE granulates, and airing of the material decreases the off-flavor to a large extent.

Temperature Effect.   Most studies on the analysis of volatile compounds originating from PE were performed at elevated temperatures, typically 100–350 °C (*3, 9, 11, 12, 15, 17, 21, 22, 27*). The use of high temperatures enables substantial amounts of products to be produced (*5*) and reduces analysis times by facilitating the removal of the volatiles. Also, the exposure of the polymer to high temperatures probably simulates the conditions encountered by the polymer during processing (*11, 12*). Only a few studies (*13, 20*) involved the isolation and characterization of odor-active compounds that are present at ambient temperatures.

The nature of the volatile degradation products and the amount of each product that is produced during the oxidative pyrolysis of PE strongly depend on the temperature of pyrolysis as well as the heating time (*3*). Thus, to identify the compounds in PE that are responsible for any odor or off-flavor imparted to foodstuffs at ambient conditions, the volatile residuals should be trapped and concentrated under the same conditions.

Despite the diversity of the available techniques for characterizing odor in PE, little success has been achieved thus far in correlating the results of instrumental methods of odor assessment with the results obtained from SEPs (*18*). Furthermore, there has been very little research into the factors that control the migration of odorous compounds found in PE into packed food products. These factors include diffusion and solubility constants of the compound in both the packaging material and the packed product, the equilibrium constant associated with such partitioning, the concentration of the migrated compound in the packaging material, temperature, contact time, and surface-to-volume ratio of the packet (*26*).

The experimental work discussed in this chapter illustrates the application of relatively low-temperature sampling techniques that can be successfully utilized to study odorous compounds in LDPE used for food-packaging applications. Three areas are of particular interest:

1. Odor activity of compounds found in commercial mineral water packed in LDPE-lined aluminium–cardboard, as determined by a combined computerized GC and sniffing port analysis technique.
2. Application of an SnO_2 semiconductor device as a means of routinely monitoring, at ambient temperatures, odor levels in a typical commercial film-grade LDPE material.

3. Low-temperature dynamic headspace trapping of odorous volatiles from film-grade LDPE followed by thermal desorption and analysis as a means of identifying those compounds responsible for odor at ambient conditions.

Materials and Methods

Combined Computerized GC and Sniffing Port Analysis.

Selection and Training of Assessors. The selection and initial training of panel members were carried out simultaneously (*40, 41*). Assessors were selected on the basis of their general health, with particular regard to their immunity against headcolds and allergies and their ability to perform well in a series of tests designed to assess odor memory, odor sensitivity, middle long-term memory, and odor recognition.

The odor memory test required the subject to correctly recall and recognize the odors of various essential oils that were previously identified to him/her. The odor sensitivity test required the subject to correctly rank different concentrations of odorous compounds dissolved in water. The middle long-term memory test required the subject to correctly recall and recognize, at a later date, the odors of various essential oils that were presented to him/her during the previous odor memory test. The odor recognition test involved noting and assessing the subject's response to odorous compounds, some of which were known to exist in mineral water packed in PE. Of the 100 persons who originally responded to a newspaper advertisement requesting participants for the experiments, 14 assessors were finally selected for sniffing-port analysis, 10 of which participated in the final panel.

Sample Preparation. The test product, mineral water packed in LDPE-lined aluminium–cardboard laminate, was obtained commercially from a local store. The product was incubated in its sealed package for 24 h at 40 °C to enhance its volatile content ("PE40" sample). Water from the same source packed in glass bottles was used as a reference ("glass" sample).

A 2-L sample of each packaged mineral water was decanted into separate glass flasks, and the flasks were equilibrated in a water bath at 40 °C. A dynamic headspace sample was prepared immediately for analysis by purging each flask with 6 L of purified nitrogen at a flow rate of 100 mL/min. The volatile compounds were trapped on Tenax-TA (35/60 mesh; Alltech Nederland BV, Zwijndrecht, The Netherlands). A cold trap maintained at −10 °C was used to condense the water vapor in the exit purge stream.

Apparatus. Figure 1 is a schematic diagram of the apparatus used for the combined computerized GC and sniffing port analysis. A thermal desorption–cold trap device (Chromopack TCT injector, model 16200) was used for desorbing the volatile compounds from the Tenax trap. These compounds

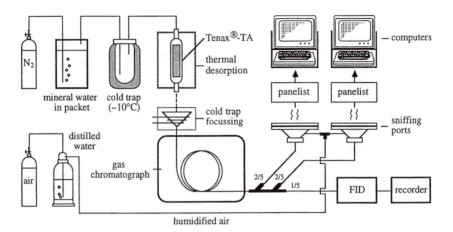

Figure 1. Schematic diagram of the apparatus used for combined computerized GC and sniffing port analysis.

were focused via a heated transfer line maintained at 250 °C onto a fused-silica capillary column (DB-1, 0.32 mm × 30 m, 1.0-μm film thickness; J & W Scientific) installed in a gas chromatograph (Carlo Erba, Type 6000 VEGA series). The chromatograph was equipped with an FID system and two sniffing ports (SGE, Milton Keynes, Great Britain). At the end of the chromatographic column the effluent was split into three streams with a split ratio of 1:2:2 for the FID system and each of the sniffing ports, respectively. Humidified air, generated by bubbling clean air through distilled water contained in a gas scrubbing bottle, was passed through stainless steel tubing to each of the sniffing ports to prevent the "drying out" of the assessors' nasal cavities during a run.

Data Collection. A Maxima 820 system (Dynamic Solutions, Ventura, California) was used to record the FID signal. This system was positioned outside the room in which the assessors were located so that their response would not be influenced by the FID response.

The perceptions of the assessors were recorded by using microcomputers (Toshiba T1000) with original software written for this purpose. The retention time and duration of an odor perceived by an assessor was registered by pushing a random key on the computer keyboard from the start of the perception until the end. After releasing the key, the assessor was given 10 s in which to record either an intensity score of the perceived odor on a random scale of 1 to 9, or a descriptor of the perceived odor chosen from a menu of 14 set descriptors.

The list of odor descriptors was generated from preliminary sniffing port analyses of the PE40 sample; assessors were asked to write down a description of the odors that were perceived. Directly after each analysis the notes were discussed by the two assessors and a panel leader. A careful selection of the fixed descriptors was necessary because of the limited time available to choose a descriptor from the menu during an analysis and to keep the list from becoming too lengthy. The final list of descriptors comprised the following: musty, metallic, artificial (as in flavor), synthetic (as in plastic), glue-like, sickly, sweet, fruity, fresh, green, mushroom-like, astringent, candy-like, cocoa, and "other". A previous study of water packed in PE–aluminium test pouches identified similar sensory descriptors for the overall flavor (20). Bravo et al. (21) also reported the descriptors "fruity" and "metallic" in relation to odorous compounds isolated from PE.

Training Assessors at the Sniffing Ports.   For the purposes of training the assessors at the sniffing ports, a standard solution of 3 ppm of pentanal (Alltech Europe, Eke, Belgium) and 2 ppm each of pyridine (Merck, Darmstadt, Germany), ethylbutyrate (Merck), hexanal, styrene, α-pinene (Janssen, Beerse, Belgium), and linalole (Janssen) in pentane (Merck) was made. A 0.1-μL aliquot of the solution was adsorbed onto a Tenax medium and later thermally desorbed onto the chromatographic column for the training run. The temperature program used for the training runs comprised a 3-min hold at 40 °C, an increase to 50 °C at 2 °C/min, an increase to 80 °C at 5 °C/min, an increase to 270 °C at 10 °C/min, and a hold at 270 °C for 10 min. The temperature of the detector was maintained at 275 °C during these runs.

The training of each of the assessors involved analyzing the standard solution desorbed from a Tenax trap twice. During the training runs the assessors were asked to use the appropriate option in the computer program to record the intensity of a perceived odor on a scale of 1 (weak) to 9 (strong) so that a chromatogram could be constructed. On the second training run, the response of each assessor matched that of the FID to within acceptable limits.

Analysis of Volatiles in Packaged Mineral Water.   The two dynamic headspace samples (PE40 and glass) taken from the packaged mineral waters were analyzed in random order by each of the 10 assessors using the combined computerized GC and sniffing port analysis apparatus. To prevent assessor fatigue, the oven temperature program was optimized for good separation of the compounds in a minimum analysis time. The temperature program for these analyses comprised a 4-min hold at 60 °C, an increase to 140 °C at 2 °C/min, an increase to 250 °C at 10 °C/min, and a 5-min hold at 250 °C. All other conditions were the same as described previously.

Each assessor was engaged for one session per week, and at the end of the session a conditioned Tenax tube containing no adsorbed volatile compounds ("dummy") was analyzed to establish the signal-to-noise level associated with the experiment. After the analyses, the response chromatograms of the assessors were accumulated. At each point in time a 0 was assigned for each assessor who did not perceive an odor, and a 1 was assigned to those who did perceive an odor. The summation of these data results in a chromatogram if the number of assessors perceiving an odor is plotted as the ordinate and the time at which the odor is perceived (retention time) is plotted as the abscissa. A method similar to this was described by Acree et al. (*42*).

Studies on Film-Grade LDPE.   *Polymer Additives and Formulations.*   LDPE with density 0.92 g/mL, melt index of 20 g/min (2.16 kg, 190 °C), and melt ratio 60.0 (21.6 kg/2.16 kg, 190 °C) (*43*) was obtained commercially. The additives were erucamide (*cis*-13-docosenamide, a fatty acid amide; Croda Chemicals Limited), silica (diatomaceous SiO_2; Celite Corporation, California), and the antioxidant Irganox 1076 (Ciba Australia Limited). The additives were used as received. Two polymer formulations representing extremes in odor properties were examined. Each formulation contained 1000 ppm erucamide and 1000 ppm SiO_2. LDPE(1) contained no thermal antioxidant, whereas LDPE(2) contained 500 ppm Irganox 1076. Before pelletization, the polymers and additives were compounded in a Brabender Plasticorder at 60 rpm and 180 °C for 5 min.

Sensory Evaluation Panel and Odor Meter Data.   A panel of 8 trained persons evaluated the odor intensity of LDPE(1) and LDPE(2) samples (50 g) that were previously sealed in separate air-tight, glass jars. The sealed samples were previously aged in an air oven at 70 °C for various times up to 21 days to produce a range of odor intensities. Members of the SEP assessed three or four samples in each sitting and described the odor on a scale of 1 to 9; 5 was equivalent to a reference sample of virgin polymer, and 1 was much less than the reference.

Analyses of the polymers by using the odor meter involved placing 50 g of polymer into a glass flow-through chamber and pumping purified air through the chamber onto the surface of the SnO_2 semiconductor in the odor meter.

Low-Temperature Trapping of Volatiles.   Samples (50 g) of LDPE(1) and LDPE(2) were sealed in separate glass jars and thermally aged at 70 °C for 14 days in an air-circulating oven. Each jar was removed from the oven and cooled to 22.5 °C before removing its lid to allow the contents to equilibrate with the atmosphere. Blank samples were run under identical conditions to those used for LDPE(1) and LDPE(2).

Volatile compounds present in the headspace of each jar were concentrated at 22.5 °C onto separate Tenax-GC (Air-Met Scientific Limited, Australia) collection tubes by pumping 13 L (flow rate, 220 mL/min) of purified air through the jar via the collection tube. After the concentration step, the contents of the collection tube were carefully transferred to a 20-mL head-space vial. The volatiles were thermally desorbed at 200 °C by using a Varian-Genesis automated headspace sampler and were concentrated into a 5-mL sample loop at 10 psi. The contents of the loop were transferred to the column of a Varian 3400 GC instrument via a heated transfer line.

The GC instrument was equipped with a fused-silica column (BP-5, 0.32 mm × 60 m containing a cross-linked methylsilicone film of 1-μm thickness; Alltech Associates). The temperature program comprised a 5-min hold at 30 °C and an increase to 250 °C at 5 °C/min. The injector and transfer line were maintained at 210 °C and 275 °C, respectively. Mass spectra were recorded on a Varian Saturn-2 mass spectrometer with an ion source temperature of 250 °C and an electron multiplier voltage of 1850 eV. Helium (flow rate, 1.0 mL/min) was used as the carrier gas.

The identity of each major volatile substance trapped from the LDPE samples was confirmed by reference to a series of chromatographic standard compounds (Alltech Associates) that were also analyzed on the GC–MS system. The standards were injected onto the GC column under conditions identical to those used for the analysis of the desorbed volatiles originating from the polymers.

Results and Discussion

Odor Map.   Because a multitude of volatile compounds were isolated from oxidized PE and identified, a classification of these compounds according to their relative severity in causing odor problems would be advantageous in the study of odor. A detailed analysis of the literature dealing specifically with volatile compounds found in oxidized PE reveals that, of the 44 compounds identified by Hoff and Jacobsson (11), certain ones were observed by researchers more frequently than others. The compounds identified originally by Hoff and Jacobsson (11) can be used as a basis for the typical compounds found in oxidized PE, and an odor map can be constructed. Such a map helps to identify those compounds whose presence in the polymer must be considered to be serious when addressing the question of odor.

Figure 2 shows the frequency with which various volatile compounds in PE were reported independently in the literature plotted against the decadic logarithm of the reciprocal of the odor threshold (44) for each given compound. Compounds located in the top right-hand corner of the plot are those that were found frequently in oxidized PE and have low odor thresholds. Therefore, hexanone, butanal, heptanone, hexanal, nonanal, and ethanal must be included in the list of compounds that cause the most serious odor problems in degraded PE resins. However, in assessing the contribution of a given compound to the overall odor of the polymer, the concentration of that compound is clearly an important factor that must be taken into account.

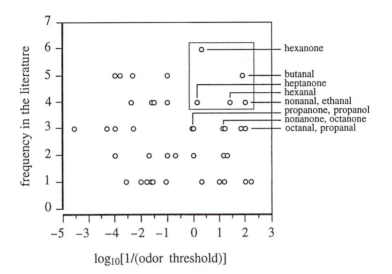

Figure 2. Frequency of independent identifications reported in the literature of various compounds in oxidized PE (11) plotted against the decadic logarithm of the reciprocal of the odor threshold (44) for the respective compounds.

Combined Computerized GC and Sniffing Port Analysis. The initial cumulative results of the sniffing port analysis of the PE40 and glass dynamic headspace samples indicated that an estimation of the signal-to-noise level was necessary. For this reason dummy samples were analyzed. Figure 3 is a plot of the percentage of the total sniffing time (ordinate) during which a given number of assessors (abscissa) perceived an odor simultaneously for the dummy, glass, and PE40 samples. When the number of assessors was less than three, the assessors simultaneously perceived an odor as frequently during the dummy run as during the sample runs. On the other hand, five or more assessors simultaneously perceiving an odor during the run of the dummy sample was never observed. However, five or more assessors did respond simultaneously to the glass and PE40 samples. In consideration of these observations, the noise level was taken to be four assessors. A noise level of greater than four assessors was considered to be above the appropriate noise level for this study. Thus, the sniffing port chromatograms indicate all possible compounds that were present and not simply those that could only be perceived by the most sensitive assessors.

Figure 4 shows the cumulative sniffing port analysis results of the 10 assessors for the PE40 and glass dynamic headspace samples run under identical experimental conditions. The noise area is shaded in each case and there are 20 peaks (numbered) above the noise level in the PE40 chromatogram. The identification of these peaks, as established by GC–MS, is detailed elsewhere (24). The "glass" chromatogram may be used to identify those peaks

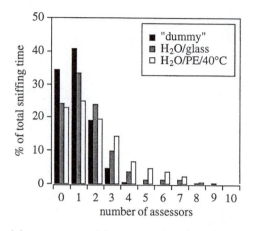

Figure 3. Plot of the percentage of the total sniffing time (ordinate) during which a given number of assessors (abscissa) perceived an odor simultaneously for the dummy, glass, and PE40 samples.

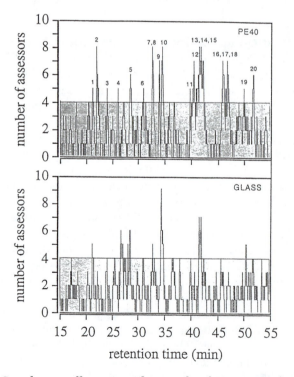

Figure 4. Cumulative sniffing port analysis results of 10 assessors for the PE40 and glass samples.

in the PE40 chromatogram that are due to the background. The peaks in the PE40 chromatogram that are not background peaks are listed in Table I along with their identities. Many of the compounds responsible for these peaks are degradation products originating from the PE lining of the mineral-water package or additives present in the PE. Of these compounds, it is particularly interesting to note the presence of hexanal (peak 2), whose position on the odor map (Figure 2) indicates that it is one of the more serious odor-causing compounds found in oxidized PE. Octanal (peak 9) is another compound of this type and is also indicated on the odor map (Figure 2). The most frequently assigned olfactory descriptors assigned during the sniffing port analyses of the odorous volatiles isolated from the PE40 sample were musty and metallic.

Even though the concentration of the trapped volatile compounds was enhanced by incubating the packed mineral water at 40 °C and by purging and collecting volatiles from a large (2-L) sample of the water, none of the compounds found was perceived by all 10 of the assessors. This finding suggests that the concentration of each volatile must have been around its threshold value. Previous sensory analyses showed that mineral water packed in PE-lined aluminium–cardboard packages and incubated at 40 °C developed an off-flavor (24). Therefore, a mixture of the odor-active compounds probably is responsible for this off-flavor; compounds of a similar chemical nature usually impart an off-flavor that is additive (45).

Studies on Film-Grade LDPE. *Odor Panel and Odor Meter Assessment of LDPE.* A total of 87 samples of LDPE, having odor characteristics ranging from low to high, were produced by thermally aging LDPE(1) and LDPE(2) formulations to different extents. These samples were presented to the trained SEP for classification and were also tested by using an odor meter. Figure 5 is a plot of the average odor panel scores versus the corresponding odor meter response for the samples. The correlation coefficient for the data is 0.605, which is statistically significant at the 0.0005 con-

Table I. Most-Probable Identities of Some Non-Background Peaks in the PE40 Chromatograms

Peak	Assigned Compound
2	Hexanal
3	4-Hydroxy-4-methylpentan-2-one
6	n-Propylbenzene
9	Octanal
11	Not assigned
12	Branched alkane(s), C_4 alkylbenzenes
15–18	Branched carbonyls
20	Not assigned

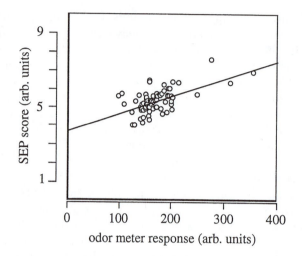

Figure 5. Plot of the average odor panel scores versus the corresponding odor meter response for LDPE samples having different odor intensities.

fidence level (r_{crit} = 0.321) and indicates that the odor meter response varies linearly with that of the odor panel. The results suggest that the SEP can be effectively replaced by the odor meter for the evaluation of the level of odor in these LDPE resins.

One explanation of the observed response of an SnO_2 semiconductor to an odoriphore was given by Fukui (*31*) and is summarized schematically in Figure 6a. A schematic diagram of the SnO_2 device showing details of its construction is given in Figure 6b. During the operation of the semiconductor, atmospheric oxygen is in equilibrium with adsorbed oxygen on the surface of the SnO_2, which is maintained at 300–400 °C. Electrons from the conduction band of the SnO_2 reduce adsorbed oxygen to produce a variety of reduced oxygen species on the surface. The reaction of an oxidizable gas at the surface releases electrons back into the conduction band of the semiconductor, thereby decreasing the semiconductor's resistance. The resulting change in resistance of the semiconductor is detected in the usual way by a Wheatstone bridge circuit.

Analysis of Volatiles in Film-Grade LDPE. The results of SEP and odor meter tests conducted on thermally aged LDPE(1) and LDPE(2) showed that each of these formulations is inherently more odorous than a reference sample containing no erucamide, silica, or thermal antioxidant (*46*). This finding suggests that the odorous nature of LDPE(1) and LDPE(2) must in some way be caused by the anti-block or slip additives present in these formulations.

In an attempt to explain the differences in the perceived odor intensities of the resins, samples of LDPE(1), LDPE(2) and the reference resin were

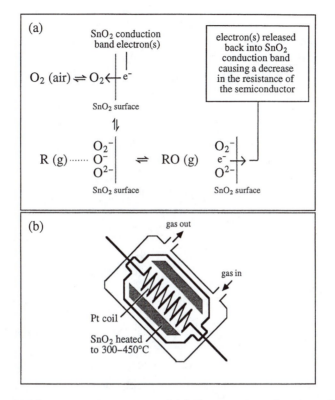

Figure 6. Schematic representations of (a) the operation of a sintered SnO₂ semiconductor device and its response to an odoriphore and (b) the SnO₂ device showing details of its construction.

aged for 18 days at 70 °C in an air-circulating oven to enhance the concentrations of volatile degradation products. A thermally degraded sample of erucamide was also produced for comparison by aging a mixture of silica and fresh erucamide in air for 10 min at 200 °C. Headspace samples of all the degraded materials were subsequently analyzed by GC–MS. Figure 7a shows synchronous portions of the chromatograms obtained during the analysis of volatiles trapped at ambient temperature from the thermally aged LDPE resins. Noticeable differences among the chromatograms were found only within the retention time interval shown. Figure 7b is a chromatogram of the thermally degraded erucamide sample that was obtained over the same retention time interval as the one selected for Figure 7a.

Table II shows the most probable identities, as determined by MS, of the peaks shown in Figure 7. Table II also indicates which of the assigned compounds are present in appreciable amounts in each of the LDPE resins and the degraded erucamide.

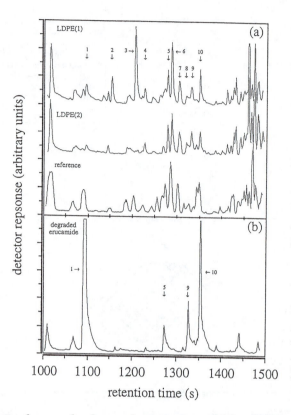

Figure 7. Synchronous headspace chromatograms of (a) volatiles trapped at ambient temperature from the LDPE(1), LDPE(2), and reference resins after 18 days thermal aging at 70 °C; and (b) erucamide–silica mixture that was thermally degraded at 200 °C for 10 min in air. Detector response for the chromatogram in (b) is 25 times that for the chromatograms in (a).

The chromatographic evidence currently available suggests that odorous compounds such as hexanal, heptan-2-one, and heptanal are produced by the oxidative degradation of the base LDPE because these compounds are present in the degraded reference resin. These odorous species are also major degradation products of erucamide, and so the incorporation of erucamide in LDPE may increase the level of odor in the formulation. The odorous ketone heptan-3-one is also present in each of the degraded LDPE resins. A very small amount of this compound can be detected in the degraded erucamide sample.

The presence of C_8 and C_9 branched alkanes (peak 4) was detected in each of the degraded LDPE resins, but the precise identities of the compounds could not be determined with a high degree of certainty. These compounds were not detected in the chromatogram of degraded erucamide, but a minute peak does appear in the chromatogram of this sample at a retention

Table II. Most-Probable Identities of Peaks Shown in Figure 7 as Determined by MS

Peak No.	Assigned Compound	Degraded Erucamide	Reference	LDPE(1)	LDPE(2)
1	Hexanal	+	+	+	+
2	2-Methyl-2-pentenal	−	−	+	−
3	2-Hexenal	−	+	+	−
4	C_8, C_9 branched alkanes	−	+	+	+
5	C_8 cyclic hydrocarbons (cyclooctane)	+	+	+	+
6	System peak	−	+	+	+
7	System peak	−	+	+	+
8	Heptan-3-one	−	+	+	+
9	Heptan-2-one	+	+	+	+
10	Heptanal	+	+	+	+

NOTE: + means compound was present in an appreciable amount; − means compound was not present in an appreciable amount.

time slightly greater than that of the C_8 and C_9 branched alkanes. MS identifies the peak in question as being due to 4-methyl-3-pentenal.

The presence of what appears to be C_8 cyclic hydrocarbons such as cyclooctane (peak 5) can be detected in each of the degraded LDPE resins as well as in the degraded erucamide sample. Such aliphatic compounds probably will not contribute greatly to the perceived odor intensities of the resins.

A much more significant contribution to the level of odor intensity may be attributable to the presence of 2-hexenal and its structural isomer 2-methyl-2-pentenal. Peaks due to these compounds feature strongly in the LDPE(1) chromatogram. 2-Hexenal is also present in the degraded reference resin; therefore, this odorous compound might be produced directly from the oxidation of the base LDPE. However, peaks due to 2-hexenal and 2-methyl-2-pentenal are not found in the chromatogram of degraded erucamide, and only a very slight amount of the structural isomer 4-methyl-3-pentenal can be found in this chromatogram. The evidence suggests that the 2-hexenal and 2-methyl-2-pentenal found in LDPE(1) do not originate from the degradation of the erucamide but are instead produced from the base LDPE. A possible explanation for the increased levels of these aldehydes in LDPE(1) as compared with the reference material is that their formation in LDPE(1) in some way may be catalyzed by the silica in this formulation.

The lower odor intensity of LDPE(2) compared with LDPE(1) is due to the presence of the thermal antioxidant in the LDPE(2) formulation, and this antioxidant suppresses the formation of degradation products, many of which have odorous properties.

Conclusions

Combined computerized GC and sniffing port analysis is a technique that can be applied successfully to the analysis of volatile compounds in mineral water packed in PE-lined aluminium–cardboard containers. The technique enables odor-active compounds that migrate from the package into the mineral water to be isolated and descriptors of these to be obtained. Hexanal and octanal are among the compounds that were detected, and both of these are serious odor-causing compounds, as revealed by the odor map.

A statistically significant, linear relationship exists between the response of the SnO_2 semiconductor device of the odor meter and the SEP response. This relationship suggests that the odor meter can be implemented effectively to replace the SEP for odor testing of LDPE resins.

The trapping of odor-active volatiles at ambient temperature can be effectively achieved by using Tenax-GC collection tubes. Compounds that can be isolated from LDPE under these conditions include C_6 unsaturated aldehydes such as 2-hexenal and 2-methyl-2-pentenal. These species probably will have similar odor characteristics to hexanal, which is shown by the odor map to be one of the more serious odor-causing compounds in oxidized PE. The evidence currently available suggests that C_6 unsaturated aldehydes may be produced in LDPE as a result of undesirable reactions involving the polymer, antiblock, or slip additives. Work is currently in progress to establish more clearly the nature of these interactions.

Safety Considerations

All odorous materials used in this work for sensory panel evaluations and computerized GC sniffing port analyses are present at trace levels that are considered to lie below the limits that are injurious to the assessor's health. The reader is advised, however, to consult relevant safety information on these materials if attempting to reproduce the experiments.

Acknowledgments

S. W. Bigger and M. J. O'Connor are grateful to Kemcor Australia Limited for financial support for this project and the Royal Australian Chemical Institute and Secomb Conference and Travel Fund for providing financial assistance to M. J. O'Connor to attend the 206th ACS National Meeting. S. W. Bigger would like to thank Elizabeth Bigger for her continued devotion and support.

References

1. *Encyclopedia of Packaging Technology*; Wiley-Interscience: New York, 1984; p 523.
2. Peled, R.; Mannheim, C. H. *Mod. Packag.* **1977,** 50, 45.
3. Bravo, A.; Hotchkiss, J. H. *J. Appl. Polym. Sci.* **1993,** 47, 1741.

4. Caul, J. F. *Proceedings of the Odor and Packaging Conference*; London, Nov., 1960, pp 39–49.
5. Bailey, W.; Liotta, C. *Polym. Prepr. (Am. Chem. Soc. Div. Polym. Chem.)* **1964,** *5*, 333.
6. Tsuchiya, Y.; Sumi, K. *J. Polym. Sci., Part A1: Polym. Chem.* **1968,** *6*, 415.
7. Kiermeier, F.; Stroh, A. Z. *Lebensm. Unters.* **1969,** *141*, 208.
8. Spore, R. L.; Bethea, R. M. *Ind. Eng. Chem. Prod. Res. Dev.* **1972,** *11*, 36.
9. Barabas, K.; Iring, M.; Kelen, T.; Tudos, F. *J. Polym. Sci.* **1976,** *57*, 65.
10. Vom Bruck, C. G.; Rudolf, F. B.; Figge, K.; Eckert, W. R. *Food. Cosmet. Toxicol.* **1979,** *17*, 153.
11. Hoff, A.; Jacobsson, S. *J. Appl. Polym. Sci.* **1981,** *26*, 3409.
12. Hoff, A.; Jacobsson, S.; Pfaffi, P.; Zitting, A.; Frostling, H. *Scand. J. Work Environ. Health* **1982,** *8* (*Suppl.* 2), 60.
13. Fales, N. J.; Stover, L. C.; Lamm, J. D. *Polym. Plast. Technol. Eng.* **1983,** *21*, 111.
14. Kozinowski, J.; Piringer, O. *Dtsch. Lebensm. Rundsch.* **1983,** *79*, 6.
15. Chuang, C. R. *Soc. Plast. Eng. J.* **1984,** 246.
16. Anselme, C.; N'Guyen, K.; Bruchet, A.; Mallevialle, J. *Environ. Technol. Lett.* **1985,** *6*, 477.
17. McGorrin, R. J.; Pofahl, T. R.; Croasmun, W. R. *Anal. Chem.* **1987,** *59*, 1109a.
18. Brodie, V., III *Tappi J.* **1988,** *Dec.*, 160.
19. Solin, F. C. F.; Fazey, A. C.; McAllister, P. R. *J. Vinyl Technol.* **1988,** *10*, 30.
20. Linssen, J. P. H.; Janssens, J. L. G. M.; Roozen, J. P.; Posthumus, M. *J. Plast. Film Sheeting* **1991,** *7*, 294.
21. Bravo, A.; Hotchkiss, J. H.; Acree, T. E. *J. Agric. Food Chem.* **1992,** *40*, 1881.
22. Dalbey, W. E.; Bynum, L. M.; Mooney, J. K.; Pulkowski, C. H. *ANTEC Proc.* **1992,** 92.
23. Marin, A. B.; Acree, T. E.; Hotchkiss, J. H.; Nagy, S. *J. Agric. Food Chem.* **1992,** *40*, 650.
24. Linssen, J. P. H.; Janssens, J. L. G. M.; Roozen, J. P.; Posthumus, M. A. *Food Chem.* **1993,** *46*, 367.
25. Berg, N. *Proceedings of the 3rd International Symposium on Migration;* Unilever: Hamburg, Germany, 1980; p 266.
26. Bojkow, E. *Osterreichische Milchwirtschaft* **1992,** *Sept.*, 37.
27. Anselme, A.; Suffet, J. H.; Mallevialle, J. *J. Am. Water Works Assoc.* **1988,** *Oct.*, 45.
28. Culter, J. D. *Pack Alimentaire '91*, New Orleans, April 1991, p 23.
29. Nielson, R. C. *J. Liq. Chromatogr.* **1991,** *14*, 503.
30. *Manual on Sensory Testing Methods*; ASTM, 1968, Method STP434.
31. Fukui, K. *J. Odor Res. Eng.* **1989,** *20*, 1.
32. ASTM Method D4526.
33. Koszinowski, J.; Piringer, O. *J. Plast. Film Sheeting* **1986,** *2*, 40.
34. Wilks, R. A.; Gilbert, S. G. *Food Technol.* **1969,** *23*, 47.
35. Culter, J. D. *J. Plast. Film Sheeting* **1992,** *8*, 208.
36. Fernandes, M. H.; Gilbert, S. G.; Paik, S. W.; Stier, E. F. *J. Food Sci.* **1986,** *51*, 722.
37. Rojas-DeGante, C.; Pascat, B. *Packag. Technol. Sci.* **1990,** *3*, 97.
38. Holmstrom, A.; Sorvic, E. M. *J. Polym. Sci. Part A: Polym. Chem.* **1978,** *16*, 2555.
39. Acree, T. E.; Butts, R. M.; Nelson, R. R.; Lee, C. Y. *Anal. Chem.* **1976,** *48*, 1821.
40. Meilgaard, M.; Civille, G. V.; Carr, B. T. *Sensory Evaluation Techniques*; CRC Press: London, 1991.
41. Sensory Evaluation Division, Institute of Food Technology, *Food Technol.* **1981,** *35*, 11.
42. Acree, T. E.; Barnard, J. B.; Cunningham, D. G. *Food Chem.* **1984,** *14*, 273.

43. ASTM Method D1277-79.
44. Verschueren, K. *Handbook of Environmental Data on Organic Chemicals;* Van Nostrand Reinhold: New York, 1983.
45. Guadagni, D. G.; Buttcry, R. S.; Okano, S.; Burr, H. K. *Nature (London)* **1963,** *28,* 1288.
46. O'Connor, M. J.; Bigger, S. W.; Scheirs, J.; Janssens, J. L. G. M.; Linssen, J. P. H.; Legger-Huysman, A. *Polym. Prepr. Am. Chem. Soc. Div. Polym. Chem.* **1993,** *34(2),* 247.

RECEIVED for review January 26, 1994. ACCEPTED revised manuscript October 10, 1994.

STABILIZATION

The use of stabilizer additives and other approaches for increasing the useful lifetime of polymeric materials are described in this section. Chapters 18–22 are concerned primarily with additives and protective coatings used for applications requiring photostabilization. Included in this group are a review of the complex action modes of the amine class of stabilizers and also a chapter discussing the photostability of UV screener type additives. Another chapter describes progress in development of protective coatings for polymeric materials used in UV environments. Chapters 23–26 are about stabilizers used for melt processing and for long-term applications that may involve thermal exposure. Several of these studies discuss the mechanisms and utility of phosphite, phosphonite, and related costabilizers used in conjunction with hindered phenols. One chapter describes the nature and characterization of reaction products arising from stabilizer additives in polymers. Chapters 27–30 cover the problem of migration and retention of stabilizer additives in polymer matrices. These chapters highlight, with respect to stabilizer efficiency, the implications of additive compatibility, diffusion rates, and volatility. Innovative approaches to the problem of stabilizer loss are described. The section ends with three chapters (31–33) that are concerned with compositional effects on stability, including the influence of copolymers, blends, and fillers.

18

Activity Mechanisms of Amines in Polymer Stabilization

Jan Pospíšil

Institute of Macromolecular Chemistry, Academy of Sciences of the Czech Republic, 16206 Prague, Czech Republic

Aromatic and heterocyclic amines have a crucial importance for stabilization of rubbers, plastics, and coatings. A detailed examination of their chemistry revealed processes accounting for antioxidant, antifatigue, and photostabilizing activities. Aspects of chemical transformations, characteristic for the generation of radical intermediates, and products from amines under conditions of polymer degradation were exploited to explain the scavenging activities of R· and ROO· as the key stabilization pathway. The data are interpreted in a way that is applicable to developing new ideas and systems for effective polymer stabilization.

TRENDS IN THE STABILIZATION OF CONVENTIONAL POLYMERS in the 1990s have responded to several problems: prevention of degradation caused by residues of polymerization catalysts of new generations remaining in the polymer bulk; expansion of technologies exploiting more drastic processing conditions; broader use of macromolecular systems filled with inorganics; use of polymer products in environments with increased chemical and physical aggressiveness in the household, machinery, energetics, automotive, or agricultural industries; and application of recycled polymer waste. Research and development has been aimed at innovative technologies for processing and long-term stabilization of plastics, rubbers, and coatings. Data obtained in studies of influences of structural factors governing the "inherent chemical efficiency" of stabilizers have been reinterpreted by using up-to-date knowledge (1–3).

An elucidation of physical factors governing stabilizer efficiency (4) accounts for a new insight into the necessary conditions for efficient polymer stabilization. Adjusting the molecular architecture of stabilizers to maximize

0065–2393/96/0249–0271$12.00/0

their physical persistency and compatibility with the polymeric matrix is of a top importance (5). The mechanistic knowledge has been exploited profitably to meet increasing demands of polymer end-use customers and environmental rules.

This chapter concerns stabilizers containing an amino group as a functional moiety assuring the inherent chemical efficiency. Two important chemical types are involved:

- aromatic and heterocyclic amines ranking among the classical commercialized antidegradants applied mainly in rubbers as antioxidants, antifatigue agents, or antiozonants
- cyclic hindered amines introduced in the market in the 1970s and ranking among the most extensively studied additives applicable as light stabilizers for most polymers and light and heat stabilizers in polyolefins

New features based on scavenging radicals $R^{\cdot}$ and $ROO^{\cdot}$ by amines contribute to the development trends in stabilizer chemistry.

Active Application Sites of Amines in Radical Degradation of Polymers

Most mechanisms for the degradation of polymers during processing, storage, and long-term use include free radicals. These radicals are involved in chain initiation, propagation, branching, and termination steps of thermally, catalytically, mechanochemically, or radiation-induced processes. Two kinds of free radicals are of key importance: carbon-centered macroalkyl $R^{\cdot}$; and oxygen-centered alkylperoxyl $ROO^{\cdot}$, alkoxyl $RO^{\cdot}$, and acylperoxyl $RC(O)OO^{\cdot}$. $R^{\cdot}$ and $ROO^{\cdot}$ can be scavenged by chain-breaking antioxidants. An electron-acceptor process fits to scavenging of $R^{\cdot}$. The electron donors like aromatic amines scavenge $ROO^{\cdot}$ (6). Some amine transformation products are able to scavenge both $R^{\cdot}$ and $ROO^{\cdot}$ (3).

Hydroperoxides and peroxyacids rank among the most dangerous species formed in polymers in an aerobic environment. Generally, hydroperoxide decomposing antioxidants, light stabilizers, or metal deactivators are applied to prevent peroxide homolysis (7).

The macroalkyl $R^{\cdot}$ radicals are formed in the initiation and chain-transfer steps by breaking C–H or C–C bonds. $R^{\cdot}$ radicals are the main species formed during polyolefin processing and in solid polymers treated by γ-radiation in an oxygen-deficient atmosphere. The reactivity of $R^{\cdot}$ is influenced by the microenvironment (8, 9): Recombination and addition to C=C double bonds, disproportionation, and fragmentation proceed in oxygen-deficient systems. According to the chemical character of the degrading polymer, cross-linking or chain scission changes physical properties of the polymer. In contact with oxygen, R˙ radicals are converted into ROO˙, and the chain oxidation governs polymer degradation. Therefore, scavenging of the primarily formed R˙ radicals, immediately after their formation (Scheme I), is of top importance for polymer stabilization.

None of the aminic stabilizers in the original chemical form is able to scavenge R˙. The sacrificial fates of amines due to scavenging ROO˙ and reactivity with hydroperoxides result in the formation of compounds with a site capable of scavenging R˙ (*2, 3, 10, 11*). This result implies that R˙ can be scavenged only by amine transformation products.

Aromatic and Heterocyclic Amines

The secondary amino group in α-position to the aromatic nucleus is the key functionality. This amino group assures the inherent chemical efficiency of classical antidegradants having structures of substituted diphenylamines (DPA), *N*-phenyl-1(or 2)-naphthylamines (PNA), *N,N'*-disubstituted 1,4-phenylenediamines (PD), 6-substituted or oligomeric 2,2,4-trimethyl-1,2-dihydroquinolines (DHQ), 2,2-disubstituted 1,2-dihydro-3-oxo(or phenylimino)-3*H*-indoles (DHI), or substituted phenothiazines (PT). Only these types of amines are considered in this section. Other nitrogen-containing heterocyclic compounds possess antioxidants properties as well.

The activity spectra of DPA, PNA, PD, DHQ, and DHI are broad. Their antioxidant, antifatigue-agent, or antiozonant efficiencies include reactivities of derived free radicals and quinoimines (*2, 3, 10, 12*). Even though the chemistry of transformation products is complicated, data dealing with reactivities of isolated and identified principal products create a base for understanding the role of amines in polymer stabilization.

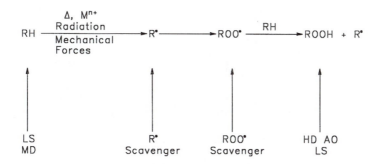

Scheme I. Sites of effective stabilizer application in polymer degradation. LS is light stabilizer; MD is metal deactivator; HD AO is hydroperoxide decomposing antioxidant.

The structure of amines affects their properties. Monoamines like DPA, PNA, and DHQ are antioxidants having more or less apparent antifatigue-agent activity. PDs are antioxidants, antifatigue agents, and mild-to-strong anti-ozonants. Sometimes, PDs are considered to be weak metal deactivators. Blends of amines account for a broadened activity spectrum due to the complementary activities of various transformation products. Discoloration and staining observed in aromatic amine-doped polymers are definitely disadvantages limiting the commercial application. Although colored products invariably arise in the service transformation of amines [i.e., processes contributing to the stabilization effect (13)], their formation is not directly correlated to the integral efficiency. Stabilizers with a reduced tendency to discoloration were synthesized (10), and most possess an improved physical persistence. For example, compounds like 4,4'-bis(α,α-dimethylbenzyl)diphenylamine are applicable in homosynergistic combinations with hindered phenols for the stabilization of polyolefins (12).

Secondary aromatic and heterocyclic amines (1, R = aryl, aminosubstituted aryl, a part of a heterocyclic moiety; see Scheme II) form the short-lived aminyl, 2a (having properties of a weak oxidation agent), in the primary step of the chain-breaking antioxidant mechanism characteristic for the H-transfer to ROO˙ (2, 10, 12). The formation of transition states characteristic for a loose π-complex or states with partial or complete charge separation is considered to precede formation of aminyls. The mesomeric forms of 2a having the character of carbon-centered radicals (2b and 2c) augment the variety of coupling products formed (2).

The aromatic aminyls, >N˙, are the primary radical species involved in the amine-stabilizing mechanism. They are able to scavenge R˙ and ROO˙ radicals (Scheme III).

Scavenging R˙ can also contribute to the regeneration of the >NH function through the thermal decomposition of the alkylated amines >NR (10, 12,

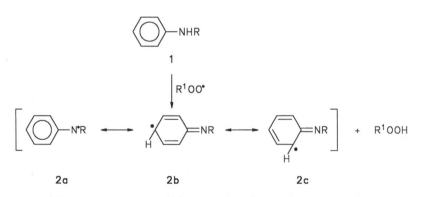

Scheme II. Formation of aminyl radical from secondary aromatic amines.

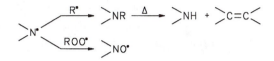

Scheme III. Reactivity of aromatic aminyls (2a) with radicals R˙ and ROO˙.

14). Aminyls such as **2a** in the DPA, DHQ, or DHI series react with ROO˙ to form nitroxides >NO˙ (e.g., **3–5**) or the corresponding nitrone when the nitroxide bears a hydrogen atom in the α-position to the nitroxide function (*2, 12, 15–17*).

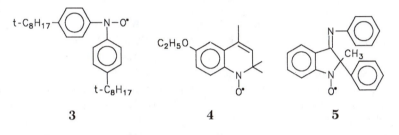

3　　　　　　　　　　**4**　　　　　　　　　　**5**

N,N'-Disubstituted 1,4-benzoquinodiimine (BQDI, **6**) is formed directly as the major product via aminyls in the PD series (*2, 10*). The respective bis(nitroxides) theoretically arise from PD and ROO˙, are unstable and react in their mesomeric bis(nitrone) forms (e.g., **7**). No electron spin resonance (ESR) signals of >NO˙ were detected in the oxidized PD (*18*). However, the generation of nitroxide signals cannot be excluded in the presence of hydrogen donors. The conjugated bis(nitrones) like **7** may retract their hydrogen atoms and form mononitroxides.

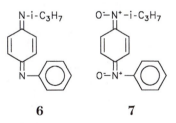

6　　　　　　　　　**7**

The delocalization of free electrons in **2a** accounts for the mesomeric forms **2b** and **2c** and controls the structure of products formed after trapping R˙. The recombination of **2b** and **2c** with R˙ irreversibly yields alkylates. The process was proved by using low-molecular weight R˙ (*12, 14*). Structure **8** is a typical alkylate (*10*). The highest spin densities in the DHQ series are in the 6- and 8-positions (*19*). Consequently, an alkylate, **9**, is formed (*14*). The formation of polymer-bound species after scavenging of macroalkyls may be extrapolated from model experiments.

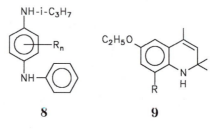

8   **9**

Oxidation is a consequence of ROO˙ trapping by C-centered radicals such as **2b** and **2c**. Most products are colored (*3, 12, 17*). N-Phenyl-1,4-benzo-quino-4-imine (**10**) is formed from unsubstituted DPA. 1,2-Benzoquinoimines such as structures **11** or **12** and more complicated quinoimines derived from the coupling products of aminyls were formed from DPA substituted in the 4,4'-positions with tertiary alkyls or aralkyls (*20*). The analogous quinoimines are formed in the PNA series. In the DHQ series, quinoimines **13** and **14** are formed from the aminyl; quinoimine **15** arises from the C–N coupling product of **14** (*17*). Quinoimines are also formed from substituted phenothiazines.

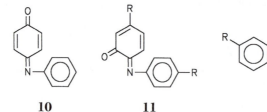

10   **11**   **12**

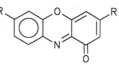

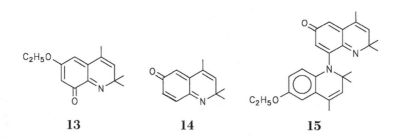

13   **14**   **15**

The nitroxides, >NO˙, represent an important radical intermediate in the stabilization mechanism of amines (*2, 10, 12*). Some mechanistic conclusions obtained in the DPA series were extrapolated to other amines. Efficiency of the participation of >NO˙ in scavenging R˙ and ROO˙ is determined by the chemical structure of >NO˙ and the tendency to participate in competitive reactions. Stability of >NO˙ drops from hindered amine stabilizers (HAS) to DHQ or DHI and, more explicitly, to the DPA series (*12*). Mechanistic features dealing with HAS are described in the next section.

The formation of mesomeric forms influences the reactivity of aromatic >NO˙. This influence is exemplified in Scheme IV for the DPA series.

A stepwise oxidation of a C-radical form of >NO˙ results in various compounds, including **16**. Benzoquinone and nitroso- and nitrobenzene are generated in ultimate phases of the lifetime of diphenylnitroxide (*16*).

Disproportionation observed with aromatic and heterocyclic >NO˙ radicals contributes to their instability and diminishes the extent of the active contribution of >NO˙ to the radical scavenging mechanism. Among other compounds, the original amine and compounds like *N*-phenyl-1,4-benzoquino-monoimine-*N*-oxide (**16**) or 2,2,4-trimethyl-6-quinolone-*N*-oxide (**17**) were formed by disproportionation in the DPA and DHQ series (*2, 16, 17, 21*).

Thermolysis may also contribute to the transformations of >NO˙. This contribution is demonstrated by the formation of substituted 2-phenyl-3*H*-indol-3-one-*N*-oxide (**18**) from >NO˙ in the DHI series (*22*).

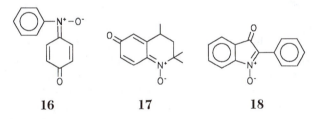

16	**17**	**18**

The ability of aromatic and heterocyclic >NO˙ to trap R˙ under the formation of the respective *O*-alkylhydroxylamine >NOR accounts for anti-fatigue-agent effects in the oxygen-deficient environment. The process was proved by using model alkyls in the DPA and DHQ series (*10, 14, 22*). Compounds such as **19** or **20** are formed. In the case of reaction with macroalkyls, polymer-bound analogues of **19** or **20** are formed.

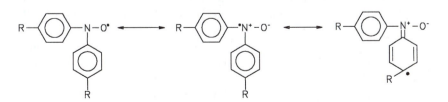

Scheme IV. Forms of nitroxide in diphenylamine series.

Aromatic >NOR groups are thermolyzed at >100 °C to hydroxylamine >NOH groups (16). These hydroxylamines react with ROO·, and the original >NO· groups may be regenerated. Therefore, aromatic >NOH are considered to be strong chain-breaking antioxidants.

Data obtained in model studies with diphenylnitroxide reveal its ability to recombine with substituted cyclohexadienonyl or 4-hydroxybenzyl radicals (i.e., the species generated from hindered phenolic antioxidant), thiyl, and sulfinyl radicals (2, 23). This result indicates a variety of >NO· reactions in the stabilized polymer matrix.

The ESR signals of >NO· observed in aged polymers doped with aromatic amine may arise from compounds other than 3–5 (i.e., "primary" >NO·). The "secondary" nitroxides may arise after R· trapping by nitrones (21). This kind of secondary >NO· (e.g., 21) is responsible for ESR signals observed in flex-cracked rubbers doped with PD (10).

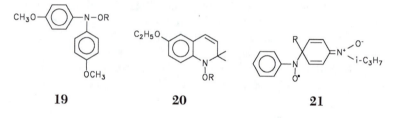

19   **20**   **21**

Regardless of the oxidation agent used (O₂, ROO·, ROOH, O₃, or ozonide), BQDI derivatives such as structure 6 are formed from antidegradants having the structure of PD (12). Wurster's cation radicals (e.g., 22 and its mesomeric forms) are expected to be formed as intermediates. BQDI derivatives retard oxidation of hydrocarbon polymers and shorten the scorch time of rubbers. Because of a strong absorption in the visible part of light between 440 and 480 nm, BQDI and some quinoide products of BQDI hydrolysis, substitution, cyclization, or disproportionation cause discoloration of rubbers.

The structural characteristics of N,N'-substituents affect the reactivity of BQDI (2, 10). Secondary alkyls C₃–C₈ present in one of the amino groups in BQDI are easily hydrolyzed. Weak organic acids and acid fillers catalyze the process yielding N-phenylbenzoquinomonoimine (BQMI, 10), which is a relatively stable compound having good oxidation power. The reaction mixture contains reduced forms of BQDI and BQMI such as structures 8 (n = 0) and 23. BQDI derivatives substituted with two N-sec-alkyl groups are hydrolyzed in two steps, and benzoquinone is the final product.

N,N'-Diphenylbenzoquinodiimine is deaminated very slowly (2, 10). The reaction product contains compounds characteristic of a quinoide moiety substituted with phenylimino groups. Compounds like azophenin, 24, are formed. The cyclization of the phenylimino-substituted quinoimines provides colored heterocyclic compounds like 1,8-diphenylfluorindine, 25 (2, 10).

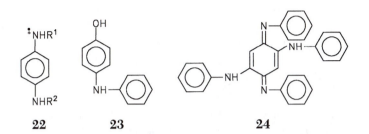

22   **23**   **24**

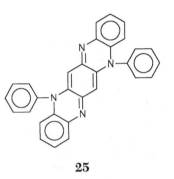

25

BQMI (**10**) formed among the transformation products of antidegradants of the DPA and PD series and BQDI (**6**) arising in the PD series are very reactive species. These species contribute to the complexity of chemical processes taking part in the aging of unsaturated vulcanized rubbers (*3, 12*). The partial regeneration of the antioxidant-active PD or 4-hydroxyldiphenylamine species (a part of which is in a polymer-bound form) was observed in the reactions of quinoimines with monohydric and dihydric phenols, alkylmercaptans, olefins (including olefinic moieties in diene-based rubbers), and alkyl radicals (*2, 10*).

The redox pairs of quinoimines with the respective reduced aminic forms enhance reactivity with R·. Stable carbon- (R· is linked to the arylene ring, e.g., **8**; *n* = 1,2) and oxygen-alkylates are formed from BQDI and BQMI, respectively, as a consequence of R· scavenging.

N-Alkylates may be thermolyzed and the parent amines regenerated. The alkyl is split off in the form of an olefin (*10, 12*). The R· scavenging ability of quinoimines represents a substantial contribution of *sec*-aromatic amines to the polymer stabilization mechanism. The process may also be extrapolated for quinoimines derived from various model amines.

A high efficiency of aromatic amines in rubbers and R·/ROO· scavenging capacities of some products of amine transformations encourage further use of amines in the stabilization of polyolefins for applications where discoloration and contact staining are not a limiting problem and in the stabilization of vinylic monomers (*12*). As a consequence, combinations of aromatic amines with phenols or quinones are of potential importance.

Hindered Aliphatic Amines

HAS rank among the most fortunate developments in additives for plastics. The protection against actinic–solar and high-energy radiations is their main application (*11*). According to this activity, they have been classified in the literature as hindered amine light stabilizers (HALS). New findings based on the proper selection of testing conditions revealed (*24*) an outstanding efficiency of HAS in the temperature range below 130 °C as the long-term heat stabilizers. This property was exploited in new, efficient, stabilizing formulations for polyolefins characterized by a reduced content of chain-breaking antioxidants. The interpretation of stability of the HAS-doped plastics in natural and accelerated weathering and oven tests revealed that combinations of oligomeric HAS (mainly structure **26**) with HAS having molecular weight lower than 1000 provide a synergistic protection in the light- and heat-induced processes (*11, 25*). Physical factors probably cooperate with the inherent chemical efficiency in these HAS combinations (*12*). The importance of HAS in the commercial stabilization of polyolefins is enhanced by their ecological acceptability. Theoretical interest in deciphering their actual activity mechanism is, therefore, enormous.

The commercialized HAS have structures of substituted piperidines. Derivatives of piperazinone are rarely used. HAS functional moieties are represented by secondary (>NH, e.g., **26**) or tertiary amino groups (>NR where R is mostly methyl, e.g., **27**; other alkyls or aralkyls are involved in developmental HAS) and by *O*-alkylhydroxylamine moieties (>NOR where R is isooctyl, e.g., **28**). The molecular architecture of some HAS is rather complicated. Hindered amine moieties form parts of polymer-bound, oligomeric, and bifunctional systems (*5, 11*).

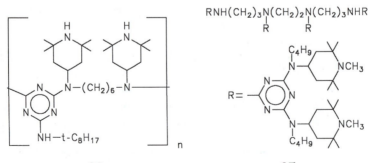

26   **27**

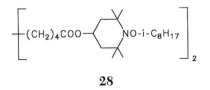

28

The primary steps in the stabilizing mechanism of HAS were interpreted in detail (*11, 26*). The activity of HAS is unequivocally based on in situ formation of the respective nitroxides, >NO·, which generally have structures **29** and **30**, respectively (X represents various substituents, including bridges containing additional HAS or other stabilizing functions). Nitroxides are formed as a consequence of the hydroperoxide-depleting pathway due to the reactivity of HAS with ROOH or ROO· (*11*) or the peracid-depleting pathway due to the reactivity with peracyl radicals or peracids formed in oxidized polymers (*27*). Recent interpretations of results obtained with polyolefins (*11*) demonstrate that charge–transfer complexes (e.g., HAS and O_2, polymer and O_2, or ROO· and HAS) play an important role in the early stages of the stabilization mechanism.

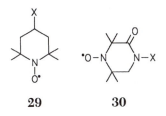

29   **30**

Nitroxides are formed directly by oxidation of secondary HAS. Tertiary HAS (>NCH$_3$) are transformed in the first step into salts of secondary amines and organic acids [>NH$_2$ $^-$OC(O)R]. Oxygen and RO· and ROO· participate in this process (*11*). Nitroxides are formed from salts in the subsequent step. This mechanism explains an equivalency in the stabilizing efficiency of >NH and >NCH$_3$ observed in commercial polyolefins.

Species >NO· derived from HAS rank among the very stable free radicals. The lack of any mesomeric form accounts for the principal difference between them and aromatic >NO· and enhances the importance of R·-scavenging efficiency in the stabilizing mechanism of HAS. O-Alkylhydroxylamines >NOR are formed as a consequence (Scheme V). A process proceeding without >NOR formation has increasing importance in the heat stabilization of polyolefins at >100 °C (*11*).

The recombination of R· with >NO· is affected by polarity of the environment (*28*). The reactions characteristic for R· in oxygen-deficient systems—

disproportionation, recombination, fragmentation, and addition to olefinic bonds (8, 9)—compete with the process in Scheme V.

In an aerobic environment, the rapid reaction of R· with oxygen accounts for ROO· formation. The relative rates of oxidation, recombination, and >NO· scavenging of various R· (pentyl, *tert*-butyl, and benzyl) were compared (28). The competitive processes may effectively limit the HAS efficiency formulated by Scheme V in terms of the R·-scavenging mechanism. Any quantitative rating of the importance of the individual competitive reactions of R· in photo- or thermooxidized solid polyolefins is still lacking.

HAS were used effectively for the stabilization of polyolefins having ethylene or propylene construction units. This success is reflected in the interest in defining properties and reactivity of the relevant >NOR expected to be formed with various R· groups derived from these polymers. Primary, secondary, and tertiary macroalkyls may be bound theoretically in >NOR. A model study (29) with three various isolated >NOR compounds derived from iso-octane revealed an easy thermolysis of *sec-* and *tert*-alkyl-containing >NOR compounds (typical for PE and PP, respectively) below 100 °C. The differences among products of >NOR thermolyses performed in nitrogen or oxygen atmospheres were observed (11, 29). Either the respective hydroxylamine >NOH was formed or >NO· was regenerated.

Thermolysis and oxidation of >NOR are key mechanisms involved in the "nitroxide regenerative cycle" (Scheme VI). This cycle has been used tradi-

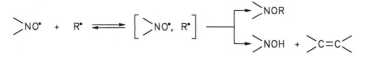

Scheme V. Formation of O-alkylhydroxylamines.

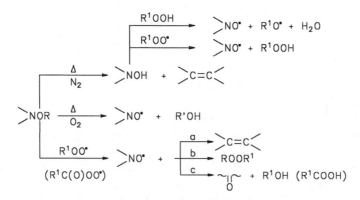

Scheme VI. Regeneration pathways of nitroxide from O-alkylhydroxylamine in HAS series.

tionally for explaining the high (also called *overstoichiometric* or *catalytic*) stabilizing efficiency of HAS.

The regeneration of >NO˙ via reactivity of >NOR with ROO˙ is accompanied by transformation of the bound alkyl group. An olefin and hydroperoxide, an alkylperoxide, or a mixture of a carbonyl compound with an alcohol were proposed as the products (*11, 27*). The carbonyl compound and the alcohol mechanism also explain the reactivity of >NOR with acylperoxy radicals. The differences among the reactivities of various alkyls in >NOR with ROO˙ were mentioned (*11*). These differences are reflected in the efficiency of the HAS regenerative cycle and may influence the service concentration of HAS necessary to reach an effective stabilization effect.

HAS fail as the processing stabilizers of polyolefins. Therefore, they have been used in most commercial applications in combinations with hydrolytically resistant phosphites and physically persistent phenols (*24*). No mechanistic data have been published on chemical interactions in the HAS–phosphite combinations. A complementary mechanism is expected. The possibilities of interactions between phenolic antioxidants and HAS were reported (*3, 12*). The oxidation of phenols by nitroxides and recombination of radicals formed from both stabilizers seem to be crucial reactions.

Phenols are transformed into phenoxyls by scavenging ROO˙ groups [a so-called *service* transformation (*13*)]. The same phenoxyl and, consecutively, the products formed via phenoxyl are generated by nitroxides. Processes such as this rank among "depleting" transformations of phenols in the stabilized polyolefins.

The oxidation of phenols with >NO˙ derived from HAS was reported in the very early stage of HAS application (*30*). The >NO˙ is reduced to the corresponding hydroxylamine >NOH. According to the chemical structure of the phenol, substituted benzoquinone, diphenoquinone, stilbenequinone (StQ), or more complicated quinomethides are formed (*3, 13*). Quinoide or quinomethinoide compounds do not diminish the stabilizing effect of secondary or tertiary HAS in the sensitized photooxidation of heptane (*31*). StQ, used as a common model of oligomeric quinomethides, enhanced the protective effect of HAS. We consider that the light-screening ability of StQ enhances the final effect.

Substituted alkylperoxy- and hydroperoxycyclohexadienones are generated from the phenolic antioxidants in sites of the augmented concentration of ROO˙ and in the presence of excited sensitizers (*13*). Thermolysis and photolysis of the peroxidic species substantially enhance the oxidation of hydrocarbon substrates. According to model experiments (*31*), the peroxidic products of phenol transformation do not diminish the stabilizing efficiency of HAS.

The recombination of free carbon radical species derived from oxidized phenols with >NO˙ was observed (*3, 12, 13*). Compounds such as **30** or more complicated analogs formed from polynuclear phenols or the C–C coupling products of phenoxyls are instable (3). Compounds like **31** are most likely decomposed into the respective quinomethide and >NOH compounds, or they undergo a recombination to cyclohexadienonyls and >NO·.

Compound **32** or its analogs were obtained from phenolic antioxidants and various >NO· (30, 32). Either recombination of hydroxybenzyl radicals with >NO· or 1,6-addition of >NOH on quinomethides generated from phenols is considered a formation pathway (12). The estimated importance (12) of O-substituted hydroxylamines in the HAS mechanism and findings concerning the efficiency of new categories of HAS with structure **28** create the prospect that compounds like **32**, which are formed from HAS and phenols during polyolefin stabilization, cannot be considered as species depleting the integral stabilizing efficiency in commercial combinations of HAS–phenol.

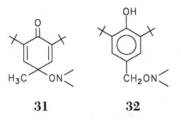

31   32

Conclusions

An efficient scavenging of R· and ROO· is conditioned by chemical activity of aminic additives. Most data used for explaining the stabilizing mechanisms of amines were obtained under model conditions. The chemical transformations of amines and properties of formed products construct the basis of understanding processes that are characteristic of stabilized polymers. The interpretation of model data for processes taking place in solid polymers must be done very carefully. An uncertainty exists in the verification of some chemical interactions due to difficulties in the isolation of trace amounts of amine transformation products from the aged polymer matrix. The synthesis of more efficient and less discoloring R· and ROO· scavengers, better understanding of amine application, and proper conditions for testing of their performance in polymers represent the practical exploitation of the knowledge of amine chemistry.

References

1. Pospíšil, J. *Pure Appl. Chem.* **1973**, 36, 207.
2. Pospíšil, J. *Developments in Polymer Stabilization;* Scott, G., Ed.; Elsevier Applied Science: London, 1984; Vol. 7, p 1.
3. Pospíšil, J. *Polym. Degrad. Stab.* **1991**, 34, 85.

4. Billingham, N.C. In *Oxidation Inhibition in Organic Materials;* Pospíšil, J.; Klemchuk, P., Eds.; CRC: Boca Raton, FL, 1990; Vol. 2, p 249.
5. Pospíšil, J. *Adv. Polym. Sci.* **1991,** *101,* 65.
6. Scott, G. In *Developments in Polymer Stabilization;* Scott, G., Ed.; Elsevier Applied Science: London, 1984; Vol. 7, p 65.
7. Pospíšil, J. In *Oxidation Inhibition in Organic Materials;* Pospíšil, J.; Klemchuk, P., Eds.; CRC: Boca Raton, FL, 1990; Vol.1, p 33.
8. Knobloch, G. *Angew. Makromol. Chem.* **1990,** *176/177,* 333.
9. Hinsken, H.; Moss, S.; Pauquet, J.-R.; Zweifel, H. *Polym. Degrad. Stab.* **1991,** *34,* 279.
10. Pospíšil, J. *Proc. 11th Int. Conf. Stabil. Control. Degrad. Polym;* Lucerne, Switzerland, 1989; p 163.
11. Gugumus, F. *Polym. Degrad. Stab.* **1993,** *40,* 167.
12. Pospíšil, J. *Proc. 15th Int. Conf. Stabil. Control. Degrad. Polym.;* Lucerne, Switzerland, 1993; p 151.
13. Pospíšil, J. *Polym. Adv. Technol.* **1992,** *3,* 443.
14. Taimr, L.; Šmelhausová, M.; Prusíková, M. *Angew. Makromol. Chem.* **1993,** *206,* 199.
15. Andruzzi, R.; Trazza, A.; Berti, C.; Greci, L. *J. Chem. Res. (S)* **1982,** 1837; *(M)* **1982,** 1840.
16. Berger, H.; Bolsman, T. A. B. M.; Brower, D. M. In *Developments in Polymer Stabilization;* Scott, G., Ed.; Applied Science: London, 1983; Vol. 6, p 1.
17. Taimr, L.; Prusíková, M.; Pospíšil, J. *Angew. Makromol. Chem.* **1991,** *190,* 53.
18. Adamic, K.; Ingold, K. U. *Can. J. Chem.* **1969,** *47,* 295.
19. Gunstone, F. D.; Mordi, R. C.; Thorisson, S.; Walton, J. C. *J. Chem. Soc. Perkin Trans. 2,* **1991,** 1955.
20. Zeman, A.; Trebert, Y.; von Roenne, V.; Fuchs, H.-J. *Tribol. Schmierungstech.* **1990,** *37,* 158.
21. Brownlie, I. T.; Ingold, K. U. *Can. J. Chem.* **1967,** *45,* 2419, 2427.
22. Alberti, A.; Carloni, P.; Greci, L.; Stipa, P.; Neri, C. *Polym. Degrad. Stab.* **1993,** *39,* 215.
23. Scott, G. *Rubber Chem. Technol.* **1985,** *58,* 269.
24. Drake, W. O. *Proc. 14th Int. Conf. Stab. Control. Degrad. Polym.;* Lucerne, Switzerland, 1992; p 57.
25. Mayer, W.; Zweifel, H. *Proc. Conf. EURETEC Paris* **1988,** p 20.
26. Gijsman, P.; Hennekens, J.; Tummers, D. *Polym. Degrad. Stab.* **1993,** *39,* 225.
27. Klemchuk, P. P.; Gande, M. E.; Cordola, E. *Polym. Degrad. Stab.* **1990,** *27,* 65.
28. Bowry, V. W.; Ingold, K. U. *J. Am. Chem. Soc.* **1992,** *114,* 4992.
29. Neri, C.; Malatesta, V.; Constanzi, S.; Riva, R. *Proc. 15th Int. Conf. Stab. Control. Degrad. Polym.,* Lucerne, Switzerland, 1993; p 119.
30. Murayama, K. J. *Synth. Org. Chem. (Japan)* **1971,** *29,* 366.
31. Scheim, K.; Pospíšil, J.; Habicher, W. *Proc. 34th IUPAC Macro Prague* 1992; p 8-P16.
32. Carloni, P.; Greci, L; Stipa, P.; Rizzoli, C.; Sgatabotto, P.; Ugozzoli, F. *Polym. Degrad. Stab.* **1993,** *39,* 73.

RECEIVED for review January 26, 1994. ACCEPTED revised manuscript September 29, 1994.

Photostability of UV Screeners in Polymers and Coatings

James E. Pickett[1] and James E. Moore[2]

[1]General Electric Company, Corporate Research and Development, Schenectady, NY 12301
[2]GE Structured Products, Mt. Vernon, IN 47620

All major classes of commercially available UV screeners undergo photodegradation at quantum yields on the order of 1×10^{-6} in polar matrices. The rates obey zero-order kinetics in the high-absorption range and are highly dependent on the nature of the matrix. Typical rates of photodegradation for screeners in cast polymethyl methacrylate films are on the order of 0.3 to 0.7 absorption units per year of outdoor exposure and 0.2 to 0.5 absorption units per 1000 h of xenon-arc or QUV exposure. Concentration of the screener appears to play little role. The presence of hindered amine light stabilizers and the nature of the light source might affect the rate of degradation of the matrix. Rapidly degrading matrices seem to cause rapid destruction of the screeners.

MOST PLASTICS DEGRADE WHEN EXPOSED OUTDOORS, and UV screeners often are added either to the bulk polymer or in coatings to protect the plastics. The screeners work by absorbing UV light, dissipating the energy harmlessly as heat, and thereby reducing the amount of UV light that can be absorbed by the polymer. However, to the extent that the quantum yield of this process is less than exactly one, the screeners are also subject to photodegradation, and their effectiveness as stabilizers can be lost. Much attention has been paid to the loss of additives by extraction or blooming, but relatively little work has been directed at the inherent photostability of UV screeners and the factors that affect this photostability.

The degradation of both benzophenone and benzotriazole UV screeners was observed in polypropylene by several groups and was attributed to free radical attack at the critical phenolic hydroxyl group (1–6) or through energy transfer from excited chromophores (7). The UV screener was depleted from

0065–2393/96/0249–0287$12.00/0
© 1996 American Chemical Society

the surface of weathered polycarbonate samples, but weathered polycarbonate is known to erode and the screener could have been physically lost (8). The photostability of benzophenone and benzotriazole screeners was studied by flash photolysis and exposure to mercury arcs (9). Recently, the photodegradation of benzotriazole screeners was reported in polymethyl methacrylate (PMMA) under conditions where leaching could be ruled out (10). Because the findings were fragmentary and the kinetics of the process were ill-defined, we undertook a systematic study of UV screener stability in several matrices with emphasis on the consequences for coatings. Some of this work was reported previously (11, 12).

Experimental Details

All additives were commercial samples obtained from the suppliers. Spectra were taken on a Shimadzu UV-240 spectrophotometer. Rates of degradation were determined by plotting the absorbance at a maximum vs. time of exposure, drawing a straight line through a portion of the curve, and dividing the slope by the correction factor as described subsequently to arrive at a zero-order rate. The rates are expressed as loss of absorption units per 1000 h of exposure (A/1000 h) unless otherwise specified.

Cast Films.   PMMA (DuPont Elvacite 2041) or crystal polystyrene were cast from chloroform solution as described previously (11).

UV-Cured Acrylics.   Resins consisting of hexanedioldiacrylate and a silylated colloidal silica were prepared according to the process described by Lewis and Katsamberis (13). UV screeners were added at 6–7% loadings, and the resins were applied as ca. 5-μm coatings onto unstabilized 15-mil (368-μm) polycarbonate films and cured under anaerobic conditions using commercially available photoinitiators and Ashdee mercury lamps.

Silicone Coatings.   Silylated UV-screener derivatives were prepared by hydrosilylating the corresponding allyl derivatives with triethoxysilane and a Pt catalyst or by treatment of aliphatic alcohol derivatives with 3-trimethoxysilylpropyl–isocyanate. Coating solutions were prepared by hydrolyzing methyltrimethoxysilane and the silylated UV screener in the presence of aqueous colloidal silica and diluting with a mixture of 2-propanol and 1-butanol. The coating solutions were flow-coated onto glass microscope slides and baked at 130 °C for 1 h to give 4–5-μm coatings.

Xenon-Arc Conditions.   An Atlas Ci35a weatherometer was run at 0.77 W/m^2 irradiance measured at 340 nm with Type S borosilicate inner and outer filters. The lamp operated in a cycle of 160 min light at 45 °C dry-bulb temperature and 50% relative humidity (ca. 65 °C black panel temperature) and 20 min dark. The final 15 min of the dark cycle was with a front and back side water spray. Under these conditions, the samples accumulated 2700 kJ/m^2 at 340 nm in 1100 h of exposure. This value is approximately equivalent to one year of Florida exposure.

QUV Conditions. Q-Panel QUV instruments were equipped with FS-40 fluorescent sunlamps. The cycle was 8 h light at 70 °C black panel followed by 4 h of dark condensation at 50 °C. The lamps were rotated or changed at 400–450 h of service according to manufacturer's specifications. The samples were mounted on the back side of a $4 \times 4 \times 1/16$ in. quartz plate by using spring clips so that the samples did not experience the condensation cycle.

Mercury Lamp Exposure. The outer glass envelope of a commercial GE HA100 high-pressure mercury street lamp was removed, and the lamp was suspended in a Pyrex photolysis immersion well. The samples were cast onto quartz plates and hung on a merry-go-round to ensure uniform exposure.

Results and Discussion

UV Screener Structure and Function. Even though many dozens of UV screeners are commercially available, they are all based on just a few chromophores. Representative structures and UV spectra are shown in Chart I and Figures 1 and 2, respectively. If these compounds are to be effective as UV stabilizers they must have high absorption at the wavelengths that cause degradation of the polymer or coating, and they must harmlessly dissipate the energy that they have absorbed. The benzophenone, benzotriazole, and oxanilide classes of screeners are thought to be photostable because their excited states can undergo a rapid internal hydrogen transfer to make higher energy ground-state species as shown in Scheme I (*14*). The reverse reaction is exothermic, and the heat is dissipated through the matrix. The triazines presumably operate through a similar mechanism.

The internal hydrogen bond is key to this process. In a polar environment, some of the hydrogen bonding may be intermolecular with the matrix, and this arrangement would interfere with the mechanism for internal hydrogen transfer. Other reactions leading to destruction of the chromophore could result (*15*). A second pathway for destruction results if free radicals due to photooxidation of the matrix abstract the phenolic hydroxyl hydrogen of the UV screener leading to oxidation of the chromophore (*2, 3*). Oxanilides are reported to be unreactive toward free radicals (*16*). To our knowledge, the photochemistry of cyanoacrylates has not been investigated. One would expect a charge-separated species to be involved in the process at some point (Scheme II). This species, if present, would be subject to attack by nucleophiles such as water. The ground-state species may also be subject to addition of free radicals across the double bond. Either process would result in loss of the chromophore.

We subjected Cyasorb 531 (Cyasorb is a trademark of Cytec) to photolysis in air-saturated ethyl acetate solution by using a Pyrex-filtered high-pressure mercury lamp. Whereas many products were observed by gas chromatography, most in very small amounts, benzoic acid was identified as the major product and accounted for about half of the degraded mass. This result is consistent

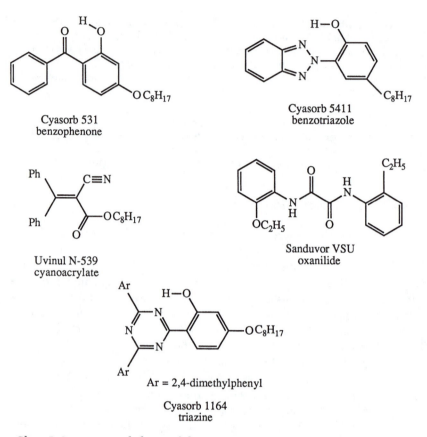

Cyasorb 531
benzophenone

Cyasorb 5411
benzotriazole

Uvinul N-539
cyanoacrylate

Sanduvor VSU
oxanilide

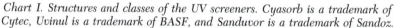

Ar = 2,4-dimethylphenyl

Cyasorb 1164
triazine

Chart I. Structures and classes of the UV screeners. Cyasorb is a trademark of Cytec, Uvinul is a trademark of BASF, and Sanduvor is a trademark of Sandoz.

with homolysis of the aryl–carbonyl bond as a major pathway for the degradation of the screener. Photolysis of a benzotriazole screener under similar conditions gave a myriad of products in tiny quantities that we have not yet identified.

Kinetics of UV Screener Photodegradation. We described (*11*) a computer model for the degradation kinetics. In doing the modeling one is faced with the choice of a simple model using a single wavelength of light in a highly absorbing region or a more complicated model including the longer wavelengths of the lesser absorbing "tail". The simple monochromatic model adequately fits the experimental results, at least for the benzophenone and benzotriazole screeners, probably because the longer wavelength light in the tail has insufficient energy (<80 kcal/einstein) to cause homolysis. Figure 3 shows the photodegradation of Cyasorb 531 in a cast PMMA film upon exposure to a xenon-arc lamp and the calculated curve in which the initial and

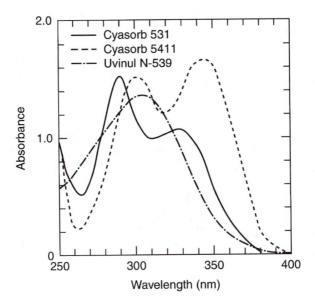

Figure 1. *Absorbance spectra of Cyasorb 531, Cyasorb 5411, and Uvinul N-539 in 1 × 10⁻⁴ M chloroform solution.*

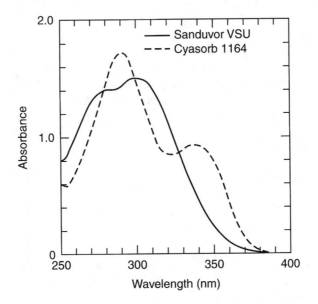

Figure 2. *Absorbance spectra of Sanduvor VSU (1 × 10⁻⁴ M) and Cyasorb 1164 (4 × 10⁻⁵ M) in chloroform solution.*

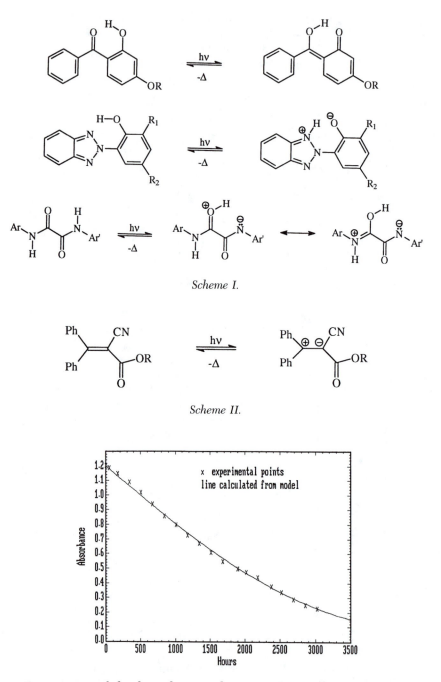

Scheme I.

Scheme II.

Figure 3. Loss of absorbance by Cyasorb 531 in a PMMA film on exposure to xenon arc. The solid line was calculated from the monochromatic model (11), and the time axis was normalized to match the observed absorption at 3040 h.

final absorbances are forced to match the experimental results. The shape of the calculated curve matches the experimental points very well. Similarly good fits were obtained for a benzotriazole screener, Cyasorb 5411.

Close examination of the results in Figure 3 shows why there has been confusion as to whether the photodegradation is a zero-order or a first-order process; the order depends on the absorbance of the film. Figure 4 shows a plot of the calculated instantaneous rates of UV screener degradation in a film as a function of absorbance. When the absorbance is greater than 1, the loss of screener is essentially zero order; that is, the rate does not depend on the initial absorbance. At absorbances less than about 0.5, the rate is highly dependent on absorbance, and the loss can be described adequately by first-order kinetics. First-order kinetics were reported (*17*) in systems with absorbance <1.

In the highly absorbing range, the light does not penetrate very deeply into the sample, so only the screener near the surface is degraded. This phenomenon can be shown in an experiment by using a stack of thin films containing Cyasorb 531. Each film was approximately 12-μm thick and had an initial absorbance of about 0.50 at 325 nm. When irradiated by a xenon-arc lamp, the screener was lost primarily from the films nearest the lamp, whereas the deeper ones remain unchanged (Figure 5). The loss of absorbance is approximately first order in the first film in the stack, but it is essentially zero order for the stack as a whole.

A practical implication of this finding is that the rate measured at any absorbance can be normalized to the zero-order rate by dividing by the factor

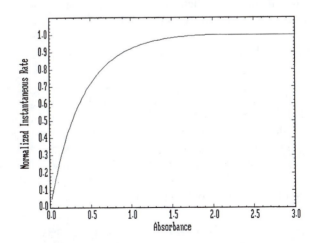

Figure 4. Normalized instantaneous rates of photodegradation for UV screeners in a film as a function of absorbance as calculated by using the monochromatic model. An empirical fit to the curve can be made with the function: Normalized Rate = 1.0 − exp(−2.5A).

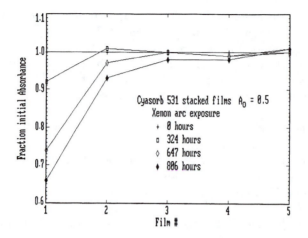

Figure 5. Depletion of Cyasorb 531 in a series of stacked films at various times on exposure to xenon arc. Film #1 was the outermost. (Reproduced with permission from reference 11. Copyright 1993 Elsevier Science Publishers, Ltd.)

found from Figure 4. For example, a rate determined by the slope at an absorbance around 1.0 in Figure 3 would be divided by 0.92 to find the limiting, zero-order rate of 0.43 A/1000 h.

Optical Consequences in Coatings.   Often the purpose of a coating is to reduce the dose of UV light reaching the surface of the substrate. The dose is the integral of the light flux over time. As a screener degrades in a coating, more light is allowed through, but the effect is nonlinear with time because of the inverse logarithmic relationship of absorbance with transmission, $A = \log(1/T)$. Figure 6 shows calculations for a coating with an initial absorbance of 3.0 at λ_{max}; that is, it initially transmits only 0.1% of the incident light. As the screener degrades over time, at first very little happens to the transmission. When two-thirds of the screener has degraded, the transmission is still only 10%, and the integrated light dose is still very small. However, further degradation of screener results in much higher transmission, and the light dose rapidly builds. The UV dose could cause yellowing of the substrate, for example. We have often observed this "hockey stick" effect on coated polycarbonate samples; yellowness will change little for a long time followed by a rather sudden onset of rapid yellowing.

Structure and Matrix Effects in Photostability.   The rates of degradation of UV screeners in various matrices exposed in the xenon-arc weatherometer are shown in Table I. The results show a wide variation in the rates among screeners and among the three matrices. In PMMA, Cyasorb 531 was the least stable, whereas Cyasorb 1164 degraded at about half the rate.

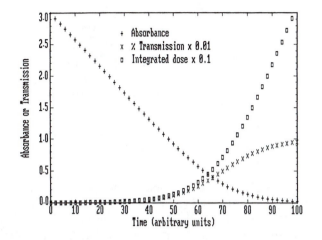

Figure 6. Calculated absorbance, transmission, and integrated transmission (light dose reaching a substrate) for a coating containing a degrading UV screener.

Table I. Rates of UV Screener Degradation in Various Matrices on Exposure to Xenon Arc

Screener	Class	PMMA	Silicone Hardcoat[a]	UV-Cured Acrylic
Cyasorb 531	Benzophenone	0.44	0.24	1.04
Cyasorb 5411	Benzotriazole	0.28	0.80	0.69
Uvinul N-539	Cyanoacrylate	0.34	0.76	1.04
Sanduvor VSU	Oxanilide	0.35	0.76	N/A
Cyasorb 1164	Triazine	0.21	2.0	0.53

NOTE: Rate values are $A/1000$ h.
[a] Screener with a trialkoxy group in place of an alkyl group, but otherwise the same chromophore.

However, in the silicone hardcoat matrix, the benzophenone-type was the most stable, and the triazine photodegraded eight times faster. In the UV-cured acrylic, all of the rates were considerably faster than in PMMA. The oxanilide was not soluble enough in the UV-cured acrylic to test.

Polystyrene films containing Cyasorb 531 and Cyasorb 5411 degraded too quickly under xenon or QUV conditions to give useful data due to rapid photooxidation of the polystyrene. These films, along with PMMA controls, were exposed to a Pyrex-filtered mercury lamp. The results (Table II) show that the rate of UV screener degradation is much slower in the nonpolar polystyrene matrix. In polystyrene, the benzophenone screener appears to be more stable than the benzotriazole, but this value is near the error limits of the experiment.

The fast rates of degradation in the UV-cured acrylic can be attributed to the higher levels of free radicals that are surely present when the coating

**Table II. Degradation of UV Screeners
when Exposed to Pyrex-Filtered Hg Lamp**

Screener	In PMMA	In Polystyrene
1% Cyasorb 531	0.44	0.04
2% Cyasorb 531	0.40	0.03
1% Cyasorb 5411	0.24	0.07
2% Cyasorb 5411	0.24	0.10

NOTE: Rate values are $A/1000$ h.
SOURCE: Reproduced with permission from reference
11. Copyright 1993 Elsevier Science Publishers, Ltd.

is exposed to UV light. There is a substantial amount of unsaturation remaining after cure (18), and the coating degrades fairly rapidly. The effect of hindered amines in reducing the rate of UV screener degradation in this matrix is described subsequently. Polystyrene is a nonpolar matrix, and the internal hydrogen bonding of the screeners should not be disrupted in any way. The very slow rate of degradation probably is close to the inherent stability of the screeners in the absence of any matrix effects. Dramatically slower loss rates were reported (17) in a stable fluoropolymer matrix compared with a reactive UV-cured acrylic urethane.

The rate differences between PMMA and the silicone hardcoat are more difficult to rationalize. The hardcoat still contains a considerable amount of hydroxyl groups after cure, and these groups may interfere with the internal hydrogen bonding that is essential to the stability of most of the screeners (15). The cyanoacrylate screener contains no hydrogen-bonded group, but it may be subject to the photoinduced hydrolytic cleavage described previously. The ranking of the rates in the silicone hardcoat are nearly reversed from that in PMMA. We have no explanation for this remarkable effect.

The results show that the stability of the screener is highly dependent on the matrix composition and that the relative rankings can change as the type of matrix changes. Unfortunately, there seems little recourse but to test all types of screeners to find the best for any particular application.

Effect of Light Source. PMMA films containing 1% of various UV screeners were exposed in a QUV apparatus and to a xenon-arc lamp in a weatherometer. The rates are shown in Table III. The second entries in the QUV column are for a duplicate run. The results show about the same relative rankings; the benzotriazole is comparatively slower to degrade in the QUV, whereas the triazine and oxanilide are comparatively faster. In general, however, the type of tester does not dramatically affect the relative ranking of UV screener stability, at least in the relatively nonreactive PMMA matrix. If the shorter wavelengths of light in the QUV cause rapid degradation of the matrix and thereby create a large concentration of free radicals, then the results could be different. This result is the case with the exposure of Cyasorb 5411 in a

Table III. Rates of UV Screener Degradation in Cast PMMA Films (1%) Exposed to Different Light Sources

Screener	Xenon Arc	QUV[a]
Cyasorb 531	0.46	0.45, 0.57
Cyasorb 5411	0.29	0.18, 0.20
Uvinul N-539	0.33	0.36, 0.45
Sanduvor VSU	0.35	—, 0.50
Cyasorb 1164	0.21	0.25, —
Cyasorb 5411[b]	0.33	0.56

NOTE: Rate values are $A/1000$ h.
[a] Duplicate measurements were tested at different times. There is considerable variation in the QUV depending on sample position and the age of the lamps.
[b] Screener in polycarbonate.
SOURCE: Reproduced with permission from reference 11. Copyright 1993 Elsevier Science Publishers, Ltd.

polycarbonate film, as shown in the last entry of Table III. The polycarbonate degrades slowly in xenon arc, and the rate of UV screener degradation is about the same as in PMMA. However, the polycarbonate itself degrades rapidly during QUV exposure and presumably makes peroxidic or radical species that contribute to the rapid degradation of the screener.

We have limited data concerning loss of UV screeners on exposure to sunlight. Results of an experiment showing absorbance loss on exposure of PMMA films containing Cyasorb 531 and Cyasorb 5411 at the South Florida Test Service in Miami are shown in Figure 7. The rate of loss of Cyasorb 5411 is about as predicted from xenon-arc exposure. An exposure of 1100 h in xenon arc under our conditions gives about the same UV dose as a year in Florida. Thus, the predicted rate was 0.32 A/year, whereas the observed rate was 0.35 A/year. However, the predicted rate from xenon arc for Cyasorb 531 is 0.51 A/year, whereas the observed rate in Florida was 0.74 A/year. Similar results were obtained on exposure in Mt. Vernon, Indiana. We have no explanation for the unexpectedly fast loss of the benzophenone screener in these experiments. Exposure on a roof rack in Schenectady, New York (where there is somewhat less sunshine), gave zero-order rates of about 0.25 A/year for Cyasorb 531 and 0.26 A/year for Tinuvin P, a benzotriazole screener closely related to Cyasorb 5411. (Tinuvin is a trademark of Ciba-Geigy.)

Extraction is not likely to contribute to the loss. PMMA films containing 1% of Cyasorb 5411 or 1% of 2,4-dihydroxybenzophenone were soaked in a circulating water bath at 65 °C for 500 h with no significant loss of screener. In general, there is little mobility of large molecules in a glassy matrix below the glass-transition temperature (*19*). Our experiments in xenon arc gave es-

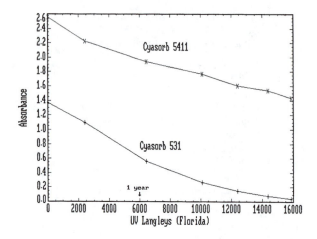

Figure 7. Exposure of PMMA films containing 1% UV screener in Florida. A dose of about 6000 UV langleys corresponds to one year of exposure. (Reproduced with permission from reference 11. Copyright 1993 Elsevier Science Publishers, Ltd.)

sentially the same results when the films were protected by quartz plates. Extraction or leaching can be a major factor in UV-cured coatings where the glass-transition temperature is low or in nonpolar semicrystalline polymers such as polyolefins.

Effect of Concentration. Free-standing films of PMMA containing up to 5% of Cyasorb 531 and Cyasorb 5411 were cast from chloroform. In addition, very thin films of PMMA containing up to 20% UV screener were cast from dilute chloroform solution onto quartz slides. The thicknesses of the films were adjusted so that the absorbances were all 0.8 to 1.4. These films were exposed in the xenon-arc weatherometer, and the rates of degradation, corrected to the zero-order rates, are shown in Table IV.

No effect of concentration on the rate of photodegradation occurs for these two screeners in the range of 1 to 20%. At concentrations around 10% the chance that a particular screener molecule will have another screener as a nearest neighbor is high, and self-quenching could occur for molecules in which the internal hydrogen bond was disrupted. However, this effect apparently is not important for these compounds, at least in PMMA. The absorbance of the entire film is an important factor, not the absolute molar concentration of the screener. A screener in a thin, highly concentrated film degrades at the same rate as in a thicker, more dilute film.

Effect of Hindered Amine Light Stabilizers. Hindered amine light stabilizers (HALS) are often added as UV stabilizers in conjunction with

UV screeners, especially in polyolefins. One possible mechanism for their action is to stabilize the UV screener from photodegradation. PMMA films containing UV screeners plus 1% Tinuvin 770 (**1**) or Goodrite 3034 (**2**) were cast and exposed to QUV or xenon arc (Goodrite is a trademark of BF Goodrich). The results (Table V) clearly show that neither HALS had any significant effect on the rate of UV screener degradation in PMMA under either xenon-arc or QUV exposure.

Table IV. Rates of UV Screener Degradation in Cast PMMA Films at Different Loadings

Form	Loading (%)	Cyasorb 531	Cyasorb 5411
Free Standing:	1	0.42	0.27
	2	0.42	0.29
	3	0.38	N/A
	5	0.41	0.29
Supported:	5	0.45	0.33
	10	0.30	0.31
	15	0.30	0.31
	20	0.45	0.33

NOTE: Rate values are A/1000 h.
SOURCE: Reproduced with permission from reference 11. Copyright 1993 Elsevier Science Publishers, Ltd.

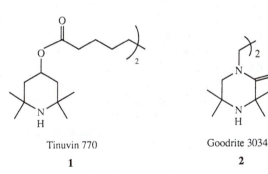

Tinuvin 770

1

Goodrite 3034

2

Table V. Rates of UV Screener Degradation in Cast PMMA Films (1% Screeners) in Presence of 1% of Tinuvin 770 or Goodrite 3034 HALS

	Xenon Arc			QUV		
	No	With	With	No	With	With
Screener	HALS	T-770	G-3034	HALS	T-770	G-3034
Cyasorb 531	0.47	0.43	0.47	0.57	0.53	0.57
Cyasorb 5411	0.29	0.29	0.27	0.21	0.19	0.18
Uvinul N-539	0.33	0.29	0.30	0.47	0.45	0.44

NOTE: Rate values are A/1000 h. For structures of Tinuvin 770 (**1**) or Goodrite 3034 (**2**), *see* text.
SOURCE: Reproduced with permission from reference 11. Copyright 1993 Elsevier Science Publishers, Ltd.

Quite different results were obtained in a rapidly degrading matrix such as the UV-cured acrylic. Samples were prepared from the UV-cured acrylic coating containing 6% of Cyasorb 5411 with and without HALS and exposed to xenon arc. The zero-order rate in the control was 0.61 A/1000 h, whereas the addition of 1% HALS reduced the rate to 0.31 A/1000 h, which is about the rate in PMMA for this screener. Nearly identical rates were obtained when the samples were exposed behind quartz, and this result indicated that leaching of the screener was not occurring. The HALS apparently help to stabilize the matrix from photooxidation and reduce the concentration of free radicals that can attack the screener. Cyasorb 1164 gave the same results in this coating. Decker and co-workers (17, 20) reported that the rate of loss of a benzotriazole UV screener in a UV-cured acrylic urethane matrix was reduced on addition of HALS, and they attributed the effect both to reduced physical loss by retention of the cross-linking and to decreased photooxidation arising from slower matrix degradation. We believe that physical loss is only a minor factor in our coating.

Estimation of Quantum Yield. We can make an order-of-magnitude estimate of the quantum yield for the degradation process by taking the case of a rate of 0.3 A/1000 h in PMMA on exposure to xenon arc. Our conditions give an exposure of about 1×10^5 J/cm^2 per 1000 h in the wavelength range of 290 to 355 nm, where the UV screeners are highly absorbing. With an average photon energy of 3.6×10^5 J/einstein in this range, the dose is approximately 0.03 einstein. This dose causes the destruction of about 3×10^{-8} moles of UV screener/cm^2 (the amount of screener with extinction coefficient of 10,000 required to make an absorbance of 0.3 in 1 cm^2) and leads to an estimated quantum yield of 1×10^{-6} for UV screener destruction in this case. Thus, for every million photons that the screener absorbs, one leads to its decomposition. A similar quantum yield was estimated by using outdoor exposure data. Of course, the rates and quantum yields are highly dependent on the matrix, as shown previously.

Conclusions

We have shown that all of the major classes of UV screeners photodegrade at appreciable rates in coatings and thin films both in accelerated testing and during outdoor exposure. The quantum yields are on the order of 1×10^{-6}. In PMMA films, the loss is 0.3 to 0.7 absorbance units per year of outdoor exposure, although the rates are highly dependent on the structure of the screener and the nature of the matrix. Strongly hydrogen bonding, hydroxylic, or photo-unstable matrices contribute to rapid photodegradation of UV screeners. At high absorbances, the loss is described by zero-order kinetics. This result initially limits the loss of screener to the surface regions of samples.

HALS and the nature of the light source have little effect on the degradation rate except in how they might affect the degradation of the matrix. Intermediates from polymer degradation such as radicals and peroxides can accelerate the degradation of the screeners. There seems to be no way to predict just what type of screener will be the most stable in any particular matrix. The rates of screener photodegradation are sufficiently high that the outdoor lifetimes of many coatings and articles could be limited by the stability of the screeners that were added to protect them.

Acknowledgments

We thank Arnie Factor, Greg Gillette, and Gautam Patel for providing samples and for helpful discussions.

References

1. Rabek, J. F. *Photostabilization of Polymers*; Elsevier Applied Science: London, 1990; p 221.
2. Allen, N. S.; Gardette, J. L.; Lemaire, J. *Polym. Degrad. Stab.* **1981,** *3*, 199.
3. Allen, N. S.; Gardette, J. L.; Lemaire, J. *Polym. Photochem.* **1983,** *3*, 251.
4. Carlsson, D. J.; Grattan, D. W.; Suprunchuk, T.; Wiles, D. M. *J. Appl. Polym. Sci.* **1978,** *22*, 2217.
5. Chakraborty, K. B.; Scott, G. *Eur. Polym. J.* **1979,** *15*, 35.
6. Hodgeman, D. K. C. *J. Polym. Sci. Polym. Lett.* **1978,** *16*, 161.
7. Vink, P. *J. Polym Sci. Symp.* **1973,** *40*, 169.
8. Bolon, D. A.; Irwin, P. C. *Makromol. Chem. Macromol. Symp.* **1992,** *57*, 227.
9. Gupta, A.; Scott, G. W.; Kliger, D. In *Photodegradation and Photostabilization of Coatings*; Pappas, S. P.; Winslow, F. H., Eds.; ACS Symposium Series 151; American Chemical Society: Washington, DC, 1981; pp 27–42.
10. Bell, B.; Bonekamp, J.; Maecker, N.; Priddy, D. *Polym. Prepr.* **1993,** *34(1)*, 624.
11. Pickett, J. E.; Moore, J. E. *Polym. Degrad. Stab.* **1993,** *42*, 231.
12. Pickett, J. E.; Moore, J. E. *Polym. Prepr.* **1993,** *34(2)*, 153.
13. Lewis, L. N.; Katsamberis, D. *J. Appl Polym. Sci.* **1991,** *42*, 1551.
14. Rabek, J. F. *Photostabilization of Polymers;* Elsevier Applied Science: London, 1990; pp 209–230 and references therein.
15. Lamola, A. A.; Sharp, L. J. *J. Phys. Chem.* **1966,** *70*, 2634.
16. Allan, M. A.; Bally, T.; Haselbach, E.; Suppan, P.; Avar, L. *Polym. Degrad. Stab.* **1986,** *15*, 311.
17. Decker, C.; Zahouily, K. *ACS Polym. Mater. Sci. Eng.* **1993,** *66(1)*, 70.
18. Factor, A.; Tilley, M. G.; Codella, P. J. *Appl. Spectrosc.* **1991,** *45*, 135.
19. Olson, D. R.; Webb, K. K. *Macromolecules* **1990,** *23*, 3762.
20. Decker, C.; Moussa, K.; Bendaikha, T. *J. Polym Sci. Part A Polym. Chem.* **1991,** *29*, 739.

RECEIVED for review December 6, 1993. ACCEPTED revised manuscript September 21, 1994.

Light Stabilization of Bisphenol A Polycarbonate

T. Thompson and P.P. Klemchuk

Additives Division, Ciba-Geigy Corporation, Ardsley, NY 10502

Bisphenol A polycarbonate (BPA-PC) is a copolymer of bisphenol A and phosgene with primarily carbonate linkages. Monofunctional phenols are added to the polymerization for molecular weight control and are incorporated as end groups. BPA-PC is a major commercial polymer with unique properties such as high impact strength, glass-like clarity, and high glass transition temperature. Its aromatic content contributes these excellent properties and makes it resistant to most environments. It absorbs terrestrial sunlight sufficiently to undergo Fries phototransformations, photooxidation, and chain scission. Discoloration that accompanies exposure to sunlight detracts from the polymer's appearance. Ultraviolet-light-absorbing additives are usually added to the polymer to reduce the rate of discoloration.

THE PRIMARY PHOTOTRANSFORMATIONS of bisphenol A polycarbonate (BPA-PC) reported in the literature are Fries phototransformations, chain scission, and photooxidation (Schemes I–III). These reactions are wavelength-dependent, and unfortunately many studies reported in the literature were conducted at wavelengths below those of terrestrial sunlight. Therefore, the findings are not necessarily relevant to what may occur during natural weathering.

Review of Literature on BPA-PC Photodegradation

Photoprocesses. In one of the earliest investigations of the photodegradation of BPA-PC, Bellus and co-workers (1) exposed chloroform solutions of polymer (no characterization information was provided) to unfiltered light from a 100-W medium-pressure mercury arc. On the basis of UV and IR spectra they postulated the formation of polymeric phenylesters of salicylic

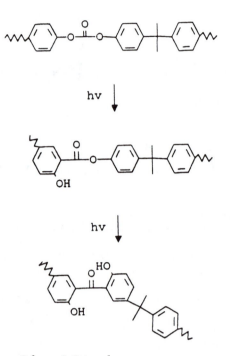

Scheme I. Fries photorearrangement.

acid (absorbance at 315 nm) and polymeric dihydroxybenzophenones (absorbances at 355 nm and 1630 cm⁻¹, UV and IR, respectively). By using alkaline hydrolysis of the photolyzed polymer, they isolated a bright yellow substance with 7% yield after 120-h exposure. This material was thought to be a substituted dihydroxybenzophenone on the basis of UV and IR spectra peaks at 360–362 nm and 1635 cm⁻¹, respectively, and on the basis of the UV and IR spectra of low molecular weight 2,2'-dihydroxybenzophenones at 335–360 nm and 1630 cm⁻¹ . These results constitute one of the first findings of Fries rearrangements in the photolysis of BPA-PC. The relevance of these results to the natural weathering of BPA-PC is questionable in view of the light source used. However, it is a beginning to understanding polycarbonate photochemistry.

In a subsequent study, Mullen and Searle (2) investigated the wavelength sensitivity of 0.1-mil solution-cast and 10-mil extruded PC films by using spectrally dispersed xenon light from 230 to 630 nm. The films were scanned with a UV spectrophotometer at predetermined wavelengths of 320, 360, and 400 nm, which are the absorbance wavelengths of polymeric phenyl salicylates, dihydroxybenzophenones, and yellow products, respectively. Activation spectra

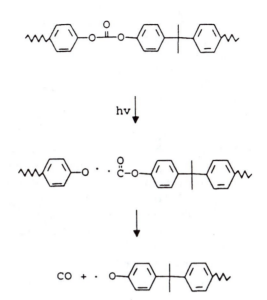

Scheme II. Chain scission.

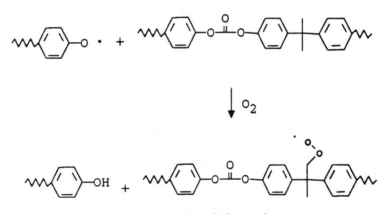

Scheme III. Initiation of photooxidation.

derived from the data showed the solution-cast film was sensitive to wavelengths from 230 to 320 nm, and especially to the region 280–290 nm.

The range of wavelengths causing photodegradation of the 10-mil extruded film was 230–430 nm, which was considerably greater than for the thinner film and was attributed to the greater thickness. The increased film thickness caused a red shift of the activation peak by about 5 nm and increased the photodegradative effect of long-wavelength UV radiation. The results with both films indicated sensitivity to UV radiation in the vicinity of 280–295 nm and supported a two-stage Fries photorearrangement: first to polymeric phenyl salicylates and subsequently to dihydroxybenzophenones. The results provided evidence that other products were also formed.

Although direct correlation between the formation of dihydroxybenzophenones and yellowing was not found, the activation spectra of the solution-cast films for absorbance at 360 and 400 nm were very similar. That similarity and the postulation of Bellus and co-workers regarding the identity of the alkaline hydrolysis product from photorearrangement (substituted dihydroxybenzophenones) suggest the dihydroxybenzophenone functionality may play a role in the yellowing of BPA-PC on exposure to UV light.

In addition to those already mentioned, many papers (3–11) were published in which evidence was presented for chain scission and Fries photorearrangements occurring simultaneously in BPA-PC undergoing exposure to UV radiation. Nearly all reports indicated the Fries photorearrangements were favored at lower wavelengths (e.g. 254–290 nm). Both processes appear to arise from C–O bond scission in the carbonate groups on absorption of light. Chain scission appears to take place more frequently toward the ends of polymer chains and suggests that when a terminal carbonate group absorbs a photon and cleaves, the fragments can move apart because they are not as restricted as when they are within the polymer chain. Scissions that occur within the polymer chain in glassy regions have a greater chance to recombine.

Many studies dealing with the photooxidation of PC were published. In early stages of photooxidation the geminal dimethyl groups were believed to be involved; in later stages ring oxidation was found to take place. Papers by Clark and Munro (12, 13), Factor and co-workers (14, 15), and Rivaton and co-workers (7) are among the most informative and provide a body of information that is essential for understanding the photooxidation of BPA-PC.

Influence of End Groups.   Chain terminators such as phenol and *t*-butylphenol are used to control the molecular weight of BPA-PC and serve to cap the polymer chains. Different manufacturers most likely have their own proprietary practices; therefore, commercial polymers will vary in the type and degree of capping. The degree of capping of the end groups is of significance to the polymer because during the high temperature processing that is mandatory with PC, terminal phenolics react with carbonate linkages and cause polymer transformations. The photostability of BPA-PC is also dependent on the degree of capping of the end groups (the more capping, the more stable the polymer), because free phenolic groups absorb UV radiation at about 290 nm, which is known to cause degradation of the polymer. Polymer terminal groups are measured by IR spectroscopy: terminal phenolic groups absorb at 3595 cm^{-1} and terminal phenyl groups absorb at 1383 cm^{-1}.

Webb and Czanderna (*16–18*) are virtually the only investigators who have looked at the influence of end groups on the photostability of PCs. They worked mainly with three polymers differing in capping and defined as follows: I, uncapped polymer of low molecular weight (M_n, 2500) with 100% free phenolic groups; IV, a commercial polymer with 12% free phenolic groups and capped with 88% phenyl end groups (M_n, 18,360); and V, acetylated I with only 1% free phenolic groups (M_n, 3120). The wavelength dependence of the photodegradation of thin films was investigated with monochromatic laser UV radiation at 265, 272, 285, 287, and 308 nm.

Changes in the vibrational spectra of the capped and uncapped films were measured quantitatively by in situ Fourier transform IR and reflection–absorption spectroscopy. The spectra showed that phenolic end groups in the uncapped polymer, if present in concentrations exceeding the water content of the polymer, were hydrogen-bonded to the backbone carbonyl groups. The correspondence of changes in molecular weight to changes in the vibrational spectra of the exposed films was investigated by size exclusion chromatography.

The results indicated that free phenolic end groups sensitized PC to some photodegradation reactions (such as cross-linking) at 287 and 265 nm while inhibiting Fries photorearrangements. High concentrations of terminal phenolic groups in I induced a cross-linking reaction that predominated at 287 nm and competed with chain scission at 308 nm. The quantum yield for chain scission in I was lower than for IV and V, evidently because of the competing cross-linking reactions of terminal phenolic groups.

Surface Photodegradation.   BPA-PC strongly absorbs UV light below 290 nm in 3-mil films and 125-mil thick plaques (Figure 1). The absorbance tails into the near-UV toward the visible region. Because 290 nm is at the tail end of terrestrial sunlight, the polymer absorbance in that region means UV light-induced reactions will occur in the polymer during natural weathering. The intensity of the polymer's absorbance establishes competition for photons with UVAs. The UVAs can mitigate the harmful effects of UV light on BPA-PC only in proportion to the fraction of UV light of relevant wavelengths that they absorb. Even though the additive's extinction coefficients are likely to be much greater than the polymer's in most of the UV region, the concentration differentials on surfaces and in thin sections are likely to favor photochemistry of the polymer on exposure to sunlight.

In addition to phototransformations, Webb and Czanderna (*16–18*) also made observations regarding the photooxidation of BPA-PC, its impact on properties, and photostabilization. They expressed the view that many properties such as transparency, tensile strength, impact resistance, and rigidity are adversely influenced by reactions at the surface of the solid polymer, where solar UV absorption and uptake of oxygen and water are highest. In their opinion, the additions of UV stabilizers and antioxidants to the bulk polymer

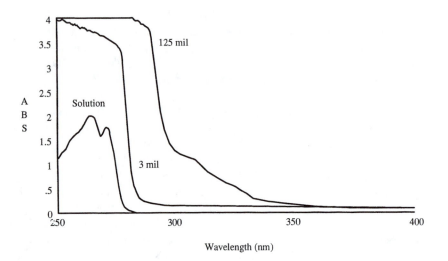

Figure 1. UV absorbance of BPA-PC: 500 mg/mL chloroform solution, 3-mil film, and 125-mil plaque.

have been generally unsuccessful in extending outdoor service life beyond three years.

Hydrolysis. Because BPA-PC is a carbonate ester, it is sensitive to degradation by hydrolysis, mainly at elevated temperatures. The polymer is resistant to hydrolysis by water at ambient temperatures. The solubility of water in the polymer is very low. However, enough water dissolves in BPA-PC when immersed in boiling water so that, when cooled to ambient temperature, the polymer appears hazy due to small droplets of water that are released. Ram et al. (*19*) examined BPA-PC for hydrolytic stability. They found immersion in water at room temperature for one year had no effect on the mechanical properties of Lexan PC143, a UV-stabilized grade polymer with M_w of 27,300 and dispersity of 1.63. However, after 30 days in water at 40 °C this polymer lost 16% of initial elongation and after the same period at 60 °C it lost 55% of initial elongation. Testing at higher temperatures revealed total tensile breakdown and severe impact loss after immersion for 2 weeks in water at 80 °C. Boiling water had a catastrophic effect on chain length and mechanical properties. Although not believed to be an important mechanism in normal circumstances, hydrolysis plays a significant role if moisture is not excluded during processing and if high temperatures are a feature of end use.

Evaluation of Light Stabilizers

Little information has been published about the effectiveness of additive classes other than UVAs in protecting the polymer against photodegradation. This study was undertaken to investigate the photostabilization of BPA-PC by the main classes of stabilizing additives. Several chemical classes of UVAs, hindered amine light stabilizers (HALS), a hindered phenolic benzoate, and a nickel-containing stabilizer were evaluated.

The major objective of this investigation was to determine the influence of stabilizing additives on several polymer properties during exposure of 1-mil thick solution-cast films to artificial light sources or sunlight. The properties of interest were color; tensile strength and elongation; UV, IR, and NMR spectra; molecular weight; and rate of oxygen consumption.

Evaluating stabilizers in solution-cast films offers two distinct advantages: First, thermal degradation of the additive or the polymer will not lead to erroneous light-stabilizer activity. Second, a thin film mimics what would be expected to happen at the surface of the polymer, which is the most difficult area to stabilize and arguably the most important.

Experimental

Materials.   The additives used were of commercial quality and were used without additional treatment. The polymer used for most of the work was Lexan 141–111, which was used without additional treatment.

Stabilizers.   The stabilizers used were as follows: UVA-1, 2-(2'-hydroxy-3',5'-di(dimethylbenzyl)benzotriazole;   UVA-2,   4-octadecyl-4'-ethoxyoxanilide; UVA-3, 2-ethylhexyl-2'-cyano-3',3'-diphenylacrylate; HALS-1, bis(2,2,6,6-tetramethyl-4-piperidinyl)sebacate; HALS-2, bis(1,1-octyloxy-2,2,6,6-tetramethyl-4-piperidinyl)sebacate; NiStab, nickel bis[ethyl-(3,5-di-*tert*-butyl-4-hydroxybenzyl) phosphonate];and benzoate, hexadecyl-3,5-di-*tert*-butyl-4-hydroxybenzoate.

Preparation of Specimens for Exposure.   BPA-PC solutions were prepared containing 20% BPA-PC in spectrograde methylene chloride. The appropriate amounts of additive were added to the polymer solutions. Films were cast with an 8-path wet-film applicator (Gardco, 2-in. width, 8-mil wet thickness) on 8 × 10 × 1/4 in. glass plates to provide 10 × 2 × 0.001 in. specimens. The films were mounted in infrared cards that were modified with 1.5 × 1.5 in. openings.

Property Measurements.   Yellowness index (ASTM1925) measurements were made on the films with the Applied Color Systems, Inc. model CS-5 spectrophotometer using the 2° observer and the large area opening with specular reflector included. The UV measurements were made with a Gilford RESPONSE UV-vis spectrophotometer; IR spectra were run with the Perkin Elmer model No. 683 IR spectrophotometer; NMR spectra were run with the Varian Unity 500 NMR system. Molecular weight analysis was carried out by gel permeation chromatography using PC standards (Aldrich). In all cases duplicate determinations were made.

Tensile testing was carried out with an Instron model No. 1123 on $6 \times 0.25 \times 0.001$-in. specimens (4–6 pulls were averaged). The following conditions were used: gauge length, 50.8 mm (2 in.); crosshead speed, 10 mm/min; and full-scale load, 5.0 kg.

Exposures. Most of this work was carried out in a custom-built exposure device (referred to as 340 FL) in which forty 48-in. UV-A fluorescent lamps (340 nm) were mounted vertically in the form of a cylinder with a diameter of 34 in. Samples were exposed vertically 2 in. away from and parallel to the bulbs on aluminum shelves that were rotated at 1.25 rpm. A limited number of samples were exposed in an Atlas Xenon Arc Weather-ometer (XAW), model Ci-65, 0.35 W/m² irradiance, Corad filters, and no spray cycle. Films supported on glass plates were exposed in Florida facing south at a 45° angle for one year.

Irradiance measurements were made with an International Light IL1700 radiometer for the two accelerated exposure devices at two relevant wavelengths of 290 and 365 nm. The XAW had a greater proportion of far-UV light than did the 340 FL exposure. The ratio of UV-B to UV-A for the XAW was consistent with published data (Table I).

Oxygen Uptake Procedure. BPA-PC solutions were prepared with spectrograde methylene chloride in Pyrex test tubes, $8 \times 3/4$ in. o.d. Each tube was rolled horizontally for 2 h to form a uniform film of BPA-PC on the inner wall of the tube. The tubes were filled with pure oxygen after a minimum of six complete evacuations. An initial pressure of 500–600 mm Hg was established to bring attention to any leaks in the closed system. The tubes were exposed in an Applied Photophysics Multilamp photoreactor equipped with fluorescent lamps centered at 320 nm and modified for concurrent exposure of six samples. Oxygen uptake was monitored with pressure transducers interfaced with an IBM-XT personal computer. Samples were run in triplicate for each formulation. The oxygen versus nitrogen exposure was carried out in a Rayonet photoreactor equipped with 310-nm fluorescent lamps.

Table I. Irradiance of Light Source

Irradiance	Sun-light	340 FL	XAW
Measured[a]			
290 nm	ND	0.173	2.41
365 nm	ND	13.14	38
Ratio, 290/365	ND	1:76	1:16
Published[b]			
250–320 nm (UV-B)	1.4	ND	3.3
320–400 nm (UV-A)	25.3	ND	54.2
Ratio, UV-B/UV-A	1:18	ND	1:16

NOTE: Values are in W/m².
[a]XAW was controlled at 0.35 W/m².
[b]Data were published by Atlas Electric Devices Co. XAW was controlled at 0.55 W/m².

Results and Discussion

Yellowness Index and Tensile Property Changes, 340 FL Exposure.   Formulations containing UVAs significantly reduced yellowing of the 1-mil BPA-PC films on exposure for 1223 h in the 340 FL exposure device. Non-UV-absorbing light stabilizers, alone or in combination with UVA-1, offered little or no improvement in yellowness index (YI). The breaking strength of formulations containing UVAs changed little after the final interval of this test of 1223 h exposure in the 340 FL exposure device. Formulations containing only non-UV-absorbing stabilizers underwent a significant loss in breaking strength. Table II summarizes the results.

YI and Tensile Property Changes, Comparison of Light Exposures.   The YI measurements (7–8 for all) of BPA-PC films for a blank and 1 or 2% UVA-1 showed little differentiation after 1 year of exposure in Florida. This result was not the case for the exposures in the 340 FL and XAW; lower rates of color development resulted in the films with UVA-1 than in the blank films. The 340 FL and XAW developed about the same degree of yellowing in the samples. The YI proved to be a very useful method for following degradation on photolysis of BPA-PC.

The blank and stabilized films underwent considerable loss of elongation after 6 months of exposure in Florida. The stabilized formulations retained elongation only slightly better than unstabilized ones. These films, when compared with films of similar YIs after exposure in the 340 FL or XAW, showed much greater loss in elongation than the films exposed in the indoor devices. The greater loss of elongation during exposure in Florida may have been due

Table II. Comparison of Light Stabilizer Classes

	Yellowness Index (1925)		Breaking Strength (kg)	
Stabilizer	0 h	1223 h	0 h	1223 h
None	3.1	18.1	7.4	0.4
UVA-1	3.2	5.7	7.8	6.2
UVA-2	3.1	6.8	7.2	6.5
UVA-3	3.1	6.5	8.0	6.2
HALS-1	3.2	13.1	6.7	0.5
HALS-2	3.1	11.7	7.3	0.7
NiStab	4.7	21.8	7.1	0.3
Benzoate	3.1	15.4	7.2	0.8
UVA01/HALS-1	3.3	5.6	6.3	7.3
UVA-1/HALS-2	3.2	4.8	7.6	6.0
UVA-1/NiStab	3.6	5.5	6.6	6.2
UVA-1/Benzoate	3.2	4.6	8.1	6.0

NOTE: Samples were 1-mil BPA-PC films exposed to 340 FL. All UVAs were 2 wt% and non-UVAs were 1 wt%.

to hydrolysis effects combined with photolysis effects. Elongation losses were greater in the XAW exposure than in the 340 FL exposure. Even though the XAW was controlled at a constant relative humidity of 30%, it probably had more moisture present than the 340 FL. This difference again implicates hydrolysis as a possible contributor to loss of tensile properties, although the greater portion of far-UV light in the XAW may also have contributed. Tables III–V summarize YI data, and Tables VI–VIII summarize percent elongation data for the accelerated and outdoor exposures.

UV and IR Spectral Changes, 340 FL Exposure.

On exposure to UV light, unstabilized BPA-PC underwent a broad increase in absorption

Table III. Yellowness Index of 1-mil BPA-PC Film on Exposure in 340 FL

Formulation	0 h	285 h	535 h	797 h	973 h	1254 h	1473 h
Blank	3.1	4.5	6.5	8.1	10.7	15.4	NA
1% UVA-1	3.2	3.9	4.5	4.9	5.4	6.6	NA
2% UVA-1	3.2	3.7	4.0	4.6	4.9	NA	6.5

NOTE: NA is not available.

Table IV. Yellowness Index of 1-mil BPA-PC Film on Exposure in XAW

Formulation	0 h	308 h	538 h	780 h	1019 h	1277 h
Blank	3.1	4.9	6.0	7.9	11.0	NA
1% UVA-1	3.1	3.8	4.6	NA	6.2	9.1
2% UVA-1	3.1	3.6	3.9	NA	4.9	6.5

NOTE: NA is not available.

Table V. Yellowness Index of 1-mil BPA-PC Film on 45 °South Exposure in Florida

Formulation	0 mo	1 mo	2 mo	3 mo	4 mo	5 mo	6 mo	9 mo	12 mo
Blank	3.1	3.3	3.8	4.2	4.6	6.1	5.7	8.2	Fail.
1% UVA-1	3.1	3.3	3.5	3.9	4.3	5.4	5.5	7.4	10.4
2% UVA-1	3.1	3.3	3.4	3.9	4.1	5.2	5.2	7.2	9.5

Table VI. Percent Elongation of 1-mil BPA-PC Film on Exposure in 340 FL

Formulation	0 h	285 h	535 h	797 h	973 h	1254 h	1473 h
Blank	90	36	18	6	1	NA	NA
1% UVA-1	96	97	64	NA	16	6	NA
2% UVA-1	95	118	70	NA	63	NA	11

NOTE: Five replicates pulled for each formulation. NA is not available.

Table VII. Percent Elongation of 1-mil BPA-PC Film on Exposure in XAW

Formulation	0 h	308 h	538 h	780 h	1019 h	1277 h
Blank	116	12	15	4	1	NA
1% UVA-1	99	77	37	NA	3	3
2% UVA-1	95	49	56	NA	5	3

NOTE: Four replicates pulled from each formulation. NA is not available.

Table VIII. Percent Elongation of 1-mil BPA-PC Film on 45 °South Exposure in Florida

Formulation	0 mo	1 mo	2 mo	3 mo	4 mo	5 mo	6 mo	9 mo	12 mo
Blank	116	67	29	11	5	3	2	Fail.	Fail.
1% UCA-1	99	50	45	27	23	12	6	3	2
2% UVA-1	95	52	47	35	14	6	9	4	3

NOTE: Six replicates pulled for each formulation.

in the near-UV region and tapered off into the visible region. This increase is in direct relation to increases in YI, which is a valuable technique for following photolysis of PC films that do not contain UVAs. The unstabilized films also underwent a significant broad increase in absorbance in the 3500 cm^{-1} region, presumably due to the formation of phenolic species. The IR spectra changed little for the films stabilized with UVA-1. IR analysis is a good qualitative technique for following photolysis of BPA-PC; however, quantification of a broad IR peak resulting from several species is not a simple task.

NMR Spectral Changes, 340 FL Exposure. Analysis by 500-MHz NMR spectroscopy was used to follow the photolysis of unstabilized PC films. Two types of protons were identified. Protons at 6.73 and 6.80 ppm in the ortho position relative to hydroxyl were a measure of the formation of phenolic species and accounted for about 2% of the aromatic protons after 1200 h exposure to 340 FL. Additionally, protons observed at 8.05 and 8.15 ppm that were ortho to a carbonyl were a measure of the amount of Fries photo-products formed. These protons accounted for about 0.5% of the aromatic protons after 1200 h exposure to 340 FL. This technique was not able to differentiate the performances of UVAs during short exposure periods: the NMR spectra of the films containing UVAs changed little during a typical accelerated exposure study. Longer exposures are needed to differentiate between UVAs by this technique.

Molecular Weight Changes, 340 FL Exposure. Films stabilized with UVA-1 maintained their molecular weights, whereas those of unstabilized BPA-PC films were reduced considerably during 1200 h exposure. (Table IX, 340 FL exposure). Difficulties in reproducibility were encountered

Table IX. Molecular Weight and Chain Scission Analysis of 1-mil Films on Exposure to 340 FL

Formulation	0 h	285 h	535 h	797 h	973 h	1254 h	1473 h
			$M_n \times 10^{-4}$				
Blank	1.60	1.34	1.31	1.17	1.04	0.74	NA
1% UVA-1	1.60	1.44	1.52	1.50	1.49	1.32	NA
2% UVA-2	1.60	1.58	1.61	1.54	1.55	NA	1.35
			Chain Scissions				
Blank	0	0.19	0.22	0.37	0.54	1.16	NA
1% UVA-1	0	0.11	0.05	0.07	0.07	0.21	NA
2% UVA-2	0	0.01	−0.01	0.04	0.03	NA	0.19

NOTE: Values reported are averages of duplicate GPC determinations. The precision is ±10% relative to polycarbonate standards. Values for M_n were calculated based on PC standards.

with the gel permeation chromatography method that was used for molecular weights. Changes in molecular weight can provide valuable insights into the photodegradation of BPA-PC.

Oxygen Uptake Results.   UVA-1 was evaluated to determine its effect on the photooxidation of BPA-PC. Film samples coated on the interior of Pyrex tubes were exposed under an oxygen atmosphere for approximately 600 h in an Applied Photophysics photoreactor equipped with 320-nm fluorescent lamps. Oxygen uptake was monitored continuously. UVA-1 was very effective at retarding photooxidation of PC: Oxygen uptake values for a blank and BPA-PC sample treated with UVA-1 (1 wt%) were 52,000 and 23,000 mmol/h, respectively.

Investigations with Poly(methyl methacrylate) Filters.   BPA-PC producers have responded to the difficulty in stabilizing the bulk polymer by focusing on stabilizing the surface of the polymer. In a limited investigation, a 10-mil poly(methyl methacrylate) filter containing 2% UVA-1 effectively stabilized the surface of BPA-PC. During 1600 h exposure in the 340 FL, results with the films behind filters without UVA-1 were reminiscent of results with blank, unfiltered, unstabilized films; they increased in YI from 3.5 at the start to 19.5. On the other hand, the films behind filters with 2% UVA-1 did not show any significant increase in YI. The same was true of changes in UV absorption spectra: The UV absorbance of the blank-filtered samples increased significantly, whereas those exposed to UV-filtered light had not changed at all in UV absorption characteristics. These results demonstrate the importance of stabilizing the surface of the polymer and the feasibility of a filtering approach to the light stabilization of BPA-PC.

Comparison of Results with Oxygen and Nitrogen Atmospheres. Unstabilized 1-mil BPA-PC films were exposed for 500 h in a Rayonet photoreactor equipped with fluorescent lamps centered at 310 nm, under both nitrogen and oxygen atmospheres, to compare the color development attributed to photooxidation with color development from Fries photorearrangements. On the basis of changes in UV-vis absorbance spectra, the color formation under nitrogen was slightly greater than under oxygen; therefore, photooxidation does not play a major role in the yellowing of BPA-PC (Figure 2).

Conclusions

- UVAs were found to be the most effective stabilizers against the photodegradation of BPA-PC. Other classes of stabilizing additives were generally ineffective when used alone and they contributed only marginally to the stabilization provided by a UVA.
- YI measurements of 1-mil BPA-PC films proved to be a reliable and reproducible measure of photodegradation of the polymer; difficulties with reproducibility were encountered with tensile testing of the 1-mil films.

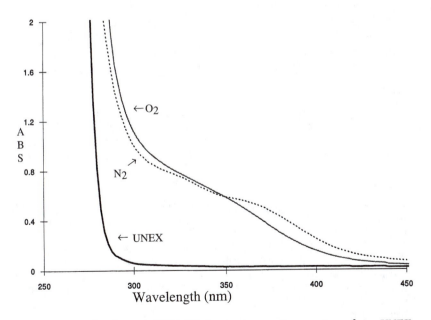

Figure 2. UV absorbance of BPA-PC for oxygen vs. nitrogen atmosphere. UNEX is unexposed polymer.

- The 340 FL exposure device was suitable for evaluating light stabilizers in BPA-PC.
- Monitoring changes in UV absorption spectra of BPA-PC on light exposure was useful for evaluating the photodegradation of films not containing UVAs.
- Natural weathering in Florida proved to be more severe than the indoor exposure devices. Percent elongation dropped faster, relative to YI, for the films exposed to natural weathering than for the films exposed in the 340 FL and the Xenon Arc weatherometer.
- NMR analysis at 500 MHz was useful for following photodegradation of unstabilized films. This technique was not useful for differentiating between UVAs during our typical exposure periods.
- IR spectroscopy was useful for qualitative monitoring of hydroxyl formation in BPA-PC during light exposure.
- Similar typical increases of UV absorption in spectra of BPA-PC were found with unstabilized films exposed to light in both oxygen and nitrogen. However, the exposure of the films in a nitrogen atmosphere caused slightly more color development than exposure in oxygen.
- Limited oxygen uptake results indicated UVAs significantly reduced the rate of photooxidation of BPA-PC.
- The absorbance of BPA-PC in the far-UV region of terrestrial sunlight made its photostabilization challenging because the polymer competed with UVAs for UV photons.
- The most effective stabilization of BPA-PC was obtained by exposing specimens behind PMMA filters containing a UVA.

References

1. Bellus, D.; Hrdlovic, P.; Manasek, Z. *Polym. Lett.* **1966**, *4*, 1–5.
2. Mullen, P. A.; Searle, N. Z. *J. Appl. Polym. Sci.* **1970**, *14*, 765–776.
3. Gupta, A.; Rembaum, A.; Moacanin, J. *Macromolecules* **1978**, *11*, 1285–1288.
4. Ong, E.; Bair, H. E. *Polym. Prepr. (Am. Chem. Soc. Div. Polym. Chem.)* **1979**, 945–948.
5. Gupta, A.; Liang, R.; Moacanin, J.; Goldbeck, R. *Macromolecules* **1980**, *13*, 262–267.
6. Moore, J. E. In *Photodegradation and Photostabilization of Coatings;* Pappas, S. P.; Winslow, F. H., Eds.; ACS Symposium Series 151; American Chemical Society: Washington, DC, 1981; pp 97–107.
7. Rivaton, A.; Sallet, D.; Lemaire, J. *Polym. Photochem.* **1983**, *3*, 463–481.
8. Torikai, A.; Murata, T.; Fueki, K. *Polym. Degrad. Stab.* **1984**, *7*, 55–64.
9. Pryde, C. A. In *Polymer Stabilization and Degradation;* Klemchuk., P. P., Ed.; ACS Symposium Series 280; American Chemical Society: Washington, DC, 1985; pp 329–351.
10. Gupta, M. C.; Tahilyani, G. V. *Colloid Polym. Sci.* **1988**, *266*, 620–623.
11. Gupta, M. C.; Pandey, R. R. *Makromol. Chem., Macromol. Symp.* **1989**, *27*, 245–254.
12. Clark, D. T.; Munro, H. S. *Polym. Degrad. Stab.* **1982**, *4*, 441–457.

13. Clark, D. T.; Munro, H. S. *Polym. Degrad. Stab.* **1984,** 8, 195–211.
14. Factor, A.; Chu, M. L. T *Polym. Degrad. Stab.* **1980,** 2, 203–223.
15. Factor, A.; Ligon, W. V.; May, R. J. *Macromolecules* **1987,** 20, 2461–2468.
16. Webb, J. D.; Czanderna, A.W. *Sol. Energy Mater. Sol. Cells* **1987,** 15, 1–8.
17. Webb, J. D.; Czanderna, A.W. *Macromolecules* **1986,** 19, 2810–2825.
18. Webb, J. D.; Czanderna, A. W. *Polym. Prepr. (Am. Chem. Soc. Div. Polym. Chem.)* **1987,** 28, 29–30.
19. Ram, A.; Zilber, O.; Kenig, S. *Polym. Eng. Sci.* **1985,** 25, 535–540.

RECEIVED for review December 6, 1993. ACCEPTED revised manuscript November 28, 1994.

Photostabilization of Macromolecular Materials by UV-Cured Protective Coatings

Christian Decker

Laboratoire de Photochimie Générale (Unité de Recherche Associée, Centre National de la Recherche Scientifique 431), Ecole Nationale Superieure de Chimie, Université de Haute Alsace, 68200 Mulhouse, France

The light stability of poly(vinyl chloride) (PVC) and wooden materials has been increased greatly by means of photocurable acrylic coatings containing benzotriazole and hindered amine light stabilizers. A 1-s exposure to the radiation of a powerful mercury lamp transformed the liquid resin into a highly cross-linked insoluble polymer. The UV-cured polyurethane–acrylate films show great flexibility, good resistance to scratching and abrasion, and excellent weatherability. When coated onto PVC panels, such protective films showed up to a sevenfold increase in the weathering resistance of both clear and pigmented PVC samples exposed to accelerated photoaging. The same procedure proved equally efficient for improving the light stability and surface properties of wooden materials.

POLYMER MATERIALS ARE KNOWN TO UNDERGO some photodegradation, when they are exposed to outdoor weathering, due to the combined action of solar radiation, atmospheric oxygen, humidity, and heat (*1, 2*). As a result, chain scissions, cross-links and oxidation products are formed in UV-exposed polymers. These chemical processes are mainly responsible for the changes observed in the mechanical and optical properties of photoaged polymers such as yellowing, hazing, cracking, and embrittlement. The outdoor lifetime of macromolecular materials can be extended substantially by the addition of UV absorbers, radical scavengers such as hindered amine light stabilizers (HALS), antioxidants, or pigments (*3, 4*). Both accelerated and outdoor weathering experiments have shown that light stabilizers remain efficient only for a certain period of time, dependent on both the type of stabilizer used and the chemical

0065–2393/96/0249–0319$12.00/0

structure of the polymer. The limited outdoor durability of stabilized polymers primarily is due to the slow disappearance through exudation, photolysis, and oxidation of the UV absorber (UVA) during photoaging (5).

Another effective method (6–9, 10a, 10b) used to increase the weathering resistance of macromolecular materials involves protecting their surfaces with highly resistant coatings typically made of strongly cross-linked polymers containing a UVA or pigment. The coating's main roles are to filter the most harmful solar radiation and, by acting as a physical barrier, to reduce the deleterious effects of moisture and oxygen. These coatings may also improve the treated materials' surface properties, such as, gloss, wettability, smoothness, and resistance to scratching and abrasion. Clear coats, paints and lacquers commonly are used for outdoor protection of a large variety of materials such as plastics, woods, papers, metals, and composites (6).

Although most organic coatings are made from solvent-based formulations, in recent years there has been a growing interest in UV-curable coatings, which have a number of distinct advantages (11, 12). The most striking advantage is the cure speed; usually less than 1 s of UV exposure is required to transform the liquid resin into a totally insoluble polymer. Moreover, no polluting solvent vapors are being released during the curing of these all-solid formulations. Finally, the reaction can be carried out at near ambient temperature, thus allowing application to heat-sensitive substrates.

In this chapter, we discuss the performance of some newly developed UV-curable acrylic coatings, which showed great resistance to accelerated weathering. They proved particularly effective in protecting macromolecular materials such as poly(vinyl chloride) (PVC) or wood against photodegradation and markedly improved their lifetime in outdoor applications.

Photocuring of Acrylic Coatings

Resin Formulation.   Photocurable resins consist of multifunctional monomers and oligomers, usually acrylates, and a photoinitiator that begins the cross-linking-polymerization following UV exposure. The basic principle of UV curing is represented in Scheme I.

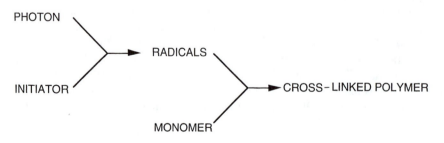

Scheme I. Basic principle of UV curing.

The UV-curable varnishes used to protect PVC or wood against photo-degradation were made of four basic components:

- a photoinitiator, which consists of an aromatic ketone, known to undergo a fast cleavage reaction upon photolysis to generate free radicals (Darocur 1173 from Ciba-Geigy or Lucirin TPO from BASF Corp.)
- a functionalized oligomer consisting of a short-chain aliphatic polyurethane [Actilane 20 from Societé Nationale de Poudres et Explosifs (SNPE)] or polyester chain [Ebecryl 830 from Union Chimique Belge (UCB)], end-capped by acrylate functions
- an acrylic monomer designed to reduce the resin viscosity, such as ethyldiethyleneglycol–acrylate (UCB), hexanediol diacrylate (UCB), or a carbamate monoacrylate (Acticryl CL-960 from SNPE)
- a light stabilizer that will effectively absorb the UV solar radiation without being destroyed; a hydroxyphenylbenzotriazole (Tinuvin 900 from Ciba-Geigy) was selected because of the good light stability of this class of UVA. A HALS radical scavenger (Tinuvin 292 from Ciba-Geigy) was also added to increase the light stability of the coating and the lifetime of the UVA (*13*)

The formulas of the various products are given in Charts I (photoinitiators, acrylate oligomers, and monomers) and II (light stabilizers). A typical formulation consisted of 3% photoinitiator, 1% photostabilizer, and equal parts of functionalized oligomer and reactive diluent. The UV-curable white lacquer also contained 20% of rutile titanium oxide.

Polymerization Kinetics. The resin was applied onto the support, in a uniform layer of controlled thickness (50 μm), by using a calibrated wire-wound applicator. Samples were exposed in the presence of air to the UV radiation of a 2-kW medium-pressure mercury lamp (International Scientific Technologies) for various durations up to 1 s. The extent of polymerization was determined quantitatively by IR spectroscopic monitoring of the decrease in absorption of the acrylate double bond at 812 cm^{-1}.

In the absence of a light stabilizer, the cross-linking polymerization develops rapidly and reaches >90% conversion after 0.1 s of exposure, as shown by the polymerization profiles in Figure 1. A 0.5-s exposure was sufficient to reach the maximum conversion value (93%) and achieve a nearly complete cure. No postirradiation polymerization was observed after storage of the UV-cured sample at ambient temperature in darkness or even in daylight. In the presence of 1% Tinuvin 900, the curing reaction proceeds about half as fast. This decrease is not due to reaction of the UVA with the photoinitiator but is due to the UVA internal-filtering effect, which partly prevents the incident photons from being absorbed by the photoinitiator. This effect leads to a profile of decreasing cure in the irradiated film from the surface to the in-

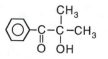
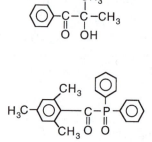

Darocur 1173

Lucirin TPO

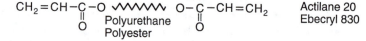

$CH_2=CH-\overset{\overset{\displaystyle O}{\|}}{C}-O \, \text{\small ⁓⁓⁓⁓} \, O-\overset{\overset{\displaystyle O}{\|}}{C}-CH=CH_2$ Actilane 20
Polyurethane   Ebecryl 830
Polyester

$CH_2=CH-\overset{\overset{\displaystyle O}{\|}}{C}-O\left(CH_2-CH_2-O\right)_2 CH_2-CH_3$ EDGA

$CH_2=CH-\overset{\overset{\displaystyle O}{\|}}{C}-O-(CH_2)_6-O-\overset{\overset{\displaystyle O}{\|}}{C}-CH=CH_2$ HDDA

$CH_2=CH-\overset{\overset{\displaystyle O}{\|}}{C}-O-CH_2-CH_2-NH-\overset{\overset{\displaystyle O}{\|}}{\underset{\underset{\displaystyle CH_3}{|}}{C}}-\overset{\overset{\displaystyle CH_3}{|}}{CH}$ Acticryl CL-960

Chart I. Formulas of photoinitiators, acrylate oligomers, and monomers.

nermost layers. The exposure time has to be doubled, therefore, to achieve a deep, through-cure of the coating and to ensure a good adhesion at the substrate interface. By contrast, HALS like Tinuvin 292 had no effect on the kinetics of such radical-induced polymerization (*13*). This unexpected result can be accounted for by considering that the polymerization of acrylic monomers can proceed only in an oxygen-depleted medium (*14*); under these conditions, the nitroxyl radicals can no longer be formed by photooxidation of HALS.

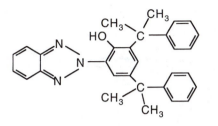

Tinuvin 900

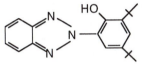

Tinuvin 326

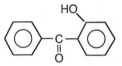

Chimasorb 81

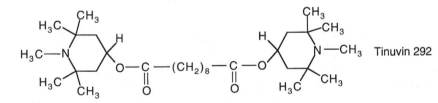

Tinuvin 292

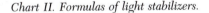

Chart II. Formulas of light stabilizers.

In the white lacquer, where the pigment particles act as an efficient screen, the cure-depth gradient and rate retardation were even more pronounced. Little difference in the curing kinetics was found between the unstabilized and UVA-stabilized coatings. By selecting the proper photoinitiator system (Darocur 1173 and Lucirin TPO from BASF Corp.), 50-μm-thick coatings could be cured at a belt speed of 10 m/min by using a 200-W/in. medium-pressure mercury lamp similar to those commonly used in UV-curing applications.

No detectable decay of the UVA occurred during the 1-s UV cure. This result is not surprising because the number of photons received during this

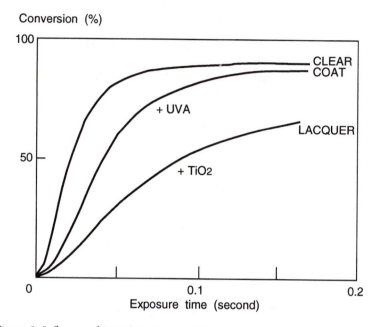

Figure 1. Influence of UVA (1% Tinuvin 900) and pigment (20% titanium oxide) on the UV-induced polymerization of a polyurethane–acrylate.

exposure was calculated to be the same as that received after only 7 min of photoaging in an accelerated weatherometer of the Q-Panel Company (QUV).

Properties of UV-Cured Coatings. The mechanical properties of UV-cured polymers are very much dependent on the functionality and chemical structure of both the oligomer and the monomer. Aliphatic polyurethane–acrylates associated to monoacrylate monomers give soft and highly flexible coatings, whereas aromatic polyethers or polyesters associated to di- or triacrylates give hard and scratch-resistant glassy polymers. By properly combining various monomers and oligomers, we succeeded (15) in producing hard but still flexible clear coats and lacquers that exhibit good resistance to both scratching and shocks. In addition, all of these coatings have a smooth, glossy surface with low wettability and good resistance to moisture and weathering.

Photoaging of Coated Polymers

Light Stability of UV-Cured Acrylic Coatings. The weathering resistance of clear coats and lacquers was tested by exposing the samples in a QUV weatherometer operated at 40 °C under continuous illumination with fluorescent UVB-313 or UVA-340 lamps. The chemical modifications occurring in the irradiated film were monitored by IR spectroscopy. The loss of UVA and photoyellowing were followed by UV–visible spectroscopy.

The stabilized coating containing both Tinuvin 900 and Tinuvin 292 remained perfectly clear and glossy after 2000 h of QUV-B aging. At that stage of the degradation, one could detect only minor changes in the IR spectrum, mainly a small increase of the OH absorbance and a 10% loss of the NH group (Figure 2). Such highly cross-linked polymers appear much more resistant to photodegradation than the linear homologous polymers and have quantum yield values on the order of 10^{-4} mol E^{-1} for the NH consumption and OH formation (13). Improved light stability can be accounted for by considering of the restricted segmental mobility in cross-linked polymers, which favors cage recombination of primary radicals over chain propagation.

Tinuvin 900 steadily disappeared according to a single exponential law and had a lifetime of about 3000 h. This value strongly depends on the overall light absorbance of the coating and, therefore, on film thickness, UVA concentration, and UVA type (5). A direct consequence of the internal filter effect is that the UVA disappears much more slowly in the bottom layer of the coating than in the top layer exposed to UV radiation, and this disparity leads to a sharp UVA-distribution gradient within the irradiated film (5). By retard-

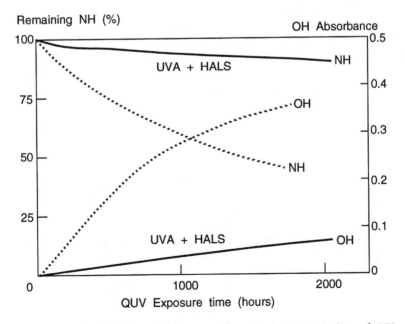

Figure 2. Accelerated photoaging by continuous irradiation at 40 °C with UV-B-313 bulbs of a stabilized, UV-cured polyurethane–acrylate (solid line) and unstabilized coating (broken line). UVA, 1% Tinuvin 900; HALS, 1% Tinuvin 292.

ing the degradation at the coating-substrate interface, the UVA contributes to the maintenance of good adhesion during weathering.

Three processes are mainly responsible for UVA loss during photoaging (5):

- direct photolysis
- radical-induced photooxidation
- exudation

The relative importance of these processes, as well as the stabilizing effect of HALS on UVA lifetime, are illustrated in Figure 3 for a UV-cured poly-urethane–acrylate containing 1% Tinuvin 900.

As stated by a basic law of photochemistry, a stabilized polymer will be better protected against UV radiation if the UVA is concentrated in a top coat rather than distributed uniformly in the whole sample. This effect is clearly illustrated in Figure 4. The same amount of UVA was used to stabilize a 90% transparent polymer by introducing it in the bulk polymer (internal filter) and by concentrating it exclusively in a top coat (external filter). When the UVA absorbs 90% of the incident light, the stabilization efficiency, based on the amount of photons absorbed by the polymer, is 4 times greater for the coated polymer. This ratio rises to 20, simply by doubling the UVA concentration or the film thickness (Figure 4). This approach constitutes the basis of the method used here to increase the light stability of macromolecular materials by means of photo-cured coatings containing a UVA.

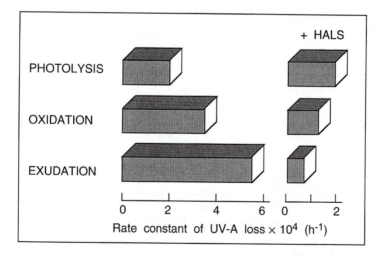

Figure 3. Influence of HALS (1% Tinuvin 292) on the various processes responsible for the UVA loss during QUV aging of a UV-cured polyurethane–acrylate coating.

Photostabilization of Poly(vinyl chloride).   Among the important industrial polymers, PVC is probably the one most sensitive to weathering. Its photodegradation kinetics and mechanism have been investigated thoroughly (for a review, *see* reference 16). The main effects of UV radiation and oxygen on PVC are represented in Scheme II.

Even a well-stabilized, clear PVC sheet undergoes oxidation, discoloration, hazing, and loss of impact strength after a few years of outdoor exposure or a few weeks of accelerated weathering (Figure 5). After a given induction period, the length of which depends on the stabilizer used, the PVC becomes discolored and loses its transparency and impact strength.

Because of their thermoplasticity, PVC-based compounds cannot be coated with baked finishes loaded with UVAs. By simply covering a well-stabilized, clear PVC panel (Lucoflex) with an acrylic coating containing 1% Tinuvin 900, the plate's resistance to QUV accelerated weathering can be

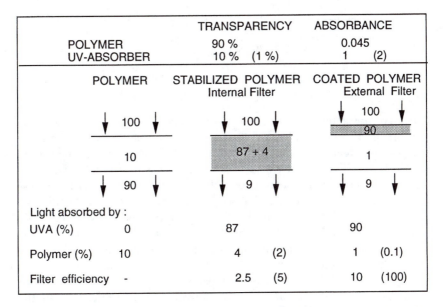

		TRANSPARENCY	ABSORBANCE
POLYMER		90 %	0.045
UV-ABSORBER		10 % (1 %)	1 (2)

POLYMER	STABILIZED POLYMER Internal Filter	COATED POLYMER External Filter
100	100	100 90
10	87 + 4	1
90	9	9

Light absorbed by :		
UVA (%) 0	87	90
Polymer (%) 10	4 (2)	1 (0.1)
Filter efficiency -	2.5 (5)	10 (100)

Figure 4. Internal vs. external filter effect in photodegradation of stabilized polymer.

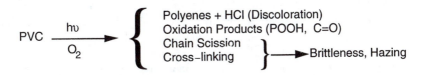

$$PVC \xrightarrow[O_2]{h\upsilon} \begin{cases} \text{Polyenes + HCl (Discoloration)} \\ \text{Oxidation Products (POOH, C=O)} \\ \left. \begin{array}{l} \text{Chain Scission} \\ \text{Cross–linking} \end{array} \right\} \longrightarrow \text{Brittleness, Hazing} \end{cases}$$

Scheme II. Effects of UV radiation and oxygen exposure on PVC.

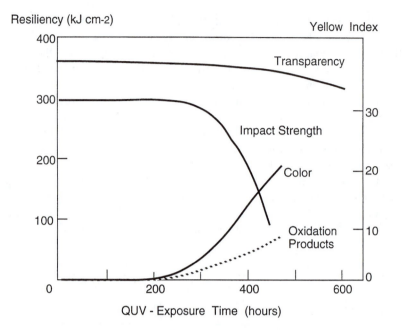

Figure 5. Accelerated QUV-B-313 weathering of a stabilized, clear PVC.

improved substantially (*see* reference 17 and Figure 6). Diacrylate monomers were more efficient for PVC stabilization than monoacrylates, most probably because the higher cross-link density of the UV-cured coating reduces the UVA loss rate. The coated PVC's weatherability was further improved by reinforcing the polyurethane elastomer with stiffer polyester units (Ebecryl 830). Finally, the greatest increase in the PVC durability was achieved by adding to the coating formulation 1% Tinuvin 292, a HALS that markedly retarded UVA loss (*see* reference 5 and Figure 6). It should be mentioned that a coating containing no UVA or Tinuvin 292 alone had no effect on the PVC photostability, as expected from the high transparency of such films to the radiation emitted by UVB-313 and UVA-340 lamps.

Because the photostability of PVC is directly related to the UVA lifetime in these tests, this system can be used as a probe to test the photostability of various UVAs in a given coating (Figure 7). The same order of UVA light stability was obtained by monitoring either the UVA loss (5) or the discoloration of the coated PVC sample. Another consequence of this relationship is that, by recording the exponential decay of the UVA in the early stages of the photoaging, it is possible to predict the long-term durability of coated polymer materials.

The great efficiency of this method of polymer stabilization was confirmed by photoaging experiments carried out in a Xenon WeatherOmeter for up to

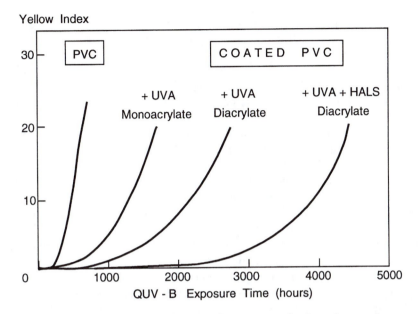

Figure 6. *Photostabilization of clear PVC by a UV-cured polyurethane–polyester acrylate coating containing a UVA (1% Tinuvin 900).*

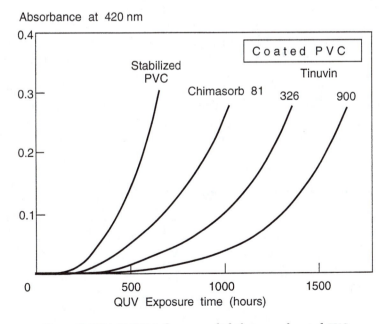

Figure 7. *UVA (0.5%) influence on lightfastness of coated PVC.*

10,000 h and by outdoor exposure. The same process was used to protect colored PVC panels against weathering. With the best-performing system, no color fading was observed after 3000 h of QUV-B exposure. Similar long-lasting protection was obtained by using UV-cured polyurethane lacquers, which can be applied to either clear or pigmented PVC panels.

Table I summarizes the stabilization performance of the various coated PVC specimens exposed to different types of accelerated weathering. With the Tinuvin 900–Tinuvin 292 combination, the service life of a well-stabilized PVC was increased by a factor of at least 5 by using a UV-cured clear coat and by a factor of over 10 by using a pigmented coating. The stabilization efficiency increased exponentially with the UVA concentration and with the film thickness, as expected from the Beer–Lambert law. Therefore, outstanding weathering resistance was achieved by protecting the PVC specimen with a 100-μm-thick coating containing 2% UVA. However, the high absorbance of this coating makes it difficult to cure the resin down to the substrate interface by using UV radiation, and adhesion failure results. In this case, electron-beam curing, although more expensive to use than UV technology, appears to be more appropriate, because the polymerization reaction would no longer be adversely affected by the UVA. Indeed, except for the initiation step (and the UVA internal filter effect), the chemistry involved in electron-beam curing of acrylate monomers was essentially the same as that involved in UV-radiation curing (18). An additional advantage of the highly cross-linked coating is that it acts as an efficient superficial barrier that prevents stabilizers and plasticizers from diffusing out of the PVC substrate.

An interesting feature observed in this study is that, for PVC panels stabilized with a UVA, the degradation process develops mainly in the 50-μm-thick top layer exposed to radiation. Therefore, highly photodegraded, clear or pigmented PVC panels can be renovated by a two-step process (see reference 19 and Figure 8). The brown, top layer is first removed by sandpapering or sandblasting, treatments that make the color disappear and restore the original impact strength as measured by the Mouton pendulum. This surface treatment was performed at ambient temperature on a 2-mm-thick PVC

Table I. Photostabilization of Clear PVC by UV-Cured Acrylate Coatings

Coating System[b]	Stabilization Efficiency[a]		
	QUV-B-313	QUV-A-340	Xenotest
Clear coat	1	1	1
Clear coat and UVA	4	3.6	3.2
Clear coat, UVA, and HALS	7	6.5	5.3
White lacquer, UVA, and HALS	>10	>10	—

[a]Stabilization efficiency is $t_{coated\ PVC}/t_{PVC}$, where t is the exposure time required to reach a Yellow Index of 10. — means test was not performed.
[b]UVA is 1% Tinuvin 900, and HALS is 1% Tinuvin 292.

panel until the top 50-µm layer was removed. A UV-curable coating is then applied to the PVC panel, which recovers its initial transparency and becomes more resistant to weathering. The main characteristics of the PVC panel at the different stages of the recycling treatment are illustrated by Figure 9.

This new method of renovating macromolecular materials that have suffered an extensive deterioration from outdoor exposure can be done with any type of clear or pigmented polymer, provided that degradation reactions are restricted to the radiation-exposed top layer. This fast and inexpensive recycling treatment presents several distinct advantages that makes it economically

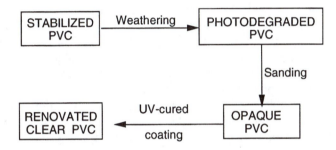

Figure 8. Two-step recycling process of photodegradable PVC specimens.

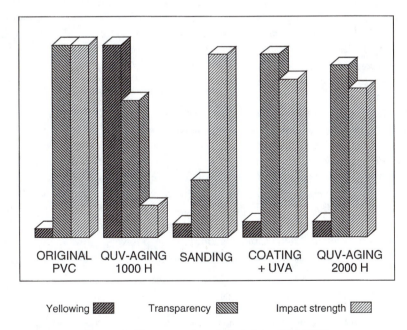

Figure 9. Characteristics of photodegradable PVC.

attractive: it is based on proven technologies, it can be carried out on-site, and it leads to a high-quality final product.

Light Stabilization of Wood Panels.   Unprotected wood is destroyed by natural weathering from the effect of both solar radiation and water; its color changes progressively to dark brown and finally to gray (6). By interacting with lignin, UV radiation generates free radicals that initiate oxidation reactions and ultimately lead to the production of hydroperoxide, carbonyl, and carboxyl groups (20). This degradation process can be prevented, to some extent, by protecting the wood surface with a coating that acts as a barrier against UV radiation and moisture. The coating must also exhibit long-term elasticity to match the dimensional instability caused by the hygroscopicity of wood. The best surface protection is obtained with paints or lacquers because of the strong screening effect of the colored pigment. With clear coats, which are often preferred to preserve the original aspect of the wood structure, long-lasting weathering protection is difficult to achieve because of the inevitable and progressive loss of UVA.

The UV-curable polyurethane–acrylate coating resins developed in this study have the high flexibility and the abrasion resistance required for application. These coatings also provide an efficient filtering of UV radiation by using Tinuvin 900–Tinuvin 292. After 3000 h of QUV aging, the coated pine- and beech-tree specimens showed no significant color change. The acrylic coating remained clear and glossy and retained good adhesion to the wooden substrate. The external filtering effect can be enhanced further by adding to the resin formulation small amounts of iron oxides (2%), which act as long-lasting UVAs but impart some coloration to the coating. The best protection against weathering was achieved by rubbing the wood panel surface with Tinuvin 292–Tinuvin 1130 in polyethyleneglycol (21) before covering the panel with a 50-µm-thick, UV-cured, clear coat containing 1.5% Tinuvin 900–1% HALS. Under these conditions, more than 4000 h of QUV aging are required before some degradation of the coated material can be observed. The preservative treatment can also be made with chromium salts such as chromated copper arsenate, which was very effective in protecting wood from UV degradation and improved the service life of semitransparent stains in outdoor exposure (22, 23). As expected from the screen effect, even greater weathering resistance (> 6000 h) was obtained by protecting the wood specimens with a white, UV-cured, polyurethane–acrylate lacquer pigmented with 20% titanium oxide and the same light-stabilizer combination. However, this high-performance wood finish does not preserve the natural color and texture of wood.

Conclusions

The weathering resistance of macromolecular materials can be enhanced by photocurable acrylic coatings containing hydroxy-benzotriazole and HALS stabilizers. A 1- s exposure was sufficient to obtain a highly cross-linked polymer showing great flexibility, good resistance to abrasion and scratching, and excellent weatherability. By filtering the harmful solar radiation, such protective coatings prevent most of the degradation of the organic substrate and the dechlamination often observed after prolonged weathering. The presence of HALS in the coating is particularly important. The HALS increases the coating durability, the UVA lifetime, and consequently, the coated material lifetime.

In PVC and wood, protecting the polymer surface by UV-cured coatings was a convenient and effective way to increase the outdoor lifetime of these polymer materials and to improve some of their surface properties. This fast and economical method of photostabilization can be extended to any kind of polymer, composite material, painted surface, or even to commercial products already well-stabilized by conventional methods, as long as one can ensure good adhesion of the coating to the substrate. The method can also be used to renovate strongly photodegraded, polymer-made structures by applying, fast-drying clear varnishes or lacquers after surface etching and cleaning. Therefore, this new approach toward photostabilization and recycling of macromolecular materials is expected to be of special interest in the building industry, where outstanding light stability and good mechanical and surface properties are essential.

References

1. Rabek, J. F. *Mechanisms of Photophysical Processes and Photochemical Reactions in Polymers*; Wiley: New York, 1987.
2. Decker, C. In *Handbook of Polymer Science and Technology*; Cheremisinoff, N. P., Ed.; Dekker: New York, 1989; Vol. 3, p 541.
3. *Development in Polymer Stabilization*; Scott, G., Ed.; Applied Science: Barking, Essex, England, 1979–1984; Vols. 1–7.
4. Gugumus, F. In *Developments in Polymer Stabilization*; Scott, G., Ed.; Applied Science: Barking, Essex, England, 1979; Vol. 1, p 261.
5. Decker, C.; Zahouily, K. *Polym. Mater. Sci. Eng.* **1993,** *68,* 70.
6. Schmidt, E. V. *Exterior Durability of Organic Coatings*; FMJ International: Redhill, Surrey, England, 1988.
7. Berner, C.; Rembold, M. In *Organic Coatings*; Parfitt, G.; Patsis, A., Eds.; Dekker: New York, 1984; Vol. 7, p 55.
8. Gerlock, J. L.; Bauer, D. R.; Briggs, L. M. *Polym. Degrad. Stab.* **1986,** *14,* 53, 73.
9. Klemchuk, P. P.; Gande, M. E. *Polym. Degrad. Stab.* **1988,** *22,* 241.
10. (a) Bauer, D. R.; Gerlock, J. L.; Mielewski, D. F.; Caputa Peck, M. C.; Carter, R. O., III. *Polym. Degrad. Stab.* **1990,** *27,* 271; (b) Bauer, D. R.; Gerlock, J. L.; Mielewski, D. F.; Caputa Peck, M. C.; Carter, R. O., III. *Polym. Degrad. Stab.* **1990,** *28,* 39.
11. Roffey, C. G. *Photopolymerization of Surface Coatings*; Wiley: Chichester, Sussex, England, 1982.
12. Decker, C. *J. Coat. Technol.* **1987,** *59* (*751*), 97.
13. Decker, C.; Moussa, K.; Bendaikha, T. *J. Polym. Sci. Polym. Chem. Ed.* **1991,** *29,* 739.
14. Decker, C.; Jenkins, A. *Macromolecules* **1985,** *18,* 1241.

15. Decker, C.; Moussa, K. *J. Coat. Technol.* **1993,** 65 (819), 49.
16. Decker, C. In *Degradation and Stabilization of PVC*; Owen, E., Ed.; Elsevier Applied Science: London, 1984; p 81.
17. Decker, C. In *Chemical Reactions on Polymers*; Benham, J. L.; Kinstle, J. F., Eds.; ACS Symposium Series 364; American Chemical Society: Washington, DC, 1988; p 201.
18. *UV and EB Curing Formulation for Printing Inks, Coatings and Paints*; Oldring, P. K. T., Ed.; SITA: London, England, 1991; Vols. 1–4.
19. Decker, C. *Makromol. Chem. Macromol. Symp.* **1992,** 57, 103.
20. Feist, W. C.; Hon, D. N. S. In *The Chemistry of Wood*; Rowell, R. M., Ed.; Advances in Chemistry Series 207; American Chemical Society: Washington, DC, 1984.
21. Forsskahl, I.; Janson, *J. Pap. Puu* **1992,** 74 (7), 553.
22. Feist, W. C.; Williams, R. S. *For. Prod. J.* **1991,** 41(1), 8.
23. Williams, R. S.; Feist, W. C. *Polym. Mater. Sci. Eng.* **1993,** 68, 216.

RECEIVED for review January 26, 1994. ACCEPTED revised manuscript September 20, 1994.

Measurements of Chemical Change Rates to Select Superior Automotive Clearcoats

J. L. Gerlock[1], C. A. Smith[1], E. M. Núñez[1], V. A. Cooper[2], P. Liscombe[2], D. R. Cummings[2], and T. G. Dusibiber[3]

[1]Food Research Laboratory; [2]Ford Central Laboratory; and
[3]Ford Body and Assembly, Paint Operations; Ford Motor Company,
Dearborn, MI 48121-2053

Hydroperoxide concentration behavior measurements and transmission Fourier transform infrared spectroscopy measurements were used to compare the photooxidation resistance of three acrylic–melamine clearcoats. Ultraviolet spectroscopy was used to compare the longevity of additives that absorb ultraviolet light for the same clearcoats. Samples were subjected to SAE J1960 JUN89 xenon-arc accelerated exposure. No dramatic difference in clearcoat photooxidative degradation rate was found, but dramatic differences in ultraviolet light absorber longevity were found. Paint systems based on the clearcoat that exhibits poor ultraviolet light absorber longevity exhibited poor Florida weathering performance.

T HE CLEARCOAT LAYER in multilayer clearcoat/basecoat/primer/electrocoat automotive paint systems can play a key role in determining weathering performance. The clearcoat must remain intact and screen underlying coating layers from UV light during exposure. The best way to determine the weathering performance of such paint systems is to expose them outdoors and follow their physical performance. The time to complete such tests, >5 years, is not always practical. Consequently, coating selections are often based on the results of accelerated weathering tests. Unfortunately, the use of harsher than natural exposure conditions can distort the chemistry of degradation during accelerated exposure relative to degradation chemistry that is driven by outdoor exposure, and these harsher conditions can yield misleading results (1).

0065–2393/96/0249–0335$12.00/0

Work in our laboratory is focused on the development of weathering tests that need not resort to harsh exposure conditions to shorten test time. It should be possible to use sensitive analytical techniques to detect chemical changes in coatings long before physical property changes can be measured. Coatings that undergo rapid chemical change should exhibit inferior weathering performance. All of our work in this area was reviewed recently by Bauer (2). Most of our work has focused on isolated clearcoats and not paint systems, because no technique was available to follow chemical changes in individual coating layers in complete paint systems as a function of exposure.

The object of the present work was to measure chemical change rates for three acrylic–melamine clearcoats, X, Y, and Z, and compare the results with Florida exposure results for paint systems based on the same clearcoats. Two measurements of chemical change rates were used; transmission Fourier transform IR spectroscopy (FTIR) (3–5) and hydroperoxide concentration behavior analysis (6–8). Florida exposure results indicate that paint systems based on clearcoat X are dramatically inferior to paint systems based on clearcoats Y and Z. Paint systems based on clearcoat X fail by clearcoat peeling.

We realized from the onset of this work that comparing Florida exposure results for paint systems with laboratory exposure results for isolated clearcoats need not be straightforward. First, laboratory clearcoat samples were weathered under SAE J1960 JUN89 xenon-arc exposure conditions, an automotive industry standard that affords shorter than natural UV light. The presence of UV light that has a shorter than natural wavelength may distort chemical degradation in the clearcoats studied. We expect that the present work will be repeated using borosilicate–borosilicate filtered xenon-arc light to examine this possibility. Also, paint-system weathering performance may not be accurately assessed by determining the weathering performance of isolated clearcoats in the absence of underlying coating layers.

The clearcoats studied in the present work are proprietary materials. Their chemical compositions and additive packages are not open for discussion. Even though it may be frustrating to discuss clearcoat weathering performance without being able to relate what is observed to polymer composition or additive effectiveness, a successful test method must assume this position in practice. Suppliers must be able to submit coatings for testing without divulging specific chemical information.

Experimental

Resin Samples.   Four versions of each clearcoat were tested; additive free clearcoat, clearcoat with hindered amine light stabilizer (HALS), clearcoat with benzotriazole UV light absorber (bz-UVA), and clearcoat with both HALS and UVA additives. The amount and nature of HALS and UVA additives used is not necessarily the same in each clearcoat.

Exposure Conditions.   Samples were exposed side-by-side in a Ci-65 Atlas Xenon arc Weather-ometer set to operate at SAE J1960 JUN89 conditions: 40 min. light at 0.55 W/m² at 340 nm, quartz inner filter, borosilicate outer filter (black panel T = 70 °C), 20 min. light with front water spray (black panel T = 70 °C, water T = 45 °C), 60 min. light (black panel T = 70 °C), 120 min. dark with front and back water spray (black panel T = 70 °C; water T = 40 °C).

Hydroperoxide Analysis.   Hydroperoxide analysis was performed by iodometric titration of cryoground clearcoat samples swelled with methylene chloride according to procedures previously described (*6–8*). Samples were prepared as 1.8–2.2-mm thick films on 5- × 7-in. glass panels. This material is sufficient for two hydroperoxide determinations per panel.

Fourier Transform IR Spectra.   FTIR spectra were recorded with a Galaxy 5020 FTIR spectrometer. Spectra were recorded as a function of exposure time for clearcoats prepared as thin (8–12-μm) films on 13-mm × 2-mm silicon disks. The silicon disks were notched to allow them to be positioned reproducibly in the FTIR spectrometer by using a sample holder with a corresponding pin.

Ultraviolet Spectra.   UV spectra were recorded with a Cary 2300 UV spectrometer. Samples were cured on 1- × 3.6-cm quartz slides. Film thickness was adjusted to obtain a starting absorbance between 0.5 and 0.8 in the 340-nm region.

Results and Discussion

Hydroperoxide analysis is a well-developed technique for polymers that can be solubilized (*9*); however, its application to insoluble cross-linked polymers is a new development. The details of the analysis procedure are described in references 6–8.

Bauer et al. (*10*) demonstrated that a kinetic analysis of the standard free-radical chain photooxidation reaction scheme, wherein the steady-state approximation is applied to free radical-intermediates and hydroperoxide, suggests that photooxidation should proceed in direct proportion to hydroperoxide concentration according to eq 1 (for a conflicting point of view, *see* ref. 11):

$$\text{Photooxidation Rate} = M + K[\text{YOOH}] \qquad (1)$$

where K is the sum of the rate constants for the photolytic decomposition of hydroperoxides into free radicals, k_i, and the rate constant for the decomposition of hydroperoxides in dark reactions, k_d, which do not initiate free-radical oxidation. The quantity M does not arise from the kinetic analysis; M arises from the observation that cross-link scission is observed (FTIR) in acrylic–melamine clearcoats that do not contain detectable hydroperoxide. Scission not driven by free radicals may be the result of photoinduced hydrolysis (*8*). This conclusion is supported by the fact that M decreases to zero for acrylic–melamine clearcoats photolyzed in the absence of humidity and is zero for acrylic–urethane clearcoats regardless of humidity.

The results of hydroperoxide concentration measurements for clearcoats X, Y, and Z are shown in Figures 1–4. Data are scattered in all cases, but trends are obvious. Hydroperoxide concentrations approached a near steady level after 500 h exposure in most cases. No evidence of an induction period or autocatalytic photooxidation was seen in any case. If K and M of eq 1 are assumed to be comparable for the three acrylic–melamine clearcoats, sus-

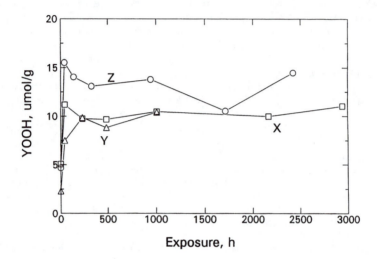

Figure 1. Hydroperoxide concentration behavior for additive-free clearcoats: X (□), Y (△), and Z (○).

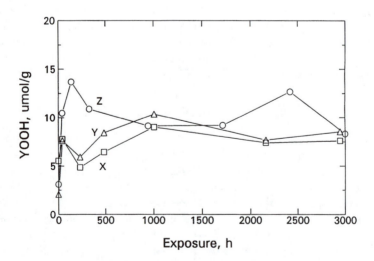

Figure 2. Hydroperoxide concentration behavior for clearcoats stabilized with HALS: X (□), Y (△), and Z (○).

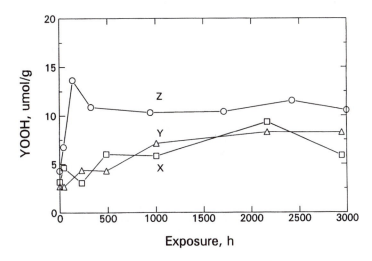

Figure 3. Hydroperoxide concentration behavior for clearcoats stabilized with UVA: X (□), Y (△), and Z (○).

tained hydroperoxide concentration should be proportional to photooxidation rates.

The curves for additive-free clearcoats suggest that clearcoat Y photooxidized 30% faster than clearcoats X and Z. The data for clearcoat Y are limited because it lost its adherence to the glass substrate after 500 h of exposure.

The addition of HALS reduced hydroperoxide concentration in all cases (Figure 2). The differences in intrinsic photooxidation resistance suggested by the curves shown in Figure 1 for additive-free clearcoats were leveled by the addition of HALS.

The addition of a UVA to a clearcoat is thought to lead to a gradient in light intensity through the clearcoat and presumably a corresponding gradient in photooxidative degradation. Therefore, hydroperoxide concentration values for clearcoats with UVA (Figures 3 and 4) cannot be compared with hydroperoxide concentration values for clearcoats without UVA (Figures 1 and 2). The hydroperoxide analysis procedure used in the present work yielded an average value that ignores any gradient in hydroperoxide concentration. Hydroperoxide concentration behavior can be compared among clearcoats containing similar bz-UVAs provided film thickness is held relatively constant. When UVA was added, clearcoat Z exhibited a higher steady hydroperoxide concentration than clearcoats X and Y.

Finally, when HALS and UVA were added, the three clearcoats did not exhibit dramatically different hydroperoxide concentrations (Figure 4). There is no case where the addition of both HALS and UVA additives dramatically reduced hydroperoxide concentration by more than the sum of the effects of both additives.

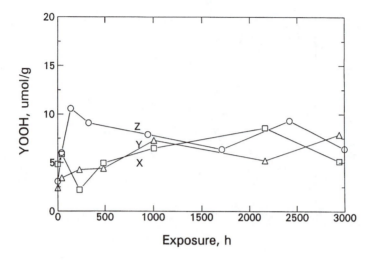

Figure 4. Hydroperoxide concentration behavior for clearcoats stabilized with HALS and UVA: X (□), Y (△), and Z (○).

Summarizing, hydroperoxide concentration behavior data is very scattered, but still clearly reflects the influence of additives. The curves for fully formulated clearcoats X, Y, and Z reveal no dramatic differences in photooxidation resistance or kinetics. No indication exists that paint systems based on clearcoat X should exhibit inferior weathering performance relative to paint systems based on clearcoats Y and Z.

The relatively high sustained hydroperoxide concentration observed for fully stabilized clearcoats, >7 μmol/g, suggests that the photooxidation resistance of each clearcoat could be improved considerably. However, the relatively large initial hydroperoxide titers observed are not understood. High initial hydroperoxide titers are not seen for laboratory-version clearcoats that do not contain all of the additives present in commercial formulations.

FTIR spectroscopy is a well-developed technique to follow functional-group changes in clearcoats as a function of weather exposure (3–5). Superior clearcoats invariably exhibit low functional group change rates. The FTIR spectra of all clearcoats studied in the present work decreased in intensity during exposure. This result clearly indicates that film erosion occurred and suggests that the film-erosion rate could be used to follow clearcoat photooxidation rate. The relationship between film-erosion rate and photooxidation rate is empirical, because no information regarding the nature of the material lost is available. An absolute measurement of film erosion is possible because spectra were consistently recorded for the same sample spot as a function of exposure. If –CH region intensity (2900–3000 cm^{-1}) is taken as a measure of FTIR spectrum intensity, plots of the –CH region peak height versus exposure are linear down to >80% intensity decrease for all samples tested. The shape of the –CH region absorbencies does not change during exposure. The –CH peak height can be converted to film thickness by using the –CH peak-height measurements for films of known thickness to yield the film-erosion rates summarized in Table I.

The effects of HALS and UVA additives on film-erosion rates strongly suggest that film erosion is linked to photooxidation. Film erosion was fastest for additive-free clearcoat, slower with the addition of either HALS or UVA, and slowest when both HALS and UVA were present.

The ability of UVAs to slow film erosion suggests that erosion may not be a simple surface process. From known absorbance per mil values and starting sample thickness, UVAs are estimated to slow photooxidation by <30%. The larger values observed for clearcoats X and Y may indicate that the UVA is behaving like an antioxidant. There is no case where the addition of both HALS and UVA additives dramatically slowed film erosion by more than the sum of either additive's ability to slow film erosion.

The film-erosion rates determined for fully stabilized clearcoats reveal no dramatic difference in the rate at which each clearcoat loses film thickness, and therefore the rates provide no clear indication that paint systems based on clearcoat X will exhibit inferior weathering performance relative to paint systems Y and Z.

The rate of accumulation of carbonyl-containing photooxidation products can be obtained by observing the ratio of carbonyl-region intensity at any time to –CH peak intensity at any time. Carbonyl intensity was determined by measuring the distance from the baseline of the spectrum to the bottom of the valley between acrylate carbonyl and subsequent absorbance peaks. Previous work has shown that this measurement yields the same results as integrating carbonyl region after subtracting initial spectrum intensity until –CH peak absorbance is zero. Plots of $>C=O$ intensity/–CH intensity are shown in Figures 5–8. It is not clear how film loss should effect carbonyl-growth measurements. For the purposes of the present work, carbonyl-growth data was limited to the point during exposure when half of the initial sample thickness was eroded away to minimize the possible effects of selectively removing specific components. The point at which this removal occurs, as calculated from the film loss data shown in Table I and the known starting film thickness

Table I. Film Loss Rates

Additive	X	Supp. (%)	Y	Supp. (%)	Z	Supp. (%)
None	9.4	—	9.8	—	7.7	—
HALS	6.7	29	5.5	44	5.6	27
UVA	5.0	47	5.1	48	5.3	31
HALS and UVA	3.5	62	4.0	59	3.7	52

NOTE: Values for X, Y, and Z are mil/h $\times$ 10^{-5} ($\pm 10\%$); Supp. is suppression.

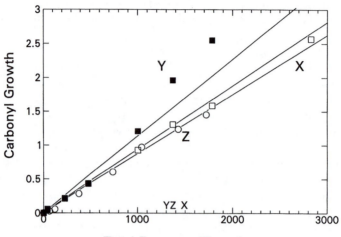

Figure 5. FTIR carbonyl growth for additive-free clearcoats. Lowercase symbols indicate point during exposure at which ~1/2 of initial sample thickness is lost: X (□), Y (■), and Z (○).

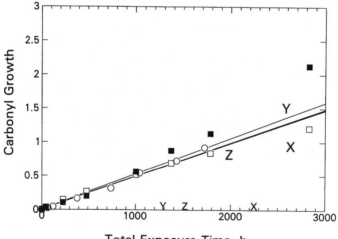

Figure 6. FTIR carbonyl growth for clearcoats stabilized with HALS. Lowercase symbols indicate point during exposure at which ~1/2 of initial sample thickness is lost: X (□), Y (■), and Z (○).

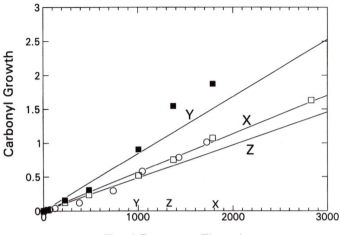

Figure 7. FTIR carbonyl growth for clearcoats stabilized with UVA. Lowercase symbols indicate point during exposure at which ~1/2 of starting sample thickness is lost: X (□), Y (■), and Z (○).

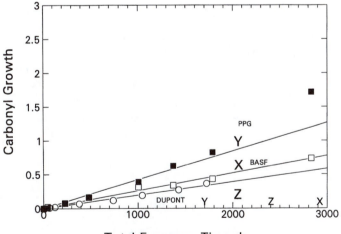

Figure 8. FTIR carbonyl growth for clearcoats stabilized with HALS and UVA. Lowercase symbols indicate point during exposure at which ~1/2 of initial sample thickness is lost: X (□), Y (■), and Z (○).

of each sample, is marked by a lowercase x, y, or z in each plot. Carbonyl-growth data is summarized in Table II.

As was the case for the film-erosion rate, carbonyl growth is fastest for additive-free clearcoat, slower with the addition of either HALS or UVA, and slowest when both additives are present. Both film-erosion and carbonyl-growth rates appear to be clearly linked to the photooxidation rate. The agreement between the ability of additives to slow film erosion and carbonyl growth, Tables I and II, is quite good in most cases.

Carbonyl-growth rates measured for fully stabilized clearcoat suggest that clearcoat Y photooxidizes considerably faster than clearcoats X and Z, and this result is in disagreement with hydroperoxide and film-loss results. It is possible that carbonyl-growth results are distorted by the selective erosion of specific components. For example, selective loss of triazine ring could increase the carbonyl content of films by leaving acrylate copolymer residue behind. With this reservation about carbonyl-growth results in mind, we concluded that neither hydroperoxide nor FTIR analysis results provide a clear indication that paint systems based on clearcoat X should exhibit inferior weathering performance relative to paint systems based on clearcoats Y and Z. This result suggests that an explanation for the poor weathering performance of paint systems based on clearcoat X must lie elsewhere.

One way that clearcoats can effect paint-system weathering performance other than by degrading and losing their physical integrity is by transmitting UV light to underlying coating layers. Even though the photophysics of bz-UVAs has been studied extensively (12), less attention has been given to their chemistry (13–16) and even less to their longevity in polymers (17–22).

The longevity of the bz-UVAs used in clearcoats X, Y, and Z was followed by measuring the decrease in the clearcoat absorbance maximum in the 344-nm region as a function of exposure. Measurements were carried out on thin films whose starting absorbance was <0.8. UVA absorbance loss follows first-order kinetics in this absorbance region, and it is necessary to compensate absorbance-loss rate for average light intensity, $<I>$. $<I>$ can be calculated using eq 2, where I_o is the intensity of the light source taken as unity (23).

$$< I > = I \, [1 - 10^{-Abs}]/2.303 \, Abs \qquad (2)$$

Table II. > C=O/−CH Ratio Increase

Additive	X	Supp. (%)	Y	Supp. (%)	Z	Supp. (%)
None	9.4	—	11.3	—	8.7	—
HALS	4.9	48	5.3	53	5.0	43
UVA	5.7	39	8.4	26	4.9	44
HALS and UVA	2.6	72	4.2	63	2.0	77

NOTE: Values for X, Y, and Z are units × 10^{-4}/h (±10%); Supp. is suppression.

UVA absorbance-loss curves for clearcoats with UVA only and UVA and HALS are shown in Figures 9 and 10, respectively, where exposure time has been multiplied by $<I>$.

The curves shown in Figure 9 indicate that all three clearcoats lose bz-UVA absorbance at a surprisingly rapid rate in the absence of HALS. The addition of HALS (Figure 10) slows bz-UVA absorbance loss by a factor of 2 for clearcoats Y and Z, but the effect is small for clearcoat X. Film loss can account for most of the bz-UVA absorbance lost by clearcoat Z, only 60% of the bz-UVA absorbance lost by clearcoat Y, and <25% of the bz-UVA absorbance lost by clearcoat X. This behavior clearly suggests that bz-UVAs are subject to chemical degradation. Prior to these observations, bz-UVAs were assumed to be chemically inert in automotive clearcoats.

The dramatic differences in the rate that clearcoat X loses bz-UVA absorbance relative to clearcoats Y and Z provides a possible explanation for the poor weathering performance of paint systems based on clearcoat X relative to paint systems based on clearcoats Y and Z, even though there is no dramatic difference in the rates at which each clearcoat undergoes photooxidative degradation. Absorbance measurements on free standing films of each clearcoat yield the following ratio of bz-UVA absorbance/mil values for clearcoats X, Y, and Z, respectively: 1.5:2.0:1.0 in the 340-nm region. Figure 10 suggests that the ratio of bz-UVA absorbance loss rates is 5:2:1 for clearcoats X, Y, and Z, respectively. If all three clearcoats were applied at the same coating thickness

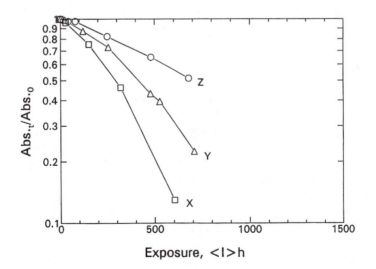

Figure 9. UVA absorbance loss for UVA containing clearcoats without HALS:
X(□), Y (△), and Z (○).

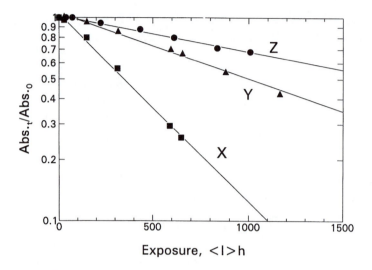

Figure 10. UVA absorbance loss for clearcoats stabilized with HALS: X(■), Y (▲), and Z (●).

in a multilayer coating system and subjected to the same exposure conditions, clearcoat X could allow UV light in the 340-nm region to penetrate to underlying coatings layers at a rate >3 times faster than clearcoats Y and Z. If all underlying coating layers were equally sensitive to UV light, paint systems based on clearcoat X could exhibit inferior weathering performance relative to paint systems based on clearcoats Y and Z.

Conclusions

Hydroperoxide concentration behavior measurements and transmission FTIR measurements of carbonyl-growth and film-loss rates provide ready means to test the effectiveness of HALS and UVA additives and to compare the relative photooxidation resistance of fully formulated clearcoats. UV spectroscopy can be used to assess the ability of HALS additives to prolong UVA longevity and determine the rate at which clearcoats lose their ability to screen underlying coating layers for UV light during exposure.

The three acrylic–melamine clearcoats studied in the present work do not exhibit dramatically different photooxidative degradation rates. However, UVA absorbance loss rates strongly suggest that clearcoat X would allow UV light to penetrate to underlying coating layers much earlier in service than clearcoats Y and Z. Paint systems based on clearcoat X exhibit inferior Florida exposure weathering performance relative to paint systems based on clearcoats

Y and Z. This apparent agreement between laboratory measurements on isolated clearcoats and Florida paint system exposure results may be fortuitous because UV light with a shorter wavelength than natural UV light was used to weather laboratory samples. All work will be repeated using borosilicate/borosilicate filtered xenon-arc light to test this possibility.

References

1. Bauer, D. R.; Paputa Peck, M.C.; Cater, R. O. III. *J. Coat. Technol.* **1987,** *59 (755),* 103.
2. Bauer, D. R. *J. Coat. Technol.* **1994,** *66 (835),* 57.
3. Bauer, D. R.; Briggs, L. M. *Characterization of Highly Cross-Linked Polymers;* Labana, S. S.; Dickie, R. A., Eds.; ACS Symposium Series 243; American Chemical Society: Washington, DC, 1984, p 271.
4. Bauer, D. R.; Gerlock, J. L.; Mielewski, D. F.; Paputa Peck, M. C.; Carter, R. O. III. *Ind. Eng. Chem. Res.* **1991,** *30 (11),* 2482.
5. Hartshorn, J. H. *Proceedings of the XIII International Conference on Organic Coating Science and Technology;* Athens, 1987; 159.
6. Mielewski, D. F.; Bauer, D. R.; Gerlock, J. L. *Polym. Degrad. Stab.* **1991,** *33,* 93.
7. Mielewski, D. F.; Bauer, D. R.; Gerlock, J. L. *Polym. Degrad. Stab.* **1993,** *41,* 323.
8. Bauer, D. R.; Mielewski, D. F. *Polym. Degrad. Stab.* **1993,** *49,* 349.
9. Carlsson, D. J.; Lacoste, J. *Polym. Degrad. Stab.* **1991,** *32,* 377.
10. Bauer, D. R.; Mielewski, D. F.; Gerlock, J. L. *Polym. Degrad. Stab.* **1992,** *38,* 57.
11. van der Ven, L. G. J.; Hoffman, M. H. *Polym. Mater. Sci. Eng. Proc.* **1993,** *68,* 64.
12. Catalan, J.; Fabero, F.; Guijarro, M. S.; Claramunt, R. M.; Santa Maria, M. D.; de la Concepcion Foces-Foces, M.; Cano, F. H.; Elguero, J.; Sastre, R. *J. Am. Chem Soc.* **1990,** *112,* 747.
13. Carlsson, D. J.; Wiles, D. M. *J. Polym. Sci. Polym. Chem. Ed.* **1974,** *12,* 2217.
14. Hodgeman, D. K. C.; Gellert, E. P. *J. Polym. Sci. Polym. Chem. Ed.* **1980,** 1105.
15. Hodgeman, D. K. C. *J. Polym. Sci. Polym. Lett.* **1978,** *16,* 161.
16. Vink, P. *J. Polym. Sci. Polym. Symp.* **1973,** *40,* 169.
17. Puglisi, J.; Schirmann, P. *J. Org. Coat. Appl. Polym. Sci. Proc.* **1982,** *47,* 620.
18. Decker, C.; Zahouily, K. *Polym. Mater. Sci. Eng.* **1993,** *68,* 70.
19. Decker, C. *Polym. Prepr.* **1993,** *34(2),* 139.
20. Pickett, J. E.; Moore, J. E. *Polym. Prepr.* **1993,** *34(2),* 153.
21. Pickett, J. E.; Moore, J. E. *Polym. Degrad. Stab.* **1993,** *42,* 231.
22. Gerlock, J. L.; Smith, C. A.; Nunez, E. M.; Cooper, V. A.; Liscombe, P.; Cummings, D. R. *Polym. Prepr.* **1993,** *34(2),* 199.
23. Bauer, D. R.; Briggs, L. M.; Gerlock, J. L. *J. Polym. Sci. Part B Polym. Phys.* **1986,** *24,* 1651.

RECEIVED for review May 13, 1994. ACCEPTED revised manuscript May 15, 1995.

23

Action Mechanisms of Phosphite and Phosphonite Stabilizers

Klaus Schwetlick[1] and Wolf D. Habicher

Institut für Organische Chemie und Farbenchemie, Technische Universität Dresden, D–01062 Dresden, Germany

Phosphite and phosphonite esters can act as antioxidants by three basic mechanisms depending on their structure, the nature of the substrate to be stabilized, and the reaction conditions. All phosph(on)ites are hydroperoxide-decomposing secondary antioxidants. Their efficiency in hydroperoxide reduction decreases in the order phosphonites > alkyl phosphites > aryl phosphites > hindered aryl phosphites. Five-membered cyclic phosphites are capable of decomposing hydroperoxides catalytically because of the formation of acidic hydrogen phosphates by hydrolysis and peroxidolysis in the course of reaction. Hindered aryl phosphites can act as chain-breaking primary antioxidants when substituted by alkoxyl radicals. Hindered aryloxyl radicals are then released and terminate the radical-chain oxidation. At ambient temperatures, the chain-breaking antioxidant activity of aryl phosphites is lower than that of hindered phenols, because the rate of their reaction with peroxyl radicals and their stoichiometric inhibition factors are lower than those of phenols. In oxidizing media at higher temperatures, however, hydrolysis of aryl phosph(on)ites takes place and produces hydrogen phosph(on)ites and phenols, which are effective chain-breaking antioxidants. Tetramethylpiperidinyl phosphites and phosphonites [HALS-phosph(on)ites] surpass common phosphites, phenols, and HALS compounds as stabilizers in the thermo- and photooxidation of polymers. Their superior efficiency is due to the intramolecular synergistic action of the HALS and the phosph(on)ite moieties of their molecules.

Oᴿɢᴀɴɪᴄ ᴘʜᴏsᴘʜᴏʀᴜs ᴄᴏᴍᴘᴏᴜɴᴅs, in particular phosphite and phosphonite esters, are used on a large scale as nondiscoloring antioxidants for the

---

[1]Current address: Canalettostrasse 32, D–01307 Dresden, Germany.

stabilization of polymers against degradation during fabrication, processing and long-term application. Chart I shows some phosphites (**I–V**) and a phosphonite (**VI**) that are commercially available. Phosphonite **VII** and various phosphites that are repeatedly studied in literature and by us are shown in Chart II.

Generally, phosphorus antioxidants are used in combination with hindered phenols and other stabilizers. However, the sterically hindered aryl phosphites and phosphonites (e.g., **III, V,** and **VI**) are, under some conditions, active by themselves. These compounds can replace phenols, especially in the processing stabilization of polyolefins. In spite of their great practical importance, the detailed mechanisms of antioxidant action of organophosphorus compounds and the relationships between chemical structure and antioxidant activity were elucidated only recently (*1–4*).

Depending on their structure, the nature of the polymer to be stabilized, and the aging conditions, phosphites and phosphonites may act as secondary and primary antioxidants. *Aliphatic* phosph(on)ite esters are only secondary hydroperoxide-decomposing antioxidants, whereas sterically hindered *ortho-tert*-alkylated aromatic compounds are capable of acting also as primary, radical chain-breaking antioxidants.

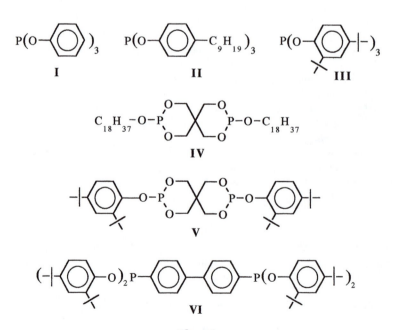

Chart I.

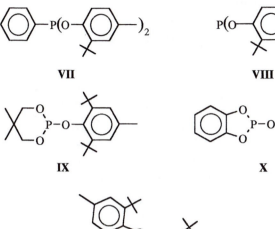

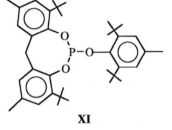

Chart II.

The stabilizing action of phosphites and phosphonites is due to three basic mechanisms:

- Oxidation of the phosphorus compound by hydroperoxides transforms these compounds into alcohols and prevents branching of the oxidation chain reaction.

$$ROOH + \overset{\backslash}{\underset{/}{P}}-OAr \rightarrow O=\overset{|}{\underset{|}{P}}-OAr + ROH \qquad (1)$$

- Substitution of hindered aryl phosphites by alkoxyl radicals formed in the reaction of peroxyl radicals with the phosphites releases hindered aroxyl radicals that are capable of terminating the oxidation chain reaction.

$$ROO^{\bullet} + \overset{\backslash}{\underset{/}{P}}-OAr \rightarrow O=\overset{|}{\underset{|}{P}}-OAr + RO^{\bullet} \qquad (2)$$

$$RO^• + \overset{\diagdown}{\underset{\diagup}{P}} - OAr \rightarrow RO - \overset{\diagup}{\underset{\diagdown}{P}} + ArO^• \tag{3}$$

- Hydrolysis of aryl phosph(on)ites gives hydrogen phosph(on)ites and phenols.

$$H_2O + \overset{\diagdown}{\underset{\diagup}{P}} - OAr \rightarrow \overset{|}{\underset{|}{H - P}} = O + ArOH \tag{4}$$

The products formed by hydrolysis are effective primary and secondary antioxidants.

Oxidation of Phosphites and Phosphonites by Hydroperoxides

Phosph(on)ites react with hydroperoxides formed in the oxidation of organic materials and give the corresponding phosph(on)ates and an alcohol (eq 1). Rates of reaction depend on the structure of the particular phosphorus antioxidant (Table I, ref. 5). The rates decrease with increasing electron-acceptor ability and bulk of the groups bound to phosphorus in the series aryl phosphonites > alkyl phosphites > aryl phosphites > hindered aryl phosphites.

Oxidation may be followed by hydrolysis and peroxidolysis of the phosph(on)ates formed. This behavior is especially true for five-membered cyclic phosphites (e.g., **X** in Scheme I); the corresponding phosphates (**XII**)

Table I. Stoichiometric Reaction of Phosphites and Phosphonites with Cumyl Hydroperoxide in Chlorobenzene at 30 °C

Phosph(on)ite	$k \times 10^3$ $(M^{-1} s^{-1})$
Triethyl phosphite	350
Triphenyl phosphite	31
III	4.9
IX	120
X	3.8
XI	0.83
VII	1500

NOTE: $[P]_o = [ROOH]_o = 0.2$ M.

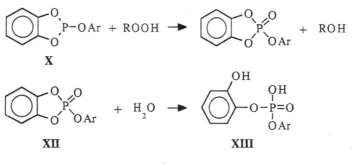

Scheme I.

of these products easily react by ring opening with water and hydroperoxides formed in the oxidation process (6). The hydrogen phosphates, **XIII,** are rather strong acids and are capable of decomposing hydroperoxides catalytically to give ketones and phenols or alcohols, which may be further dehydrated to olefins and water (6, 7).

Furthermore, if the particular cyclic phosphite is an aromatic one, like the *o*-phenylene derivative, **X,** in Scheme I, the resulting product, **XIII,** is an *o*-substituted phenol capable of acting as an effective chain-breaking antioxidant (8). Therefore, polyfunctional antioxidants are formed in the course of application of cyclic arylene phosphites, and the various hydroperoxide-decomposing and chain-breaking moieties of which enhance their antioxidative activity autosynergistically.

Reactions of Phosphites and Phosphonites with Peroxyl and Alkoxyl Radicals

Phosph(on)ites react with peroxyl radicals formed in the autoxidation of an organic compound according to eq 2 by oxidation to give phosph(on)ates. The simultaneously formed alkoxyl radicals further react with a second phosph(on)ite molecule in a way depending on the structures of the particular phosphorus compound and of the alkoxyl radical.

Phosphonites, both alkyl and aryl, and alkyl phosphites are always oxidized by alkoxyl radicals to give the corresponding phosph(on)ates and alkyl radicals that propagate the oxidation chain reaction (9, 10).

$$RO^{\cdot} + \overset{\diagdown}{\underset{\diagup}{P}}-OR' \rightarrow O=\overset{|}{\underset{|}{P}}-OR' + R^{\cdot} \tag{5}$$

Therefore, phosphonites (at least at ambient temperatures) and alkyl phosphites are unable to act as chain-breaking antioxidants.

Hindered aryl phosphites, however, react with alkoxyl radicals by substitution according to eq 3 to give a substituted alkyl phosphite and hindered aryloxyl radicals that terminate the oxidation chain reaction (10–12). Hence, only aryl phosphites are capable of acting as primary antioxidants (at any temperature).

The chain-breaking antioxidant activity of hindered aryl phosphites at ambient temperatures is lower than that of hindered phenols, because the rate constants of their reactions with peroxyl radicals are lower by a factor of 20 to 60 (Table II, reference 12). Furthermore, their stoichiometric inhibition factors, f, the number of peroxyl radicals trapped by one phosphite molecule, are less than 1. This value is lower than those of hindered phenols, which generally are 2. The low stoichiometric factors are due to the reaction of the phosphite with hydroperoxides (eq 1), which destroy the antioxidant. These factors are especially low in highly oxidizable substrates and at low phosphite concentrations (12). Therefore, at ambient temperatures, hindered aryl phosphites themselves are effective chain-breaking antioxidants only in rather high concentrations and predominantly in substrates of low oxidizability.

Hydrolysis of Phosphites and Phosphonites

Remarkably, in the autoxidation of hydrocarbons at higher temperatures and inhibited by aryl phosphites, hydrolysis of the phosphite takes place (eq 5) to give phenols and hydrogen phosphites, which may further hydrolyze to phosphorous acid [$HPO(OH)_2$] (13). Even the sterically hindered aryl phosphites and phosphonites hydrolyze at 150–190 °C (14). The mixture of antioxidants thus generated is responsible for the high stabilizing efficiency of phosphite and phosphonite esters at these temperatures.

In addition to hydrolysis, oxidation of the phosphorus compounds by hydroperoxides (eq 1) and peroxyl radicals (eq 2) takes place in the course of reaction to give the corresponding phosph(on)ates (14). The ratio of oxidation to hydrolysis depends on the oxidizability of the particular substrate and on the reaction conditions (temperature). In the inhibited oxidation of the less-oxidizable n-paraffins at 150 °C, hydrolysis exceeds oxidation, whereas at 180

Table II. Reactions of Stabilizers with Cumyl- and Tetralylperoxyl Radicals at 65 °C

Stabilizer	k_{cum} $(M^{-1} s^{-1})$	k_{tet} $(M^{-1} s^{-1})$
VIII	300	640
IX	500	1420
X	400	810
XI	230	
BHT	15000	30000

°C the two processes take place in a 1:1 ratio. In the inhibited oxidation of the more easily oxidizable aralkyl hydrocarbons (such as tetralin), oxidation of the phosphorus compound is faster than hydrolysis. In the oxidation of a polyether (polyethylene-propylene oxide) only oxidation and not hydrolysis of the phosphorus inhibitors occurs (*14*).

HALS Phosphites and Phosphonites

Phosph(on)ites modified by additional functional groups should exhibit new stabilizing properties due to action mechanisms characteristic for these functional groups or synergistic effects of these groups on the common phosph(on)ite stabilizer mechanisms. Interesting in this context are phosphite esters bearing 2,2,6,6-tetramethylpiperidinyl (hindered amine light stabilizer, HALS) groups that contain hydroperoxide-decomposing, chain-breaking, and light-stabilizing moieties in their molecules. Some cyclic representatives of such HALS-phosphites were described in patents (*15–18*).

We synthesized and studied the stabilizing efficiency of HALS-phosphite and -phosphonite esters such as **XIV** and **XV** (*19–22*). A remarkable property of these HALS-phosph(on)ites is their pronounced hydrolytic resistance. Whereas common phosphorus(III)-based stabilizers are more or less sensitive to hydrolysis, HALS-phosph(on)ites are stable even under conditions where sterically hindered aryl phosphites are completely hydrolyzed. This property may be ascribed to the amino function of the HALS group, which neutralizes existing and formed acid and therefore suppresses acid catalysis of hydrolysis.

The inhibiting efficiency of HALS-phosph(on)ites in the thermo- and photooxidation of polymers is generally much higher than that of common hindered phenols, phosph(on)ites, and HALS stabilizers. Therefore, in the thermooxidation of a polyether alcohol (SYSTOL T 154), the HALS-phosph(on)ites **XIV** and **XV** (Chart III) exhibited a lower critical antioxidant concentration and a longer induction period than phenols and phosphites (Table III). Because of the electron-donating property of the piperidinyl group, they also destroy polyether hydroperoxides with higher rates than common aryl phosphites (Table IV).

In the course of inhibited oxidation of the polyether, HALS-phosph(on)ites are exclusively oxidized to the corresponding phosph(on)ates;

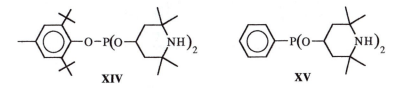

Chart III.

Table III. Critical Antioxidant Concentrations (c_{cr}) and Induction Periods (t_{ind} at 1500 ppm Antioxidant) in Thermooxidation of Polyether Alcohol SYSTOL T 154 (1:9 in o-Dichlorobenzene) at 100 °C

Antioxidant	$c_{cr}(10^{-4} M)$	t_{ind} (min)
None		11
Irganox 1010	19.2	120
Triphenyl phosphite	23.5	31
XIV	2.9	1780
XV		6000

Table IV. Rate Constants for Decomposition of Polyether Hydroperoxides by Phosph(on)ites

Phosph(on)ite	k × 10^3 ($M^{-1} s^{-1}$)
Trioctyl phosphite	110
Triphenyl phosphite	18
XIV	31
XV	104

NOTE: $[P]_O = [-OOH] = 0.15$ M, 30 °C, PEA T 154 (1:9 in o-dichlorobenzene).

Table V. Induction Periods and Relative Oxidation Rates (S_r after Induction Period) in Thermooxidation of Polypropylene at 180 °C

Antioxidant	t_{ind}/min	S_r
None	45	1
BHT	410	0.49
III	230	0.5
XIV	1020	0.015
XV	430	0.17

NOTE: Propathene HF 20 CGV 170 [ICI]; 0.1-mm films; [stabilizer] = 0.02 mol/kg.

Table VI. Relative Irradiation Times To Reach a CI of 1 in Photooxidation of Polypropylene Films

Stabilizer	Efficiency Ratio
None	1
Triphenyl phosphite	1.4
4-Hydroxy-2,2,6,6-tetramethylpiperidine	1.5
4-Hydroxy-2,2,6,6-tetramethylpiperidin-1-oxyl	3.6
XIV	4.0
XV	4.6

NOTE: Materials are the same as in Table V.

neither hydrolysis nor substitution products could be detected. On the other hand, nitroxyl radicals could be observed by electron spin resonance spectroscopy. Therefore, the superior inhibition efficiency of HALS-phosph(on)ites may be due to the synergistic cooperation of the hydroperoxide decomposing ability of the phosph(on)ite moiety and the chain breaking antioxidant activity of the hindered piperidinyl substituent.

The outstanding and wide antioxidative efficiency of the multifunctional stabilizers of the HALS-phosph(on)ite type manifests itself especially in the stabilization of polyolefins against oxidative and thermal degradation and discoloration (19–23). In the thermooxidation of polypropylene, HALS-phosph(on)ites are better inhibitors than common phenols, phosphites, and HALS compounds. They give rise to longer induction periods and lower oxidation rates after the induction periods (Table V) (19).

The efficiency of stabilizers in the photooxidation of polypropylene is shown in Table VI (22). Also in Table VI, the HALS-phosph(on)ites proved to be the most effective of the compounds studied.

The rather good efficiency of HALS-phosph(on)ites in the photo- and thermostabilization of polymers could be related to a special type of synergism that might be called *intramolecular synergism,* because mixtures of individual components with appropriate structural elements give only additive or less pronounced synergistic effects. This phenomenon was demonstrated for the photooxidation of polypropylene (23).

Whereas most HALS-type stabilizers do not efficiently act under thermooxidative and processing conditions, the HALS-phosph(on)ites just here show their advantages. These were demonstrated for polypropylene (24) and linear low-density polyethylene (LLDPE) (25); the HALS-phosph(on)ites effectively contributed to melt flow and color stabilization during processing. Furthermore, HALS-phosphites with pentamethylpiperidinyl groups exhibited an extraordinarily good compatibility with LLDPE and are not at all prone to blooming out. Their solubilities in the polymer were determined to be an order of magnitude above the usual application levels. Because of their high efficiency and wide applicability, HALS-phosph(on)ites could complete the scale of current commercial stabilizers.

References

1. Schwetlick, K. In *Mechanisms of Polymer Degradation and Stabilisation;* Scott., G., Ed.; Elsevier Applied Science: London and New York, 1990; p 23.
2. Schwetlick, K. *Pure Appl. Chem.* **1983,** *55,* 1629.
3. Kirpichnikov, P. A.; Mukmeneva, N. A.; Pobedimskii, D. G. *Usp. Khim.* **1983,** *52,* 1831.
4. Pobedimskii, D. G.; Mukmeneva, N. A.; Kirpichnikov, P. A. In *Developments in Polymer Stabilisation;* Scott, G., Ed.; Applied Science: London, 1980; p 125.
5. König, T.; Habicher, W. D.; Schwetlick, K. *J. Prakt. Chem.* **1989,** *331,* 913.
6. Schwetlick, K.; Rüger, C.; Noack, R. *J. Prakt. Chem.* **1982,** *324,* 697.
7. Humphris, K. J.; Scott, G. *J. Chem. Soc. Perkin Trans.* **1973,** *2,* 826.
8. Rüger, C.; König, T.; Schwetlick, K. *Acta Polym.* **1986,** *37,* 435.
9. Walling, C.; Rabinowitz, R. *J. Am. Chem. Soc.* **1959,** *81,* 1243.
10. Schwetlick, K.; König, T.; Rüger, C.; Pionteck, J. *Z. Chem.* **1986,** *26,* 360.
11. Schwetlick, K.; Pionteck, J.; König, T.; Habicher, W. D. *Eur. Polym. J.* **1987,** *23,* 383.
12. Schwetlick, K.; König, T.; Pionteck, J.; Sasse, D.; Habicher, W. D. *Polym. Degrad. Stab.* **1988,** *22,* 357.
13. Bass, S. I.; Medvedev, S. S. *Zh. Prikl. Khim.* **1962,** *36,* 2537.
14. Schwetlick, K.; Pionteck, J.; Winkler, A.; Hähner, U.; Kroschwitz, H.; Habicher, W. D. *Polym. Degrad. Stab.* **1991,** *31,* 219.
15. Minagawa, M.; Kubota, N.; Shibata, T.; Sugibuchi, K. Japanese Patent 52 022 578, 1975.
16. Minagawa, M.; Nakahara, Y.; Shibata, T.; Arata, R. European Patent 149 259, 1983.
17. Rasberger, M. German Patent DE 2 656 999, 1977.
18. Rasberger, M.; Hofmann, P.; Meier, H. R.; Dubs, P. European Patent 155 909, 1984.
19. Habicher. W. D.; Hähner, U.; Marquart, R.; Schwetlick, K. German Patent DD 290 906, 1988.
20. Hähner, U.; Habicher, W. D.; Ohms, G.; Schwetlick, K. German Patent DD 301 614, 1988.
21. Hähner, U.; Habicher, W. D.; Ohms, G.; Schwetlick, K. German Patent DD 301 615, 1988.
22. Hähner, U.; Habicher, W. D.; Chmela, S. *Polym. Degrad. Stab.* **1993,** *41,* 197.
23. Chmela, S.; Habicher, W. D.; Hähner, U.; Hrdlovic, P. *Polym. Degrad. Stab.* **1993,** *39,* 367.
24. Habicher, W. D.; Bauer, I.; Staniek, P. *Abstracts of Papers of the 206th ACS National Meeting;* American Chemical Society: Washington, DC, 1993; part 2, No. 0338.
25. Lingner, G.; Staniek, P.; Stoll, K. H. *Lecture at the Polyolefines VIII RETEC of the SPE;* Society of Petroleum Engineers: Richardson, TX, 1993.

RECEIVED for review December 6, 1993. ACCEPTED revised manuscript December 5, 1994.

Characterization of Conversion Products Formed during Degradation of Processing Antioxidants

J. Scheirs[1,2], Jan Pospíšil[3], M. J. O'Connor[4], and S. W. Bigger[4]*

[1] Product Technology Section, Kemcor Australia Ltd., 228-238 Normanby Rd., South Melbourne 3205, Australia
[3] Department of Degradation and Stabilization of Polymer Materials, Academy of Sciences of the Czech Republic, Heyrovsky Sq. 2., 162 06 Prague, Czech Republic
[4] Department of Environmental Management, Victoria University of Technology, PO Box 14428, Melbourne City Mail Centre, Melbourne 8001, Australia

A range of analytical techniques was used to characterize the conversion products of a common polyolefin processing stabilizer (Irgafos 168, a hindered phosphite) and a common polyolefin antioxidant (Irganox 1076, a hindered phenol). Analysis by gas chromatography (GC)–mass spectrometry revealed that the phosphite stabilizer is oxidized to a phosphate under processing conditions. The extent of conversion of the phosphite can be quantified by using GC as well as ^{31}P NMR spectroscopy. The applicability of other techniques such as IR spectroscopy, fluorimetry, and supercritical fluid–liquid chromatography are also investigated. The quantitation of Irganox 1076 in the presence of Irgafos 168 by GC is complicated because the phosphate co-elutes with Irganox 1076. The thermal degradation of Irganox 1076 in air was studied by using both high-performance liquid chromatography (HPLC)–UV-visible spectrophotometry and by preparative HPLC–NMR spectroscopy. The degradation of Irganox 1076 at 180 °C produced cinnamate and dimeric oxidation products, whereas the degradation of Irganox 1076 at 250 °C produced dealkylation products.

HINDERED PHOSPHITES AND HINDERED PHENOLS are widely used as stabilizers during the processing of polyolefins such as polyethylene (PE) and

[2]Current address: ExcelPlas Australia Ltd., PO Box 102, Moorabbin 3189, Australia.
*Corresponding author

0065–2393/96/0249–0359$12.00/0
© 1996 American Chemical Society

polypropylene (PP) (*1–4*). Indeed, the processing of PP would be almost impossible if processing stabilizers were not used. Phosphite processing stabilizers maintain the melt index and prevent polymer discoloration by decomposing hydroperoxides and by reacting with quinoidal conversion products of the hindered phenol (*5, 6*).

A synergistic effect is obtained when hindered phosphites are used in combination with hindered phenols. Even though the precise nature of this synergism is not fully understood, aromatic phosphites preserve the concentration of the hindered phenol during polymer processing (*7*). This observation led to the commercialization of the so-called "B-blends", which are mixtures of a hindered phosphite and a hindered phenol in a set ratio.

The Action of Organic Phosphites

The principal reactions of hindered organic phosphites are shown in Scheme I. Phosphites decompose polymer hydroperoxides by reduction to alcohols, and in the process the phosphites are oxidized to phosphates (*2, 8, 9*). Phosphites may also react with

- peroxyl radicals to produce the phosphate and an alkoxyl radical
- an alkoxyl radical to produce a nonpropagating free phenoxyl radical that can, in turn, react with peroxyl radicals to terminate a chain
- an alkoxyl radical to produce the phosphate and a macroalkyl radical

In addition to decomposing hydroperoxides, aromatic phosphites can also

- react with unsaturated (vinyl) groups in the polymer
- coordinate with transition metal residues
- help to preserve the hindered phenol
- prevent discoloration by reacting with quinoidal compounds

Furthermore, Schwetlick et al. (*1, 8*) showed that sterically hindered aryl phosphites may act as chain-terminating primary antioxidants.

Hindered aromatic phosphites can function as chain-terminating antioxidants by their interaction with alkoxyl radicals (*8*). A product of this reaction

$$P(OAr)_3 + ROOH \longrightarrow O=P(OAr)_3 + ROH \qquad (1)$$

$$P(OAr)_3 + ROO^{\cdot} \longrightarrow O=P(OAr)_3 + RO^{\cdot} \qquad (2)$$

$$P(OAr)_3 + RO^{\cdot} \longrightarrow RO–P(OAr)_2 + ArO^{\cdot} \qquad (3)$$

$$P(OAr)_3 + RO^{\cdot} \longrightarrow O=P(OAr)_3 + R^{\cdot} \qquad (4)$$

Scheme I.

is the di-*tert*-butylphenoxyl radical that can react subsequently with a peroxyl radical to terminate the chain in the oxidation reaction (reaction 3 in Scheme I). Aliphatic phosphites, unlike their aromatic counterparts, are not chain-breaking antioxidants because they react with alkoxyl radicals to produce macroalkyl radicals, which in turn can react with oxygen to cause chain-propagation (8) (reaction 4 of Scheme I). At polymer processing temperatures, the radical scavenging action of aromatic phosphites is comparable, if not better than, hindered phenolic antioxidants. However, phosphites provide no long-term stabilization at the lower temperatures experienced by polymers during their service.

Allen et al. (10) studied the effect of γ-irradiation on the oxidation of PP containing Irgafos 168 and found that the conversion of the phosphite to the phosphate reflects the role of Irgafos 168 in reacting with peroxyl radicals that are generated during irradiation. During the processing of high-density PE (HDPE) at 220 °C in the absence of a phosphite stabilizer, as much as 45% of the phenolic antioxidant is consumed after one extrusion pass (7). In contrast, the presence of a phosphite stabilizer reduces the consumption of phenolic antioxidant to 20% after the first extrusion pass (7). Although the mechanism is not fully understood, one of the major roles of the phosphite is in the preservation of the hindered phenol. This function may be partly due to the phosphite altering the yield of specific radicals or the phosphite regenerating the phenol. In the case of HDPE, increased amounts of phosphite provide increased processing stability at a constant phenol concentration (7).

Phosphites are known to form coordination complexes with metal ions, thereby changing the potential activity of the metal (11, 12). In addition, some phosphite–metal complexes incorporated into polymers were shown to be active stabilizers. One particular study found that in polyolefins containing high levels of catalyst residues, phosphites effectively improve the weathering resistance of the polymer by interacting with the metal residues (13).

In formulations of high-molecular-weight HDPE, where calcium or zinc stearate (processing aid or acid scavenger) and high levels of phosphite exist, an interaction may occur between the phosphite and the free calcium or zinc ions. For instance, complex formation was observed in the form of precipitates when Irgafos 168 was added to calcium stearate in solution (14). Furthermore, an early study on transition-metal coordination complexes found that aryl phosphates can coordinate with chromium ions (15).

Phosphites are well known for preventing discoloration caused by the quinoidal conversion products of primary phenolic antioxidants (2). The phosphite disrupts the conjugated π-electron systems of the quinone derivatives and renders the derivatives colorless. Scheme II shows the reaction of an aromatic phosphite with quinone and quinonemethide (both of which are highly colored) to produce the colorless compounds aryl phosphate and aryl phosphonate, respectively. Phosphite stabilizers may also react with unsaturated atoms

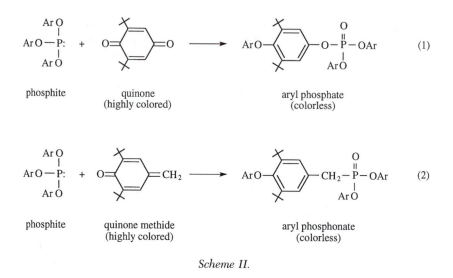

Scheme II.

and thereby disrupt the polyene sequences in the backbone of oxidized PE to minimize coloration.

Degradation Products of Processing Stabilizers

The two basic conversion products of Irgafos 168 are the phosphate that forms as a result of oxidation and the phosphonate that forms as a result of hydrolysis (Figure 1). Both the phosphate and the phosphonate are pentavalent, and this valency renders them ineffective as processing stabilizers because the lone pair of electrons on the phosphorus is involved in the stabilization process (2).

In practice, the fact that the phosphite decomposes is not of as much concern as is the extent to which it decomposes. Apparently, when Irgafos 168 is used in chromium-catalyzed HDPE, a significant amount of the phosphite is consumed in the initial compounding step (16). Factors that may be responsible for this excessive consumption are as follows:

1. The phosphite is depleted by its reaction with hydroperoxides formed when the aluminium alkyl cocatalysts are exposed to the air (17).
2. The phosphite can complex with residual chromium sites (15), thereby reducing the effective concentration of Irgafos 168.

The oxidation of Irgafos 168 to its phosphate degradation product was studied previously by using the technique of multipass extrusion. This technique simulates processing and involves repeatedly passing the polymer through an extruder and collecting samples after each pass. The oxidation of Irgafos 168 by multiple extrusion was studied in various polymers including PP (18), linear

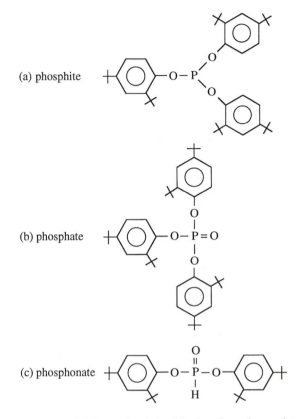

(a) phosphite

(b) phosphate

(c) phosphonate

Figure 1. Structures of (a) Irgafos 168, (b) phosphate form of Irgafos 168 (oxidation product), and (c) phosphonate form of Irgafos 168 (hydrolysis product).

low-density polyethylene (LLDPE) (4), and HDPE (6, 7, 12). In addition, the degradation products of Irgafos 168 resulting from γ-irradiation (10) and electron-beam irradiation (19) were characterized.

Although Irgafos 168 is quite resistant to hydrolysis it can be hydrolyzed under severe conditions. The hydrolysis of Irgafos 168 is shown in Scheme III. It follows a reaction sequence in which bis(2,4-di-*tert*-butyl-phenyl)phosphonate is formed initially (reaction 1 in Scheme III) along with the free-phenol by-product 2,4-di-*tert*-butylphenol (2,4-DTBP). The phosphonate formed in reaction 1 reacts further with water to produce the hydrophosphate (reaction 2). On completion of the hydrolysis reaction (reaction 3), *o*-phosphorus acid is formed (20). In each reaction 2,4-DTBP is formed as the by-product. The slight antioxidant behavior of degraded Irgafos 168 at low temperature was attributed to the presence of the 2,4-DTBP by-product (21).

The oxidation of Irganox 1076 yields a primary quinonemethide and a cinnamate, and these compounds may be transformed into the dimeric bis(quinonemethide) and the bis(cinnamate) species, respectively (22, 23). The structures of these species are shown in Figure 2. A recent publication by Klemchuk and Horng (24) discussed the various transformation products of Irganox 1076 that form as a consequence of its antioxidant activity.

Techniques Used To Study Antioxidant Degradation

The experimental techniques used for the analysis of processing antioxidants are numerous and varied. Johnston and Slone (21) used Fourier transform IR (FTIR) spectroscopy to monitor the consumption of Irgafos 168 in LLDPE with progressive processing. Mass spectrometry (MS) was also used for detecting the presence of the degradation products of Irgafos 168 (10).

Phosphorus NMR spectroscopy is an ideal method for determining the extent of conversion of phosphite to phosphate because of the high degree of specificity of this technique. Allen et al. (10) successfully used ^{31}P NMR spectroscopy to study the decomposition of Irgafos 168 in PP during γ-irradiation. Klender et al. (20) also used this technique to examine the hydrolysis of organophosphorus stabilizers. Supercritical fluid extraction was used in various chromatographic studies (25–28) aimed at determining the concentration of organophosphites in polymeric materials.

$$(ArO)_3P + H_2O \longrightarrow (ArO)_2\overset{\overset{\displaystyle O}{\|}}{P}H + ArOH \qquad (1)$$

$$(ArO)_2\overset{\overset{\displaystyle O}{\|}}{P}H + H_2O \longrightarrow ArO\overset{\overset{\displaystyle O}{\|}}{-P}-OH + ArOH \qquad (2)$$

$$ArO\overset{\overset{\displaystyle O}{\|}}{-P}-OH + H_2O \longrightarrow H_2\overset{\overset{\displaystyle O}{\|}}{P}-OH + ArOH \qquad (3)$$

Scheme III.

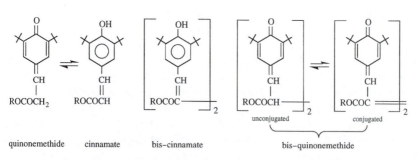

quinonemethide   cinnamate   bis–cinnamate   bis–quinonemethide

Figure 2. Transformation products of Irganox 1076.

The accurate quantitation of processing antioxidants in PE is an area that has gained more importance in recent years because of requirements placed on PE producers to provide their customers with certificates of analysis that verify the presence and levels of all additives in the formulation (*16*). The quantitation of polymer processing antioxidants such as hindered phosphites is of importance because it enables the polymer producer to determine the amount of antioxidant that is converted during polymer compounding. This chapter examines a range of techniques that can be used to characterize and quantify the commercial processing antioxidants Irganox 1076 and Irgafos 168 and their conversion products.

Materials and Methods

The antioxidants were obtained from Ciba-Geigy (Australia) and were used as received. Irgafos 168 was incorporated at a level of 1300 ppm into samples of HDPE powder that had increasing degrees of unsaturation to produce the following formulations: HDPE-1, HDPE-2, HDPE-3, and HDPE-4. The relative unsaturation levels in the virgin polymers were determined by the intensity of the IR absorbance at 910 cm^{-1}, which is due to terminal vinyl groups. The samples were extruded up to five times in a twin-screw extruder at 180 °C and 60 rpm, and the concentration of phosphate was measured after each pass by gas chromatography (GC) (*17*).

The GC analysis involved taking an accurately known mass of approximately 3 g of HDPE and extracting it with 20 mL of analytical-grade chloroform. The samples were filtered and injected onto a chromatographic column. Analysis was performed by using a Hewlett Packard 3890 GC with flame ionization detection and a 10 m × 0.31 mm BP-5 capillary column with a stationary phase thickness of 0.52 μm. The following temperature conditions were set in the chromatograph: inlet, 280 °C; oven, 280 °C; and detector, 300 °C.

GC–MS analyses were performed by using a Hewlett Packard 5890 GC with a 5971 series mass selective detector. The chromatographic column was a J&W DB-5MS, 25 m × 0.2 mm, and stationary phase thickness of 0.33 μm. The following conditions were set in the instrument: oven, 280 °C; inlet, 280 °C; transfer line, 300 °C; detector, 190 °C; electron impact energy, 70 eV; and scan mode, 35–650 amu.

Analysis of the samples by high-performance liquid chromatography (HPLC) was accomplished by using a Varian 9010 HPLC pump fitted with a Waters Resolve 12.5 cm × 5 μm C$_{18}$ column connected to a Varian 9065 Polychrom diode array detector. A detection wavelength of 280 nm and a flow rate of 1 mL/min were used. The ethyl acetate–methanol mobile phase in the chromatograph was held constant at 10% ethyl acetate for 3 min and then increased linearly to 100% ethyl acetate over a 12-min period.

The ³¹P NMR experiments were performed in hexane on a Bruker MSL 300-MHz instrument. Fluorimetric studies were performed using a Perkin-Elmer model LS-50B spectrofluorimeter. Supercritical fluid chromatography of the various antioxidants was performed by Tony Tikuisis, Novacor Chemicals, Canada (25–27).

Results and Discussion

Analysis of Irganox 1076 and Irgafos 168 by GC.   During routine GC analyses of PE containing a 1:1 Irganox 1076–Irgafos 168 B-blend, the level of the Irganox 1076 analyte is seemingly higher than the amount originally added (16). Furthermore, the amount of Irgafos 168 is correspondingly much lower than the amount originally added. Indeed, the relative amounts of these antioxidants remaining in the polymer as determined by GC is typically 2:1. This finding may suggest that during compounding the phosphite decomposes to form a phosphate product that co-elutes with the Irganox 1076 and causes falsely high Irganox 1076 readings (16).

The phosphate peak overlaps with the Irganox 1076 peak during GC analysis as well as during HPLC analysis when using an RP8 column (29, 30). However, the use of an RP18 column and acetonitrile–dioxane (70:30, v/v) eluent purportedly (29) enables these compounds to be separated by HPLC. A recent publication by Nielson (31) described the use of an acetonitrile–water mobile phase that prevented Irgafos 168 from hydrolyzing and also prevented the hydrolysis products from interfering with the quantitation of Irganox 1076. Moreover, good separation can be achieved by HPLC using a newly developed method (30) that involves using an octadecylsulfonate column with a ethyl acetate–methanol–water (60:30:10, v/v/v) mixture as the mobile phase. The elution order is Irgafos 168 (phosphate), Irganox 1076, and Irgafos 168 (phosphite) at 7.12, 12.2, and 13.45 min, respectively.

Multipass Extrusion Experiments.   Figure 3 is a plot of the phosphate concentration in HDPE as a function of the number of extruder passes at 230 °C. The plot shows that the phosphate concentration increases with increasing number of passes, which, in turn, reflects the decrease in the phosphite concentration. Two important observations are that most of the phosphite stabilizer is consumed during the first two extruder passes, and progressively less is consumed during subsequent passes; and as the initial vinyl content of the polymer becomes higher, the observed phosphite consumption also increases.

These results can be explained by the fact that in the postpolymerization stage of the polymer production process, the nascent, unstabilized polymer is exposed to air, and peroxidation of the allylic hydrogen atoms occurs. During the melt-compounding stages the hydroperoxides are reduced by the phos-

phosphate conc. (ppm)

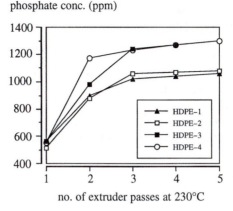

no. of extruder passes at 230°C

Figure 3. Plot of phosphate concentration as a function of the number of extruder passes at 230 °C.

phite: hence the higher the unsaturation, the greater the hydroperoxide concentration of the polymer and the higher the degree of phosphite consumption. This statement is consistent with the observation reported in the literature that the consumption of Irgafos 168 is higher when the material is processed in air compared with processing under nitrogen (6).

Despite reports in the literature (*12, 32*) that the degree of conversion of phosphite to phosphate can be measured by FTIR, limitations of this method make it unsuitable for quantitative work. The major limitation is that the phosphate P–O stretching absorption at 968 cm^{-1} is in the same region as the *trans*-vinylene group absorption in PE (*11*). The disappearance of the phosphite P–O absorption at 850 cm^{-1} is indicative of complete oxidation of the phosphite. However, changes in the vinyl concentration with processing of LLDPE, which contains a partially degraded phosphite antioxidant, cannot be followed accurately by FTIR. This limitation is due to small absorbance contributions by the phosphite P–O stretch at 908 cm^{-1} and the phosphate P–O stretch at 895 cm^{-1}. Indeed the phosphate and phosphite absorbances produce a single peak whose intensity is a weighted average of the two absorbances. Johnston and Slone (*21*) found that only by studying samples of LLDPE in which the phosphite was completely oxidized could accurate determinations of the vinyl group concentration be made.

Analysis of Irganox 1076 by HPLC–UV-Vis Spectrophotometry.
Figure 4 shows a chromatogram of a sample of Irganox 1076 that was degraded for 1 day at 180 °C in an air-circulating oven. The main product of thermal degradation at this temperature is the cinnamate, which elutes at 9.3 min and is slightly later than Irganox 1076 (8.1 min). The degradation of Irganox 1076 was also performed at 250 °C in air, and at this higher temper-

ature the cinnamate was not produced in appreciable quantities. Instead, both the mono- and didealkylated derivatives of Irganox 1076 were produced. These products can be readily identified by NMR spectroscopy following preparative HPLC.

At 250 °C, a retro-Friedel–Crafts reaction takes place, and isobutylene is liberated from the Irganox 1076 molecule (33). This dealkylation reaction (Scheme IV) can also occur in the presence of a Lewis acid, such as $MgCl_2$, which is a catalyst component in the polymerization system of some polyolefins (33). Therefore, it is important that polymers containing residual acid moieties

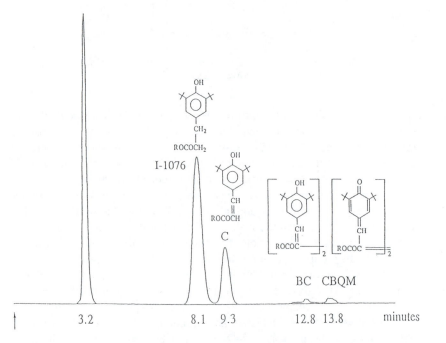

Figure 4. Chromatogram of conversion products of Irganox 1076. Peak identification: BHT, 3.2 min, internal standard; Irganox 1076, 8.1 min; cinnamate, 9.3 min; bis(cinnamate), 12.8 min; conjugated bis(quinonemethide), 13.8 min.

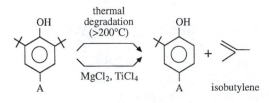

Scheme IV.

be formulated with an acid-acceptor such as calcium stearate or hydrotalcite. This formulation prevents dealkylation of the phenolic antioxidants in the additive package.

The quinonemethide degradation product of Irganox 1076 (Figure 2) is unstable and can readily isomerize to the more stable cinnamate. Alternatively, it can dimerize to the bis(quinonemethide) (BQM), of which there are conjugated and unconjugated isomers. Similarly, the cinnamate can dimerize to the bis(cinnamate). Fortunately, all of these products are spectrally distinct (Table I) and have appreciable extinction coefficients, and therefore their detection by UV-vis spectrophotometry is very convenient. BQM has an interrupted conjugated system that makes it a weaker chromophore than the strongly yellow-colored stilbene quinone of butylated hydroxytoluene (BHT). To cause similar discoloration of the polymer substrate, BQM must be present at approximately 250 times the concentration of the BHT stilbene quinone.

Analysis of Irgafos 168 by GC–MS.   Figure 5 shows the mass spectra of Irgafos 168 and its phosphate degradation product. MS is a very specific technique for detecting the presence of the phosphate degradation product because an abundant ion peak occurred at m/z 316 in the mass spectrum of the phosphate (Figure 5b). This peak is totally absent in the spectrum of the phosphite (Figure 5a). The m/z 316 ion peak was also observed by Allen et al. (*10*) in extracts from irradiated PE containing Irgafos 168. The origin of this peak was assigned to a doubly charged molecular ion (m/z 662) minus two methyl groups. Also, an intense peak occurred at m/z 647 in the spectrum of the phosphate and was assigned to the quasimolecular ion formed by the loss of a methyl group from the molecular ion. The peak at m/z 207 in the phosphate spectrum is due to the 2,4-di-*tert*-butylphenol species. The peak at m/z 57 is prominent in both the phosphite and phosphate spectra and is due to a *tert*-butyl fragment. This sequence suggests that some dealkylation has occurred. The m/z 91 ion peak present in both spectra is attributable to the tropylium ion.

By using advanced versions of MS such as Fourier-transform cyclotron-resonance MS and time-of-flight secondary-ion MS, Asamoto et al. (*34*) and

**Table I. Wavelengths of Maximum
Absorbance of the Degradation Products of
Irganox 1076**

Degradation Product	λ_{max} (nm)
Quinonemethide	300
Cinnamate	320
Bis(cinnamate)	305
Unconjugated bis(quinonemethide)	314
Conjugated bis(quinonemethide)	322

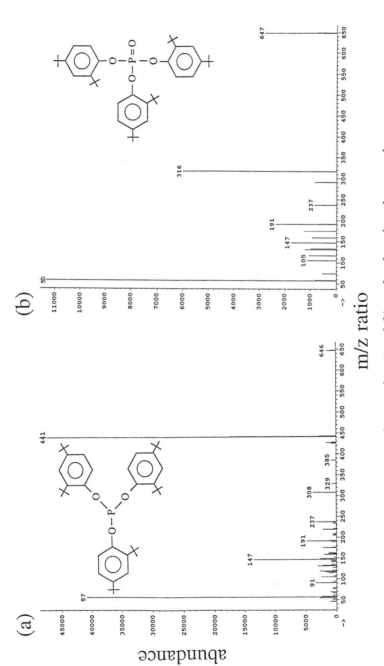

Figure 5. Mass spectra of (a) Irgafos 168 and (b) its phosphate degradation product.

Mawn et al. (35) observed an m/z 662 ion and attributed this ion to the phosphate degradation product of Irgafos 168.

Analysis of Irgafos 168 by Supercritical Fluid Chromatography. Supercritical fluid chromatographic (SFC) analyses revealed that the phosphate degradation product of Irgafos 168 eluted slightly before the phosphite, whereas the free phenol hydrolysis product (2,4-DTBP) eluted 3 min earlier, near the elution time for BHT (25, 26). The similarity in elution times can be explained by the chemical similarity between 2,4-DTBP and BHT.

Figure 6a shows the SFC chromatogram of an extract of HDPE film containing both Irgafos 168 and Irganox 1010. The peak at 5.5 min is the phosphate degradation product of Irgafos 168. The integrated area of the phosphate peak may be added to that of the phosphite peak so that the original concentration of Irgafos 168 can be calculated (26). Such a calculation is based on the premise that the response factor for the phosphate degradation product

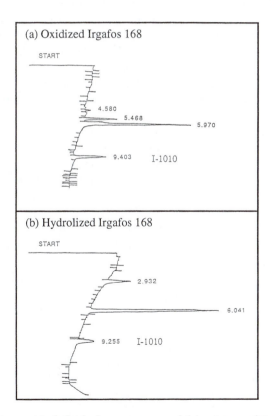

Figure 6. Supercritical fluid chromatograms of (a) mixture of Irganox 1010, Irgafos 168, and the phosphate degradation product of Irgafos 168; and (b) mixture of Irganox 1010, Irgafos 168, and 2,4-DTBP.

of Irgafos 168 is similar to that of pure Irgafos 168. In the second chromatogram (Figure 6b), the peak at 2.9 min is due to the hydrolysis by-product of Irgafos 168, namely 2,4-DTBP (25, 26).

Quantitation of Irgafos 168 by SFC has been complicated by its degradation and hydrolysis products (27, 28). For example, Raynor et al. (28) reported a shoulder on the Irgafos 168 peak due to a degradation product of Irgafos 168 that forms during polymer processing. Similarly, Tikuisis and Cossar (27) showed that during SFC analysis of additives in HDPE, there is normally a secondary peak associated with the Irgafos 168 that is either due to degraded or hydrolyzed Irgafos 168 and also forms as a result of polymer processing.

Analysis of Irgafos 168 by ^{31}P NMR Spectroscopy and Spectrofluorimetry. The ^{31}P nuclear magnetic resonances of Irgafos 168 are shown in Figure 7. Irgafos 168 gave a discrete ^{31}P NMR signal at $\delta = 128.3$ ppm (relative to H_3PO_4; $\delta = 0$ ppm). The phosphonate signal ($\delta = -0.1$ ppm and $\delta = -6.0$ ppm) is split into a doublet because of the coupling of the phosphorus nucleus to the adjacent proton. The phosphate degradation product produces a single resonance at $\delta = -21$ ppm. The presence of a weak signal at $\delta = -21$ ppm in the ^{31}P NMR spectrum of "pure" Irgafos 168 standard is presumably due to some phosphate impurity.

The fluorescence emission spectra of Irgafos 168 and its degradation product were recorded in hexane by using an excitation wavelength of 270 nm. Although the phosphite and phosphate are not spectrally distinct, the fluorescence quantum yield of the phosphate is much greater than that of the phosphite. This difference should, in principle, enable the quantitation of the phosphate concentration to be achieved using this technique.

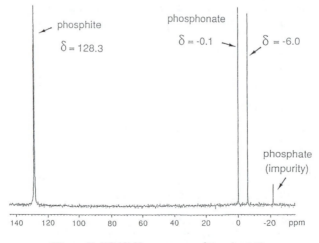

Figure 7. ^{31}P NMR spectrum of Irgafos 168.

Conclusions

Both the literature and the GC–MS work demonstrate that the phosphate conversion product of Irgafos 168 co-elutes with the Irganox 1076 peak. This co-elution can result in erroneously high concentrations of Irganox 1076 obtained during routine quantitation of PE containing a "B-blend" processing stabilizer package. Some recent HPLC methods overcome this problem by the judicious selection of mobile phases.

The extent of conversion of the phosphite to phosphate can be quantified readily by using GC and ^{31}P NMR spectroscopy. MS, fluorimetry, and SFC are also useful techniques for studying the degradation of Irgafos 168. During multipass extrusion of HDPE, a higher proportion of phosphite stabilizer is consumed on the first extruder pass, and as the initial vinyl content of HDPE is increased, the overall phosphite consumption during compounding is increased.

The thermal degradation of Irganox 1076 in air was studied by using both HPLC–UV-vis spectrophotometry and preparative HPLC–NMR spectroscopy. Degradation of Irganox 1076 at 180 °C produced the cinnamate and dimeric oxidation products, whereas the degradation of Irganox 1076 at 250 °C produced mainly dealkylation products.

Acknowledgments

We are grateful to David Allen (Sheffield Hallum University, UK), Peter Klemchuk (Ciba-Geigy Corp.), Gerald Klender (Ethyl Corp.), Gilbert Ligner (Sandoz Huningue, France), Paul Roscoe (General Electric Corp.), Roberto Tedesco (Ciba-Geigy Corp., Switzerland), Tony Tikuisis (Novacor Chemicals, Canada), and the Kemcor laboratory staff. The help of Sandra Hopkins (Kemcor Ltd., Australia) during the preparation of the manuscript is also acknowledged.

References

1. Schwetlick, K. *Pure Appl. Chem.* **1983**, 55, 1629.
2. Ray, W. C.; Isenhart, K. *Polym. Eng. Sci.* **1975**, 15, 703.
3. Lemmen, T.; Roscoe, P. *J. Elast. Plast.* **1992**, 24, 115.
4. Ligner, G.; Stoll, K. H.; Wolf, R. *Proceedings of the Annual Technical Conference of the Society of Plastics Engineers;* Society of Plastics Engineers: Palatine, IL, 1991; p 1920.
5. Pospisil, J. *Polym. Degrad. Stab.* **1993**, 40, 217.
6. Moss, S.; Zweifel, H. *Polym. Degrad. Stab.* **1989**, 25, 217.
7. Drake, W. O.; Pauquet, J. R.; Tedesco, R. V.; Zweifel, H. *Angew. Makromol. Chem.* **1990**, 176/177, 215.
8. Schwetlick, K.; Pionteck, J.; Koenig, T.; Habicher, W. D. *Eur. Polym. J.* **1987**, 23, 383.

9. Pobedimskii, D. G.; Mukmeneva, N. A.; Kirpichnikov, P. A. *Developments in Polymer Stabilization-2*; Scott, G., Ed.; Elsevier: London, 1980; p 125.
10. Allen, D. W.; Leathard, D. A.; Smith, C. *Chem. Ind.* **1987**, 854.
11. Tannahill, M. M.; Enlow, W. P. "Limitations of Liquid Chromatography and FTIR Determination for Ultranox 626 in Polyolefins"; GE Specialty Chemicals Technical Bulletin; General Electric, 1988. (Available from GE Specialty Chemicals, PO Box 1868, Parkersburg, WV, 26102.)
12. Reeder, M.; Enlow, W.; Borkowski, E. In *Polyolefins VI International Conference*; Society of Petroleum Engineers: Houston, TX, 1989; p 181.
13. Masaki, Y. F.; Akiyoshi, O. T. U.S. Patent 4 261 880, 1981.
14. Klender, G. J., private communication, 1993.
15. Podall, H. E.; Iapalucci, T. L. *J. Polym. Sci. Polym. Lett. Ed.* **1963**, *1*, 457.
16. Scheirs, J.; Pospisil, J.; O'Connor, M. C.; Bigger, S. W. *Polym. Prep. Am. Chem. Soc. Div. Polym. Chem.* **1993**, *34*, 203.
17. Mulhaupt, R. *New Trends in Polyolefin Catalysis and Influence on Polymer Stability*; 12th Annual International Conference on Advances in the Stabilization and Controlled Degradation of Polymers; Technomic Press, 1990.
18. Tobita, E.; Haruna, T. *Proceedings of the Annual Technical Conference of the Society of Plastics Engineers*; Society of Plastics Engineers: Palatine, IL, 1993; p 2544.
19. Bourges, F.; Bureau, G.; Pascat, B. *Packag. Technol. Sci.* **1992**, *5*, 197.
20. Klender, G. J.; Kolodchin, W.; Gatto, V. J. *Proceedings of the Annual Technical Conference of the Society of Plastics Engineers*; Society of Plastics Engineers: Palatine, IL, 1993; p 180.
21. Johnston, R. T.; Slone, E. J. *The Effect of Stabilizers on the Crosslinking vs. Scission Balance During Melt Processing of LLDPE*; SPE Polyolefins VII; Society of Petroleum Engineers: Houston, TX, 1991; p 207.
22. Bartelink, H. J. M.; Beulen, J.; Bielders, G.; Cremers, L.; Duynstee, E. F. J.; Konijnenberg, E. *Chem. Ind.* **1978**, 586.
23. Jonas, R. O.; Parsons, B. A. G.; Wilkinson, R. *Proceedings of the Conference on Chem. and Phys. Phenomena in the Ageing of Polymers*; Chemtec, 1988; p 84.
24. Klemchuk, P. P.; Horng, P. L. *Polym. Degrad. Stab.* **1991**, *34*, 333.
25. Tikuisis, T., Novacor Chemical Co., Canada; private communication, 1993.
26. Tikuisis, T.; Cossar, M. *ANTEC '93*; p 270.
27. Tikuisis, T.; Cossar, M. In *Suprex Technology Focus Notes*; 1992.
28. Raynor, M. W.; Bartle, K. D.; Davies, I. L.; Williams, A.; Clifford, A.; Chalmers, J. M.; Cook, B. W. *Anal. Chem.* **1988**, *60*, 427.
29. Ligner, G., Sandoz Chemicals, France; private communication, 1993.
30. Spatafore, R., private communication, 1993.
31. Nielson, R. *Millipore-Waters Polym. Notes* **1993**, *4*, 6.
32. Wieboldt, R. C.; Kempfert, K. D.; Dalrymple, D. L. *Appl. Spectrosc.* **1990**, *44*, 1028.
33. Pospisil, J. *Polym. Degrad. Stab.* **1993**, *39*, 103.
34. Asamoto, B.; Young, J. R.; Citerin, R. *J. Anal. Chem.* **1990**, *62*, 61.
35. Mawn, M. P.; Linton, R. W.; Bryan, S. R.; Hagenhoff, B.; Jurgens, U.; Benninghoven, A. *J. Vac. Sci. Technol.* **1991**, *A9*, 1307.

RECEIVED for review December 6, 1994. ACCEPTED revised manuscript February 21, 1995.

Effect of Stabilization of Polypropylene during Processing and Its Influence on Long-Term Behavior under Thermal Stress

Hans Zweifel

Ciba-Geigy AG, Additives Division, CH-4002 Basel, Switzerland

Sterically hindered phenols are efficient stabilizers against the degradation of polymers upon processing of the melt. They also protect the polymer during end-use and impart protection against the loss of chemical, physical, and aesthetic properties. Combinations of sterically hindered phenols with hydroperoxide decomposers bring about a further contribution to the stabilization of polymers in general, and polypropylene in particular. Optimal contribution is achieved with a blend of one part phenol and two parts phosphite. Such a combination also protects the phenol from excessive consumption during melt processing and thus contributes to improvement of long-term behavior. However, long-term behavior is most strongly influence by combinations of sterically hindered phenols with thiosynergists, such as dilaurylthiodipropionate. Such combinations do not contribute to stabilization of the melt, because sulfenic acid has to be formed as an active precursor. The use of combinations of sterically hindered phenols with hindered amine stabilizers contributes to long-term stabilization of polypropylene. The effect is strongly influenced by the temperature during use.

P LASTICS UNDERGO DEGRADATION in the course of their processing as melts during extrusion or injection molding, and this degradation may lead to changes in the original mechanical properties. Furthermore, plastic end products are exposed during their use to external effects such as heat or weathering. These effects cause deterioration of mechanical properties and have a

0065–2393/96/0249–0375$12.50/0

detrimental influence on the aesthetic aspect as a result of chalking. Suitable stabilizers and stabilizer systems can inhibit or delay degradation.

Polypropylene (PP) is a material that cannot be processed without adequate stabilization, and products made of this material have to be particularly well stabilized against thermo- and photooxidative degradation. Studies concerned with the stabilization of this polymer are, therefore, eminently well-suited to gain insight into the effectiveness of different stabilizers and stabilization systems.

Submitting unstabilized PP to several successive extrusions results in a degradation of the macromolecules because of chain scission. Figure 1 shows the IR spectra (measured as differential spectra related to virgin material) and the molecular weights (determined by gel permeation chromatography) of an unstabilized PP homopolymer after various extrusion steps at 280 °C. The bands of oxidation products such as γ-lactones, esters, and aldehydes are visible in the carbonyl region (1800–1700 cm^{-1}) rather weakly because the amount of oxidation products is low. Clearly visible are the absorption bands at 1645 cm^{-1} assigned to C=C bonds. Oxidation products are the result of thermal oxidation of the polymer caused by residual oxygen dissolved in the polymer. The unsaturated C=C molecules arise from thermooxidative decomposition of hydroperoxides (e.g., β-scission) and from thermomechanical chain scission. Alkyl radicals are generated and then undergo disproportionation.

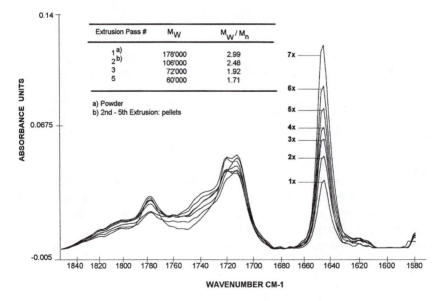

Figure 1. Spectral changes (FTIR) in the carbonyl region and in M_w (GPC) of PP after multiple extrusion at 280 °C related to untreated sample.

If the same unstabilized PP-homopolymer is subjected to aging in a circulating air oven for several hours at 135 °C, then molecular weight degradation is again observed. Figure 2 shows the IR spectra (compared with unaged material) and the resulting molecular weights after each aging period. The strong absorption bands in the carbonyl region between 1800–1700 cm^{-1} indicate a substantial portion of oxidation products, whereas the proportion of molecules with C=C bonds is comparable with that of thermomechanically aged samples (multiple extrusion).

These results are in good agreement with earlier experiments (1–3). Under processing conditions as they exist in an extruder, very little oxygen is available for thermal oxidation, and the concentration ratio of ROO˙ to R˙ is less than 1. During thermooxidative aging in a circulating air oven, the polymer undergoes thermooxidative degradation. The concentration of peroxy radicals (and hydroperoxides) is, because of autoxidation, much greater than that of alkyl radicals (ROO˙:R˙ >> 1), which react immediately with oxygen dissolved in the polymer and form further peroxy radicals.

The stabilization of plastics in general and PP in particular has to take into account both aspects. It is based on intercepting or transforming reactive radicals arising during processing and thermal stress into stable transformation products. Figure 3 shows the available possibilities.

In the course of the past 30 years numerous chemically different compounds have been tested for their suitability as stabilizers, and some of them

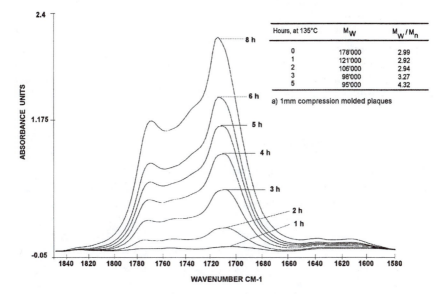

Hours, at 135°C	M_W	M_W / M_n
0	178'000	2.99
1	121'000	2.92
2	106'000	2.94
3	98'000	3.27
5	95'000	4.32

a) 1mm compression molded plaques

Figure 2. Spectral changes (FTIR) in the carbonyl region and in M_w of PP after aging in a draft-air oven at 135 °C related to untreated sample.

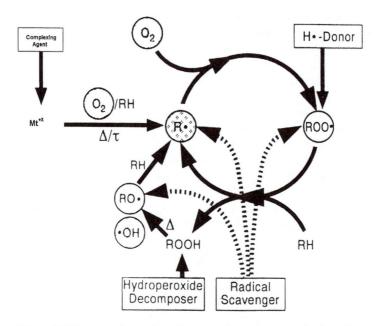

Figure 3. Thermooxidative degradation and stabilization of polyolefins.

were developed as commercial products. The chemistry of inhibiting autoxidation by means of amines and phenols (*primary antioxidants*) was investigated exhaustively by Scott and Al-Malaika (*4*), Pospisil (*5*, *6*), and Henman (*7*). The transformation of hydroperoxides into stable products by means of suitable hydroperoxide decomposers (*secondary antioxidants*) is still the subject of experimental work. The objective of this chapter is to discuss the stabilization of PP during processing and thermal stress during its lifetime and to point out the stabilizers and stabilizer systems exhibiting the greatest effect.

Phenolic Antioxidants

Phenolic antioxidants are the most widely used stabilizers for polyolefins. Unlike stabilizers based on aromatic amines, phenolic antioxidant derivatives are approved for applications in contact with food. For the user of such stabilizers this means that there is no danger of cross-contamination with nonapproved additives whenever there is a change in formulations. Furthermore, the use of aminic stabilizers leads to pronounced discoloration of the substrate; and for that reason, their application is essentially limited to black rubbers (vulcanizates) and elastomers.

Investigations were conducted by Gugumus (*8*) concerning the physical and chemical properties of phenolic antioxidants and their effect on polyolefins. Recently, Pospisil and co-workers (*9–11*) summarized the state of the art regarding the mechanism of antioxidant action. The key reaction in the stabilization of polyolefins by phenolic antioxidants is the formation of hydroperoxides by transfer of an electron and proton from the phenolic moiety to the peroxy radical. This transfer results in the phenoxyl radical (eq 1).

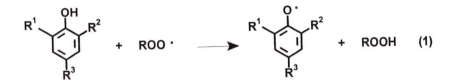

The steric hindrance of substituents such as *tert*-butyl groups in the 2- or 6-positions influences the stability of the phenoxyl radical or the mesomeric cyclohexadienonyl radicals. Sterically hindered phenols can be classified according to the substituents in the 2-, 4-, and 6-positions, as follows.

Fully Sterically Hindered Phenols. Phenols that are fully sterically hindered (**A** in Scheme I) have substituents in the 2-, 4-, and 6-positions that have no H atom on the α-carbon (no tautomeric benzyl radical formation possible). The contribution of such phenols to stabilization consists essentially of the stoichiometric reaction between the phenol and the peroxy radical. The cyclodienonyl radicals can add to ROO• radicals (Scheme I); however, this reaction is reversible.

Partially Hindered Phenols. The partially hindered phenols (**B–D**; Schemes II–IV) have substituents at least in the 4-position (or the 2- or 6-position) having H atoms on the α-carbon. The original phenol **B, C**, or **D** is reformed by a disproportionation reaction resulting in the corresponding quinonemethide **11** (Scheme II), **20** (Scheme III), or **27** (Scheme IV). Quinonemethides react with alkyl, alkoxy, and peroxy radicals. Inter- and intramolecular recombinations lead to the generally irreversible C–C products **13, 14**, and **15** (Scheme II), **21** and **22** (Scheme III), and **29** and **30** (Scheme IV).

Reactions between O- and C-centered radicals lead to reversible coupled products and are not noted in the schemes. Sterically hindered phenols of the **B, C**, and **D** type are, therefore, also radical scavengers. They can contribute to stabilization by stepwise reactions resulting in stable transformation products in an "over-stoichiometric" way (related to the equivalents of available phenolic groups). In this case, reference is made to a stoichiometric factor, f, larger than one (*12*). (De Jonge and Hope (*13*) erroneously referred to such antioxidants as "regenerating".) In any event, phenolic antioxidants are thus consumed. Currently used phenols for the stabilization of polyolefins are usually of type **D**. Table I shows the results from multiple extrusions at 270 °C

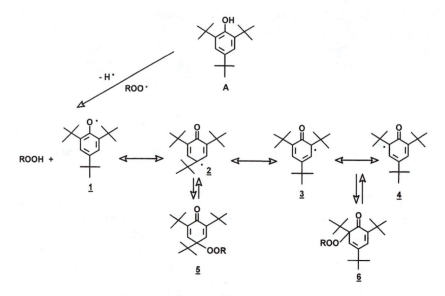

Scheme I. Fully sterically hindered phenol **A**.

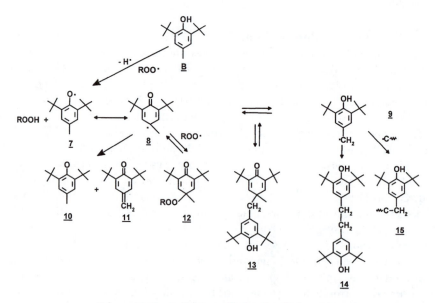

Scheme II. Partially sterically hindered phenol **B**.

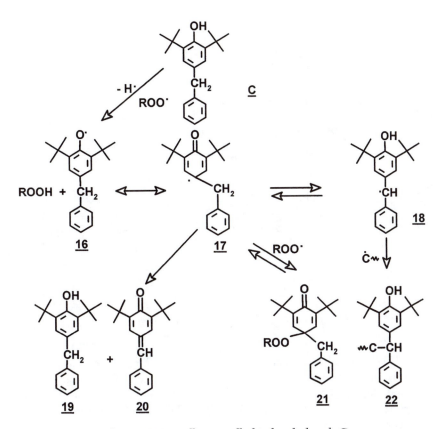

Scheme III. Partially sterically hindered phenols **C**.

Table I. PP Homopolymer, Multiple Extrusions at 270 °C

Compound	M_w	Mol OH/ kg AO	Multiple Extrusion[a]		
			1×	3×	5×
No antioxidant	—	—	7	>30	>30
AO-1	222	4.5	5.2	6.1	8.0
AO-2	531	1.9	6.8	14	25
AO-3	639	3.1	7.0	12.5	20
AO-4	1178	3.4	5.6	10	15

NOTE: All samples contain 0.075% calcium stearate and 0.075% antioxidant (phenol). For structures of antioxidant compounds, *see* Chart I.
[a]Values are MFR g/10 min (230 °C, 2.16 kg), where MFR is melt mass-flow rate according to ISO 1133.

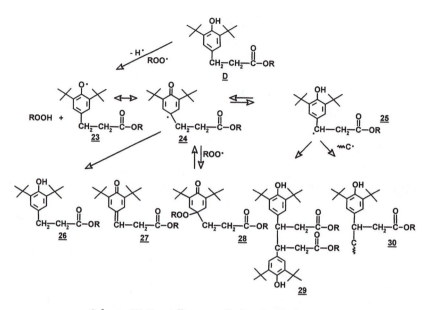

Scheme IV. Partially sterically hindered phenol **D**.

of PP [PP-homopolymer; melt mass-flow rate according to ISO 1133 (MFR) of the virgin polymer was ~3.5 at 230 °C and 2.16 kg]. The stabilization of PP melt during processing essentially depends on the available phenolic groups (mol OH/kg antioxidant [AO]). (*See* Chart I for structures of compounds from Tables I–VI.)

Table II summarizes the results obtained with PP-homopolymer with regard to its aging behavior in a circulating air oven at 135 °C and 149 °C and using various phenols. Obviously, the low-molecular weight phenol AO-1 (type **B**) does not contribute to the stabilization of the polymer because of its high volatility. The contribution to long-term thermal stabilization within the homologous series of phenols of type **D**, AO-2, AO-3, and AO-4, depends on the content of phenolic groups and on the diffusion behavior of phenols influenced by their molecular weight. This relationship is particularly pronounced in aging at 149 °C. The phenol AO-5 (type **C**) is markedly different compared with the phenol AO-6 (Type **A** similar) even though they have nearly the same molecular weights. The much higher contribution to thermal stabilization of the polymer by AO-5 compared with AO-6 may be because phenols of type **C** can form quinonemethides (**11**, Scheme III) by disproportionation (*f* greater than 1). Analogous reaction with phenol AO-6 is not possible because of the lack of H-substitution on the α-carbon atom (Scheme I).

Phenols such as AO-6 belong to the "cryptophenols"; that is, those phenols that are not substituted in the 2- or 6-position (phenol type **F**; *see* Scheme

Table II. PP Homopolymer, Oven Aging at 135 °C and 149 °C

Compound	M_w	Mol OH/ kg AO	Time to Embrittlement (days) 135 °C	149 °C
No antioxidant	—	—	<1	<1
AO-1	222	4.5	2	<1
AO-2	531	1.9	51	11
AO-3	638	3.1	141	36
AO-4	1178	3.4	187	51
AO-5	775	3.9	124	29
AO-6	795	5	44	12

NOTE: Samples were 1-mm compression-molded plaques. All samples contained 0.1% calcium stearate and 0.2% antioxidant (phenol). For structures of antioxidant compounds, *see* Chart I.

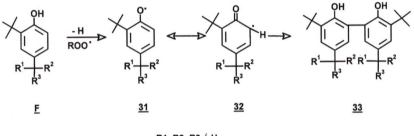

R1, R2, R3 ≠ H

*Scheme V. Intermolecular C-C coupling reaction with cryptophenols such as **F**.*

V). The intermolecular C–C bond of two cyclohexadienonyl radicals (**32** in Scheme V) mentioned in the literature (9) and forming the bisphenyldiol dimer **33** does not appear to contribute to stabilization because of the bulky AO-6 molecule.

These examples demonstrate how stability of a polymer is influenced during processing and particularly during thermal aging (long-term behavior). The reactions of the various phenols are shown in the Schemes II–V. Because of the complexity of the reactions and their kinetics, the temperature necessary for accelerated aging seems to play an important role (*14*).

Phenolic Antioxidants in Combination with Other Stabilizers

In recent years the use of phenolic antioxidants in combination with suitable cooperating stabilizers has become state of the art. Physical blends of stabi-

AO-1

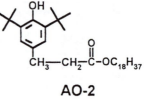

AO-2

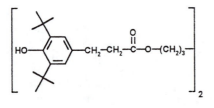

AO-3

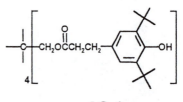

AO-4

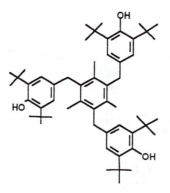

AO-5

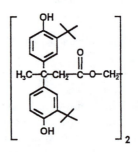

AO-6

Chart I.

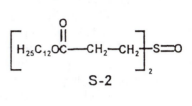

S-1

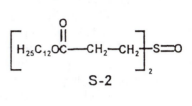

S-2

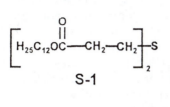

S-3

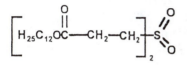

P-1

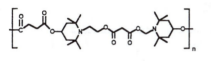

HAS-1

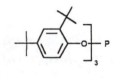

HAS-2

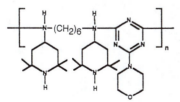

HAS-3

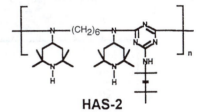

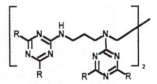

HAS-4

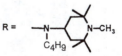

Chart I. Continued.

lizers with the same effective mechanism are referred to as homosynergistic combinations. Combinations of sterically hindered phenols with different reactivity are mentioned in the literature (*10*) but are hardly used in practice.

Much more important are the heterosynergistic combinations of stabilizers that have different effective mechanisms and complement each other. A blend of two additives is referred to as synergetic if the observed effect is greater than that of simple addition of the effects of the individual components (Figure 4). Because phenolic antioxidants do not contribute directly to the transformation of hydroperoxides arising in the course of autoxidation, the use of combinations with hydroperoxide decomposers together with sterically hindered phenols is advantageous (Figure 3).

Phenolic Antioxidants in Combination with Organophosphorus Derivatives

The use of trivalent organophosphorus compounds in combination with sterically hindered phenols is the general way of stabilizing plastics and polyolefins. A variety of phosphites and phosphonates are mentioned in the literature

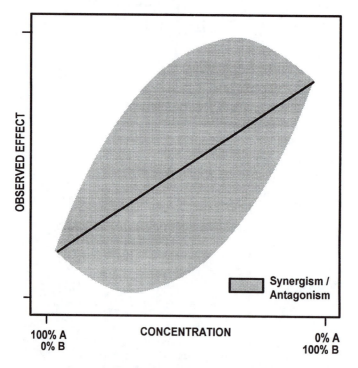

Figure 4. Synergistic and antagonistic effects.

(*15*). The hydroperoxide reacts exactly stoichiometrically, forms the corresponding alcohol, and simultaneously oxidizes the phosphite to the corresponding phosphate (eq 2, Scheme VI).

Table III summarizes the results from multiple extrusions of PP at 270 °C (PP-homopolymer; MFR of the virgin polymer was ~2.5 at 230 °C and 2.16 kg) and stabilized with the blends AO-4–P-1 and AO-5–P-1. The results show that phenolic antioxidants of type **D** (AO-4) and of type **C** (AO-5) combined with the arylphosphite P-1 are considerably more effective than the corresponding phenols alone with respect to preserving the original MFR and, thus, the molecular properties of the PP-homopolymer. Drake and Cooper (*16*) found that a blend of approximately 1 part AO and 2 parts phosphite displayed a generally optimal synergistic effect in stabilizing PP during processing. Figure 5 shows the results from multiple extrusions of PP after first extrusion at 280 °C, and Figure 6 shows results after five extrusions at 280 °C. The MFR of the virgin PP-homopolymer was ~3.5 at 230 °C and 2.16 kg. Stabilizer blends were used of AO-4–P-1, and the different ratios of AO-4:P-1 varied at total concentrations from 0.075 to 0.15% w/w with respect to the polymer.

The effectiveness of P-1 alone, that is without phenolic antioxidant, in stabilizing PP during processing indicates a complex stabilization mechanism.

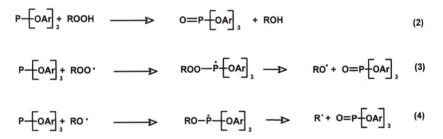

Scheme VI. *Hydroperoxide decomposition and reactions of peroxy- and alkoxy radicals in presence of aromatic phosphites.*

Table III. PP Homopolymer Multiple Extrusions at 270 °C

Sterically Hindered Phenol		Phosphite		Multiple Extrusion[a]		
Compound	*Conc. (%)*	*Compound*	*Conc. (%)*	*1×*	*3×*	*5×*
AO-4	0.05	P-1	—	9.4	18.4	>30
	0.05		0.1	3.2	5.1	7.7
AO-5	0.05	P-1	—	7.3	18.4	>30
	0.05		0.1	3.1	4.9	7.0

NOTE: Samples contained 0.05% calcium stearate. For structures of antioxidant compounds and P-1, *see* Chart I.
[a]Values are MFR g/10 min (230 °C, 2.16 kg).

1. Extrusion

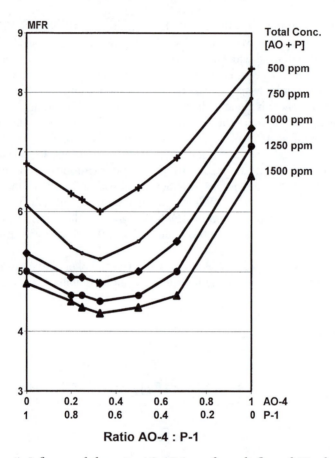

Figure 5. Influence of the ratio AO-4:P-1 on the melt flow of PP after first extrusion at 280 °C.

Reactions of peroxy- (ROO•) and alkoxy radicals (RO•) (Scheme VI, eqs 3 and 4) are mentioned in the literature (17, 18). Furthermore, rapid oxidation of the phosphite to phosphate by the residual oxygen dissolved in the polymer may contribute to stabilization during processing (19). Because of the kinetics of the oxidation from phosphite to phosphate, no contribution to stabilization of PP can be expected on aging at temperatures of 135 °C or 150 °C. However, the use of synergistic blends of sterically hindered phenols with trivalent or-ganophosphorus compounds protects the phenol from excessive consumption. Consequently, P-1 can fully contribute to stabilization during thermal aging (19). For this reason, phenol-phosphite can occasionally show an improved

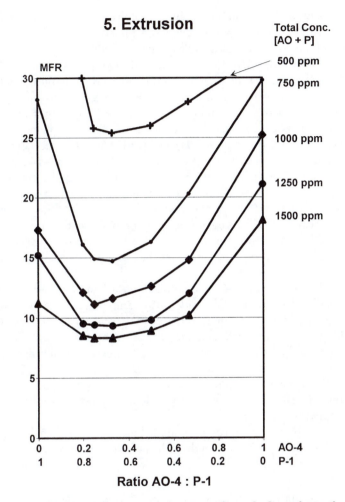

Figure 6. *Influence of the ratio AO-4:P-1 on the melt flow of PP after fifth extrusion at 280 °C.*

effect (2) with regard to thermal aging of polyolefins as compared with formulations containing the same amount of phenol alone.

Phenolic Antioxidants in Combination with Thiosynergists

Organosulfur compounds such as sulfides, dialkyl dithiocarbamate, or thiodipropionates are well-known hydroperoxide decomposers and have been used for many years for the stabilization of various plastics. Dioctadecyl-3,3-thiodipropionate (DSTDP) and its didodecyl homolog (DLTDP) bring about substantial increase in long-term heat stability of polyolefins in general and PP in particular in blends with sterically hindered phenols. Table IV summarizes some data obtained from heat-aging experiments with PP at 135 °C and 149 °C and stabilized with 0.1% sterically hindered phenol with 0.3% DSTDP.

Even in the absence of phenolic antioxidants, thiosynergists contribute significantly to heat-aging stability. In blends with 2,6-di-*tert*-butyl-*p*-cresol (AO-1), aging behavior is determined solely by the thiosynergist: because of its volatility, this phenol is quickly lost physically. The two phenolic antioxidants AO-2 or AO-4 (phenyl type **D**) in blends with thiosynergists lead to improved heat-aging stability because of an accumulating effect (compare in this respect the values in Table II). Synergism is manifested by blends of the phenols AO-5 (phenol type **C**) or, in a pronounced way, AO-6 (phenol Type A) with thiosynergists. However, the phenolic AO-6 and the thiosynergist exist *associated* because of a hydrogen-bridge bond (*20*), so that hydroperoxide formation and decomposition proceed in a quasi-concerted way. The mechanism of hydroperoxide decomposition by thiosynergists is shown in Scheme VII.

The sulfide (**34**) reacts stoichiometrically in a first step with a hydroxyperoxide molecule forming the oxide (**35**). Sulfenic acid (**37**) is formed through thermal decomposition. Another possible reaction is the formation of the dioxide (**36**). Starting with sulfenic acid (**37**), the key product in the reaction chain, further oxidation with hydroperoxides eventually may lead to sulfuric acid and other sulfur-containing oxidation products (*4, 21*). The overall reaction sequence contributes overstoichiometrically with respect to the used thiosynergist, because sulfur-containing acids act catalytically in the decomposition of hydroperoxides.

The effect of thiosynergists is limited, however, to thermal, long-term aging. DSTDP or DLTDP in blends with the phenolic antioxidant AO-4 do not contribute to the stabilization of polyolefins during processing of the melt.

Table IV. PP Homopolymer, Oven Aging at 135 °C and 149 °C

Compound	M_w	Mol OH/ kg AO	Time to Embrittlement (days) 135 °C	Time to Embrittlement (days) 149 °C
No antioxidant	—	—	42	8
AO-1	222	4.5	44	7
AO-2	531	1.9	79	18
AO-4	1178	3.4	219	56
AO-5	775	3.9	211	54
AO-6	795	5	152	47

NOTE: Samples were 1-mm compression-molded plaques. All samples contained 0.1% calcium stearate, 0.1% antioxidant (phenol), and 0.3% DSTDP. For structures of antioxidant compounds, *see* Chart I.

Table V summarizes the results from multiple extrusions with PP at 280 °C (PP-homopolymer; MFR of the virgin polymer was ~2.5 at 230 °C and 2.16 kg) stabilized with blends of AO-4–DLTDP (**34**), AO-4–DLTDP oxide (**35**), or AO-4–DLTDP dioxide (**36**).

In contrast to the blends AO-4–DLTDP and AO-4–DLTDP dioxide (36) only the blend AO-4–DLTDP oxide (**35**) contributes significantly to the processing stability of PP (22). This finding confirms that the formation of sulfenic acid as reactive precursor is the rate-determining step. The long-term behavior during thermal aging of the polymer stabilized with a blend of AO-4–P-1–DLTDP [or DLTDP oxide (**35**) or a corresponding combination with DLTDP dioxide (**36**)] is, as expected, improved by using the blend with DLTDP or DLTDP oxide (**35**).

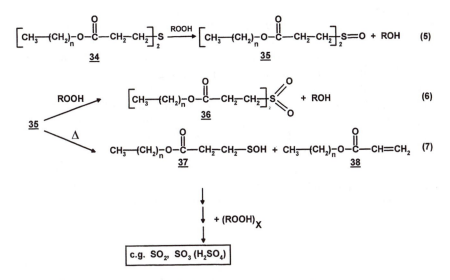

Scheme VII. Hydroperoxide decomposition with thiodipropionate esters.

Table V. PP Homopolymer Multiple Extrusions at 280 °C

Sterically Hindered		Thiosynergist		Multiple Extrusion[a]			t^a
Compound	Conc.	Compound	Conc. (%)	1×	3×	5×	
AO-4	0.01	None	—	9.4	18.4	>30	69
AO-4	0.01	S-1	0.5	9.8	20.1	>30	90
AO-4	0.01	S-2	0.5	3.1	4.9	6.8	87
AO-4	0.01	S-3	0.5	10.5	22	>30	69

NOTE: Samples contained 0.05% calcium stearate. For structures of antioxidant and thiosynergist compounds, *see* Chart I.
[a]Values are MFR g/10 min (230 °C, 2.16 kg).
[b]t is time to embrittlement in days at 135 °C. These formulations contained additional 0.1% P-1.

Phenolic Antioxidants in Combination with Sterically Hindered Amines

The discovery of sterically hindered amines based on tetramethyl-piperidine derivatives, the hindered amine light stabilizers (HALS), as stabilizers against photooxidative degradation of polyolefins led to an extensive change in the stabilization of these plastics. Numerous publications dealt with a variety of mechanism aspects related to HALS effect as inhibitors of the photooxidation of polymers (23).

Such sterically hindered amines are, however, also effective stabilizers against thermal degradation of polyolefins (24, 25).[1] The activity of these amines as antioxidants is based on their ability to form nitroxyl radicals. The reaction rate of nitroxyl radicals with alkyl radicals appears to be insignificantly lower than that of alkyl radicals with oxygen (26). For this reason, nitroxyl radicals are extremely efficient alkyl radical scavengers. Scheme VIII summarizes the reactions. It can be seen that the intermediary N-O-R, formed by the reaction with a peroxy radical ROO·, is returned to the reactive nitroxyl radical. It follows that in this cycle, there is a regenerating process (27) (or "Denisov" cycle).

Nitroxyl radicals are formed only in the course of polymer autoxidation. Stabilization of the polymer during processing of the melt, such as in the extruder, is unavoidable. For this study of PP a blend was chosen of the phenolic antioxidant AO-4 with P-1 and various HAS derivatives.

Table VI depicts the results obtained from thermal long-term aging of PP stabilized with the phenolic antioxidant AO-4 and from different HAS derivatives under long-term thermal exposure conditions. At temperatures below

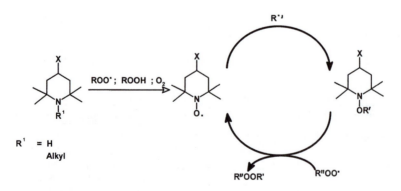

Scheme VIII. Stabilization mechanism of hindered amines.

[1] The designation "hindered amine light stabilizer (HALS)" should be replaced by "hindered amine stabilizer (HAS)", which reflects the ability of scavenging of radicals generated by both ways, light and heat.

120 °C, all blends containing sterically hindered amines display clear (120 °C) to significant (110 °C) improvement with regard to long-term thermal stability compared with samples without HAS. Long-term aging at higher temperatures appears to lead to antagonistic effects depending on the structure of the sterically hindered piperidine. Pospisil and Vyprachticky (*10, 28*) demonstrated that phenols (**39**) can react with nitroxyl radicals to form hydroxylamines (**44**) and with phenoxyl radicals (or the mesomeric and tautomeric forms, **41–43**) (Scheme IX). These compounds are, in turn, radical scavengers and add ni-

Table VI. PP Homopolymer Oven Aging at 110 °, 120 °, and 135 °C

Sterically Hindered Phenol	Sterically Hindered Amine	Time to Embrittlement (days)		
		110 °C	120 °C	135 °C
AO-4	—	205	170	69
AO-4	HAS-1	447	194	44
AO-4	HAS-2	377	173	45
AO-4	HAS-3	282	181	42
AO-4	HAS-4	461	220	85

NOTE: Samples were 1-mm compression-molded plaques. All samples contained 0.05% calcium stearate, 0.05% antioxidant (phenol), 0.1% phosphite (P-1), and 0.1% hindered amine (HAS). For structures of compounds, *see* Chart I.

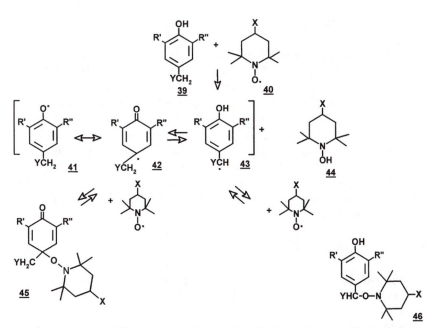

Scheme IX. Possible reactions of nitroxyl radicals with sterically hindered phenols.

troxyl radicals to the reaction products **45** and **46**. The kinetics of the retro-reaction determine the efficiency of such stabilizer blends.

At present there are no fundamental studies concerning these complex reactions, the products formed, and their reactions kinetics. On the strength of available data (25), there is great potential for stabilizer systems consisting of phenolic antioxidants, trivalent phosphites (protection against degradation during processing), and sterically hindered amines as stabilizers for long-term aging. This assumption is valid, at least, for specific applications of polyolefins, such as fibers. Of decisive importance is the proper choice of the individual components, their concentration, the processing method, and thermal long-term aging conditions.

Acknowledgments

I express my gratitude for data supplied by my colleagues K. Cooper and R. E. King III (Ciba-Geigy Additives Division, Ardsley, N.Y.), W. O. Drake, F. Gugumus, J. R. Pauquet, R. V. Todesco, and J. Zingg (Ciba-Geigy Additives Division, Basle, Switzerland) and M. Bonora and P. Canova (Ciba-Geigy Additives Division, Italy). Particular thanks are due to J. Pospisil, Czech Republic Academy of Sciences, Prague, Czechoslovakia, for his valuable contributions in discussions in the course of recent years.

List of Stabilizers

Sterically Hindered Phenols.   AO-1: Phenol, [2,6-bis(1,1-dimethylethyl)-4-methyl]-, CAS Reg. No. 128-37-0 (trade name, Ionol). AO-2: Benzenepropanoic acid, 3,5-bis(1,1-dimethylethyl)-4-hydroxyoctadecyl ester, CAS Reg. No. 2082-79-3 (trade name, Irganox 1076). AO-3: Benzenepropanoic acid, 3,5-bis(1,1-dimethylethyl)-4-hydroxy-1,6-hexanediyl ester, CAS Reg. No. 35074-77-2 (trade name, Irganox 259). AO-4: Benzenepropanoic acid, 3,5-bis(1,1-dimethylethyl)-4-hydroxy-, 2,2-bis[[3-[3,5-bis(1,1-dimethylethyl)-4-hydroxyphenyl]-1-oxopropoxy]methyl[-1,3-propanediyl ester, CAS Reg No. 6683-19-8 (trade name, Irganox 1010). AO-5: Phenol, 4,4',4"-[(2,4,6-trimethylethyl-1,3,5-benzenetriyl)-tris(methylene)]-tris[2,6-bis(1,1-dimethylethyl)-, CAS Reg No. 1709-70-2 (trade name, Irganox 1330). AO-6: Ethylene glycol, bis(3,3-bis(3'-dimethylethyl-4'-hydroxyphenyl) butyrate), CAS Reg No. 32509-66-3 (trade name, Hostanox 03).

Phosphites.   P-1: Phenol, 2,4-bis(1,1-dimethylethyl)-phosphite, CAS Reg. No. 31570-04-4 (trade name, Irgafos 168).

Thioethers.   DLTDP: Propanoic acid, 3,3'-thiobis-didodecyl ester, CAS Reg. No. 123-28-4 (trade name, Irgafos PS 800).

Hindered Amine Stabilizers. HAS-1: Butanedioic acid, polymer with 4-hydroxy-2,2,6,6-tetramethyl-1-piperidineethanol, CAS Reg. No. 65447-77-0 (trade name, Tinuvin 622). HAS-2: Poly[[6-[(1,1,3,3-tetramethylbutyl)-amino]-1,3,5-triazine-2,4-diyl][(2,2,6,6-tetramethyl-4-piperidinyl)imino]-1,6-hexane-diyl-[(2,2,6,6-tetramethyl-4-piperidinyl)imino]], CAS Reg. No. 71878-19-8 (trade name, Chimassorb 944). HAS-3: Poly-[[4-[(4-morpholinyl)imino]-1,3,5-triazine-2,4-diyl][(2,2,6,6-tetra-methyl-4-piperidinyl)imino]-1,6-hexane-diyl-[(2,2,6,6-tetramethyl-4-piperidinyl)imino]], CAS Reg. No. 82451-48-7 (trade name, Cyasorb 3346). HAS-4: 1,3,5-Triazine-2,4,6-triamine, N,N'''-[1,2-ethane-diyl-bis[[[4,6,-bis-[butyl-(1,2,2,6,6-pentamethyl-4-piperidinyl)amino]-1,3,5-triazine-2-yl]imino]-3,1-propanediyl]]-bis[N,N''-dibutyl-N',N''-bis-(1,2,2,6,6-pentamethyl-4-piperidinyl)-, CAS Reg. No. 106990-43-6 (trade name, Chimassorb 119).

References

1. Drake, W. O.; Pauquet, J. R.; Todesco, R. V.; Zweifel, H. *Angew. Makromol. Chem.* **1990,** *176/177,* 215.
2. Moss, S.; Pauquet, J. R.; Zweifel, H. In *Proceedings of the 13th International Conference on Advances in the Stabilization and Degradation of Polymers;* Patsis, A. V., Ed.; Lucerne, 1991; p 203.
3. Hinsken, H.; Moss, S.; Pauquet, J. R.; Zweifel, H. *Polym. Degrad. Stab.* **1991,** *34,* 279.
4. Scott, G.; Al Malaika, S. In *Atmospheric Oxidation and Antioxidants;* Scott, G.; Ed.; Elsevier Science: London, 1993; Vol. I.
5. Pospisil, J. In *Developments in Polymer Stabilization;* Scott, G., Ed.; Applied Science: London, 1979; Vol. 1, p 1.
6. Pospisil, J. *J. Adv. Polym. Sci.* **1980,** *36,* 69.
7. Henman, T. J. In *Developments in Polymer Stabilization;* Scott, G., Ed.; Applied Science: London, 1979; Vol. 1, p 39.
8. Gugumus, F. *Angew. Makromol. Chem.* **1985,** *137,* 189.
9. Pospisil, J. *Polym. Degrad. Stab.* **1993,** *40,* 217.
10. Pospisil, J. *Polym. Degrad. Stab.* **1993,** *39,* 103.
11. Pilar, J.; Rotschova, J.; Pospisil, J. *Angew. Makromol. Chem.* **1992,** *200,* 147.
12. Denisov, E. T. In *Developments in Polymer Stabilization;* Scott, G., Ed.; Applied Science: London, 1980; Vol. 3, p 1.
13. De Jonge, C. R. H. I.; Hope, P. In *Developments in Polymer Stabilization;* Scott, G., Ed.; Applied Science: London, 1980; Vol. 3, p 21.
14. Glass, R. D.; Valange, B. M. *Polym. Degrad. Stab.* **1988,** *20,* 355.
15. Gugumus, F. In *Plastic Additives Handbook*; Gächter, R.; Müller, H., Eds.; Hanser: Munich, Germany, 1990; p 1.
16. Drake, W. O.; Cooper, K. D. *Proceedings of the SPE Polyolefins VIII International Conference;* Society of Petroleum Engineers: Houston, TX 1993; p 417.
17. Schwetlick, K.; Könih, T.; Rüger, C.; Pionteck, J.; Habicher, W. D. *Polym. Degrad. Stab.* **1986,** *15,* 97.
18. Schwetlick, K. *Pure Appl. Chem.* **1983,** *55,* 1629.
19. Pauquet, J. R. In *Proceedings of the 42nd International Wire and Cable Symposium;* International Wire and Cable Symposium: Eatontown, NJ, 1993; p 77.
20. Yachigo, S.; Sasaki, M.; Kojima, F. *Polym. Degrad. Stab.* **1992,** *35,* 105.

21. Pospisil, J. In *Oxidation Inhibition in Organic Materials;* Pospisil, J.; Klemchuk, P. P., Eds.; CRC: Boca Raton, FL, 1990; Vol. I, p 40.
22. Drake, W. O.; Pauquet, J. R.; Zingg, J.; Zweifel, H. *Polym. Prepr.* **1993**, *2*, 174.
23. Sedlar, J. In *Oxidation Inhibition in Organic Materials;* Pospisil, J.; Klemchuk, P. P., Eds.; CRC: Boca Raton, FL, 1990; Vol. II, p 1.
24. Shlyapintokh, V. Y.; Ivanov, V. B. In *Developments in Polymer Stabilization;* Scott, G., Ed.; Applied Science: London, 1982; Vol. 5, p 41.
25. Drake, W. O. In *Proceedings of the 14th International Conference on Advances in the Stabilization and Degradation of Polymers;* Patsis, A. V., Ed.; Lucerne, 1992; p 57.
26. Bowry, V. W.; Ingold, K. U. *J. Am. Chem. Soc.* **1992**, *114*, 4992.
27. Shilov, Y. B.; Denisov, Y. T. *Vysokomol. Syed.* A16, **1974**, *10*, 2316. (*See* Shilov, Y. B.; Denisov, Y. T. *Polym. Sci. USSR* **1974**, *16/8*, 2686.)
28. Vyprachticky, D.; Pospisil, J. *Polym. Degrad. Stab.* **1990**, *27*, 227.

RECEIVED for review January 26, 1994. ACCEPTED revised manuscript October 7, 1994.

Fluorophosphonites as Co-Stabilizers in Stabilization of Polyolefins

G. J. Klender

Albemarle Corporation, Albemarle Technical Center, PO Box 14799, Baton Rouge, LA 70898

Fluorophosphonites, a family of phosphorus co-stabilizers, are discussed in regard to their preparation, properties, and effectiveness in polyolefins, particularly when combined with phenolic stabilizers to control melt flow by multipass extrusion. The presence of a fluorine–phosphorus bond contributes to improved hydrolytic stability in aryloxy trivalent phosphorus compounds with and without the phosphorus atom in a seven- or eight-membered ring system. The superior hydrolytic stability of the fluorophosphonite FP-1, its excellent thermal stability, and competitive stabilization effectiveness compared with other phosphorus stabilizers are clearly demonstrated. Improvements in process color control and long-term heat aging with FP-1 are also shown. Some of the excellent heat aging characteristics of FP-1, particularly under gamma radiation or high temperature moisture exposure conditions, are exceptional. These fluorophosphonites may open up new application areas that were considered unattainable because of the limited hydrolytic stability of other phosphorus containing stabilizers.

THE SYNERGISTIC COMBINATION OF PHOSPHITES OR PHOSPHONITES with phenolic antioxidants can result in a more cost effective system for the melt stabilization of polyolefins than a phenolic antioxidant alone (1). The benefits of these phosphorus stabilizers are not limited to melt stability; other advantages such as minimizing polymer color during processing (2) and improving resistance to gas yellowing (3) have been shown in some polyolefin resins. Along with the advantages of these phosphorus stabilizers come some major concerns to users. As a class, phosphites, particularly alkyl or mixed alkyl-aryl phosphites, have poor hydrolytic stability (4). Aromatic phosphites provide improved hydrolytic stability but do not equal the antioxidant performance of

0065–2393/96/0249–0397$13.75/0

alkyl or mixed alkyl-aryl phosphites (5). Phosphite hydrolysis can result in caking problems, which can cause difficulties in handling the additive, as well as inferior performance of the additive in the polymer. The addition of small amounts of organic bases such as triisopropanolamine resulted in marginal improvement in the hydrolytic stability of phosphites (6). This approach was extended recently (7, 8) to the incorporation of amine functionality into phosphite molecule and resulted in a significant improvement in the hydrolytic stability of these phosphites. Poor thermal stability can accompany the poor hydrolytic stability of phosphites and result in a characteristic appearance of "black specks" during the processing of the polymer. These black specs have been attributed to the breakdown of the phosphorus stabilizer.

The goal of developing hydrolytically and thermally stable phosphorus compounds, with the stabilizer performance of alkyl-aryl phosphites, was achieved with the synthesis of a new family of phosphonite antioxidant, fluorophosphonites (9). In general, the fluorophosphonites exhibit hydrolytic stability superior to the commercially available aromatic phosphites, while providing performance in polyolefins that approaches that of the best mixed alkyl-aryl phosphite. This chapter will describe the preparation of fluorophosphonites, some observations regarding the effect of structure on performance, and a comparison of a fluorophosphonite to some of the commercially available phosphorus stabilizers in respect to their performance and behavior in various environments.

Experimental

Stabilizers used in this work are identified by their chemical structure in Chart I, and a commercial name is listed where possible. The commercial materials were used without further purification, and experimental materials had a minimal purity of 95% by gas chromatography and phosphorus NMR spectroscopy. The polypropylenes (PPs) used in this study were homopolymers produced on a supported Ziegler–Natta catalyst system by bulk processes. All samples were compounded with acid neutralizers except where otherwise specified. Generally, 0.05 wt% of calcium stearate (CaSt) was determined to be adequate for neutralization of catalyst residues and maintenance of good process color in these PPs (10). In some instances, only 0.01 wt% of CaSt was used in a model system to demonstrate the contribution of the phosphorus stabilizer to color suppression. Sometimes the inorganic acid neutralizer, synthetic dihydrotalcite (DHT4a), was used in place of calcium stearate. This is a magnesium–aluminum salt with the nominal formula $Mg_6Al_2(OH)_{16}CO_3 \cdot 4H_2O$. Other polyolefins used in this study were high-density polyethylene (HDPE) from a chromium-based slurry process that was used without neutralizer, and gas-phase, linear, LLDPE from a Ziegler–Natta catalyst system that was used with zinc stearate (ZnSt).

Procedures for compounding and for multipass extrusion as well as some polymer test methods are described elsewhere (11). Temperature profiles for these extrusions are as follows:

Profile	Zone Temperature (°C)	Stock Temperature (°C)	Polymer
TP-1	288, 288, 288, 288	295–298	PP
TP-2	218, 218, 232, 246	253–255	HDPE, LLDPE
TP-3	260, 260, 260, 260	268–270	PP

Melt flow index measurements were made to determine melt viscosity by using method ASTM D 1238. Standard test conditions of load and temperature were used: condition L (2160 g at 230 °C) for PP and condition P (5000 g at 190 °C), E (2160 g at 190 °C), or F (21,600 g at 190 °C) for PE. A measure of a broadening of the molecular weight distribution can be made by dividing the high load melt index (F) by a low load melt index. During processing, an increase in melt index ratio indicates a broader molecular weight distribution caused by long-chain branching.

The details of the hydrolysis studies and ^{31}P NMR shifts after hydrolysis were previously published (*12*). Thermal stability testing, fiber spinning conditions, heat aging, and gamma radiation test procedures are described in detail in another publication (*13*).

Results and Discussion

Preparation of Fluorophosphonites.   The reaction of alcohols and phenols with phosphorus trichloride can yield organophosphites or partially substituted chlorophosphonites (*14*). Aryloxy mono- or di-halo phosphonites can be isolated in good yields by the reaction of the appropriate phenol with phosphorus trichloride or tribromide in an aprotic solvent because of the reduced reactivity of these partially substituted compounds (*15, 16*). The formation of the chlorophosphonite, CP-1 as seen in Scheme I, is followed by the replacement of the chlorine atom by fluorine by the use of various fluorinating agents to yield the fluorophosphonite, FP-1 (*9, 17*).

In this manner, a variety of mono- and difluorophosphonites have been prepared from mono- and poly-nuclear phenols. Some of these are shown in

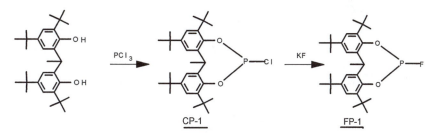

Scheme I.

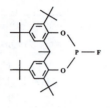

FP-1

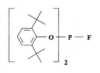

FP-2

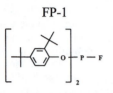

FP-3

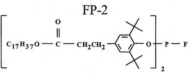

FP-4

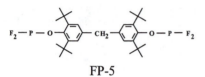

FP-5

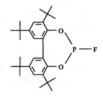

FP-6

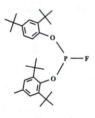

FP-7

FP-8

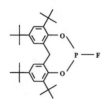

FP-9

FP-10

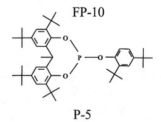

P-5

Chart I.

Abbreviation	Structure	Trade Name

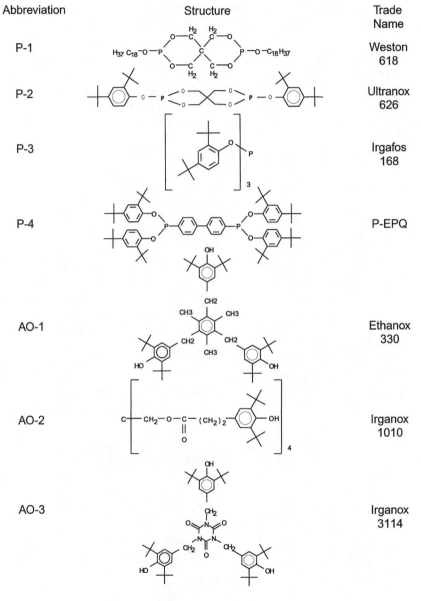

Chart I. Continued

Chart I, and they illustrate the variety of structures that are available with this technology.

Thermal Stability of Fluorophosphonites.   In Table I, several of the fluorophosphonites are listed by increasing molecular weight. They provide a range of melting points from liquid at room temperature to 200 °C and a varying phosphorus content. Some of these compounds show weight losses under nitrogen that reflect their relative molecular weights. However, some of the higher molecular weight fluorophosphonites that are based on monophenols show higher volatility than expected and suggest that they are degrading rather than volatilizing in their evaluation by thermogravimetric analysis.

Some phosphorus stabilizers produce a char when melted in a test tube open to the air at 288 °C. Even though such a small-scale test may not represent the same thermal degradation that can occur in processing, it is encouraging to observe that the order ranking for char formation in this test is the same order of ranking that is recognized by many polymer producers for commercial phosphorus stabilizers in regard to the formation of black specks in polyolefins during processing. In Table II, neat phosphites and phosphonites were held at 288 °C for 30 min. The fluorophosphonites, for the most part, discolored but did not produce char, whereas the other compounds discolored and produced char. After cooling, the percent of each remaining phosphorus compound was determined by ^{31}P NMR spectroscopy. The commercial phosphorus stabilizers showed significant decomposition, whereas many of the fluorophosphonites remained intact (Table II).

Because fluorophosphonites have the potential of releasing hydrogen fluoride, there is a legitimate concern that fluorophosphonites could contribute to the corrosion of processing equipment as well as produce a severe health and safety problem. To determine if HF was evolved during processing, FP-1

Table I. Properties of Some Fluorophosphonites
Determined at 10 °C/min under Nitrogen

Fluorophos- phonite	M_r	P (%)	DSC mp (°C)	10% TGA Weight Loss (°C)
FP-8	376	8.2	~20	220
FP-6	458	6.8	188–190	225
FP-3	460	6.7	68–69	240
FP-9	472	6.6	194–196	237
FP-7	474	6.5	152–153	210
FP-1	486	6.4	202–204	257
FP-10	514	6.0	67–71	276
FP-5	560	11.0	125–126	200

was exposed to the most adverse conditions that we could envision: a combination of high heat and moisture. Table III summarizes the results of these studies where samples of neat FP-1 or compositions with PE were placed in a 1-in. diameter stainless steel tube in a furnace, swept with wet helium, and heated up to 350 °C. The helium was passed through a train of scrubbers containing sodium hydroxide where any HF released could be detected as NaF by ion chromatography. In an evaluation test of the system, quantitative amounts of HF were recovered at the 250 (μg/mL level with an accuracy of about 1%. No detected HF evolved from FP-1 alone or from FP-1 compounded in PP, even at 350 °C. After each test, the apparatus was taken apart and inspected. Although some of the FP-1 sublimed into a cooler portion of the tube, there was no evidence of corrosion in the tube. The P–F bond appears to be very stable and FP-1 should not contribute to equipment corrosion during compounding. To determine the decomposition point of FP-1,

Table II. Phosph(on)ite Remaining in Samples Heated at 288 °C for 30 min in Air Determined by ^{31}P NMR Spectroscopy after Cooling

Compound	Remaining (%)	Appearance
P-1	0	Severe char
P-2	15	Char
P-4	40	Some char
P-3	50	Slight char
FP-5	61	No char
FP-3	73	Slight char
P-5	78	Slight char
FP-1	93	No char
FP-2	99	No char
FP-4	99	No char

Table III. Ion Chromatographic Detection of Fluoride Ion Resulting from Wet Helium Passed Over Sample for 30 min at Each Temperature

Sample	Temp. (°C)	Observed [NaF] (μg/mL)	Theoretical [NaF] (μg/mL)	Corrosion
FP-1	250	0	854	No
FP-1	300	0	780	No
FP-1	350	0	871	No
FP-1 (6% in PP)	250	0	163	No
FP-1 (6% in PP)	300	0	160	No
FP-1 (6% in PP)	350	0	261	No
HF solution (250 μg/mL)	100	248.5	250	Yes

it was compounded into polycarbonate and heated in a Paar bomb to 350 °C over an 80-min period without decomposition; after 3 min at 350 °C, 0.7% of HF was detected (18).

Moisture Effects on Neat Stabilizers Containing Phosphorus. All of the phosphorus stabilizers currently used in polyolefins are free flowing particles that can stick together, encrust, or solidify when exposed to heat and humidity. In Table IV, comparisons among some phosphites, phosphonites, and fluorophosphonites are made in regard to caking under different conditions of temperature, relative humidity (RH), and surface area. Temperature appears to be a critical factor in the acceleration of caking. The phosphite P-3 cakes in 42 days at 36 °C, whereas it took only 7 days to cake at 50 °C with the same surface area. The fluorophosphonites in Table IV did not cake under conditions of exposure.

These phosphorus stabilizers that caked also picked up moisture. Although the pickup of moisture does not necessarily indicate hydrolysis of the phosphorus compound, those compounds that did not gain weight when exposed can be assumed to be stable and resistant to hydrolysis. This assumption was verified in Table V; the solid fluorophosphonites FP-1, FP-7, and FP-9 did not pick up water when exposed to moisture at 50 °C and also did not show hydrolysis by ^{31}P NMR spectroscopy. The alkyl and aryl-alkyl phosphites P-1 and P-2 showed rapid moisture pickup within the first day under these conditions, and these results indicated that hydrolysis was occurring by the presence of insoluble solids when these exposed phosphites were dissolved in dry methylene chloride. These phosphites were completely soluble in methylene chloride before exposure to moisture. A semiquantitative, phosphorus-NMR analysis was made on the soluble portion of P-2 on the assumption that any insoluble portion was completely hydrolyzed. It appears that 50% of P-2 was hydrolyzed in nine hours in this analysis. The appearance of insolubles

Table IV. Time to Caking for Neat Phosphorus Stabilizers Exposed at a Surface Area of 106 cm²/10 g and Relative Humidity from 75 to 90%

Compound	36 °C[a]	50 °C	55 °C
P-1	< 1	< 1	
P-2	2	< 1	< 1
P-3	42	4[b]	3
P-4	40	3	< 1
FP-1	> 380	> 120	> 6
FP-7	> 365	> 120	
FP-9	> 150	> 120	

NOTE: Values are in days.
[a]Surface area was 57 cm²/10 g.
[b]Caked in 7 days at 57 cm²/10 g.

Table V. Time to Moisture Pick Up or Hydrolysis of Neat
Phosphorus Stabilizers Exposed to 50 °C/86% RH at a
Surface Area of 106 cm²/10 g of Sample

Compound	2% Weight Gain	10% Weight Gain	10% Hydro- lysis	50% Hydro- lysis
P-1	0.08	0.25		
P-2	0.25	0.60		0.38
P-3	5	7	5	6
P-4	2	5	2	3
FP-1	> 120	> 120	> 120	> 120
FP-7	> 120	> 120	> 120	> 120
FP-8	4	8		
FP-9	> 120	> 120	> 120	> 120

NOTE: Values were determined by ³¹P NMR spectroscopy and are in days.

with P-1 was faster than with P-2 and it is assumed that the hydrolysis of P-1 is faster than P-2. In either case, hydrolysis products were only seen by ³¹P NMR spectroscopy in the portion of the samples soluble in methylene chloride after 24 h of moisture exposure.

The solid phosphonite P-4 and the phosphite P-3 were more hydrolytically stable. Moisture pickup for P-4 began on the second day and was accompanied by significant hydrolysis. By the third day, P-4 was 50% hydrolyzed, and it was completely hydrolyzed by the fifth day. In a similar manner, P-3 hydrolyzed rapidly once moisture pickup began. Virtually no moisture pickup occurred until the fourth day, but once the process began it was accompanied by rapid hydrolysis. P-3 was completely hydrolyzed in 8 days. The ³¹P NMR spectrum of P-3 in Figure 1 illustrates the chemical shift and hydrolysis reactions that are occurring. The ³¹P NMR spectrum of FP-1 (Figure 2) was taken after four months of exposure and shows no P–H structures associated with hydrolysis products. FP-1 was also completely soluble in methylene chloride after exposure. The liquid fluorophosphonite FP-8 appears to be less hydrolytically stable than the other fluorophosphonites based on moisture pickup and does show some hydrolysis. Its hydrolytic stability could be similar to P-3, which has almost the same affinity for water in this test.

Effect of High Temperature and Humidity on PP with Phosphorus Stabilizers.

The effect of partially hydrolyzed phosphorus stabilizers in polyolefins was a loss of effectiveness as a processing stabilizer and, in some cases, resulted in greater discoloration of the polymer during processing (11). To determine if partial hydrolysis of the phosphorus stabilizer can occur during pellet storage, PE containing AO-2, calcium stearate (CaSt), and different phosphorus stabilizers were evaluated. One-half of the pellets were exposed to 80 °C and high humidity for 6 weeks, and the other portion

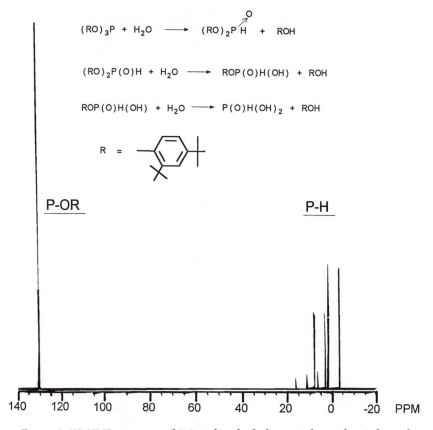

Figure 1. ³¹P NMR spectrum of P-3 and its hydrolysis products after 6 days of exposure to 50 °C and 86% RH (surface area was 106 cm²/10 g).

was kept at the controlled 20 °C and low humidity. The high temperature/ high moisture exposure was an accelerated test that attempted to simulate several months of storage under the most adverse atmospheric conditions such as might be found in hopper car storage on the Gulf coast of the United States. After exposure, the melt flow indexes (MFIs) of the pellets after multipass extrusion were compared to the MFIs of the pellets stored under controlled conditions. The phosphite P-2 and the phosphonite P-4 both showed a significant loss of effectiveness because of exposure of the pellets to high humidity. This effect is most evident after fifth-pass extrusion, as seen by the higher melt indexes of the exposed samples compared with the control samples containing these two phosphorus stabilizers in Figure 3. This figure shows that these stabilizers have apparently hydrolyzed to some degree to less effective compounds. The fluorophosphonite FP-1 does not show a significant difference in melt stabilizer effectiveness after fifth-pass extrusion between

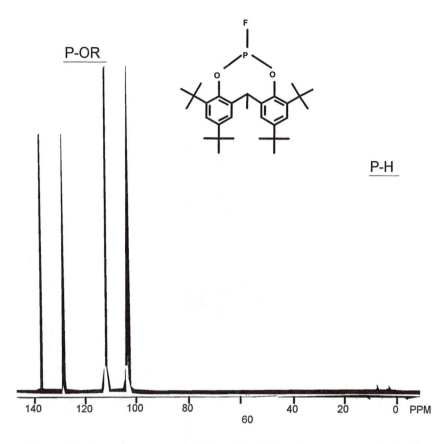

Figure 2. [31]*P NMR spectrum of FP-1 after 120 days of exposure to 50 °C and 86% RH (surface area was 106 cm^2/10 g); no hydrolysis is shown.*

the moisture exposed and the unexposed pellets; therefore, it has resisted hydrolysis in the pellet as expected based on the data from exposure of the neat stabilizer.

Factors That Influence Hydrolytic Stability of Phosphorus Stabilizers. The resistance of the phosphorus–fluorine bond to reaction with water is at least partially responsible for the hydrolytic stability of the fluorophosphonites. In solution, the chlorophosphonite (CP-1) is readily converted into the phosphonate (HP-1) in several hours at room temperature (Scheme II).

Replacement of the chlorine–phosphorus bond by a fluorine–phosphorus bond enhances the room temperature solution stability of the compound as seen in Table VI. There was no hydrolysis of FP-1, which is the fluoro-derivative of CP-1, for at least two weeks. This relatively good hydrolysis resistance

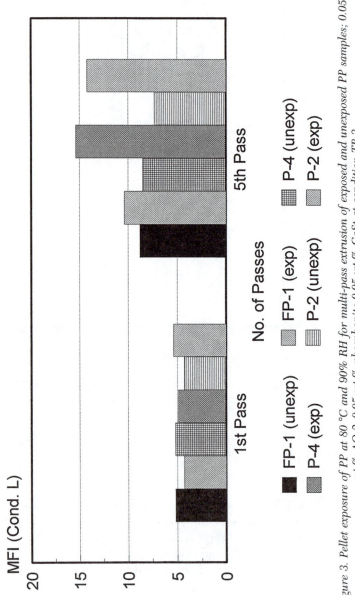

Figure 3. Pellet exposure of PP at 80 °C and 90% RH for multi-pass extrusion of exposed and unexposed PP samples; 0.05 wt.% AO-2–0.05 wt.% phosphonite-0.05 wt.% CaSt at condition TP-3.

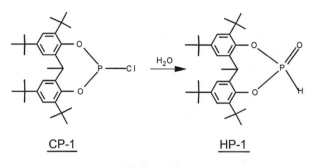

CP-1   HP-1

Scheme II.

**Table VI. Percent of Sample
Hydrolyzed in Tetrahydrofuran
Containing 8% by Weight H_2O at 25 °C
Determined by ^{31}P NMR Spectroscopy**

Compound	After 1 Day	After 2 Weeks
P-2	32	
P-4	78	
FP-1		0
FP-2		0
FP-3		0
FP-4		0

of FP-1 is also shared by several fluorophosphonites that do not have the phosphorus atom bonded in an eight-member ring structure. On the other hand, the phosphite, P-2, and the phosphonite, P-4, in solution were readily hydrolyzed in the first day, and this result further supports the importance of the P–F bond to the hydrolytic stability of fluorophosphonites.

Solid derivatives of seven- and eight-membered rings containing a trivalent phosphorus atom ([1,3,2]dioxaphosphocine ring systems), such as compound P-5, have good hydrolytic stability and stabilizer effectiveness in polyolefins (*19*). These hydrolytic stability tests were performed on the solid compounds. Some of the fluorophosphonites have the same structure as these solid phosphites, and as solids they show good hydrolytic stability (FP-1 and FP-9 in Table V). It is important from a practical standpoint to determine the effect of moisture on the solids for handling purposes. It is also impossible to separate the effects that particle size, crystal structure, and the ability of water to wet the crystal will have on the hydrolytic stability of the solid; therefore, it is not possible to determine the effect of the [1,3,2]dioxaphosphocine ring structure on hydrolytic stability from these results. To make this kind of a determination, it would be necessary to compare similar ring and non-ring structures in solution. The importance of these solid phase characteristics on the hydrolytic stability of the compound can be seen in the comparison of FP-7 and FP-8 (Table V). Both are fluorophosphonites based on mononuclear phenols. The solid fluorophosphonite, FP-7, is much more hydrolytically stable than the liquid fluorophosphonite FP-8. The solid-state characteristics of the phosphite P-2 and the phosphonite P-4 are probably important in accounting for the greater hydrolytic stability of P-4 in comparison to P-2 in the solid state (Table V) because P-4 will hydrolyze faster than P-2 in solution (Table VI).

Processing Stability of Fluorophosphonites in PP. The mechanisms of the antioxidant action of phosphorus stabilizers were reviewed (20). The reduction of polymer hydroperoxides by phosphites is considered to be the important mode of stabilization during melt processing. This phenomenon is certainly the case in the action of the cyclic fluorophosphonites, and the fluorophosphonate oxidation product was identified (18). Even though the fluorophosphonate oxidation product is expected to be less hydrolytically stable than the fluorophosphonites, no hydrolyzed phosphonate was found by ^{31}P NMR spectroscopy in the polymer after multipass extrusion.

A comparison of PP containing AO-1 and various fluorophosphonites shows that these compounds are effective synergists with the phenolic during multi-pass extrusion processing. They add to the stability of the system and are generally better or equal to the effectiveness of P-3 when compared on an equal weight basis (Table VII). This result is confirmed in Table VIII, which contains data that were obtained in another lot of the same PP but had a reduced concentration of acid neutralizer (0.01%).

The data in Table VIII also show that there are relatively small differences in stabilizer effectiveness among the various types of cyclic fluorophosphonites after multi-pass extrusion. The size of the phosphorus-containing ring, the appendant groups on the aromatic rings, and the aliphatic bridge have relatively small effects on the melt stability of PP during melt processing under

Table VII. Effect of Stabilizer Composition on Melt Stability of PP Containing 0.10% Calcium Stearate

Composition	First Pass	Third Pass	Fifth Pass
AO-1	8.9	13.2	18.8
AO-1/FP-1	5.5	9.3	15.1
AO-1/FP-3	5.5	9.5	15.4
AO-1/FP-4	7.9	11.5	16.2
AO-1/P-3	5.9	10.7	16.6
AO-1/P-5	6.5	11.9	18.9

NOTE: Multipass extrusion at condition TP-1 (see Experimental) in a single-screw extruder at 30 rpm screw speed. Values are MFI at condition L-ASTM D 1238. Concentrations for all compositions were 500 ppm each.

**Table VIII. Comparison of Phenolic–
Phosphorus Stabilizer Blends in PP Containing
0.01% Calcium Stearate**

Composition	First Pass	Third Pass	Fifth Pass
AO-1/FP-1	5.4	8.7	13.1
AO-1/FP-6	5.1	9.7	14.2
AO-1/FP-7	6.0	10.8	16.8
AO-1/FP-8	5.5	9.0	13.9
AO-1/FP-9	5.0	8.8	14.1
AO-1/P-3	5.7	10.5	15.5

NOTE: Multipass extrusion at condition TP-1 (*see* Experimental) in a single-screw extruder at 30 rpm screw speed. Values are MFI at condition L-ASTM D 1238. Concentrations for all compositions were 500 ppm each.

these oxygen-starved conditions. This behavior has been notice with the other cyclic fluorophosphonites.

Processing Stability: FP-1 Versus Commercial Phosphorus Stabilizers. There have been comparisons run between FP-1 and other commercial phosphorus stabilizers in many different PPs. The data in Figure 4 summarize the results of three phenolic antioxidants with P-2, P-3, P-4, and FP- I. The results are typical of melt flow data in other PPs that are made on supported Ziegler–Natta catalyst systems. With each phenolic, FP-1 is always a superior melt stabilizer when compared with P-3, generally a little better than P-4, and not quite as good as P-2.

In HDPE, cross-linking is a primary mechanism of degradation but chain scission can also occur. One method of measuring stability is to look at the ratio of melt flow at two different shear rates that will detect a broadening of the molecular weight distribution due to cross-linking. The greater change in ratio of the high to low melt index is indicative of greater degradation. Figure 5 shows that FP-1 with AO-2 is more effective than P-2, P-3, or P-4 in this fractional-melt HDPE polymerized with a chromium catalyst system.

The primary mechanism of degradation in LLDPE is also cross-linking. In a fractional-melt LLDPE polymerized by a gas-phase Ziegler–Natta process, the melt index was reduced with each pass through the extruder with stabilizer systems comprising AO-1, zinc stearate, and a phosphorus stabilizer (Figure 6). The percent change in melt index from the powder melt index after five passes through the extruder was smallest or best for P-2. FP-I and P-4 gave slightly larger reductions of the melt index. The poorest results of the phosphorus stabilizers were yielded by P-3.

Processing Color Stability: FP-1 versus Commercial Phosphorus Stabilizers. Processing discoloration in PP has become less critical in polymers made by high-activity-supported catalysis if adequate neu-

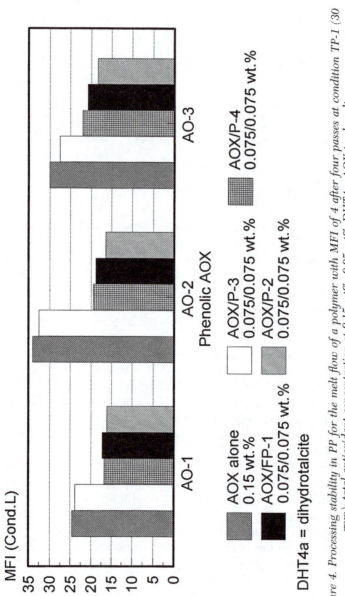

Figure 4. Processing stability in PP for the melt flow of a polymer with MFI of 4 after four passes at condition TP-1 (30 rpm); total antioxidant concentration at 0.15 wt%; 0.05 wt% DHT4a. AOX is phenolic.

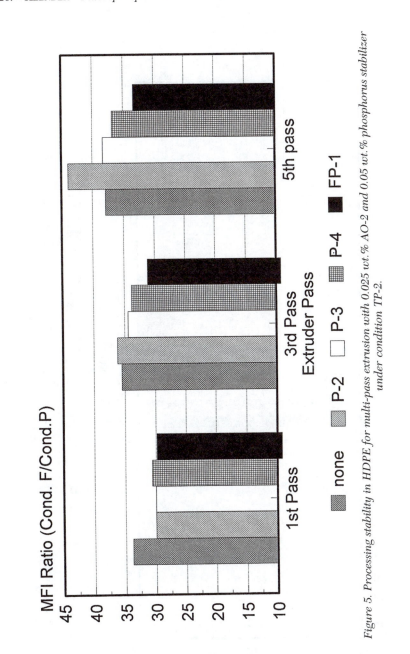

Figure 5. Processing stability in HDPE for multi-pass extrusion with 0.025 wt. % AO-2 and 0.05 wt. % phosphorus stabilizer under condition TP-2.

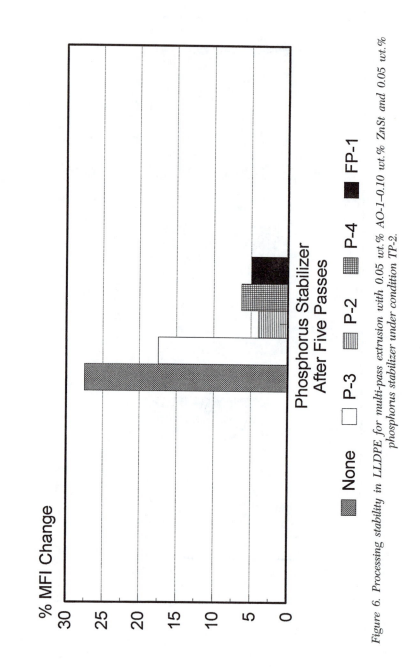

Figure 6. Processing stability in LLDPE for multi-pass extrusion with 0.05 wt.% AO-1–0.10 wt.% ZnSt and 0.05 wt.% phosphorus stabilizer under condition TP-2.

tralization of any acid catalyst residues is achieved (*10*). This result is primarily due to the low levels of titanium metal residue in the product. To see a significant effect of phosphorus stabilizers on color control in these polymers, it is necessary to use an artificial model and minimize the amount of acid neutralizer in the system. The color suppression effect of phosphorus stabilizers in PP is seen in Figure 7. With AO-1, the aryl phosphite P-3 does not show much color-suppression effect even after fifth-pass extrusion, whereas FP-1, P-2, and P-4 all exhibit strong color control particularly in minimizing color increases on multi-pass extrusion.

HDPEs produced on supported chromium oxide catalysts are not generally compounded with acid neutralizers, although zinc stearate has provided some processing color suppression with phenolic antioxidants in these polymers (*10*). Multi-pass extrusions with AO-2 in this HDPE were conducted without the use of an acid neutralizer, and the effect of adding the various phosphorus stabilizers is shown in Table IX. P-2 is the best in controlling both initial color and color development on multi-pass extrusion. FP-1 and P-4 provide similar control of color but are not as good as P-2, whereas P-3 gives the poorest color control of the phosphorus stabilizers tested in this polymer.

The LLDPE used in this study was compounded with AO-1, zinc stearate, and the various phosphorus stabilizers. P-2 and FP-1 provide the best initial color and color control during processing of this polymer (Table X). Even though P-4 provided some initial color suppression, it did not maintain the effect on processing, and P-3 showed essentially no benefit in either initial or processing color control.

Effects of Gamma Radiation on PP Containing Phosphorus Stabilizers.   Gamma radiation has become an important method of sterilization, particularly for medical devices. Unfortunately, PP is severely degraded by gamma radiation and must be protected especially from postirradiation deterioration on storage (*21*). Because phenolic antioxidants tend to severely color on exposure to gamma radiation, it is desirable to use a minimum amount of phenolic as a processing stabilizer. The processing stabilization can be accomplished by the use of a phosphorus co-additive with a minimum amount of phenolic. However, some phosphorus stabilizers are destroyed by gamma radiation and can have adverse effects on the polymer (*22, 23*).

To evaluate their resistance to gamma radiation, the various phosphorus stabilizers were blended with AO-3 in PP, and the PP powder with an MFI of 4 was cracked to a polymer of 30 MFI with peroxide during compounding. These materials were made into tensile bars and plaques then irradiated with 3.0 megarads of radiation. The plaques were aged at 60 °C for up to a year, and color measurements were made periodically. In Figure 8, the data show that P-4 rapidly yellowed and was worse than a composition without a phosphorus stabilizer after 11 weeks. P-3 began accelerated yellowing at 21 weeks,

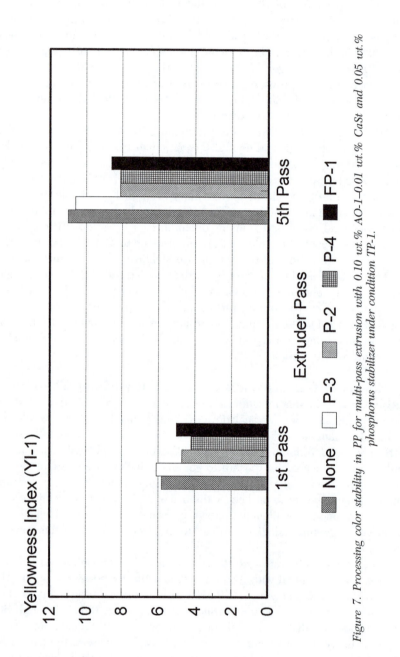

Figure 7. Processing color stability in PP for multi-pass extrusion with 0.10 wt.% AO-1–0.01 wt.% CaSt and 0.05 wt.% phosphorus stabilizer under condition TP-1.

Table IX. Process Color for 0.025% AO-2 and 0.05% Phosphorus Stabilizer in HDPE after Multipass Extrusion

Compound	First Pass	Third Pass	Fifth Pass
None	8.5	11.5	12.8
P-2	−0.6	0.2	1.3
FP-1	2.0	4.0	5.0
P-4	2.4	4.0	5.1
P-3	3.2	5.7	7.0

NOTE: Multipass extrusion at condition TP-2 (*see* Experimental) at 30 rpm screw speed. Values are yellowness index (YI-1).

Table X. Process Color for 0.05% AO-1 in LLDPE with 0.10% Zinc Stearate and 0.05% Phosphorus Stabilizer after Multipass Extrusion

Compound	First Pass	Fifth Pass
None	0.44	4.8
P-2	−0.56	3.5
FP-1	−0.60	3.3
P-4	−0.50	5.2
P-3	0.43	4.5

NOTE: Multipass extrusion at condition TP-2 (*see* Experimental) at 30 rpm screw speed. Values are yellowness index (YI-1).

and P-2 began at 31 weeks. FP-1 gained 3 units of yellowness index over its original value of 6.6 after exposure to 51 weeks of oven aging, and this result shows significant color stability in comparison to the drastic changes in the other samples.

The tensile bars were also aged at 60 °C, and the change in tensile strength of the bars was examined periodically over the same time period. Figure 9 shows that the composition containing FP-1 maintained 98% of its tensile strength, whereas the composition with P-2 fell to 91%. The samples containing P-3 and P-4 fell in tensile strength to values below the sample that did not contain any phosphorus stabilizer.

Heat Aging of PP with Phosphorus Stabilizers. Phosphorus stabilizers in combination with phenolics do not show the same synergistic effect on the long-term heat aging (LTHA) of polymers that are shown by phenolic/thioester combinations (24). Phosphorus stabilizers, however, may improve the heat-aging stability of the polymer because of the effectiveness that they may have in preserving polymer integrity during processing by de-

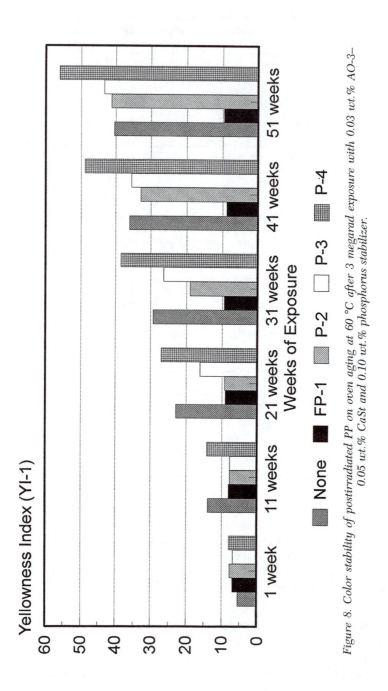

Figure 8. Color stability of postirradiated PP on oven aging at 60 °C after 3 megarad exposure with 0.03 wt.% AO-3– 0.05 wt.% CaSt and 0.10 wt.% phosphorus stabilizer.

stroying hydroperoxides or because, in the presence of a phosphorus stabilizer, less phenolic would be consumed during processing and would be available to protect the polymer during LTHA (25). The improvement in LTHA would be highly dependent on the processing conditions and the efficiency of the phosphorus stabilizer. In Table XI, the LTHA time-to-failure data are shown for a 0.6-mm thick samples of PP containing AO-1 with phosphorus stabilizers both with and without a thioester synergist [distearylthiodiproprionate (DSTDP)] after 150 °C oven aging. These samples were compounded by pass-

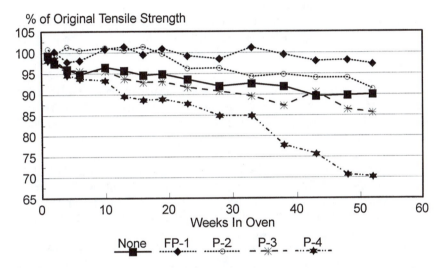

Figure 9. Tensile strength retention of irradiated PP on oven aging at 60 °C after 3 megarad exposure with 0.03 wt.% AO-3–0.05 wt.% CaSt and 0.10 wt.% phosphorus stabilizer.

**Table XI. Phosphorus Stabilizer Effect on
150 °C Oven Aging of 0.6-mm PP Plaques
Containing 0.05% Calcium Stearate**

Additives	Concen-tration (%)	Days to Failure
AO-1[a]	0.1	21
AO-1/P-2[a]	0.1/0.1	26
AO-1/FP-1[a]	0.1/0.1	25
AO-1/P-3[a]	0.1/0.1	21
AO-1/DSTDP	0.06/0.25	46
AO-1/P-2/DSTDP	0.06/0.06/0.25	50
AO-1/FP-1/DSTDP	0.06/0.06/0.25	50
AO-1/P-3/DSTDP	0.06/0.06/0.25	48

[a]Samples contained 0.1% calcium stearate.

ing them quickly through a twin screw extruder at 245 °C. Consequently, the phosphorus stabilizers did not contribute much to the overall LTHA performance of the system because the mild processing conditions did not stress the polymer sufficiently to show the effectiveness of these compounds.

The processing of fibers was done under more severe conditions than those used in Table XI. The data in Figure 10 show the LTHA effect on fibers of PP at a reduced temperature of 105 °C with three phenolic antioxidants. In all cases, the addition of FP-1 to the phenolic resulted in an increase in the time to failure for the fiber. FP-1 was the most effective phosphorus stabilizer of the formulations containing AO-2. Both FP-1 and P-2 showed good oven-aging characteristics with AO-1 and AO-3. Both P-3 and P-4 had lesser and perhaps negative effects on LTHA stability in these phenolic formulations.

Boiling Water Applications with Phosphorus Stabilizers. Many applications exist for polyolefins where aqueous extraction of the additives is a concern of the fabricator. The use of phosphites in some of these applications becomes questionable because of the potential hydrolysis of the phosphorus stabilizer. One severe test to determine how well a stabilizer system will stand up to extraction is to expose PP samples that were boiled for 7 days in water to 150 °C oven aging. Data in Table XII show the effect of boiling on AO-1 and AO-2 with phosphorus stabilizers and DSTDP in PP. AO-2 (26) and DSTDP (27) can be extracted by water from PP. Our data appear to support these conclusions.

Also, samples containing P-3 and P-4 are less stable to LTHA after exposure to boiling water than the same compositions without a phosphorus stabilizer. This result suggests that these two additives help reduce the concentration of the phenolic/sulfur compounds in these polymers. The phosphorus stabilizers themselves would not contribute much to oven aging under the mild processing conditions used in the study. Because the hydrolytically stable FP-1 does not show this LTHA loss, the hydrolysis of P-3 or P-4 may be accelerating the loss of the other additives. The processing advantages of a combination of AO-1 with FP-1 over a phenolic alone could be used in polyolefin applications where severe conditions such as exposure to hot or boiling water are used.

Conclusions

Fluorophosphonites represent a new development in polymer stabilization because they combine superior hydrolytic stability and exceptional thermal stability with the processing stabilization characteristics of the more effective phosphorus co-stabilizers. A variety of structures are possible, and some contain phosphorus in 7- or 8-membered ring systems. A spectrum of melting

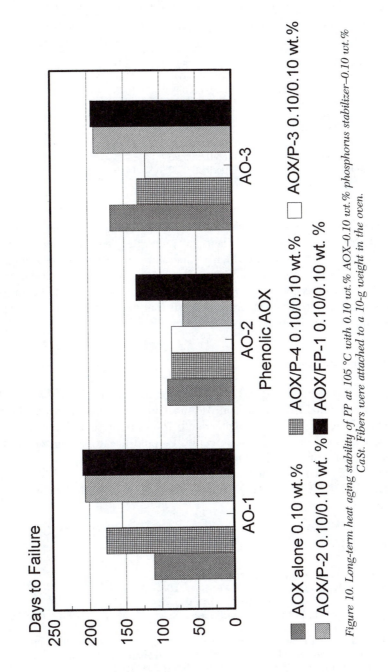

Figure 10. Long-term heat aging stability of PP at 105 °C with 0.10 wt.% AOX–0.10 wt.% phosphorus stabilizer–0.10 wt.% CaSt. Fibers were attached to a 10-g weight in the oven.

Table XII. Effect of Boiling on Oven Aging of 0.6-mm PP Plaques Containing Phosphorus Stabilizers and DSTDP

Additives	Concentration (%)	Hours to Failure	
		Untreated	Boiled 7 days
AO-1/FP-1	0.06/0.06	264	264
AO-1/DSTDP	0.06/0.25	1104	456
AO-1/FP-1/DSTDP	0.06/0.06/0.25	1008	528
AO-1/P-3/DSTDP	0.06/0.06/0.25	1416	240
AO-2/P-3	0.06/0.06	360	48
AO-2/DSTDP	0.06/0.25	1608	360
AO-2/FP-1/DSTDP	0.06/0.06/0.25	1416	360
AO-2/P-3/DSTDP	0.06/0.06/0.25	1416	96
AO-2/P-4/DSTDP	0.06/0.06/0.25	1416	96

NOTE: All samples contained 0.05% calcium stearate.

points and stabilities can be achieved. So far, the dioxaphosphocine ring compounds have shown the best balance of properties. The resistance of these compounds to hydrolysis and high temperature may open up new areas of application to phosphorus stabilizers. The unique resistance to discoloration of PP formulations containing FP-1 after gamma radiation coupled with good retention of mechanical properties suggests that fluorophosphonites may provide superior stabilizers for this application.

Acknowledgments

The author presented part of the information in this chapter at the 13th Annual International Conference on Advances in the Stabilization and Degradation of Polymers, Lucerne, Switzerland. The work presented in this chapter is the combined effort of many people. I would like to recognize the pioneering efforts of L. B. J. Burton in developing fluorophosphonites and the following major contributors: S. E. Amos, M. S. Ao, K. P. Becnel, V. J. Gatto, R. D. Glass, K. R. Jones, M. K. Juneau, K. A. Keblys, W. Kolodchin, and L. S. Simeral.

References

1. Mitterhofer, F. *Polym. Eng. Sci.* **1980,** *20,* 692.
2. Asay, R. E.; Wilson, R. A. *Plast. Technol.* **1986,** *November,* 73.
3. Reed, G. V. *Plast. Comp.* **1982,** *September/October,* 63.
4. Schwarzenbach, K. In *Plastics Additives;* Gachter, R.; Muller, K., Eds.; Carl Hanser Verlag: New York, 1985; p 15.
5. Lemmen, T.; Enlow, W.; Dietz, S. *Proceedings of the SPE 48th Annual Tech. Conference;* Society of Petroleum Engineers: Richardson, TX, 1990; p 1186.
6. Hodan, J. J. ; Schall, W. L. U.S. Patent 3,553,298, 1971.
7. Hahner, U.; Habicher W. D.; Chmela, S. *Polym. Degrad. Stab.* **1993,** *41,* 197.

8. Lingner, G.; Staniek, P.; Stoll, K. K. *SPE Polyolefins VIII Proceedings;* Society of Petroleum Engineers: Richardson, TX, 1993; p 386.
9. Burton, L. P. J. U.S. Patent 4,912,155, 1990.
10. Klender, G. J.; Glass, R. D.; Juneau, M. K.: Kolodchin, W.; Schell R. A. *SPE-Polyolefins V RETEC Proceedings;* Society of Petroleum Engineers: Richardson, TX, 1987; p 225.
11. Juneau, M. K.; Burton, L. P. J.; Klender, G. J.; Kolodchin, W. *SPE Polyolefins VI RETEC Proceedings;* Society of Petroleum Engineers: Richardson, TX, 1989; p 347.
12. Klender, G. J.; Kolodchin, W.; Gatto, V. J. *SPE 48th Annual Tech. Conference Proceedings;* Society of Petroleum Engineers: Richardson, TX, 1990; p 1178.
13. Klender, G. J.; Jones, K. R. *SPE-Polyolefins VII RETEC Proceedings;* Society of Petroleum Engineers: Richardson, TX, 1991; p 293.
14. Kosolapoff, O.N. *Organophosphorus Compounds;* Wiley: New York, 1950; pp 180–189.
15. Odoriso, P. A.; Pastor, S. D.; Spivak, J. D.; Steinhuebel, L. *Phosphorus Sulfur* **1983,** *15*, 9.
16. Spivak, J. D. U.S. Patent 4,182,704, 1980.
17. Keblys, K. A.; Ao, M.-S.; Burton, L. P. J. U.S. Patent 4.929,745, 1990.
18. Klender, G. J.; Gatto, V. J.; Jones, K. R.; Calhoun, C. W. *Polym. Prepr.* **1993,** *34(2)*, 156.
19. Spivack, J. D.; Pastor, S. D.; Patel A.; Steinhuebel, L. P. *Polym. Prepr.* **1981,** *25*, 72.
20. Schwetlick, K. In *Mechanisms of Polymer Degradation and Stabilizaation;* Scott, G., Ed.; Elsevier Applied Science: London, 1990; p 23.
21. Carlsson, D. J.; Dobbin, C. J. B.; Jensen, J. P. T.; Wiles, D. M. *Polymer Stabilization and Degradation;* Klemchuk, P. P., Ed.; ACS Symposium Series No. 280; American Chemical Society: Washington, DC, 1985; p 359–371.
22. Allen, D. W.; Leathard, D. A.; Smith, C. Chem. Ind. **1987,** 198.
23. Horng, P.; Klemchuk, P. P. Plast. Eng. **1984,** *40*, 35.
24. Bartz, K. W. *SPE-Polyolefins IV RETEC Proceedings;* Society of Petroleum Engineers: Richardson, TX, 1984; p 151.
25. Drake, W. O. *Speciality Plastics Conference Proceedings;* 1988; p. 339.
26. Pfahler, G.; Lotzsch, K.; Holifield, P. J. *Proceedings of First European Conference on High Performance Additives;* London, 1988; p. 119.
27. Rysavy, D. *Donaulaendergespraech* 1976, 9, 243.

RECEIVED for review December 6, 1993. ACCEPTED revised manuscript January 5, 1995.

Processing Effects on Antioxidant Transformation and Solutions to the Problem of Antioxidant Migration

S. Al-Malaika* and S. Issenhuth

Polymer Processing and Performance Group, Department of Chemical Engineering and Applied Chemistry, Aston University, Birmingham B4 7ET, England

Polymer oxidation takes place inadvertently throughout the life cycle of polymers. Antioxidants are generally incorporated in polymers to inhibit or minimize oxidative degradation. Many of these antioxidants are depleted by undergoing chemical transformations while performing their antioxidant function at various stages of polymer processing and fabrication. In addition, antioxidants are subject to loss from polymers during processing and in service. Migration of antioxidants is a major concern in applications involving polymers in direct contact with food and human environment. This concern is compounded by the realization that very little is known about the nature and the migration behavior of antioxidant transformation products. In this chapter two approaches are advocated to minimize risk attached to the migration of antioxidants and their transformation products: the use of the biological antioxidant vitamin E and the efficient grafting of reactive antioxidants on polyolefin backbones during processing.

THE WELL-KNOWN FREE-RADICAL OXIDATIVE DEGRADATION PROCESS of hydrocarbon polymers (1) is accelerated, to varying extent, during all phases of the polymer life cycle. The stages of converting monomers to polymers (polymerization) and polymers to finished products (processing and fabrication) contribute significantly to accelerating the oxidative degradation process of polymers due to the presence of adventitious chemical impurities. During polymerization, for example, residues of metal ions from catalysts play a main role in accelerating oxidation (2). During processing and fabrication, the build

*Corresponding author

0065–2393/96/0249–0425$12.00/0

up of small concentrations of hydroperoxides, unsaturation, and carbonyl compounds, all due to the effect of high processing temperatures, shearing forces, and the inevitable presence of a small oxygen concentration, contribute to undesirable chemical changes in the polymer such as cross-linking and chain scission (3, 4).

These chemical changes, together with the effects of outdoor exposure to sunlight, heat, stresses, leaching solvents, and detergents, lead to premature oxidation of the product with concomitant loss of useful properties. The reprocessing of in-house rejects or fully pledged recycling operations for reuse of polymers in second and possibly subsequent lives, followed by reexposure to environmental conditions during reuse, contribute further to oxidative degradation. These approaches may cause major changes to the additive system used in the polymer and lead to inferior performance of the recycled product. Figure 1 gives a schematic representation of oxidation during various steps of the life cycle of polymers.

The Role of Antioxidants and Problems of Their Migration

Most polymers require the use of stabilizers and antioxidants to inhibit the oxidative degradation reactions that occur at the different stages of the polymer life cycle. Antioxidants are normally incorporated into polymers during the high-temperature processing operation to serve as melt stabilizers or to provide protection during service as thermal and UV stabilizers. However, under the normal conditions of processing, most antioxidants undergo oxidative transformation as a consequence of their antioxidant function (5-7). This transformation results in a chemical loss of antioxidants, which can occur during processing, reprocessing, and recycling or during service life. The overall stabilization afforded to the polymer, therefore, depends not only on the in-

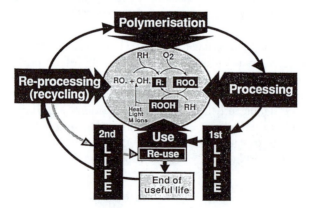

Figure 1. Oxidation during polymer life cycle.

itially added antioxidants but also on the behavior of the antioxidant transformation products. This behavior can either be beneficial (when transformation products are themselves antioxidants) or harmful (when products exert pro-oxidant effects) to the overall polymer stability. Physical losses of antioxidants occur because of volatilization (especially at the high processing temperatures), poor solubility, diffusion, and leachability when in contact with aggressive solvents during service (8, 9).

For antioxidants to be effective, therefore, they must be inherently efficient (based on their chemical structure and activity) and physically retained by polymers throughout their life cycles. Migration of antioxidants and their chemical transformation products can occur from polymer articles into the surrounding medium. Even at small concentrations, this migration gives rise to major concerns when antioxidants are used in applications that involve direct contact with the human environment (e.g., food-packaging plastics, children's toys, textiles, packaging materials for pharmaceuticals and other medical applications). Migration of this sort can lead not only to premature failure of the plastics article but also to associated health hazards. In the case of food-packaging applications migration of antioxidants and their transformation products into food represent a major source of contamination of the packaged food.

Health authorities in Europe, the United States, and many other countries have strict regulations to control the use of additives in plastics used for food packaging (10, 11). Existing regulations stipulate that packaging materials must not alter the quality of food and that additives must have toxicity clearance and approval for their use. Recent mechanistic studies on antioxidant action (12, 13), aided by advances in analytical techniques, have led to a better understanding of the antioxidancy role of parent antioxidants and their transformation products. Further, the problem of migration of antioxidants and transformation products has received greater attention. The shortcomings of existing regulations are that although parent antioxidants may have toxicity clearance, the chemical nature, migration behavior, and toxicity effects of their transformation products (mainly formed during processing) still remain either unknown or uncertain.

Effect of Processing on Antioxidant Performance and Antioxidant Transformations

Two approaches are advocated to reduce risks associated with the migration of antioxidants and their transformation products: use of a fat-soluble biological antioxidant such as vitamin E (α-tocopherol), which may be considered as an acceptable migrant in food packaging; and the immobilization of antioxidants by tying them down to the polymer backbone to eliminate the problem of migration. In both cases the processing operation plays an important role: in the use of vitamin E, processing is the major step that leads to the formation of transformation products (not only from vitamin E but from other antioxidants as well); whereas in the immobilization case, processing is used in a novel way to affect high levels of grafting of reactive antioxidants on polymer backbones.

Use of Vitamin E in Polyolefins. The structure of vitamin E is essentially based on a chroman nucleus and a phytyl chain containing 3 chiral carbon atoms on 2-, 4'-, and 8'-positions. The hindered-phenol-type structure of the chroman nucleus is responsible for the intrinsic antioxidant activity of vitamin E; the side chain plays only a minor role and acts to enhance solubility. Natural vitamin E is a mixture of tocopherols (α, β, γ, and δ) that differ only by the number and position of the aromatic methyl groups on the benzene ring. The most bioactive of these is $2R,4'R,8'R$ α-tocopherol, **1**, which occurs in only one stereochemical form, the R,R,R configuration (*14*). Synthetic vitamin E, on the other hand, is a *dl*-α-tocopherol that is an all-racemic mixture of the eight possible stereoisomers.

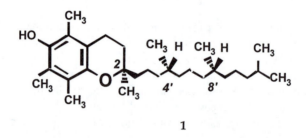

1

The antioxidant action of synthetic hindered phenols such as Irganox 1076 (**2**), Irganox 1010 (**3**), and BHT (**4**) is due to their chain-breaking donor activity; they donate H to ROO· to give a stable phenoxyl radical (Scheme I) (*12*). Likewise, α-tocopherol was shown (*15*) to be a very effective alkyl-peroxyl radical trap leading to the formation of a very stable tocopheroxyl radical (Scheme II). The reactivity of α-tocopherol toward ROO· in styrene was

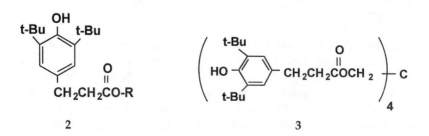

2   3

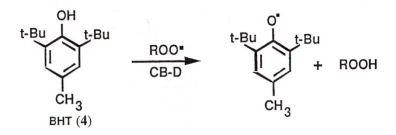

Scheme I.

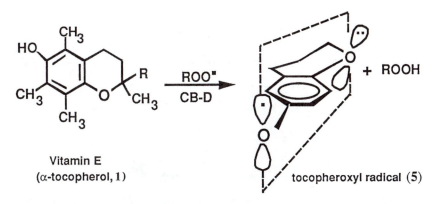

Scheme II.

found (*15*) to be 250 times greater than BHT. This finding was attributed to stereoelectronic effects exerted by the chroman structure giving additional stabilization to the tocopheroxyl radical (**5**) through interaction between p-orbitals of the two *para*-oxygens.

Synthetic antioxidants containing a hindered phenol function such as Irganox 1076 and 1010 are known to be good melt stabilizers for polyolefins. Figure 2 shows the melt-stabilizing effect of 0.2% of Irganox 1010, Irganox 1076, and α-tocopherol in polypropylene (PP) and low-density polyethylene (LDPE) as functions of processing severity. PP undergoes chain scission during processing (reflected in the increase in the melt-flow index [MFI]), whereas LDPE undergoes cross-linking (evidenced by a decrease in MFI) (*3*). Figure 2 shows clearly that, under these conditions, α-tocopherol-containing polyolefin samples exhibit higher melt stability than the best commercially available synthetic melt stabilizers such as Irganox 1010. The superiority of α-tocopherol is also evident at very low concentration, such as 0.01% in PP (*16a*), and under severe processing and reprocessing (multiple-extrusion) conditions (unpublished work).

The effect of polyethylene extrusion and multiple extrusions (reprocessing at 180 °C) on the retention and chemical transformations of the hindered phenol antioxidants α-tocopherol, Irganox 1076, and Irganox 1010 was further

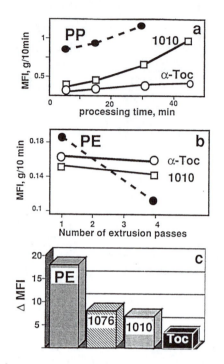

Figure 2. Effect of processing on melt stability (measured by melt flow index) of PP and PE containing 0.2% of different hindered phenol antioxidants: (a) PP processed in a closed chamber of an internal mixer at 180 °C; (b) LDPE extruded (up to four extrusion passes) at 180 °C; and (c) rate of change of MFI (ΔMFI, between first and fourth passes) for PE in the presence and absence of antioxidants.

examined (*16b*). Single and multiple extrusions of antioxidant-containing LDPE was accompanied by chemical transformations.

High-performance liquid chromatography (HPLC) of extracts of polymer samples containing 1% antioxidants separated by exhaustive extraction in a good solvent demonstrated that (Figure 3a) that the amount of α-tocopherol retained in the polymer, after successive extrusion passes, is higher than that of Irganox 1076 and Irganox 1010. Further HPLC analysis of polymer samples containing 0.2% antioxidants revealed that the higher amount of α-tocopherol retained in the polymer is paralleled by lower overall concentration of trans-formation products when compared to Irganox 1076 (Figure 3b). For example, after the first extrusion pass, the total amounts of transformation products obtained from α-tocopherol and Irganox 1076 were about 5 and 33%, respectively (Figure 3c). At this low initial antioxidant concentration (0.2%), the physical loss of the antioxidants during processing was minimal and their sol-

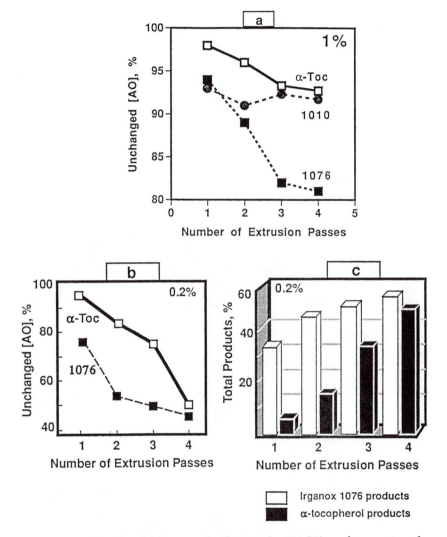

Figure 3. Effect of multiple extrusion of LDPE (at 180 °C) on the retention of antioxidants processed at (a) 1% and (b) 0.2%. The total weight percent of transformation products (c) formed from 0.2% α-tocopherol and Irganox 1076 are shown (all concentrations were measured by HPLC).

ubility in the polymer presented no problem. In all cases, chemical transformations accounted entirely for the loss of the antioxidants.

Optimization of HPLC conditions led to a good separation of the different transformation products formed from the antioxidants after the different extrusion passes. Six major products were isolated from α-tocopherol and Irganox 1076 by preparative HPLC. Figure 4 shows that, in both cases, multiple

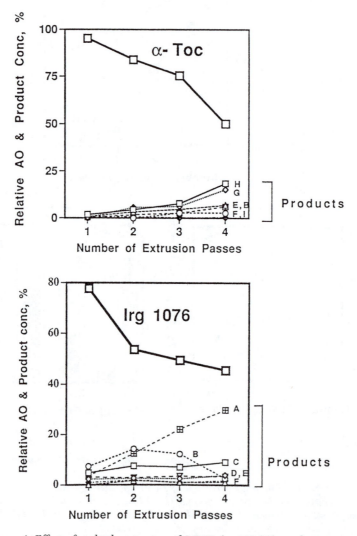

Figure 4. Effect of multiple extrusion of LDPE (at 180 °C) on the retention of 0.2% α-tocopherol, 0.2% Irganox 1076, and their transformation products (A-I; weight %). All concentrations were measured by HPLC; transformation products were isolated by preparative HPLC.

extrusions affects not only the level of retention of the antioxidant but also the distribution of transformation products. Further, the initial amount of antioxidant (added to the polymer during processing) had a dramatic effect on the relative distribution of transformation products of α-tocopherol (Figure 5). Polymer samples containing Irganox 1076 and Irganox 1010 showed similar behavior. Detailed characterization of each of the isolated transformation products revealed differences in chemical and physical behavior and significant contributions from dimeric and trimeric structures (*16b*).

Migration of Hindered Phenol Antioxidants and Their Transformations. Migration is an umbrella term that encompasses a whole range of physical processes and interactions involving the polymer surroundings and its constituents. These processes and interactions include the rate of additive diffusion, additive solubility in the polymer and in the contacting media, and volatilization of the additive from the polymer surface. Although the effect of each of these parameters for a number of antioxidants was studied (*8, 9*), the internationally accepted migration tests for polymer additives in foodstuffs refer only to the overall migration. Both static and dynamic migration tests are normally carried out. The static test, which is generally accepted by many countries with some exceptions (e.g., the United States), is used to measure the amount of additive migrating during a specified time interval. The dynamic test is used to measure the amount of additive migration as a function of time.

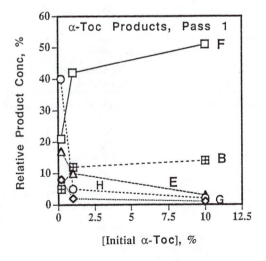

Figure 5. Effect of initial concentration of α-tocopherol on the relative distribution of transformation products formed during extrusion (single pass) in LDPE.

Overall, the quantity of the migrant additive (usually expressed in milligrams per squared decimeter) and its toxicity will determine its suitability for use in plastic food-packaging materials. Current regulations for the approval of antioxidant use in food packaging refer only to migration and toxicity of the parent antioxidant; these regulations do not deal with aspects of migration and toxicological behavior of the major transformation products that are formed, in most cases, during polymer processing and fabrication.

Migration tests are not usually carried out in the presence of food because of the analytical difficulty of separating and measuring low concentrations (parts per million or below) of additives, hence the use of specific solvents as food simulants. The type of food simulants and the test conditions are standardized in most countries and differ only slightly. For example, a range of test conditions and simulants for oily, alcoholic, and aqueous foods specified by a European Economic Community directive (17) were used (except where indicated) in the studies presented in this chapter (see Table I). The U.S. Food and Drug Administration (FDA) proposal of 1982 (19) assumes that about 41% of all packaged food are packed in plastics materials, and about 49% of that total are in contact with aqueous food, 16% with acidic food, 34% with fat, and less than 0.01% with alcohol.

The importance of processing history and parent antioxidant concentration on transformation products was clearly established from results presented in the previous section (see Figures 3–5). Furthermore, these results add a dimension to the question of migration of antioxidants from polymers into contact media such as food. The formation of different transformation products during polymer processing and reprocessing and the relationship among the products' relative distribution and the parent antioxidant concentration are

Table I. Some Food Simulants and Test Conditions Used for Our Migration Studies

Experimental Materials and Conditions Used	Materials and Conditions Simulated
Food simulant used	**Type of food simulated**
Distilled water	Aqueous Food
Acetic acid (3% by wt)	Acidic aqueous food
Aqueous ethanol (50% by vol)[a]	Alcoholic food (ethanol > 0.5%)
Rectified olive oil	Fatty food
n-Heptane[b]	Fatty food
Migration test conditions	**Intended contact and use**
$t = 10$ days; $T = 40$ °C	$t > 24$ h; 5 °C $< T \leq 40$ °C
$t = 2$ h; $T = 70$ °C	$t \leq 2$ h; 40 °C $< T \leq 70$ °C

NOTE: Unless otherwise noted, experimental conditions were as recommended by the European Council (17). T is temperature; t is time.
[a]Simulant recommended by U.S. FDA (19).
[b]Fat simulant recommended by U.S. FDA (19) and used to overcome the problems of detection of very small quantities of additives in fats, which is only possible by using radioactive labeling.

further manifestation of the complex nature of migration of antioxidants in polymers. The inevitable variations in the chemical behavior, migration characteristics, and toxicological effects of transformation products pose a real problem with antioxidant migration when it is in contact with leaching media, especially in applications that involve direct contact with the human environment.

Figure 6 shows results from static migration tests carried out at 70 °C for 2 h on extruded PE films (after first extrusion pass) containing 0.2% w/w each of α-tocopherol and Irganox 1076 in different food simulants. For both anti-

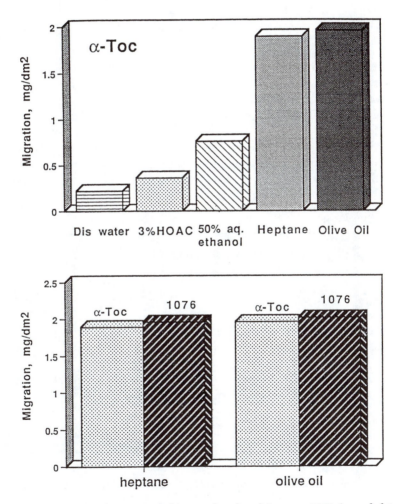

Figure 6. Extent of migration of α-tocopherol and Irganox 1076 (extruded in LDPE, single pass, at 0.2% w/w) in food simulants distilled water, 3% acetic acid, 50% aqueous ethanol, rectified olive oil, and n-heptane at 70 °C for 2 h.

oxidants, the overall migration levels (i.e., for the antioxidant and its transformation products) are highest in olive oil and in the fatty food simulant, heptane; α-tocopherol migrates to a slightly lower extent in both cases. However, structural and chemical differences of the antioxidant transformation products would be expected to create differences in migration behavior in the presence of different food simulants. Figure 7 compares the relative extent of migration at 40 °C for 10 days of the individual transformation products from α-tocopherol containing PE film in *n*-heptane. The extent of migration of the individual products appears to follow the order of their polarities; for example, G and B (least polar) show higher migration levels than the other more polar products. The relative distribution of transformation products obtained by extraction in dichloromethane of identical films (Figure 5, 0.2% α-Toc) is very different.

Grafting of Antioxidants on Polyolefin.   A number of approaches deal with the problem of loss and migration of synthetic antioxidants from polymers. The two earlier approaches involve the use of oligomeric antioxidants for rubbers and plastics (*20*) and the incorporation of polymerizable antioxidants (e.g., ones containing vinyl group) during polymer synthesis (*21, 22*) to produce polymers containing randomly copolymerized antioxidant functions. Both these approaches yield antioxidants that show great persistence in the polymer, especially at high temperatures. However, there are drawbacks

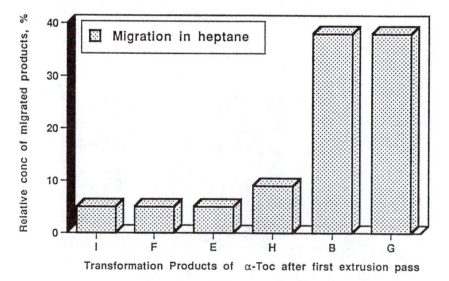

Figure 7. Relative extent of migration at 40 °C for 10 days of transformation products of α-tocopherol-containing PE (at 0.2%) in heptane tested at 40 °C for 10 days. Concentrations were measured by HPLC.

in both cases: oligomeric antioxidants suffer from leaching by solvents, whereas the method of copolymerization that is more suitable for amorphous polymers is more expensive because new specialty polymers must be produced for each end-use product.

The use of suitably functionalized antioxidants that may attach (graft) chemically to the polymer melt during reactive processing is an alternative approach. The grafted antioxidant cannot be detached from the polymer matrix except through the severance of chemical bonds. Grafting of antioxidants should lead to highly substantive and effective antioxidant system specially designed for use under aggressive leaching environments. The use of standard processing machineries to achieve targeted grafting reactions is an added advantage of this method. Essentially two different types of polymer-reactive functionalities on antioxidants can be used in reactive processing: nonpolymerizable and polymerizable functions.

An early example of the nonpolymerizable type was based on antioxidants containing a thiol function that were exploited in mechanochemically (shear-) initiated reactions for polymers containing double bonds (23), for example, rubbers and rubber-modified plastics. Although this method yields highly effective binding of the antioxidant, very high levels of binding are difficult to achieve because of the formation of disulfides, which are a major by-product and are easily lost from the polymer under leaching environments.

The second type of reactive antioxidants, which are based on monomeric antioxidants containing a polymerizable group (e.g., a vinyl, acryloyl, or methacryloyl group), have been used in the presence of very small concentrations of a free-radical initiator to give polymer bound-antioxidant functions. This use is referred to as the conventional reactive-processing method (RP). A limitation of this approach is the general low level of grafting as a result of major competitive side reactions (e.g., homopolymerization) of the antioxidant during processing (24, 25). These limitations were the object of our research into reactive processing for a number of years.

Three reactive processing procedures were developed to address these problems. The first procedure relies on the use of antioxidants containing a maleate function that is known to be much less susceptible to polymerization (26). The second procedure uses antioxidants containing two polymerizable functions that would yield minimum homopolymerization of the antioxidant by a careful choice of processing parameters (27). The third procedure involves the use of another functional monomer in a novel reactive processing (NRP) procedure to enable the grafting of simple monomeric polymerizable antioxidants at very high levels with minimum homopolymer formation (28). These same antioxidants led to low grafting levels using RP procedures (25).

Synergistic systems based on the NRP approach with high levels of grafted antioxidants have led to high retention of stabilizing activity in polyolefins under aggressive leaching solvent environments (unpublished work). The NRP procedure is not limited to antioxidants; it is applicable to other monomeric modifiers and additives and can achieve very high levels of additive retention and polymer performance. This approach offers less hazardous (with minimum, if any, migration) and more effective polymer systems such as homopolymers, copolymers, blends, and composites designed for demanding applications and applications that are in contact with the human environment.

Acknowledgments

The work described in the sections of this chapter on (1) the use of vitamin E in polyethylene and (2) the migration of hindered phenol antioxidants and their transformations was financially supported by the Ministry of Agriculture Fisheries and Food (England), to whom we are grateful. The results in these sections are the property of MAFF and are crown copyright; the Ministry is thanked for permission to publish. We are also grateful to Hoffmann La-Roche (Switzerland), Ciba-Geigy (Switzerland), Ltd., and British Petroleum (England), Ltd., for samples of α-tocopherol, Irganox 1010 and Irganox 1076, and polyethylene, respectively.

References

1. Al-Malaika, S. In *Atmospheric Oxidation and Antioxidants;* Scott, G., Ed.; Elsevier: Amsterdam, 1993; Vol. 1, Chap. 2, and references therein.
2. Osawa, Z. In *Developments in Polymer Stabilization;* Scott, G., Ed.; Elsevier Applied Science Publications: London, 1984; Vol. 7, Chap. 4.
3. Scott, G. In *Atmospheric Oxidation and Antioxidants;* Scott, G., Ed.; Elsevier: Amsterdam, 1993; Vol. 2, Chap. 3.
4. Al-Malaika, S. *Polym. Degrad. Stab.* **1991,** *34,* 1.
5. Al-Malaika, S.; Coker, M.; Scott, G.; Smith, P. *J. Appl. Polym. Sci.* **1992,** *44,* 1297.
6. Scott, G. In *Developments in Polymer Stabilization;* Scott, G., Ed.; Elsevier Applied Science Publications: London, 1984; Vol. 7, Chap. 2.
7. Pospíšil, J. In *Developments in Polymer Stabilization;* Scott, G., Ed.; Elsevier Applied Science Publications: London, 1979; Vol. 1, Chap.; ibid 1984; Vol. 7, Chap 1.
8. Billingham, N. In *Oxidation Inhibition in Organic Materials;* Pospíšil, J.; Klemchuk, P. P., Eds.; CRC Press: Boca Raton, 1990; Vol. 2, Chap. 6.
9. Al-Malaika, S.; Goonetilleka, M. R. J.; Scott, G. *Polym. Degrad. Stab.* **1991,** *32,* 231.
10. Crompton, T. R. In *Additive Migration from Plastics into Food;* Pergamon Press: Oxford, 1979.
11. McGuinness, J. D. *Food Addit. Contam.* **1986,** *3,* 95-102.
12. Scott, G. In *Atmospheric Oxidation and Antioxidants;* Scott, G., Ed.; Elsevier: Amsterdam, 1993; Vol. 2, Chap. 4.
13. Al-Malaika, S. In *Atmospheric Oxidation and Antioxidants;* Scott, G., Ed.; Elsevier: Amsterdam, 1993; Vol. 1, Chap. 5.
14. Nilsson, J. L. G. *Acta Pharm. Suec.* **1969,** *6(1),* 1.
15. Burton, G. W.; Ingold, K. U. *J. Am. Chem. Soc.* **1981,** *103,* 6472.

16. (a) Laermer, S. F.; Nabholz, F. *Plast. Rubber Process. Appl.* **1990,** *14,* 235; (b) Al-Malaika, S.; Ashley, H.; Issenhuth, S. *J. Polym Sci. Part A Polym. Chem.* **1994,** *32,* 3099.
17. *Official Journal of the European Communities;* 1982; Article No. L 297, pp 26–30.
18. *Food and Drug Administration Code of Federal Regulations;* U.S. Government Printing Office: Washington, D.C., 1967; Title 21, Chap. 1, Part 121, Sec. 1221.2514 and 121.2546, pp 239, 269.
19. *Food Chem. News* **1982,** *23,* 34.
20. *Plastics Additives Handbook;* Gachter, R.; Muller, H., Eds.; Hanser Publishers: New York, 1987.
21. Fu, S.; Gupta, A.; Albertsson, A. C.; Vogl, O. In *New Trends in the Photochemistry of Polymers;* Allen, N. S.; Rabek, J. F., Eds.; Elsevier Applied Science: London, 1985; Chap. 15.
22. Pospíšil, J. In *Oxidation Inhibition in Organic Materials;* Pospíšil, J.; Klemchuk, P. P., Eds.; CRC Press: Boca Raton, 1990; Vol. 1, Chap. 6.
23. Scott, G. In *Developments in Polymer Stabilization;* Scott, G., Ed.; Elsevier Applied Science Publications: London, 1984; Vol. 7, Chap. 5.
24. Munteanu, D. In *Developments in Polymer Stabilization* Scott, G., Ed.; Elsevier Applied Science: London, 1987; Vol. 8, Chap. 5.
25. Al-Malaika, S.; Scott, G.; Wirjosentono, B. *Polym. Degrad. Stab.* **1993,** *40,* 233.
26. Al-Malaika, S.; Ibrahim, A. Q.; Scott, G. *Polym. Degrad. Stab.* **1988,** *22,* 233.
27. Al-Malaika, S.; Ibrahim, A. Q.; Rao, J.; Scott, G. *J. Appl. Polym. Sci.* **1992,** *44,* 1287.
28. Al-Malaika, S.; Scott, G. *Int. Pat. Appl.* PCT/GB 89/ 00909 (1989).

RECEIVED for review January 26, 1994. ACCEPTED revised manuscript January 13, 1995.

Diffusion of Benzotriazoles in Polypropylene

Influence of Polymer Morphology and Stabilizer Structure

Vincent Dudler and Conchita Muiños

Ciba-Geigy, Ltd., Materials Research, 1723 Marly 1, Switzerland

The diffusion coefficient of some commercial benzotriazole stabilizers was measured in polypropylene processed by compression molding and by extrusion. Results show that the diffusion coefficient was not greatly different in polypropylene samples presenting the same crystallinity but different morphology. In extruded films, the diffusion coefficient was measured along the three principal axes. In the film's plane, for directions parallel and perpendicular to extrusion, the diffusion coefficients were similar. However, both values were smaller than the diffusion coefficient across the film's surface, and this difference indicates that an anisotropy in the crystal organization can affect the diffusion coefficient of additives. The study of the influence of the stabilizer structure has shown that the diffusion coefficient of benzotriazoles can be predicted from the molecular weight.

THE PHYSICAL DEPLETION OF STABILIZERS from polymers is an important mechanism in polymer stabilization because it generally shortens the lifetime of articles. This loss can be modeled with three properties of the additive (1) that, when combined, are referred to as the *additive–polymer compatibility.* The compatibility is not only given by the intrinsic properties of the additive such as its polarity or its chemical structure, but it also depends on the characteristics of the substrate (Scheme I). For example, the mobility of gases, small molecules, or additives can be strongly influenced by the polymer properties such as density, crystallinity, or orientation (2–5). However, many of these investigations were carried out with polymers modified by mechanical or thermal treatment (drawing and annealing) to enhance the effect.

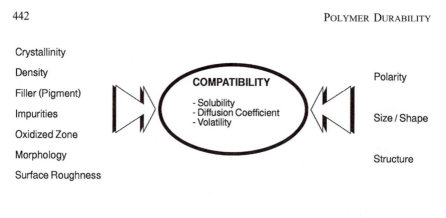

Crystallinity

Density

Filler (Pigment)

Impurities

Oxidized Zone

Morphology

Surface Roughness

COMPATIBILITY

- Solubility
- Diffusion Coefficient
- Volatility

Polarity

Size / Shape

Structure

POLYMER CHARACTERISTICS   **ADDITIVE PROPERTIES**

Scheme I. Factors influencing the additive–polymer compatibility.

In our study of the physical depletion of additives from polymers, we have been interested in looking at industrially prepared polypropylene (PP) and in determining if a diffusion coefficient measured in PP of a given morphology could characterize the diffusion behavior of additives in PPs of different morphologies. Therefore, we measured the diffusion coefficient of UV-absorbers in thick (compression-molded plates) and thin (extruded films) PP samples showing large morphology differences but having the same crystallinity and density.

Processing conditions (rapid cooling and mechanical stress) can also affect the polymer morphology and induce a crystal orientation. To assess the importance of crystal orientation on the films and to study the effect of orientation on the additive migration, the diffusion coefficient was measured along the three axes of an extruded PP film (Scheme II).

The stabilizer structure is also of importance, and changes in the substituents of an additive are known to have different affects on the diffusion coefficient (6). The diffusion rate of many different structures from the generic class of benzotriazoles was measured to investigate the rate dependence on the molecular mass.

Experimental

Materials. Chemical structures of the 12 UV-absorbers studied and some of their characteristics are included in Table I and structures **1–3**.

The isotactic PP employed was an unstabilized Propathene produced by ICI, Inc. Plates were compression molded to a thickness of 1 mm. The film strips (width, 50 mm; thickness, 100 μm) were made by extrusion. Plates and films were processed under nitrogen at 230 °C and showed an identical crystallinity of ca. 48% ($\pm 2\%$, measured by differential scanning calorimetry) and the same density of 0.899 g/cm^3 (± 0.003, measured by flotation technique, ISO R 1183).

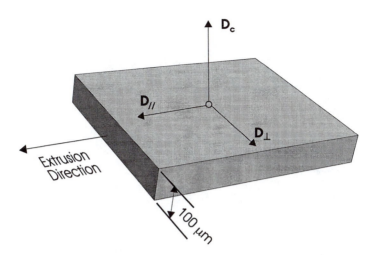

Scheme II. Definition of diffusion axes in PP film.

In polarization microscopy, the plates showed a spherulitic structure (ca. 20–30 μm large), whereas the films showed no measurable crystalline regions. The large difference in morphology between the two samples can be observed by electron microscopy. An etching with permanganic acid (7) preferentially removes the amorphous material of PP and reveals the crystalline structure. Figure 1 depicts the large spherulites of the plates and the small crystalline structure (*see* arrows) of the film.

Diffusion Experiment.   A stabilizer-free plate was clamped between two reservoir plates (supersaturated plates made by solubilizing the additive at a temperature higher than the experimental temperature). This diffusion cell was placed in a forced-air oven at the required temperature. After diffusion, the plate was removed, washed with acetone, and sectioned (10-μm thick slice); and the concentration along the diffusion path was measured by UV-microspectrometry.

For diffusion across the film, a stack of 10 films was clamped between the two reservoir plates. At the end of the diffusion, the stack was washed and the films were separated for concentration analysis. The diffusion experiment along the film's plane was carried out with the following technique.

The film was first plasma-etched (plasma conditions: O_2, 4-mbar pressure, 21 °C, 15 s; performed in the laboratory of H.-P. Haerri, Materials Research, Ciba-Geigy, Marly) and embedded in an epoxy resin (Araldite HY 951/CY 223, Ciba-Geigy) at room temperature. The piece was then trimmed with a microtome to expose only one side of the film (parallel or perpendicular to the extrusion direction) and to obtain a clean surface. Afterward, the open side was put in close contact with amorphous stabilizer powder for diffusion (the resin stops the stabilizer diffusing from other sides). After diffusion, the piece was washed with acetone and sectioned along the diffusion direction for measurement of the concentration profile.

An example of such a section after diffusion is given in Figure 2. The dark zone in the PP-film (right side) was produced by the absorption of stabilizer (UV-6).

Table I. Chemical Structures and Characteristics of UV Absorbers

Code Name	M_w (g/mol)	mp (°C)	R1[a]	R2[a]	R3[a]
UV-1	225.3	132	$-H$	$-CH_3$	$-H$
UV-2	267.3	99	$-H$	$-t\text{-}C_4H_9$	$-H$
UV-3	316.6	142	$-t\text{-}C_4H_9$	$-CH_3$	$-Cl$
UV-4	323.4	106	$-H$	$-C(CH_3)_2CH_2C(CH_3)_3$	$-H$
UV-5	323.4	88	$-t\text{-}C_4H_9$	$-CH(CH_3)CH_2CH_3$	$-H$
UV-6	323.4	156	$-t\text{-}C_4H_9$	$-t\text{-}C_4H_9$	$-H$
UV-7	351.5	83	$-C(CH_3)_2CH_2CH_3$	$-C(CH_3)_2CH_2CH_3$	$-H$
UV-8	357.9	158	$-t\text{-}C_4H_9$	$-t\text{-}C_4H_9$	$-Cl$
UV-9	447.6	141	$-C(CH_3)_2C_6H_5$	$-C(CH_3)_2C_6H_5$	$-H$
UV-10	486.0	—	$-t\text{-}C_4H_9$	$-CH_2CH_2COOC_8H_{17}$	$-Cl$
UV-11[b]	658.9	198	NA	NA	NA
UV-12[c]	760.9	114	NA	NA	NA

NOTE: mp is melting point; NA is not applicable.
[a]For UV-1 through UV-10, groups R1, R2, and R3 refer to positions in structure **1**.
[b]See structure **2**.
[c]See structure **3**.

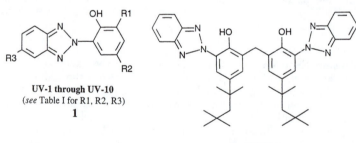

UV-1 through UV-10
(*see* Table I for R1, R2, R3)
1

UV-11
2

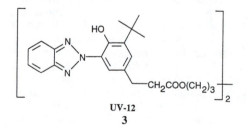

UV-12
3

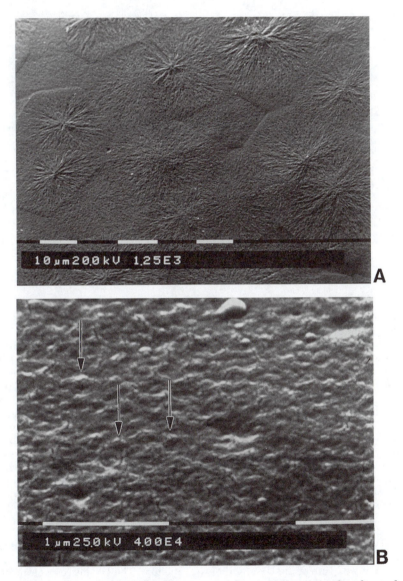

Figure 1. Surface of PP observed by scanning electron microscopy after chemical etching: A, spherulitic morphology of compression-molded plate and B, crystalline structure of extruded film. (Numbers on pictures give the value of the micron bar, the high tension voltage, and the magnification.)

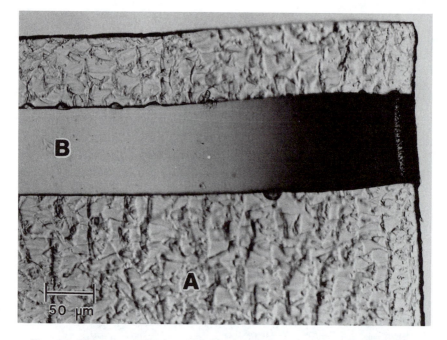

Figure 2. UV-micrograph of a microtomed section of an epoxy resin (A) with an embedded PP film (B): monochromatic illumination, 345 nm; bandwidth, 10 nm.

The black edges around the embedding medium show that diffusion of UV absorber from the other sides of the film was completely prevented.

Possible artifacts due to diffusion of unhardened epoxy resin into PP during the embedding were scrupulously checked: films immersed during 1 h at 60 °C in the individual components of the resin showed no difference in the diffusion coefficient (D_c) compared with untreated films. For technical reasons, the additive source used in these experiments was a powdered, amorphous material obtained by melting and quenching in water. Because the physical state of the additive source is known to affect the measured D values in the "diffusion-in" experiment (i.e., the experiment in which additive diffuses from an external reservoir into an initially additive-free polymer) (8), the two different supply systems were tested on the diffusion rate in plates and films. No difference in D was observed when using pure additive powder instead of the usual reservoir plate. The calculation of D was done by fitting the best theoretical profile to the experimental concentrations (least-squares error minimization).

UV Microspectrometry. The stabilizer concentration profile of plates or films was measured by UV microspectrometry (transmission) at the longest absorption wavelength of the UV absorbers (ca. 350 nm). The UV-microspectrometer used was a UMSP 80 (Carl Zeiss). This instrument is a UV-microscope (quartz optic) equipped with a monochromatic illuminating system (xenon lamp, grating monochromator) and a photomultiplier. It is a one-beam spectrometer with the possibility to adjust the size of the photometric field (down to 0.5 μm) by placing a small diaphragm in the optical path (9). A motorized XY-stage allows line scans of the samples to be taken (Figure 3).

Results and Discussion

Influence of Morphology.   The influence of morphology on the diffusion was studied for three benzotriazole stabilizers (UV-1, -6, and -9; *see* Table I) with molecular weights between 225 and 450 g/mol. From the results given in Tables II–IV and plotted in Figure 4, the difference between the diffusion coefficient measured in thick plate (D) and in thin film (D_c, across the surface) is small; the ratio D/D_c varies from 0.7 to 1.6. The 17% average difference between the two diffusion coefficients is as large as the standard deviation of the measurements, which was estimated to be 15%. Therefore, D and D_c can be considered as equal.

This result is contrary to the observations of Schwarz et al. (10) who measured the diffusion coefficient for a phenolic antioxidant in PP and found it to be four times higher in a film than in a thick plate. Although differences in the experimental procedures make a comparison difficult, these contradic-

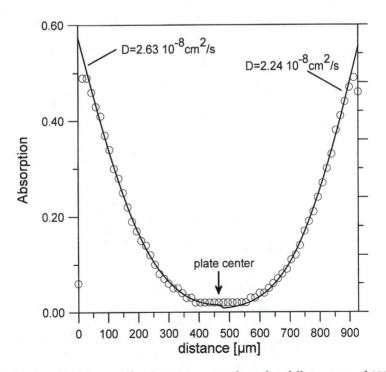

Figure 3. Absorption profile of UV-1 in a PP plate after diffusion time of 115 min at 80 °C (thickness, 930 μm). Circles are experimental data, and curves are calculated diffusion profiles for the half thickness.

Table II. Measured Diffusion Coefficients of UV-1 in PP

T (°C)	Plate	Film		
		D_c	$D_\parallel$	$D_\perp$
40	1.55×10^{-10}	1.80×10^{-10}	7.52×10^{-11}	4.23×10^{-11}
50	1.03×10^{-9}	8.53×10^{-10}	1.96×10^{-10}	1.44×10^{-10}
60	3.16×10^{-9}	4.11×10^{-9}	1.67×10^{-9}	1.16×10^{-9}
70	1.03×10^{-8}	9.91×10^{-9}	4.49×10^{-9}	4.63×10^{-9}
80	2.43×10^{-8}	2.63×10^{-8}	1.19×10^{-8}	1.61×10^{-8}
90	5.76×10^{-8}	7.24×10^{-8}	2.53×10^{-8}	3.21×10^{-8}
100	7.21×10^{-8}	7.28×10^{-8}	2.59×10^{-8}	2.08×10^{-8}
110	1.83×10^{-7}	1.87×10^{-7}	ND	ND
120	4.99×10^{-7}	4.13×10^{-7}	ND	ND

Note: Values are in cm²/s. T is temperature; ND is not determined.

Table III. Measured Diffusion Coefficients of UV-6 in PP

T (°C)	Plate	Film		
		D_c	$D_\parallel$	$D_\perp$
50	ND	5.85×10^{-11}	3.29×10^{-11}	3.91×10^{-11}
60	3.46×10^{-10}	3.37×10^{-10}	1.95×10^{-10}	2.05×10^{-10}
70	9.31×10^{-10}	1.15×10^{-9}	5.25×10^{-10}	5.58×10^{-10}
80	4.27×10^{-9}	3.48×10^{-9}	ND	ND

Note: Values are in cm²/s. T is temperature; ND is not determined.

Table IV. Measured Diffusion Coefficients of UV-9 in PP

T (°C)	Plate	Film (D_c)
60	7.22×10^{-11}	8.39×10^{-11}
70	3.00×10^{-10}	2.90×10^{-10}
80	1.33×10^{-9}	8.33×10^{-10}
90	2.92×10^{-9}	2.40×10^{-9}
100	9.56×10^{-9}	1.16×10^{-8}
105	9.18×10^{-9}	1.26×10^{-8}
110	2.25×10^{-8}	1.67×10^{-8}
120	2.46×10^{-8}	3.52×10^{-8}

Note: Values are in cm²/s. T is temperature.

tory observations could be explained by a size difference between the molecules studied. The antioxidant was quite bulky ($M_w = 774$ g/mol) compared with benzotriazoles UV-1, -6, and -9. These results could indicate that the diffusion coefficient is only affected by the PP morphology when the additive size reaches a certain level.

Figure 3 shows that the diffusion curves from both sides of a plate were not symmetrical, and a difference of 15% on calculated diffusion coefficients was usually observed. This experimental error does not reflect the accuracy

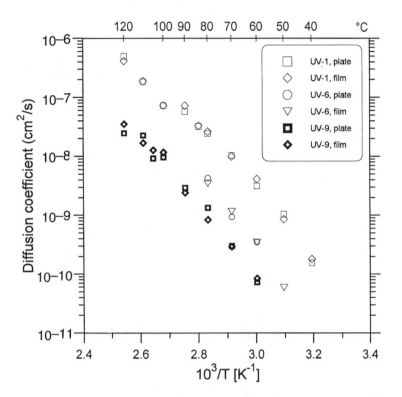

Figure 4. Arrhenius plot for the diffusion coefficient of UV-1, UV-6, and UV-9 in differently processed PP.

of the spectrometric analysis but does reflect some structural difference between the two sides of a plaque. This heterogeneity is quite common for semicrystalline material produced by compression molding.

Anisotropic Diffusion in Film. The importance of the migration direction on the diffusion coefficient was studied for two stabilizers by measuring their diffusion coefficient along the three directions of a film. The diffusion coefficients of UV-1 and UV-6 are reported in Tables II and III, respectively, and are depicted in Figure 5.

The results show that the diffusion coefficient in the plane of the film was not greatly different for the migration parallel or perpendicular to the extrusion direction ($D_{\parallel}$ or $D_{\perp}$). For UV-1, $D_{\parallel}$ seems to be slightly larger at low temperature, and the tendency was reversed at temperatures higher than 70 °C. Considering that the experimental error is slightly larger for the measurement of D in the plane of the film, $D_{\parallel}$ and $D_{\perp}$ can be considered to be equal.

In the case of UV-6, values of $D_{\perp}$ and $D_{\parallel}$ match almost perfectly. Comparing the diffusion across the plane (D_c) with the diffusion in the plane, one

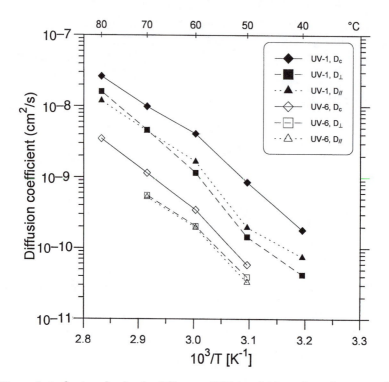

Figure 5. Arrhenius plot for the diffusion of UV-1 and UV-6 along the 3 axes of an extruded PP film.

observes that $D_\parallel$ and $D_\perp$ are always smaller than the diffusion coefficient measured across the film. This difference suggests that the extruded film presents either a one-axis anisotropy in its crystal organization or a morphology difference between the surface and the bulk (skin morphology). The processing conditions (rapid cooling with a temperature gradient perpendicular to the surface) and the fact that Fickian diffusion curves were always observed favor the first hypothesis.

The average ratio of D_c/D_{in} (D_{in} is $D_\parallel$ or $D_\perp$) is small, about 3 for UV-1 and 2 for UV-6. This small ratio can be explained by the processing conditions: the PP films were extruded without mechanical stress (draw ratio of 1) and the rapid cooling is supposed to produce only a slight anisotropy. The observation of a D_{in} smaller than D_c is reasonable: The orientation of impermeable polymer crystallites along the D_c-axis increases the tortuosity (diffusion path) in the other directions and decrease the diffusivity. However, the fact that D_c/D_{in} is smaller for UV-6 than for UV-1 is not easily explained. An increase of the additive size will decrease the diffusion rate; however, it is still unclear how this augmentation could differently affect D_c and D_{in}.

Influence of Additive Structure. The D values measured as a function of temperature are depicted in Figure 4 and indicate that the diffusion coefficient of benzotriazole stabilizers does not follow an Arrhenius relationship but shows a curvature in the Arrhenius plot. Such an observation was already made by Billingham et al. (*11*) for an homologous series of the 2-hydroxybenzophenones. They showed that the experimental data could be fitted with the Williams, Landel, and Ferry (WLF) equation. However, in a small temperature interval, the relationship can be considered as linear and the parameters of the Arrhenius equation (E_d, the activation energy for the diffusion; and D_0, the preexponential factor) can be calculated.

In the 60–80 °C range, we obtained values (average of films and plaques) for E_d of 95.2, 117.6, and 127.3 kJ/mol for UV-1, UV-6, and UV-9, respectively. Because E_d represents the energy required to generate a free space between the polymer chains large enough to allow the molecule to diffuse, its increase with the size of the side groups is normally expected. The size of the molecule is roughly proportional to its molecular weight, and authors have shown (*6, 11*) that $\log(D)$ is related to $\log(M_w)$.

To check this proportionality, the diffusion coefficients of different benzotriazoles were measured at 80 °C in PP plates (Table V). A reasonably good linear fit (considering the large differences in substituents) can be drawn between $\log(D)$ and $\log(M_w)$ (Figure 6), if one omits the stabilizers containing two benzotriazoles moieties (UV-11 and UV-12):

$$\log(D) = -3.64 \log(M_w) + 0.89$$

This equation allows one to estimate the diffusion coefficient of new structures (at 80 °C) within 30–40% confidence. This accuracy is, of course, only

**Table V. Diffusion Coefficient of
Benzotriazole Stabilizers in PP
at 80 °C**

Code name	D [cm²/s]
UV-1	2.43×10^{-8}
UV-2	1.35×10^{-8}
UV-3	5.37×10^{-9}
UV-4	3.19×10^{-9}
UV-5	6.13×10^{-9}
UV-6	4.27×10^{-9}
UV-7	6.21×10^{-9}
UV-8	3.47×10^{-9}
UV-9	1.33×10^{-9}
UV-10	1.80×10^{-9}
UV-11	7.71×10^{-10}
UV-12	1.06×10^{-9}

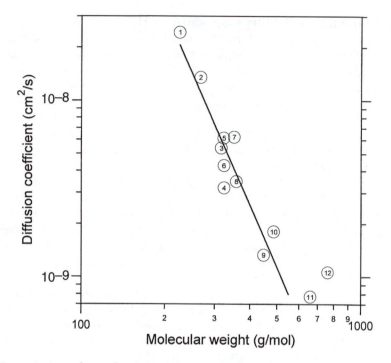

Figure 6. Dependence of D *(at 80 °C) on the molecular weight of benzotriazoles.*

partially satisfactory; therefore, we tried to correlate the diffusion coefficient with the molar volume of additives to give a better representation of the real molecular size (*12*). However, the calculated fit was not better; this result suggests that a shape factor and the effect of the substituents should be considered for better modeling. Observation of the more scattered data leads to the conclusion that the groups R1 and R2 (*see* structure **1**; Table I) can abnormally affect the diffusion coefficient (i.e., UV-4, UV-5, and UV-6). Because of their position on the molecule, a slight change of R1 or R2 is likely to greatly change the apparent size of the molecule. Conversely, R3 is in the line of the molecule and its effect is limited. Moreover, the group in R1 can modify the hydrophilicity of the molecule by hindering the phenol group.

UV-1 presents a discontinuity in its Arrhenius plot (Figure 4), which cannot be fitted with the WLF equation. In the 90–100 °C range, the diffusion coefficient stayed constant. No reasonable argument has been found presently to rationalize this observation. Nevertheless, this temperature range coincides with the temperature maximum of the exothermal recrystallization peak (96 °C by differential scanning calorimetry), and an association of molecules (competitive recrystallization) is not to be excluded.

Conclusions

We conclude that the influence of PP morphology on the diffusion behavior of the benzotriazole stabilizers is small. The D values measured in thick, compression-molded PP are relevant to describe the diffusion in samples presenting a different morphology. For diffusion in extruded film, the difference observed in the diffusion coefficients (perpendicular versus in-plane) could be rationalized by an anisotropic crystal orientation. For physical depletion problems, this observation is of limited importance because the loss occurs through the film's surface.

Compared with the PP morphology, the influence of the benzotriazole structure on diffusion is much more important. However, accurate modeling of diffusion is not simple, but a rough estimation of the diffusion coefficient of new benzotriazole structures from the molecular weight is possible.

References

1. Billingham, N. C.; Calvert, P. D. In *Developments in Polymer Stabilisation–3;* Scott, G., Ed.; Elsevier Applied Science Publishers: London, 1980; Vol. 3, p 139.
2. Peterlin, A. *J. Macromol. Sci., Phys.* **1975**, *B11*, 57.
3. Moisan, J. Y. *Eur. Polym. J.* **1980**, *16*, 989.
4. Moisan, J. Y. *Eur. Polym. J.* **1980**, *16*, 997.
5. Weinkauf, D. H.; Paul, D. R. In *Barrier Polymers and Structures;* Koros, W. J., Ed.; ACS Symposium Series 423; American Chemical Society: Washington, DC, 1990; pp 60–91.
6. Al-Malaika, S.; Goonetileka, M. D. R. J.; Scott, G. *Polym. Degrad. Stab.* **1991**, *32*, 231.
7. Olley, R. H.; Basset, D. C. *Polymer* **1982**, *23*, 1707.
8. Földes, E.; Turcsányi, B. *J. Appl. Polym. Sci.* **1992**, *46*, 507.
9. Piller, H. *Microscope Photometry;* Springer-Verlag: Berlin, 1977.
10. Schwarz, T.; Steiner, G.; Koppelmann, J. *J. Therm. Anal.* **1989**, *35*, 481.
11. Billingham, N. C.; Calvert, P. D.; Uzuner, A. *Eur. Polym. J.* **1989**, *25*, 839.
12. Saleem, M.; Asfour, A.; De Kee, D.; Harrison, B. *J. Appl. Polym. Sci.* **1989**, *37*, 617.

RECEIVED for review January 26, 1994. ACCEPTED revised manuscript January 5, 1995.

Diffusion and Solubility of Hindered Amine Light Stabilizers in Polyolefins

Influence on Stabilization Efficiency and Implications for Polymer-Bound Stabilizers

Ján Malík[1,2], Alexander Hrivík[3], and Dam Q. Tuan[3]

[1]VUCHT a.s., 836 03 Bratislava, Slovakia
[3]Slovak Technical University, Faculty of Chemical Technology, Bratislava, Slovakia

Results of diffusion and solubility measurements of hindered amine light stabilizers (HALS) in polyethylene and polypropylene are presented. These physical parameters are correlated with the efficiency of stabilizers. Moisan's empirical relationship for phenolic antioxidants in low-density polyethylene also applies to HALS efficiency. This relationship supports the idea of polymer-bound stabilizers; therefore, some results and ideas in this field are also presented.

HINDERED AMINE LIGHT STABILIZERS (HALS) were introduced to industry relatively recently but soon they took the leading role in light stabilization of polymers, especially polyolefins. Since their introduction, no new class of light stabilizers has appeared that surpasses the level of polyolefin stabilization offered by HALS.

Consequently, HALS has been one of the most actively investigated classes of polymer additives in recent years. The number of papers dealing with chemical mechanisms of HALS in polymer stabilization is indicative of their importance.

Most authors considered oxidation products of HALS (especially nitroxyl radicals) to be the key to their excellent light stabilization performance. Some recent works showed that the HALS mechanism is quite complex. In addition to the various transformation products, the charge transfer complex of the parent amine with polymer can play an important role in polymer stabilization (*1–5*).

[2]Current address: Clariant Huningue S. A., Polymer Additives, BP 149, F-68331 Huningue Cedex, France.

It is well known that the degradation of a stabilized polymer is accompanied by a loss of most of the effective stabilizer. Obviously this loss can be caused by chemical consumption of the stabilizer in stabilization reactions, but it may also be caused by physical loss of the stabilizer by such processes as evaporation, leaching, and blooming.

Contrary to the great amount of published work devoted to stabilization mechanisms of HALS, practically no data are available concerning the physical behavior of HALS in polymers. The published works on physical loss problems (6–8) do not include the data for hindered piperidine stabilizers.

In our earlier works (9, 10), we presented some results on diffusion and solubility measurements of HALS in polyolefins. In this chapter we present a summary on the diffusion and solubility measurements. We also attempt to correlate these parameters with the measured efficiency of the stabilizers. This correlation lead to the experiments with polymer-bound stabilizers, and the results and suggestions in this field are also discussed.

Diffusion and Solubility of HALS

Low-Density Polyethylene.   Diffusion coefficients (D) and solubilities (S) of HALS in polyolefins were measured by the dynamic method described by Moisan (8). The method was modified slightly to avoid formation of air bubbles between individual films of polymer (9). After the diffusion experiment the individual films were peeled off from the stack and were extracted in CCl_4 for 24 h. Each extract was analyzed for HALS content. From the concentration profile in the stack of films, D and S were computed. For this computation a mathematical approximation of the appropriate solution of Fick's second law was used (9). In the experiments where solubility values exceeded 7–8% (observed with Dastib 845 [*see* Chart I] in low-density polyethylene [LDPE] at higher temperatures), a deviation of the theoretical curve from the experimental data was observed. The reason for this deviation could be a concentration dependence of diffusion at these relatively high concentrations of diffusant.

Each diffusion experiment was performed at least twice at the same temperature. The experimental error of a single D estimation was $\pm 20\%$. The structures of stabilizers are shown in Chart I.

The results obtained in D and S measurements are presented in the form of parameters of Arrhenius equations for D and S. Table I shows the results for LDPE. With Diacetam 5 (*see* Chart I), we were not able to measure the penetration of the stabilizer into the stack of polymer films at room temperature. The unmeasurable penetration was caused by very low solubility of this stabilizer in LDPE matrix. With oligomeric Chimassorb 944 (*see* Chart I), no measurable penetration occurred after 10 months at room temperature. Also, experiments with this stabilizer at elevated temperatures were not successful.

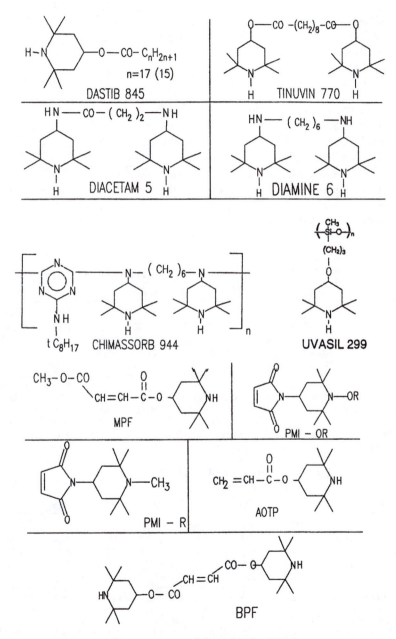

Chart I. Structural formulas for HALS.

Only one stabilizer (Diamine 6, *see* Chart I) exhibited first-order transition (melting) within the measured temperature range. With this stabilizer we observed a change in the slope of the Arrhenius plot for *S*, where the difference between change in enthalpy (ΔH) over melting point (mp) and ΔH under mp was higher than the heat of fusion of this stabilizer (71.7 kJ/mol vs. 42.9 kJ/mol fusion, respectively). No obvious change in the slope of the Arrhenius plot for *D* was found (the calculated difference was within the span of experimental error).

Polypropylene. Results obtained with polypropylene (PP) are presented in Table II. Again, no measurable diffusion of Chimassorb 944 was observed. For Diamine 6 the same effect of melting on *D* and *S* was observed with PP as with LDPE—no evident change in activation energy of diffusion E_d and relatively great change in *S*. The difference in ΔH (87.6 kJ/mol) over and under mp is again much bigger than the heat of fusion. In comparison to the other measured stabilizers, ΔH values for Diamine 6 under mp are relatively high in PP and LDPE.

Table I. Diffusion and Solubility of HALS in LDPE

HALS	mp	E_d	log D_0	D	T	ΔH	log S_0	S
D-845	28–31	57.2	1.55	2.0	30–75	18.2	7.54	39,600
T-770	81–83	83.1	5.3	0.74	25–75	43.2	10.55	3700
Diac-5	134–137	38.3	−1.6	1.6	50–80	30.6	7.21	180
D-6	61–63	75.5	4.37	1.5	30–75	96.2	19.8	17,760
D-6	—	—	—	—	>mp	24.5	8.4	—

NOTE: D-845 is Diastib 845, T-770 is Tinuvin 770, Diac-5 is Diacetam 5, and D-6 is Diamine 6 (*see* Chart I): mp is melting point (°C), E_d is activation energy of diffusion (kJ/mole), D_0 is preexponential parameter of Arrhenius equation for diffusion, D is diffusion coefficient at 50 °C ($\times 10^{-8}$ cm²/s), T is temperature (°C), ΔH is heat of solution (kJ/mole), S_0 is preexponential parameter of Arrhenius equation for solubility, and S is solubility at 50 °C (ppm).

Table II. Diffusion and Solubility of HALS in PP

HALS	mp	E_d	log D_0	D	T	ΔH	log S_0	S
D-845	28–31	109	8.2	3.8	60–90	14.7	8.65	18,780
T-770	81–83	89.5	4.9	2.7	55–80	69.7	14.5	1800
Diac-5	134–137	79.9	3.3	2.5	60–90	17.7	5.81	890
D-6	61–63	114.7	9.1	3.6	55–80	147.1	27	1670
D-6	—	—	—	—	>mp	59.5	13.4	—

NOTE: D-845 is Dastib 845, T-770 is Tinuvin 770, Diac-5 is Diacetam 5, and D-6 is Diamine 6 (*see* Chart I): mp is melting point (°C), E_d is activation energy of diffusion (kJ/mole), D_0 is preexponential parameter of Arrhenius equation for diffusion, D is diffusion coefficient at 50 °C ($\times 10^{-10}$ cm²/s), T is temperature (°C), ΔH is heat of solution (kJ/mole), S_0 is preexponential parameter of Arrhenius equation for solubility, S is solubility at 50 °C (ppm).

The relative error of E_d estimation amounted to $\pm 10\%$, and the relative error of ΔH was higher ($\pm 20\%$). Measurements with Diamine 6 were the exception, because the relative error of E_d estimation in PP was ± 24 % and the relative error of ΔH in PP and LDPE reached 35%. Later investigations showed that Diamine 6 is not a very stable compound. In the presence of water (even exposed to air moisture for an extended time), Diamine 6 forms several crystalline modifications with different heats of fusion and different mps. Therefore, the presented data for Diamine 6 should be regarded with this fact in mind. However, the observed discontinuities in the Arrhenius plots of S of Diamine 6 are outside the error.

The discontinuities in Arrhenius plots for D and S of phenolic antioxidants in LDPE at their mps were observed by Moisan (8). Al-Malaika et al. (11) also reported discontinuities for D of several antioxidants in LDPE. However, explanations of this phenomenon were different. Whereas Moisan (8) attributed the discontinuity in E_d to differences in the physical form of the additives, Al-Malaika et al. (11) suggested that the discontinuity is more likely due to morphological changes of the polymer.

In this connection, Billingham (12) pointed out that diffusion of an additive is a molecular-level process that is determined by an interaction between polymer and an isolated diffusant molecule. Therefore, no physical reason exists for the change in E_d at the mp of diffusant. Billingham (12) also suggested that the change in E_d results from morphological changes on annealing of the polymer during experiments. As we already mentioned, no obvious change in the slope of Arrhenius plot for D was observed with Diamine 6 in LDPE and in PP.

A change in Arrhenius plot for S of an additive in a polymer is a different phenomenon. When a crystalline additive was used as a source of diffusing molecules (12), a change in the ΔH was observed that corresponded to the heat of fusion of the used additives within experimental error. We used (9) pure additive as a source of diffusing molecules and observed a marked change in the slope at the mp of Diamine 6 in LDPE and PP. However, it is difficult to explain the change in the Arrhenius S plot that was reported by Moisan (8), who used a polymer with 2–10% additive as the source of the diffusing molecules. The source was prepared by processing at 150 °C, a temperature that is higher than the mps of the additives. We are not aware of any reports of a well-characterized crystallization of antioxidants inside polyolefins, and so the antioxidants after the processing were probably presented in a form of supersaturated solution. Hence a change in Arrhenius S plot connected to the heat of fusion of crystalline additive at its mp can hardly be expected.

Solubility in Model Liquids.   The D coefficient values of all low-molecular stabilizers in LDPE and PP obtained in our measurements (Tables I and II) were comparable, but the S values differed greatly. Therefore, the S results obtained from dynamic measurements were compared with S values of the stabilizers in organic liquids (see Table III). The relative ranking and ratios of S in model liquids coincide very well with the results obtained in the polymers. Surprisingly the S of oligomeric Chimassorb 944 in the nonpolar liquids was one of the best. Therefore, the reason for nonmeasurable penetration of the oligomer into the stack of polymer films was the size of stabilizer's molecules.

Permeation Experiments and Extrapolation of Oligomer Diffusion Parameters

Because the experiments with a stack of polymer films did not enable us to obtain the D parameters of oligomeric Chimassorb 944, we tried to measure the permeation of this stabilizer through swollen LDPE film (10). Two types of experimental arrangements were used (Figure 1) and both offered comparable results. In the first type, a steel chamber was divided into two parts that were separated by LDPE film. One part of the chamber contained concentrated solution of Chimassorb 944 in CCl_4 and the other part contained pure CCl_4. In the second type, the concentrated solution of the stabilizer was thermowelded into an LDPE bag, and the bag was then placed into a vessel with pure CCl_4. The experiments were done at room temperature (at higher temperatures CCl_4 dissolves polyethylene films).

The amount of penetrated stabilizer for a given experimental time (Q_t) was measured in the solvent by UV-spectroscopy of the HALS complex with iodine (9). The obtained values were used for computation of D from an appropriate form of Fick's law, which included an intercept on the time axis known as "time-lag":

$$Q_t = -DA(c_2 - c_1)(1/l)[t - (l^2/6D]$$

where A is the area of the film, l is film thickness, and t is time. The concentrations c_2 and c_1 are given as:

$$c_2 = (Q_0 - Q_t)/V_2 \qquad c_1 = Q_t/V_1$$

where Q_0 is the original amount of HALS in the solution, and V_2 and V_1 represent volume of HALS solution and volume of pure solvent, respectively.

Experimental data generated by measurements with Chimassorb 944 were not properly fitted by the theoretical curve. The experimental dependence started to settle down after permeation of only about 3% Q_0 dissolved in the concentrated solution. Apparently, only a small part of the oligomer was able to penetrate the swollen LDPE film. Therefore, in our next calculations only a fraction of the original amount of stabilizer presented in concentrated solution was used. With 4% Q_0 the experimental data fit very well with the

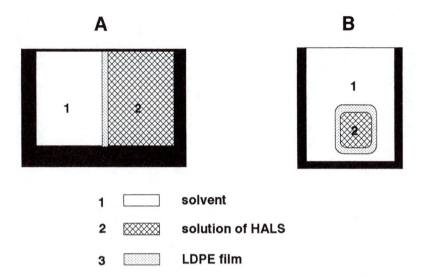

1		solvent
2		solution of HALS
3		LDPE film

Figure 1. Experimental arrangements of CCl₄ solvent (1), saturated solution of stabilizer in CCl₄ (2), and LPDE membrane (3) for permeation measurements. Part A: steel chamber. Part B: LDPE bag.

Table III. Solubility of HALS in Model Liquids

HALS	Heptane	Parafin Oil	PP Oil
Dastib 845	49.20	23.20	1.20
Tinuvin 770	5.80	2.10	0.70
Diacetam 5	0.08	0.16	0.04
Diamine 6	30	4.5	—
Chimassorb 944	53.70	14.50	3.00

NOTE: For structures of HALS, *see* Chart I. All values are in weight percent.
SOURCE: Reproduced with permission from reference 9. Copyright 1992 Elsevier Science.

theoretical curve (Figure 2). (The dotted line in the Figure 2 corresponds to the real value of Q_0.) The calculated values for D of 4% of the oligomer ranged from 2.1×10^{-10} to 2.6×10^{-10} cm²/s.

The same experiments were done with low-molecular Dastib 845. The experimental data fit well with the theoretical curve for the whole amount of stabilizer (100% Q_0), and D ranged from 1.6×10^{-9} to 2.49×10^{-9} cm²/s. The D value obtained from the measurement in a stack was 7×10^{-10} cm²/s, which means that D in swollen LDPE was 2–3 times faster.

In the next step an attempt for mathematical extrapolation of D parameters of the oligomer was done. The extrapolation was based on three equations (10):

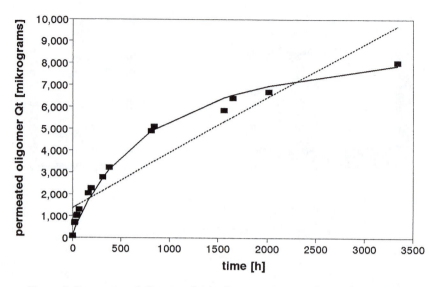

Figure 2. Permeation of Chimassorb 944 showing experimental data (solid boxes), calculated line for actual amount of Chimassorb 944 (broken line), and calculated line for $Q_0 = 4\%$ of the actual amount of Chimassorb 944.

- Arrhenius equation for D
- empirical Auerbach dependence of D on molecular weight of diffusant
- dependence of the logarithm of preexponential factor of Arrhenius equation on E_d, known also as "compensation effect equation"

Mutual substitution of these three equations leads to the dependence of E_d on molecular weight of diffusant (M_w):

$$E_d = A' + B' \times \ln M_w$$

For the computation of the constants A' and B' in the extrapolation equations, data from Table I were used. The D parameters of the oligomer were calculated, and results are presented in Table IV. Obviously, only molecules with one structural unit can diffuse through LDPE matrix at a comparable rate to that of the low-molecular weight stabilizers. The D values of molecules with two and more structural units are too small to be comparable with the rate of diffusion of low-molecular weight stabilizers. The computed D for one structural unit is about one-half the value obtained from permeation experiments; a similar difference was found for low-molecular weight Dastib 845.

On the basis of the obtained and extrapolated results, we assumed that the oligomer should contain approximately 4% low-molecular weight fraction corresponding to molecules with one structural unit. This assumption was later confirmed by gel permeation chromatographic measurement of molecular weight distribution. As seen in Figure 3, the fraction with molecular weight under 1000 is really about 5%.

According to the previously stated results, approximately 95% of the molecules of the highly effective oligomeric stabilizer are translationally immobile in the polymer matrix. Therefore, the common explanation of the decreased efficiency of oligomeric stabilizers with increased molecular weight as a result of reduced stabilizer mobility is questionable.

Table IV. Extrapolated Diffusion Parameters of Chimassorb 944 in LDPE

No. units	M_w	E_d	$ln\ D_0$	D
1	600	115.5	24.1	1.2×10^{-10}
2	1200	210.8	58.1	1.1×10^{-12}
3	1800	266.6	78.1	7.8×10^{-14}
4	2400	306.2	92.2	1.1×10^{-14}
5	3000	336.9	103.2	2.5×10^{-15}
6	3600	362.0	112.1	6.8×10^{-16}

NOTE: M_w is molecular weight, E_d is activation energy of diffusion (kJ/mole), D_0 is preexponential parameter of Arrhenius equation for diffusion, and D is diffusion coefficient at 23 °C (cm²/s).

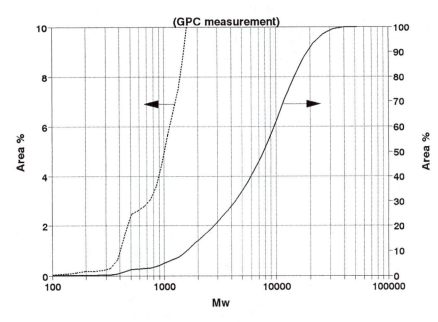

Figure 3. Molecular weight distribution of Chimassorb 944.

Research data from Ciba-Geigy (13) obtained with a polymerized acrylic derivative of HALS implied that the stabilizer gives the maximum performance at molecular weight ~2700. However, Minagawa (14) reported the maximum in HALS efficiency at a much lower molecular weight (~600), whereas the thickness of tested PP specimens was the same (50 μm). Hrdlovic and Chmela (15, 16) showed that stabilization efficiency of a copolymer of acrylate HALS in PP decreases exponentially with increasing molecular weight of the copolymer. They also (17) documented that copolymers of acrylate HALS with monomers containing long alkyl chains and having molecular weight ≤9500 are much more efficient than a homopolymer with molecular weight 1300. Experimental data from the various laboratories do not correspond sufficiently.

The observed phenomenon—decreased stabilization efficiency with increased molecular weight of a stabilizer—is usually ascribed to lower compatibility and lower mobility of the stabilizer (14–17). However, no data on the stabilizer's mobility were reported. Contrary to the previous assumption, Langlois et al. (18) studied thermooxidation of cross-linked PE and concluded that even the relatively high diffusion rate of a used phenolic antioxidant ($D \approx 2 \times 10^{-6}$ cm²/s) is too low to induce significant redistribution effects after a heterogeneous consumption of the antioxidant in a first phase of thermal oxidation of a 2.2-cm thick PE sample. This calculation was done for PE degradation at 130 °C, and the conclusion was determined to remain valid at higher temperatures because diffusivity increases less than oxidation rate with the temperature.

Therefore, taking into account our measured and extrapolated D parameters, we think that the more probable explanation for decreased stabilization efficiency of stabilizers with high-molecular weight is a decreased homogeneity of distribution of active stabilizing functionalities throughout the polymer. In the high-molecular weight stabilizers, a lot of the stabilizing functionalities are connected to one backbone of a polymeric stabilizer, and so they are concentrated in one place and other parts of polymer matrix are left unprotected. However, there are no doubts that the compatibility of the stabilizer molecules with a host polymer plays an important role in the stabilization process.

Relation of the Physical Parameters to Stabilization Efficiency

The measured and extrapolated physical parameters were used for estimation of the stabilization efficiency of individual HALS in LDPE. We tried (10) to use the theoretical model for physical loss of additives that was proposed by Billingham and Calvert (6). This model uses three parameters for the estimation of protection time: D, S, and V_0, where V_0 is the rate of evaporation of pure additive into free air or the rate of dissolution of pure additive in a contacting liquid. In our calculations (10) the rate of dissolution of stabilizers in water was used as V_0, because liquids are supposed to be much more effective extracting media than gasses (*6*) and water is the most probable contacting liquid for a stabilized polymer article that is exposed to weathering or accelerated aging in Xenotest. The calculations (*10*) led to the predictions of a very short protection time with low-molecular HALS and an unrealistically long protection with the oligomeric HALS.

Other authors (*6, 19*), however, reported good correlation of the calculated values from the model and experimental data. These authors used the rate of evaporation of additives into air as V_0. Therefore, some barrier (interlayer) seems to be formed between polymer surface and contacting water, and this barrier causes the rate of additive loss to be significantly lower than the rate of dissolution of pure additive.

In our next attempt we used the empirical model proposed by Moisan (*8*). Moisan showed that the ratio S^2/D is linked to the stabilization efficiency of phenolic antioxidants in LDPE. Therefore, this ratio was calculated for individual HALS stabilizers from the measured and extrapolated physical parameters. The S value for Chimassorb 944 was estimated to be approximately the same as for Dastib 845 on the basis of the S measurements in hydrocarbon liquids (Table III). The calculations of S^2/D with HALS were done for a temperature of 40 °C, which roughly corresponded to the test temperature of samples in Xenotest. The results versus the measured stabilization performance in Xenotest are plotted in Figure 4. The plotted curve for HALS showed

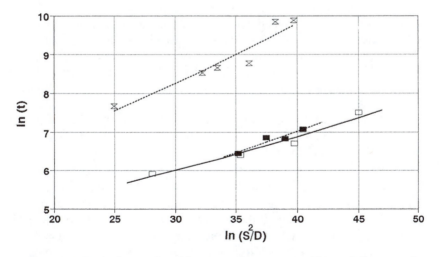

Figure 4. Dependence of stabilization efficiency on additive diffusion and solubility [ln(t) = f[ln(S²/D)]] for phenolic antioxidants in LDPE and natural weathering (data recalculated from a plot in reference 8) (double inverted triangles), HALS in LDPE and Xenotest 1200 exposure (open rectangles), and HALS in PP and QUV-A exposure (filled rectangles).

a course very similar to the Moisan curve for phenolic antioxidants (dotted line in Figure 4) even though hindered phenols and HALS are two different classes of antidegradants. Moreover, efficiency measurements of phenols were done in 1-mm thick LDPE sheets exposed to natural aging, whereas HALS efficiency was measured in blown LDPE films of 160 μm thickness exposed to accelerated aging in Xenotest 1200. Surprisingly, a similar dependence was also observed in PP when 100 μm compression molded films were exposed in a QUV tester with UV-A lamps [operated at black panel temperature (bpt) of 60 °C].

The importance of an additive's solubility in relation to its diffusion suggests that solubility is not significant only from the point of view of physical behavior of a stabilizer molecule, but this parameter probably influences the kinetics of chemical reactions of the stabilizer molecule in polymer matrix.

Polymer-Bound Stabilizers

The results presented in Figure 4 indicate that an effective stabilizer should be highly soluble in polymer, while its translation mobility is less important than was previously proposed and should be low. Therefore, polymer-bound additives could be a convenient solution. Recently, this type of HALS stabilizers has been attracting more and more attention, but the scientific community still holds two discordant opinions:

1. The increasing use of polymers in demanding applications clearly places greater demands on the performance of polymer additives. Conventional additives, because of their physical losses (which have toxicological consequences especially in medical and food applications), are not able to meet these demands. Polymer-bound additives are suggested as an attractive solution to the problem of physical losses (20, 21).

2. Doubts exist concerning polymer-bound additives. These additives are considered to be completely insoluble in the polymer and present in a metastable alloy (12).

In our opinion, the problems connected with polymer-bound stabilizers are case-by-case dependent. First of all, the grafting process depends on the type of reactive functionality in the stabilizer molecule. The situation should be different with the most commonly presented functionalities such as acrylates and maleates (20–22). Whereas acrylates are known for the tendency to form homopolymers, maleate derivatives are typical copolymerization monomers and usually do not form homopolymers. This difference means that with stabilizers having acrylate-reactive functionality, we can expect formation of homopolymer chains during a reactive processing with polyolefin. However, the reactive processing of a polyolefin with maleate-functionalized stabilizers would prefer single stabilizer molecule attachment to polymer. The grafting of acrylate can result in something like metastable polymer alloy; alternately, grafting of maleate can be regarded as a process similar to an alkylation of the stabilizer molecule to improve solubility of the additive (*12*).

In our experiments with polymer-bound stabilizers we used the structures presented in Chart I. Fumarates (BPF and MPF) and maleimides (PMI-R where R is methyl, and PMI-OR where R is octyl) are typical derivatives with low tendency to homopolymerization, whereas acrylate (AOTP) is a typical monomer that easily forms homopolymer. In the first experiments, 5% master batches of the BPF processed with different molar ratios of dicumylperoxide (DCP) were prepared. The level of BPF grafting was measured by IR spectroscopy before and after the extraction of master batches in chloroform, and the grafting was 30 and 50% for DCP:BPF ratios of 0.2 and 1.0, respectively. The master batches were used for PP stabilization, and their efficiency was compared with the efficiency of Chimassorb 944. Figure 5 shows the stabilization efficiency results (in Xenotest 450 operated at bpt of 45 °C) for chloroform extracted samples (20 h at room temperature), and Figure 6 shows the results for unextracted samples. In both cases the grafted stabilizers were more efficient than Chimassorb 944.

In these tests the stabilizers were compared on a weight basis. In the next series of experiments the weight concentrations of stabilizers were recalcu-

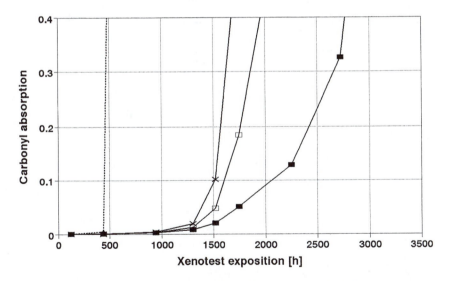

Figure 5. Stabilization efficiency of polymer-bound BPF [DCP:BPF ratios of 0.2 (open rectangles) and 1.0 (filled rectangles)] and oligomeric Chimassorb 944 (Xs) extracted samples and reference (filled triangles) in 0.1 mm PP.

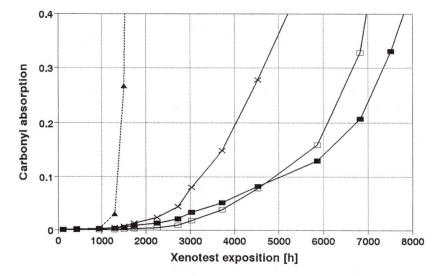

Figure 6. Stabilization efficiency of polymer-bound BPF [DCP:BPF ratios of 0.2 (open rectangles) and 1.0 (filled rectangles)] and oligomeric Chimassorb 944 (Xs) unextracted samples and reference (filled triangles) in 0.1 mm PP.

lated to piperidine content and compared with an increase of carbonyl absorption of 100-μm compression-molded, stabilized PP samples after 2200 h of Xenotest exposure (bpt, 63 °C) (Figure 7). Data were fitted easily with a simple linear relationship and were not influenced by different levels of grafting of individual stabilizers (from 25 to 70%). Standards—low-molecular Tinuvin 770 (*see* Chart I) and oligomeric Chimassorb 944—fit this relationship very well. Only AOTP showed a much higher carbonyl increase in relation to its piperidine content, and this result is consistent with the assumed homopolymerization tendency leading to a metastable alloy. In later stages of exposure, the PMI-R sample degraded somewhat faster than corresponded to the linear relationship. This increase is perhaps due to methyl substitution on piperidine nitrogen.

In later experiments with AOTP, varying grafting conditions improved the stabilization efficiency. This result suggests that under different conditions this stabilizer can give different reaction products—from homopolymeric or long-chain branching to short chains of stabilizer molecules attached to a polymer backbone. Dependence similar to that shown in Figure 7 was found for extracted samples, but the extracted standards (Tinuvin 770 and Chimassorb 944) showed no stabilization effect in this case.

Figure 8 summarizes the efficiency results from various test series obtained from several polymer-bound HALS stabilizers that are dependent on their piperidine concentration. The concentration of used master batches var-

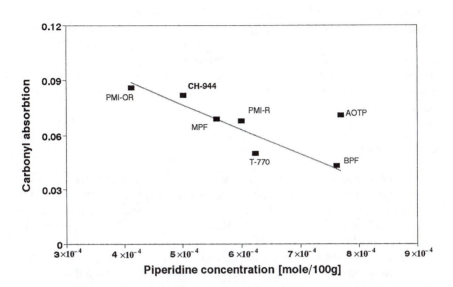

Figure 7. Stabilization efficiency of various HALS (see Chart I) according to their content of tetramethylpiperidine for unextracted samples in 0.1 mm PP after 2200 h of Xenotest-450 exposure.

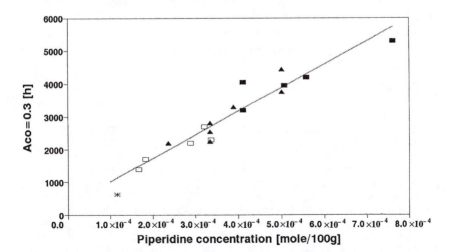

Figure 8. Stabilization efficiency of unextracted samples of polymer-bound HALS (filled rectangles), extracted samples of polymer-bound HALS (open rectangles), unextracted samples of commercial HALS (filled triangles), and extracted sample of commercial HALS (star) according to their content of tetramethylpiperidine.

ied from 5 to 15% of reactive HALS. The results for some commercial HALS (Chimassorb 944, Dastib 845, and Uvasil 299) are plotted as well. All efficiency results were obtained in thin compression-molded PP films exposed to accelerated aging in xenon arc devices operated at 63 °C bpt. Results for unextracted and extracted samples can be fitted by a simple linear relationship; the intercept on the y-axis corresponds to degradation of unstabilized PP film. The extracted samples, which fit well to the calculated line, contained virtually no mobile stabilizer.

Conclusion

The diffusion and solubility measurements as well as efficiency data of polymer-bound stabilizers suggest that the generally accepted concept of the importance of stabilizer's migration may need to be reconsidered. Concentration and especially homogeneity of distribution of the active stabilizing moieties in a polymer article should be addressed.

Increased solubility of the stabilizer improves its performance, whereas an increase in diffusion rate has the opposite effect. Polymer-bound stabilizers offer a favorable combination of these properties. However, the grafting process is a specific chemical reaction that is very sensitive to the type of monomer used and the grafting conditions. As with any other chemical reaction, the grafting reaction yields the desired product, a grafted stabilizer; the undesired side products, homopolymerized stabilizer; and unreacted monomeric stabilizer. Consequently, the resulting effect can vary from poor to excellent stabilization efficiency that outperforms the efficiency level commonly achieved by commercial stabilization systems.

References

1. Denisov, E. T. *Polym. Degrad. Stab.* **1989,** 25, 209.
2. Klemchuk, P. P.; Gande, M. E. *Polym. Degrad. Stab.* **1988,** 22, 241.
3. Klemchuk, P. P.; Gande, M. E.; Cordola, E. *Polym. Degrad. Stab.* **1990,** 27, 65.
4. Gugumus, F. *Polym. Degrad. Stab.* **1993,** 39, 117.
5. Gijsman, P.; Hennekens, J.; Tummers, D. *Polym. Degrad. Stab.* **1993,** 39, 225.
6. Billingham, N. C; Calvert, P. D. In *Developments in Polymer Stabilization-3;* Scott, G., Ed.; Applied Science: London, 1980; p 139.
7. Chalykh, A. Ye *Diffuzija v Polimernych Sistemach,* Chimija: Moscow, 1987.
8. Moisan, J. Y. In *Polymer Permeability;* Comyn, J., Ed.; Elsevier, 1985; p 119.
9. Malík, J.; Hrivík, A; Tomová, E.; *Polym. Degrad. Stab.* **1992,** 35, 61.
10. Malík, J.; Hrivík, A; Alexyová, D. *Polym. Degrad. Stab.* **1992,** 35, 125.
11. Al-Malaika, S.; Goonetileka, M. D. R. J.; Scott, G. *Polym. Degrad. Stab.* **1991,** 32, 231.
12. Billingham N. C. In *Oxidation Inhibition in Organic Materials;* Pospíšil, J.; Klemchuk, P. P., Eds.; CRC Press: Boca Raton, FL, 1990; Vol. 2; p 249.
13. Gugumus, F. *Res. Discl.* **1981,** 209, 357.
14. Minagawa, M. *Polym. Degrad. Stab.* **1989,** 25, 121.

15. Hrdlovič, P.; Chmela, Š. *Vybrané Problémy Stabilizácie Monomérov a Polymérov;* Preprints, Bratislava 1986, p 41.
16. Chmela, Š.; Hrdlovič, P. *11th Discussion Conference on Chemical and Physical Phenomena in the Ageing of Polymers;* Prague, 1988, p 9.
17. Chmela, Š.; Hrdlovič, P.; Maňásek, Z. *Polym. Degrad. Stab.* **1985,** *11,* 233.
18. Langlois, V.; Audouin, L.; Verdu, J.; Courtois, P. *Polym. Degrad. Stab.* **1993,** *40,* 399.
19. Sampers, J. T. E. H. *Physical Loss of UV Stabilizers from LDPE films;* 34th IUPAC International Symposium on Macromolecules, Prague, 13-18 July 1992, 8-P56.
20. Al-Malaika, S. *Polym. Plast. Technol. Eng.* **1990,** *29,* 73.
21. Al-Malaika, S. *CHEMTECH* **1990,** *20(6),* 366.
22. Al-Malaika, S.; Ibrahim, A. Q.; Scott, G. *Polym. Degrad. Stab.* **1988,** *22,* 233.

RECEIVED for review January 26, 1994. ACCEPTED revised manuscript September 22, 1994.

Reactive Oligomeric Light Stabilizers

Š. Chmela and P. Hrdlovič

Polymer Institute, Slovak Academy of Sciences, 842 36 Bratislava, Slovak Republic

A new class of photoreactive oligomeric light stabilizer containing a weak link in its structure has been synthesized and tested in polypropylene (PP). The light-sensitive 1-phenyl-2-propenone (VPK) was used as the weak link. The other monomers were as follows: 2,2,6,6-tetramethyl-4-piperidylacrylate (TMA), 2-hydroxy-4-(2-acroyloxyethoxy)-benzophenone (2HAB), and n-octadecylacrylate (ODA). TMA contains a sterically hindered amine as the light stabilizing unit, and 2HAB is a UV absorber. ODA improves the compatibility of oligomers with non-polar PP. For terpolymers TMA–ODA–VPK with similar molecular mass, the stabilizing efficiency depends on the VPK content. The higher the VPK content the higher the efficiency. For terpolymers 2HAB–ODA–VPK this dependence is not valid and the efficiency is extremely low. The reasons for this different behavior are discussed.

THE QUEST FOR IMPROVED RESISTANCE of plastics and fibers to degradation by outdoor weathering has resulted in the development of several classes of light stabilizers (1). The most successful and rapidly growing stabilizer class at present is the hindered amine light stabilizer (HALS) family. The majority of commercial products of this class are derivatives of 4-substituted 2,2,6,6-tetramethylpiperidine. The result of very extensive research in several countries is a reasonable understanding of the mechanism of HALS protection. HALS and their conversion products act as multifunctional additives during stabilization. Because of this complexity some details of their action are still controversial (2).

In addition to suitable chemical structure, an effective additive has to have good physical properties. First of all, it has to have good processing stability, good thermal stability, and sufficient retention in the polymer substrate. Other important aspects are good dispersion in the polymer during processing and sufficient mobility during service lifetime. Some of these requirements might

0065–2393/96/0249–0473$12.00/0

be achieved easily by increasing the molecular mass of the additive. But the mobility of the additive decreases with increasing molecular mass. Generally, the stabilizing efficiency for a particular additive decreases with increasing molecular mass (3, 4). It is necessary to find the optimal region of molecular masses for each particular oligomeric additive with respect to its application.

Alternately, we have tried to solve this problem by preparing oligomeric additives with relatively high molecular mass but with some content of statistically distributed light-sensitive units. The role of these units on the action of light is to cleave the oligomeric chain of the additive and thus produce shorter fractions with higher mobility.

Experimental Details

Starting monomers were 2,2,6,6-tetramethyl-4-piperidylacrylate (TMA), 2-hydroxy-4-(2-acroyloxyethoxy)benzophenone (2HAB), n-octadecylacrylate (ODA), and 1-phenyl-2-propanone (VPK). Their preparation is described elsewhere (5, 6). Terpolymers (TMA–ODA–VPK and 2HAB–ODA–VPK) and copolymers (TMA–ODA, TMA–VPK, ODA–VPK, and 2HAB–ODA) were prepared by radical copolymerization under nitrogen in benzene by using 2-2'-azobisisobutyronitrile (AIBN) as the initiator. The limiting viscosities (η_{sp}/c; intrinsic viscosity) were determined in benzene at 30 °C. Polypropylene (PP) powder (Tatren HPF, Slovnaft s.e., Bratislava, Slovak Republic) and the additives (0.2 wt %) were mixed and homogenized in a Brabender Plastograph (Duisburg, Germany) at 190 °C for 5 min. The bulk polymer was then pressed into 0.15–0.20-mm films in an electrically heated laboratory press at 200 °C for 1 min. Films were irradiated in a merry-go-round at 30 °C. A medium-pressure mercury lamp was used as a source of irradiation ($\lambda < 310$ nm). Absorption at the 1700–1750 cm^{-1} region was monitored by IR spectroscopy (IR-75, Carl Zeiss, Jena, Germany). UV spectra were measured on a Specord UV-VIS and M-40 (C. Zeiss, Jena, Germany).

Results and Discussion

Synthesis of Additives. Two types of terpolymers were prepared (Table I). The first ones were the terpolymers containing the HALS unit TMA (**1**). The second type consisted of terpolymers with 2HAB (**2**), which is a UV-absorber stabilizing unit. VPK (**3**) was used as a light-sensitive unit in both cases. From the structural formulae it is clear that all comonomers are rather polar compounds. To increase compatibility with nonpolar PP, ODA (**4**) was used as the third comonomer.

To verify importance and contribution of the particular comonomers, TMA–ODA, ODA–VPK, TMA–VPK, and 2HAB–ODA were prepared. Two groups of additives were prepared according to molecular mass by using different initiator concentrations. The first group (polymers A, B, and D in Table I) had higher molecular masses and η_{sp}/c values of 110–180 mL/g; the second group (polymers C and E–K in Table I) had relatively lower molecular masses and η_{sp}/c values of 40–80 mL/g.

Table I. Polymerization Characteristics of Copolymers and Terpolymers

Polymer	Structure	VPK (wt %)[a]	2HAB (wt %)[a]	AIBN (wt %)	%N[b]	η_{sp}/c^c (mL/g)
A	TMA–ODA–VPK	4.8	—	0.1	2.98	164
B	TMA–ODA–VPK	10.2	—	0.1	2.61	111
C	ODA–VPK	9.6	—	0.1	0	83
D	TMA–ODA	0	—	0.1	3.06	180
E	TMA–ODA	0	—	0.7	2.44	66
F	TMA–VPK	12.2	—	0.7	5.78	55
G	TMA–ODA–VPK	8.2	—	0.7	2.64	62
H	TMA–ODA–VPK	2.5	—	0.7	2.78	64
I	2HAB–ODA	0	53	0.1	—	80
J	2HAB–ODA–VPK	16[d]	43	0.1	—	40
K	2HAB–ODA–VPK	5[d]	47	0.1	—	52

[a]Calculated from UV spectra.
[b]Elemental analysis; %N calculated for TMA–ODA = 2.61% (ratio 1:1).
[c]Intrinsic viscosity for benzene; c = 7 g/L.
[d]Value determined from monomer mixture.

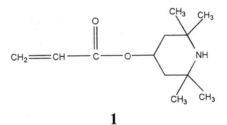

1

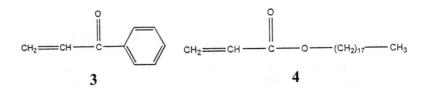

2

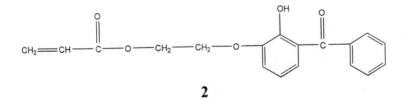

3   **4**

Compositions of the terpolymers and copolymers were roughly the same as the composition of the starting monomers mixture [copolymerization parameters for TMA and ODA are $r_1 = 0.86$ and $r_2 = 0.89$, respectively (7)]. Terpolymers and copolymers were white or slightly yellowish powders or rubbery materials. The presence of VPK units was evident from UV and IR spectra of carbonyl absorption at 1680 cm^{-1}.

Photolysis in Solution. A Norrish type II reaction leads mainly to main-chain scissions in polymer containing carbonyl groups and γ-hydrogen atoms (8, 9). Terpolymers and copolymers containing VPK units from Table I represent this type of polymer. The changes in molecular mass of prepared terpolymers and copolymers are shown in Figures 1 and 2.

The molecular masses of copolymers D, E (TMA–ODA) and I (2HAB–ODA) do not contain the light-sensitive VPK unit and do not change throughout the photolysis in benzene. On the other hand, all copolymers and terpolymers (with the exception of the terpolymers 2HAB–ODA–VPK) containing VPK units exhibited a remarkable decrease in molecular mass, especially at the first stage of photolysis. The mechanism of the main-chain scission reaction for terpolymers TMA–ODA–VPK is shown in Scheme I. The rate of decrease in molecular mass and the final molecular mass depend on the VPK content. The greater the VPK content, the higher the rate and the

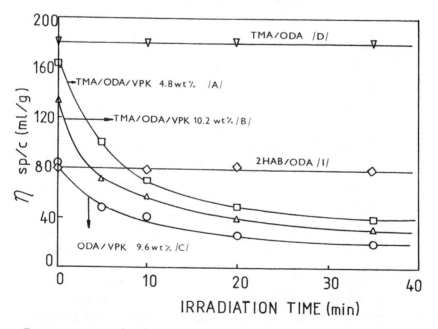

Figure 1. Course of molecular mass changes of copolymers and terpolymers during photolysis in benzene solution: □, A; △, B; ○, C; ▽, D; and ◇, I.

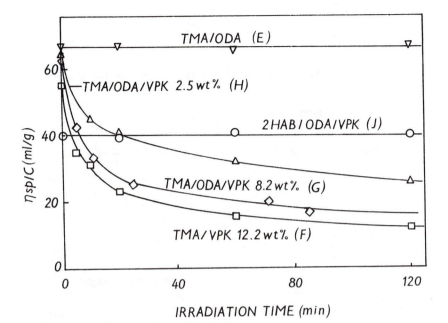

Figure 2. Course of molecular mass changes of copolymers and terpolymers during photolysis in benzene solution: ▽, E; □, F; ◇, G; △, H; and ○, J.

smaller the final molecular mass (compare the course of changes for polymers H and G in Figure 2).

The molecular mass of terpolymer J did not change despite the presence of the VPK light-sensitive unit. The explanation for this might be in the difference between the extinction coefficients for 2HAB and VPK. The 2-hydroxybenzophenone units in 2HAB that are around the VPK molecule can absorb all the light because their extinction coefficient, ε, is 100 times larger [ε_{2HAB} = 8700 L/(mol × cm) and ε_{VPK} 83 L/(mol × cm) at 330 nm]. Another possible mechanism is energy transfer from the excited triplet state of VPK (340 kJ/mol) to the 2HAB structure, which has a lower triplet energy (260 kJ/mol; references 10 and 11). On the other hand, degradation may occur at a correct ratio of VPK and 2HAB. The preparation conditions for a terpolymer of this type were not optimized.

Photostabilizing Efficiency in PP.   We investigated the action of the light-sensitive VPK unit in the copolymer without the light-stabilizing unit (copolymer C). This copolymer acts neither as an inhibitor nor as a sensitizer, and the course of the photooxidation is the same as for unstabilized PP film (Figure 3). This result means that the absorption of light by phenyl ketone groups is not sufficient to act as a UV screener or absorber. On the other

STATISTICAL TERPOLYMER

-TMA-ODA-VPK-TMA-ODA-TMA-ODA-ODA-TMA-VPK-ODA-

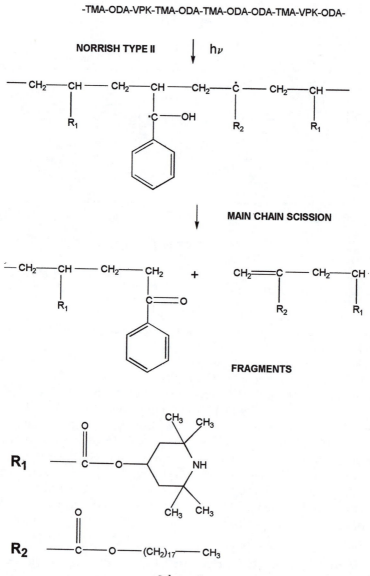

Scheme I.

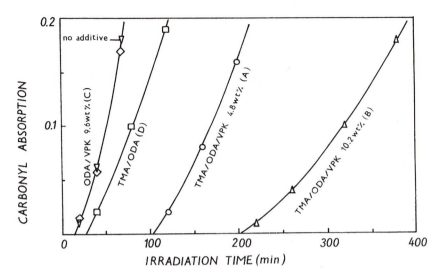

Figure 3. Rates of photooxidation of PP films (ca. 0.2 mm) containing additives (0.2 wt %): ▽, no additive; ◇ C; □, D; ○, A; and △, B.

hand the photolytic process and its products do not sensitize or inhibit photooxidation of PP under these irradiation conditions.

Among the terpolymers with higher molecular mass, A, B, and D, the highest efficiency was obtained with terpolymer B followed by terpolymer A wt% (Figure 3). The minimum efficiency was exhibited by copolymer D. The molecular mass of D did not change during irradiation in benzene solution (Figure 1). Its molecular mass is not expected to change in PP film, either. We assume that such a large molecule is not able to migrate at all, and because it is completely immobile, it remains in the same position at all times. According to Figures 1 and 3, the stabilizing efficiencies for A, B, and D increase with decreasing molecular mass and, what is more essential, with increasing VPK content. The importance of VPK content is clear from Figure 4, where the efficiencies of copolymers G and H, which have the same molecular mass but different VPK contents, and copolymer E are compared.

The increasing of VPK content from 0 to 2.5 to 8.2 wt % resulted in an increase in the stabilizing efficiency from 150 to 350 to 700 h, respectively, to reach carbonyl absorption (A_{co} = 0.2). These differences suggest a possible role of stabilizer migration. The initial distribution of the stabilizers in PP is supposed to be similar, because molecular masses are almost the same for all three additives. Copolymer E without VPK units retains its original molecular mass which is too high for effective action; its stabilizing efficiency is the smallest (4). Both terpolymers react with light by Norrish type II reactions and the result is main-chain scission. Fragments from terpolymer G, which has a higher VPK content, have smaller molecular mass than fragments of

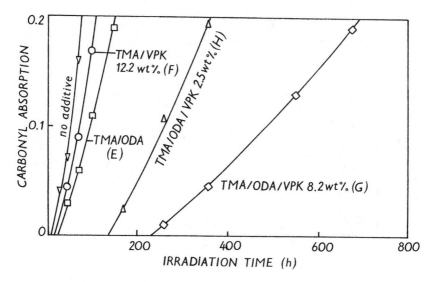

Figure 4. Rates of photooxidation of PP films (ca. 0.2 mm) containing additives (0.2 wt %): ▽, no additive; ○, F; □, E; △, H; and ◇, G.

terpolymer H, which has a lower VPK content. Consequently, the highest efficiency is exhibited by terpolymer G and is due to the highest migration ability of its smaller fragments.

Having the smallest fragment size itself is not a sufficient condition for effective action. The chemical composition of fragments with respect to their physical properties is also very essential. Copolymer F, which contains the stabilizing unit in an even higher content but does not contain the ODA unit with the long alkyl chain, exhibited the highest rate of splitting reaction in benzene (Figure 2). It also has fragments of the smallest size. Regardless, its stabilizing efficiency is extremely low (Figure 4), and the course of carbonyl product accumulation is almost the same as for unstabilized PP. The absence of ODA units increases polarity and decreases compatibility, which results in the too-polar fragments not being able to migrate in nonpolar PP to the places where photooxidation is occurring.

Quite a different situation is found in the case of terpolymers containing 2HAB (2HAB–ODA–VPK) instead of TMA (TMA–ODA–VPK). The stabilizing efficiency of terpolymers J and K is extremely low, even a bit lower than the efficiency of copolymer I (Figure 5). No interdependence exists between efficiency and VPK content, as is the case of TMA–ODA–VPK terpolymers. The reason ensues from the different behavior of 2HAB-containing terpolymers upon photolysis in benzene solution. Contrary to the terpolymers TMA–ODA–VPK, the molecular masses of terpolymers 2HAB–ODA–VPK do not change.

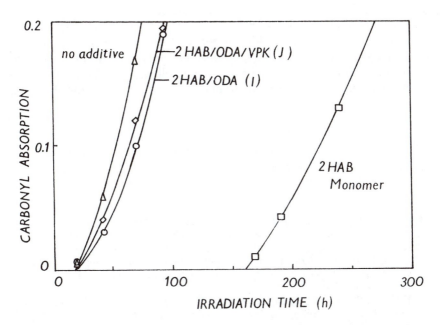

Figure 5. Rates of photooxidation of PP films (ca. 0.2 mm) containing the following additives: △ *no additive;* ◇, *J (0.2 wt %);* ○, *I (0.2 wt %); and* □, *2HAB monomer (0.1 wt %).*

Monomer 2HAB exhibited a much higher stabilizing efficiency at the same concentration of absorbing structural units. The molecular mass of terpolymer J ($M_n = 60{,}700$) determined by gel permeation chromatography using tetrahydrofuran as the mobile phase is very high in comparison with that of monomer 2HAB ($M_w = 312$). The greater dispersion of monomeric 2HAB units in PP acts as a much better UV absorber in comparison with the localized concentration in terpolymer J. Similarly, in the terpolymer case, many absorbing units are concentrated in small areas, whereas the rest of PP volume is unprotected.

Several authors (*12, 13*) concluded that the efficiency of 2-hydroxybenzophenone-type stabilizers is based not only on screening effect but also on quenching of the excited states and radical scavenging processes. Because the terpolymers 2HAB–ODA–VPK (J and K) and copolymer 2HAD–ODA (I) have rather high molecular masses, they could not be the effective quenchers. The same reasoning is valid for explaining the low effectiveness of radical scavenging by these types of additives. The mobility of the monomer 2HAB ensures the inhibition of photooxidation during the induction period up to the consumption of about 80% of the monomer (decrease of absorption at 330 nm measured by UV spectroscopy). Then, the oxygen-containing products begin to be formed. For terpolymers J and K and for copolymer I, photooxidation starts at about a 10% decrease in absorption at 330 nm. These facts indicate that the polymeric additives act in the same manner as the low-molecular weight ones, but their low mobility hinders sufficient PP protection. As a result, the induction period is very short, although the polymeric stabilizers still contain high concentrations of the original, virgin, active, absorbing units.

References

1. Allen, N. S. *Chem. Soc. Rev.* **1986,** *15,* 373.
2. Chirinos Padron, A. J. *J. Macromol. Sci. Rev. Macromol. Chem. Phys.* **1990,** *C30,* 107.
3. Gugumus, F. *Res. Discl.* **1981,** *209,* 357.
4. Hrdlovič, P.; Chmela, Š. *J. Polym. Mater.* **1990,** *13,* 249.
5. Chmela, Š.; Hrdlovič, P.; Maňásek, Z. *Polym. Degrad. Stab.* **1985,** *11,* 233.
6. Karvas, M.; Jexova, E.; Holcik, J.; Balogh, A. *Chem. Prum.* **1968,** *18,* 427.
7. Watrt, J. *Diploma Thesis;* Polymer Institute: Bratislava, Slovak Republic, 1983.
8. Guillet, J. E. *Polymer Photophysics and Photochemistry;* Cambridge University Press: 1985; p 261.
9. Hrdlovič, P.; Lukáč, I. In *Developments in Polymer Degradation;* Grassie, N., Ed.; Applied Science Publications: London, 1982; Vol. 4, p 101.
10. Klopffer, J. *J. Polym. Sci., Polym. Symp.* **1976,** *57,* 205.
11. Lukáč, I.; Hrdlovič, P. *Eur. Polym. J.* **1979,** *15,* 533.
12. Carlsson, D. J.; Suprunchuk, T.; Wiles, D. M. *J. Appl. Polym. Sci.* **1972,** *16,* 615.
13. Wink, P.; Van Ween, T. H. *J. Eur. Polym. J.* **1978,** *14,* 533.

RECEIVED for review May 14, 1994. ACCEPTED revised manuscript May 5, 1995.

Weathering of Acrylonitrile–Butadiene–Styrene Plastics: Compositional Effects on Impact and Color

D. M. Kulich and S. K. Gaggar

GE Plastics, Technology Center, Washington, WV 26181

The effects of outdoor aging on surface embrittlement and color changes were determined by using model ABS (acrylonitrile–butadiene–styrene) polymers systematically varying in rubber level, grafting, and SAN (styrene–acrylonitrile) copolymer composition. A high-speed puncture test was used to provide greater sensitivity than that typically obtained with routine pendulum and falling-dart type measurements. Color readings were also obtained to identify the color changes occurring in the time period to surface embrittlement. Elastomer type and SAN phase composition were shown to be key factors affecting time to surface embrittlement, whereas grafting and rubber level had much less or no significant effect. The primary color changes occurring in the time period to surface embrittlement were bleaching and fading. The effects of ABS composition on yellowing with and without added TiO_2 are also described.

ACRYLONITRILE-BUTADIENE-STYRENE (ABS) polymers comprise an important class of multiphase polymer blends that contain a discrete, particulate elastomeric phase dispersed in a thermoplastic matrix. In ABS the blend components consist of polybutadiene (BR) or butadiene copolymer particles grafted with a copolymer of styrene and acrylonitrile and dispersed in a matrix of styrene and acrylonitrile copolymer (SAN). ABS provides a favorable balance of properties that includes high impact strength, ease of processing, and good dimensional stability. Various grades of ABS are offered to meet specific

0065–2393/96/0249–0483$12.00/0

end-user requirements by adjusting the relative proportions of rubber, grafting, and matrix composition (1).

ABS is susceptible to photodegradation resulting in discoloration and loss of toughness. Applications involving outdoor exposure require protective measures such as the use of light stabilizers, pigments, or protective coatings (2–9). Light aging of ABS results in degradation of the BR phase and a corresponding embrittlement of the surface layer (2, 10–19) by oxidative cross-linking of the rubber particles and graft scission (15). Mechanisms describing the photooxidation of ABS were proposed (20–25). Prior thermal processing can introduce polymer hydroperoxides that can act as catalysts for photodegradation (10). Oxidation studies with singlet oxygen have also shown that initial attack on ABS involves oxidation of the BR phase (23). Studies using laminates of brittle polymer films on ABS to simulate environmental surface embrittlement demonstrated that at a critical brittle layer thickness a surface crack is able to propagate across the layer-core interface (26). Weathering studies were conducted by using both outdoor (11–13, 15–17, 24, 27–30) and artificial aging methods (2, 11, 15, 17, 21, 22, 28, 29); the effects of wavelength were recognized (11, 20–22, 25, 31, 32). Photooxidative degradation of the BR phase (vis-à-vis IR degradation) is primarily initiated by wavelengths below 350 nm (24, 25). Wavelength sensitivity studies have shown that photochemical yellowing is caused by wavelengths between 300 and 380 nm, and maximum bleaching of yellow-colored species occurs at wavelengths in the 475–485 nm region (31).

Test methods previously used to determine the effects of light aging on embrittlement of ABS include Izod impact (13, 29, 33–35), Charpy impact (3, 14, 29), flexural tests (3, 14, 27, 28), falling weight (15, 19), dynastat (24), and dynamic mechanical measurements (Rheovibron) (19). Because photodegradation occurs essentially only on the exposed surface and the interior of the sample remains essentially unaffected, a routine pendulum type of notched impact test will not be sensitive to changes in surface embrittlement. Falling dart types of testing, although more sensitive to the surface, still do not have adequate sensitivity, and a large number of samples is required.

In this study, a high-speed puncture test was used to determine the effects of outdoor exposure on crack-initiation energy values by using model rubber-modified SAN copolymers systematically varying in structure. Structural variations included rubber type, rubber level, grafting level, and SAN composition. The samples studied were not light stabilized, and comparisons were made by using unpigmented and pigmented (4.0 pph TiO$_2$) samples. Correlations are also drawn with associated weather-induced color changes. Obtaining color measurements on samples well characterized for surface embrittlement permits an identification of the types of color changes occurring in the time frame during which surface embrittlement occurs. Although discoloration phenomenon were described previously (15, 21, 31), many of the observations may relate to exposures well beyond the point required for surface embrittlement or were determined using artificial light sources raising concerns of relevance to natural weathering conditions.

Experimental

Laboratory prepared ABS samples were used to provide better control over compositional variations. Elastomers grafted with SAN were prepared in an emulsion process (*1*) and compounded using a BR (Banbury) with SANs having various styrene-to-acrylonitrile ratios or with a SAN at different rubber levels. Pigmented samples contained 4 pph rutile TiO_2. After compounding, samples were prepared for light aging by injection molding into $75 \times 126 \times 3.2$-mm plaques by using an 8-oz injection molding machine at 260 °C stock temperature.

Samples for outdoor aging were mounted on south-facing racks inclined at 45° to the horizon located at a site adjacent to the Technology Center in Washington, WV; sample exposures were in the month of July.

Color readings were on a Macbeth 1500+, specular component included, illuminant D65, 10° observer. Color changes were recorded [1976 Commission International de l'Eclairage (CIE) L*a*b*] as L* (lightness-darkness axis in color space with lower L* indicating a darker sample), a* (red-green axis with a larger value indicating a more red sample), and b* (yellow-blue axis with a larger value indicating a more yellow sample) values by using the Macbeth spectrophotometer.

Puncture impact testing on aged and unaged (control) samples was performed on a Material Test System high-rate tester at a crosshead speed of about 0.21 m/s (500 in./min.). The test specimen was positioned such that the punch with a 0.635-cm (0.25-in.) diameter hemispherical tip impacted the sample on the unexposed surface and caused tensile failure of the aged surface (Figure 1). The punch penetration depth in the specimen was controlled to determine the crack initiation energy values by recording the load–displacement curve on a Nicolet oscilloscope. The energy values (divide in lb/in. by 22.4 to get J/cm) were calculated by measuring the area under the load–displacement curve and normalizing for the sample thickness. At each punch-penetration depth, the specimen surface was visually

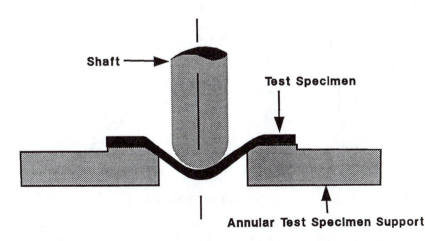

Figure 1. Puncture impact apparatus.

examined for the appearance of any surface cracks, and the punch depth was adjusted accordingly for the next impact on an untested portion of the specimen surface until the minimum depth for crack initiation was identified. A maximum of four impact points on each test specimen was used to establish the crack initiation energy value. The total fracture energy values were also established for the samples by driving the punch clear through the specimens and recording the load–displacement curves corresponding to total fracture. A typical load–displacement curve to total fracture is shown in Figure 2 for unaged and aged samples. The area under the entire load-displacement curve normalized for the specimen thickness is defined as the total fracture energy. Initial energy values can be reproduced within 5% of reported averages. Measurement accuracy will decrease as the measured values decrease due to aging; as energy values approach 50 units, differentiation becomes difficult.

Results and Discussion

Figure 3 shows the effect of exposure time on crack initiation energy values for rubber modified SAN copolymers differing in elastomer type as follows: polybutadiene in ABS, EPDM (ethylene–propylene–diene monomer) rubber in AES (acrylonitrile–EPDM–styrene), and polybutylacrylate rubber in

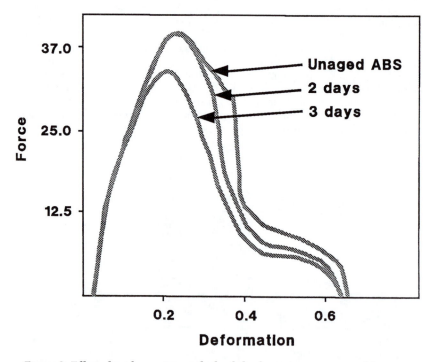

Figure 2. Effect of outdoor aging on the load-displacement curve to total fracture. The total fracture energy value (lb-in./in.) is determined from the area under the curve normalized for sample thickness.

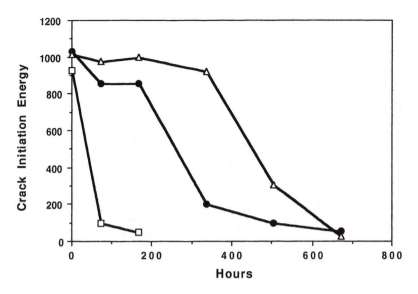

Figure 3. Effect of outdoor exposure time (hours) on crack initiation energy value (lb-in./in.) for unpigmented samples varying in elastomer type: □, ABS; ●, AES; △, ASA.

ASA (acrylate–styrene–acrylonitrile). In each case, the sample was prepared by using the same SAN as the continuous phase. Note that embrittlement occurred in all samples, and stability increased in the order: ABS<AES<ASA. After 400 h, all samples showed a decrease to very low impact values. The reduced stability of ABS is consistent with the increased lability of polybutadiene toward oxidation. Rather interestingly, in the absence of light stabilizers, even AES and ASA showed relatively rapid surface embrittlement ensuing after only 200 h of outdoor exposure for AES and 400 h for ASA.

The bulk-failure mode (the energy required to punch through the sample) is shown in Figure 4. Comparison of the percent energy change in Figure 3 with Figure 4 illustrates the greater sensitivity obtained through crack-initiation energy value measurements. The results in Figure 4 indicate that impact was reduced to almost a constant value and that a constant brittle layer thickness was formed due to light screening and a reduction in oxygen permeability by the oxidized layer. Because brittle failure does not occur in the bulk-failure mode, the brittle layer formed is not sufficiently thick for a crack to propagate through the brittle layer–core interface. Previous studies on the thermal oxidation of ABS indicated that the critical layer thickness at which surface cracks will propagate through the sample is between 0.07 and 0.2 mm for impact tests at ambient temperatures (35). The critical layer thickness will vary with the impact test temperature and test rate.

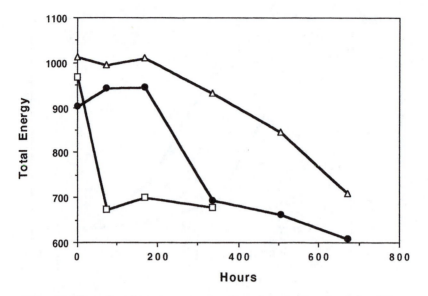

Figure 4. Effect of outdoor exposure time (hours) on total energy value (lb-in./ in.) for unpigmented samples varying in elastomer type: □, *ABS;* ●, *AES;* △, *ASA.*

The effect of pigmentation with TiO₂ on impact retention is shown in Figure 5. Note that the addition of TiO₂ has a stabilizing effect, but the same relative ranking of elastomer type is obtained. The protective effect by pigmentation is consistent with the ability of rutile TiO₂ to strongly absorb UV below 400 nm (36).

Figures 6 and 7 illustrate the effects of rubber level for pigmented and unpigmented ABS, respectively. Although initial energy values are significantly affected, complete brittle failure occurs after 72 h. Similarly, varying of the extent of grafting (Figures 8 and 9) primarily influences initial toughness.

In sharp contrast to the effects of grafting and rubber level, SAN composition does not significantly affect initial energy values but does significantly affect time to embrittlement. The effect was consistently seen both in the unpigmented and pigmented samples (Figures 10 and 11). We proposed that varying the percent acrylonitrile in SAN affects oxygen permeability and thereby affects time to embrittlement. The effect of SAN composition on oxygen permeability in SAN is shown in Figure 12 (37). The addition of TiO₂ is not expected to have a significant effect on permeability. The effect on permeability due to simply an increase in path length with added filler can be calculated from an expression by Nielson (38) and is estimated to be <2% at 4 pph TiO₂ (which is less than 1% filler by volume due to the high density of TiO₂). Although the effect of a filler on permeability can be increased further by ordering of the polymer on the filler surface (39, 40), the large screening

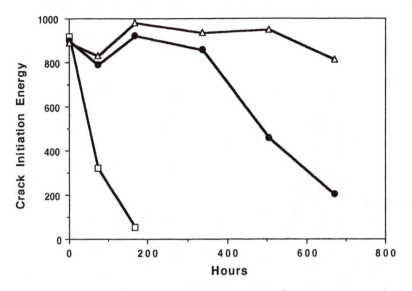

Figure 5. Effect of outdoor exposure time (hours) on crack-initiation energy value (lb-in./in.) for samples pigmented with TiO₂ and varying in elastomer type: □, *ABS;* ●, *AES;* △, *ASA.*

effect of TiO_2 is expected to be the dominant factor accounting for impact retention with TiO_2.

Changes in color during outdoor exposure were also determined. Figure 13 shows the effect of elastomer type on changes in yellowing (b* value) for unpigmented ABS; an increase in b* value corresponds to an increase in sample yellowness. Changes in L* and a* values are also recorded in Figures 14 and 15, respectively. For each composition, the samples undergo rapid initial fading (within 24 h) followed by subsequent yellowing. Initial fading (reduction in b* value) occurred irrespective of elastomer type and suggests that yellow chromophores being photooxidized are in the SAN phase. After 24 h, the ABS samples show greater yellowing (increased b* value) and darkening (decreased L* value) than AES or ASA. Changes in the red-green coordinate are relatively small (<0.5 units). We concluded that photooxidation of the rubber phase contributes to yellowing and darkening for the exposures studied, but the effect is relatively small. Jouan and Gardette (21) proposed that yellowing is derived from oxidation of the SAN phase induced by photooxidation of BR. Because no large differences in yellowing were observed between unpigmented ABS as compared with AES and ASA, which have more photolytically stable elastomeric phases, we concluded (contrary to Jouan and Gardette) that photooxidation of BR is not a significant sensitizer in unpigmented ABS under the conditions of this study. Because the contribution of the rubber phase to overall discoloration in unpigmented ABS is small, in-

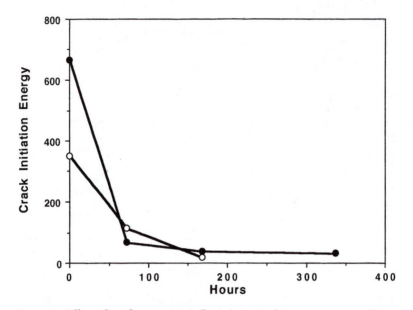

Figure 6. Effect of outdoor exposure (hours) on crack-initiation energy (lb-in./ in.) value for ABS samples pigmented with TiO2 at constant grafting but differing in rubber level: ○, *15% BR;* ●, *25% BR.*

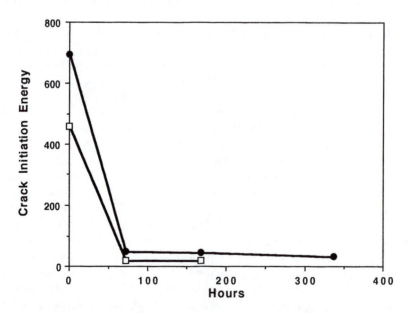

Figure 7. Effect of outdoor exposure (hours) on crack-initiation energy (lb-in./ in.) value for unpigmented ABS samples at constant grafting but differing in rubber level: □, *15% BR;* ●, *25% BR.*

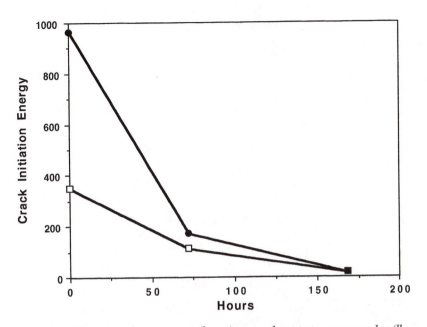

Figure 8. Effect of outdoor exposure (hours) on crack-initiation energy value (lb-in./in.) for ABS samples pigmented with TiO$_2$ at constant rubber but differing in graft/rubber ratio: □, graft/rubber = 0.43; ●, graft/rubber = 0.50.

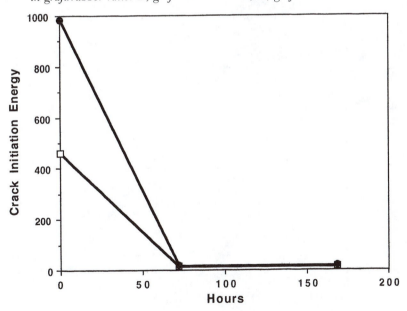

Figure 9. Effect of outdoor exposure (hours) on crack-initiation energy value (lb-in./in.) for unpigmented ABS samples at constant rubber but differing in graft/rubber ratio: □, graft/rubber = 0.43; ●, graft/rubber = 0.50.

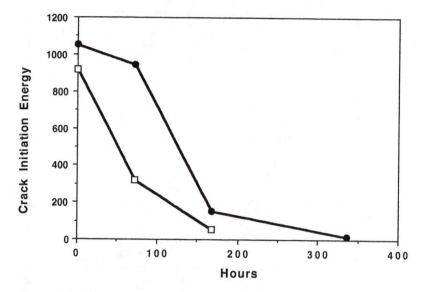

Figure 10. Effect of outdoor exposure (hours) on crack-initiation energy value (lb-in./in.) for ABS samples pigmented with TiO₂ and differing in SAN composition: □, *28% (w/w) acrylonitrile;* ●, *36% (w/w) acrylonitrile.*

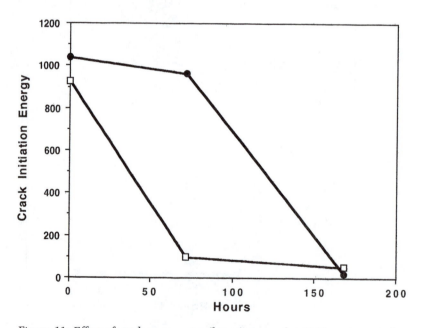

Figure 11. Effect of outdoor exposure (hours) on crack-initiation energy value (lb-in./in.) for unpigmented ABS samples differing in SAN composition: □, *28% (w/w) acrylonitrile;* ●, *36% (w/w) acrylonitrile.*

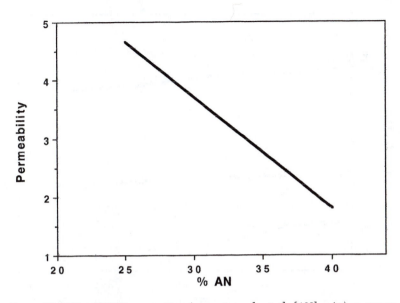

Figure 12. Effect of SAN composition (percent acrylonitrile [AN], w/w) on oxygen permeability $(cm^3 \cdot cm)/(cm^2 \cdot s \cdot cm\ Hg) \times 10^{11}$ at 25 °C. Data from reference 37.

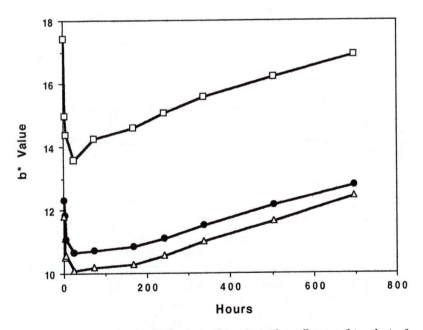

Figure 13. Effect of outdoor exposure (hours) on the yellowing (b* value) of unpigmented samples differing in elastomer type: □, ABS; ●, AES; △, ASA.

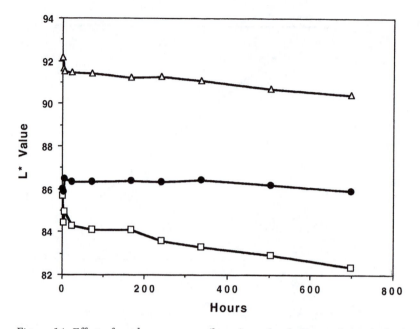

Figure 14. Effect of outdoor exposure (hours) on the darkening (L* value) of unpigmented samples differing in elastomer type: □, ABS; ●, AES; △, ASA.

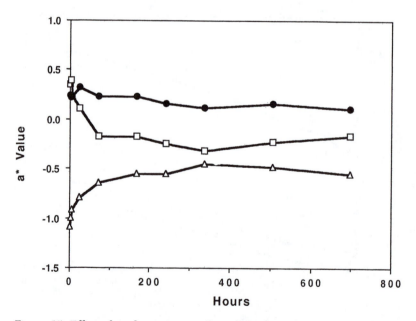

Figure 15. Effect of outdoor exposure (hours) on the red-green color shift (a* value) of unpigmented samples differing in elastomer type: □, ΛBS; ●, AES; △, ASA.

creasing BR stability by increasing acrylonitrile content (lower permeability) should have a small effect on overall discoloration. This result was observed (Figure 16).

A comparison of Figure 3 with Figure 13 shows that impact loss and initial bleaching and fading were independent phenomenon; the time scale for bleaching and fading was relatively unaffected by rubber type and stability. Yellowing that ensued subsequent to initial bleaching and fading does appear to correlate with the onset of impact loss. However, surface embrittlement precedes the point at which yellowness surpasses initial unaged values.

A comparison of the weathering-induced discoloration of the components of ABS (Figure 17) confirms that the SAN phase alone significantly fades and the BR phase alone yellows. As determined by Fourier transform IR (FTIR), BR films underwent photooxidation within the time frame studied; hence, some discoloration is not unexpected. Because the percent BR in the ABS samples was 25%, the contribution of BR in ABS to yellowing will be proportionately reduced. The ABS sample after 72 h showed evidence of $-OH$ formation at 3300 cm^{-1}, carbonyl formation at 1640–1720 cm^{-1}, ether formation at 1050–1200 cm^{-1}, and loss of *trans*-butadiene absorption at 960 cm^{-1}. A determination of the subambient glass-transition temperature via differential scanning calorimetry analysis of an ABS sample aged 400 h shows complete loss of the polybutadiene transition to a depth of at least 2 mils. Gel Permeation chromatography (GPC) area data also indicate the presence of more of the soluble polymer fraction after aging (more ungrafted SAN) and suggest degrafting may occur. Conversely, FTIR analysis indicated no change in the composition of the SAN sample after 240 h of exposure. For SAN alone, GPC analysis showed no change in molecular weight after 400 h, even in the top 1-mil layer.

With TiO$_2$ pigmented samples, rapid initial fading (decrease in b* value) was observed with all samples as with unpigmented samples, and the magnitude of fading showed little dependence on elastomer type (Figure 18). Pigmented SAN alone similarly underwent fading (Figure 19). The sources of yellow chromophores in SAN include conjugated species derived from sequences of adjacent acrylonitrile units as described by Grassie (*41*). More fading was observed with SAN versus polystyrene (PS), and this result is consistent with greater initial yellowness of SAN versus PS (PS, unaged b* value = 3.0; SAN, unaged b* value = 4.7).

However, subsequent discoloration of pigmented ABS was considerably more pronounced relative to AES and ASA. The result is unexpected based on the screening of wavelengths <400 nm by TiO$_2$ (*36*) and the expectation that wavelengths of 300–380 nm primarily contribute to yellowing (*31*). The role of the BR phase in yellowing was confirmed by demonstrating that adding increasing amounts of BR (not grafted with SAN) to SAN resulted in increased yellowing (Figure 20). Because TiO$_2$ enhances (rubber) stability as demonstrated by impact measurements, we propose that color formation is a side

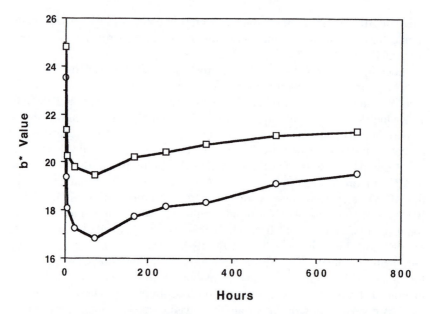

Figure 16. *Effect of outdoor exposure (hours) on the yellowing (b* value) of unpigmented ABS samples differing in SAN composition:* ○, *28% (w/w) acrylonitrile;* □, *36% (w/w) acrylonitrile.*

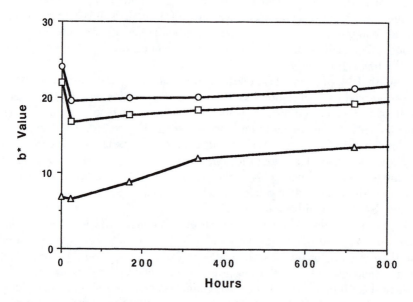

Figure 17. *Effect of outdoor exposure (hours) on the yellowing (b* value) of unpigmented ABS components:* ○, *SAN;* □, *ABS;* △, *BR.*

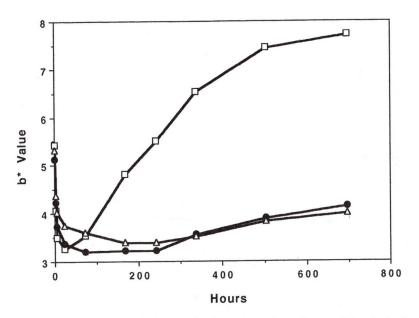

Figure 18. Effect of outdoor exposure (hours) on the yellowing (b* value) of samples pigmented with TiO$_2$ and varying in elastomer type: □, *ABS;* ●, *AES;* △, *ASA.*

reaction most probably involving interactions of TiO$_2$ with oxidized fragments from the rubber phase. This interpretation is consistent with the expected presence of TiO$_2$ in the SAN phase, and this presence makes direct rubber–TiO$_2$ interactions less likely. The absence of an increase in yellowing with SAN with added TiO$_2$ demonstrates that the increase in yellowing observed in ABS with added TiO$_2$ is not simply an optical effect due to the pigment obscuring bleached and faded material in underlying layers.

Because photooxidation of the BR phase contributes more significantly to yellowing in the presence of TiO$_2$, reduced yellowing would be expected with SAN having a higher percentage acrylonitrile (*see* previous discussion on acrylonitrile effect on impact retention in weathering). This result was observed (Figure 21) and is consistent with previous impact measurements. The effect of SAN composition on color stability is also consistent with a previous observation by Donald and co-workers (6). Again, the effect on ABS stability may be accounted for by reduced oxygen permeability with increasing acrylonitrile content. A comparison by Jouan and Gardette (21) using a long wavelength (>300 nm) light source of a spaced, multilayered sample with samples differing in thickness indicates discoloration to occur more rapidly if oxygen accessibility is permitted between each layer. This comparison supports the significance of limited oxygen permeability.

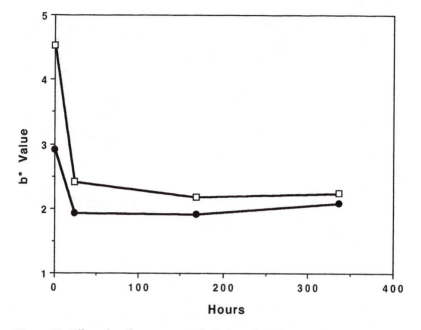

Figure 19. Effect of outdoor exposure (hours) on the yellowing (b* value) of SAN and PS each pigmented with TiO₂: $\square$, *SAN;* $\bullet$, *PS.*

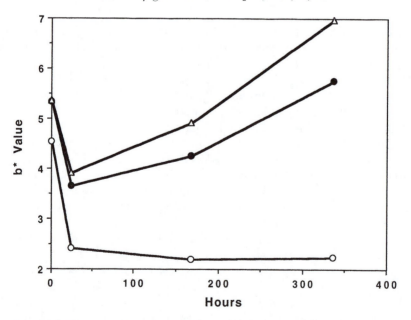

Figure 20. Effect of outdoor exposure (hours) on the yellowing (b* value) of SAN pigmented with TiO₂ and with added BR rubber: $\bigcirc$, *0% BR;* $\bullet$, *15% BR;* $\triangle$, 25% BR.*

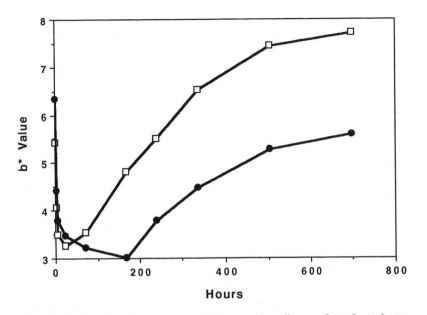

Figure 21. Effect of outdoor exposure (hours) on the yellowing (b* value) of ABS samples pigmented with TiO₂ and differing in SAN composition: □, 28% (w/w) acrylonitrile; ●, 36% (w/w) acrylonitrile.

Summary

The effects of compositional variables on impact and color of ABS exposed outdoors with no added light stabilizers were determined. High-speed puncture impact was shown to be a sensitive technique to detect surface embrittlement in the early stages of light aging. The effect of elastomer type on increasing time to embrittlement was ABS<AES<ASA. Even AES and ASA developed surface embrittlement within 500 h of outdoor exposure. Elastomer type and SAN phase composition were shown to be two key factors affecting both color and impact retention. Oxygen permeability may play a key role. Color changes involved initial fading followed by subsequent yellowing. A comparison of SAN with ABS indicated that the SAN phase is a primary contributor to initial fading and the polybutadiene phase plays a role in subsequent yellowing. FTIR analyses corroborated severe oxidation of the polybutadiene component as the cause of impact loss. In the absence of TiO₂, the contribution of the rubber phase to yellowing was minor, but a significant effect was seen in the presence of TiO₂. The effect of rubber in the presence of TiO₂ was also confirmed by demonstrating that increased levels of polybutadiene (ungrafted polybutadiene rubber added to SAN) resulted in increased levels of yellowing. The comparative studies with and without pigment illustrate the significant effects that colorants can have on oxidation phenom-

enon and the caution to be exercised in extending conclusions based on un-pigmented systems to commercial applications typically having added colorants.

Acknowledgment

We thank the staff of GE Plastics for their technical support.

References

1. Kulich, D. M.; Pace, J. E.; Fritch, L. W., Jr.; Brisimitzakis, A. In *Kirk-Othmer Encyclopedia of Chemical Technology*; 4th ed.; Kroschwitz, J., Ed.; Wiley: New York, 1991; Vol. 1, pp 391-411.
2. Kelleher, P. G.; Boyle, D. J.; Miner, R. J. *Mod. Plast.* **1969**, *46*, 188.
3. Gugumus, F. In *Developments in Polymer Stabilization-1;* Scott, G., Ed.; Applied Science: London, 1979; Chapter 8.
4. Kolawole, E. G.; Adeniyi, J. B. *Eur. Polym. J.* **1982**, *18*, 469.
5. Adenyi, J. B. *J. Polym. Mater.* **1986**, *3*, 25.
6. Donald, R. J.; Landes, S. K.; Maecker, N. L. *SAE Techn. Pap. Ser.* **1987**, 2.715.
7. Kurumada, T.; Ohsawa, H.; Yamazaki, T. *Polym. Degrad. Stab.* **1987**, *19*, 263.
8. Bullet, T. R.; Mathews, P. R. *Plast. Polym.* **1971**, *39*, 200.
9. Loelinger, H.; Gilg, B. *Die Angew. Makromol. Chem.* **1985**, *137*, 163.
10. Adeniyi, J. B.; Kolawole, E. G. *Eur. Polym. J.* **1984**, *20*, 43.
11. Davis, A.; Gordon, D. *J. Appl. Polym. Sci.* **1974**, *18*, 1159.
12. Gesner, B. D. *SPE J.* **1969**, *25*, 73.
13. Bair, H. E.; Boyle, D. J.; Kelleher, P.G. *Polym. Eng. Sci.* **1980**, *20*, 995.
14. Bucknall, C. B. In *Weathering and Degradation;* Gordon and Breach: New York, 1966; Chapter 6.
15. Scott, G.; Tahan, M. *Eur. Polym. J.* **1977**, *13*, 981.
16. Hirai, T. *Jpn. Plast.* **1970**, *4*, 22.
17. Gesner, B. D. *J. Appl. Polym. Sci.* **1965**, *9*, 3701.
18. Gilg, B.; Muller, H.; Schwarzenbach, K. In *Advances in the Stabilization and Controlled Degradation of Polymers;* SUNY Conference: New Paltz, NY, 1982.
19. Ghaemy, M.; Scott, G. *Polym. Degrad. Stab.* **1981**, *3*, 233.
20. Adeniyi, J. B. *Eur. Polym. J.* **1984**, *20*, 291.
21. Jouan, X.; Gardette, J. L. *Polym. Degrad. Stab.* **1992**, *36*, 91.
22. Jouan, X.; Gardette, J. L. *J. Polym. Sci. Part A* **1991**, *29*, 685.
23. Kaplan, M. L.; Kelleher, P. G. *J. Polym. Sci. Part A* **1970**, *8*, 3163.
24. Shimada, J.; Kabuki, K.; Ando, M. *Rev. Electr. Comm. Lab.* **1972**, *20*, 553.
25. Shimada, J.; Kabuki, K. *J. Appl. Polym. Sci.* **1968**, *12*, 671.
26. So, P.; Broutman, L. *J. Polym. Eng. Sci.* **1982**, *22*, 888.
27. Kelleher, P. G.; Boyle, D. J.; Gesner, B. D. *J. Appl. Polym. Sci.* **1967**, *11*, 1731
28. Ruhnke, G. M.; Baritz, L. F. *Kunstoffen* **1972**, *62*, 250.
29. Casale, A.; Salvatore, O.; Pizzigoni, G. *Polym. Eng. Sci.* **1975**, *15*, 286.
30. Karaenev, S.; Nikolova, S.; Dobrewa, D.; Tschilingirian, M. *Die Angew. Makromol. Chem.* **1988**, *158/159*, 177.
31. Searle, N. D.; Maecker, M. L.; Crewdson, L. F. E. *J. Polym. Sci. Part A* **1989**, *27*, 1341.
32. Davis, A.; Gordon, D. *J. Appl. Polym. Sci.* **1974**, *18*, 1173.
33. Heaps, J. M. *Rubber Plast. Age* **1968**, *49*, 967.

34. Falk, J. C.; van Fleet, J. *Plast. Rubber Mater. Appl.* **1978,** *3(4),* 123.
35. Wolkowicz, M. D.; Gaggar, S. K. *Polym. Eng. Sci.* **1981,** *21,* 571.
36. Day, R. E. *Polym. Degrad. Stab.* **1990,** *29,* 73.
37. Salame, M. In *Transport Phenomenon Through Polymer Films;* Wiley: New York, 1973; pp 1-15.
38. Nielson, L. E. *J. Macromol. Sci. Chem.* **1967,** *A-1,* 929.
39. Kumins, C. A.; Roteman, J. *J. Polym. Sci. Part A* **1963,** *1,* 527.
40. Kwei, T. K. *J. Polym. Sci. Part A* **1965,** *3,* 3229.
41. Grassie, N. In *Developments in Polymer Degradation-1;* Grassie, N., Ed.; Applied Science: London, 1977; Chapter 5.

RECEIVED for review December 6, 1993. ACCEPTED revised manuscript January 5, 1995.

32

Thermal and Photooxidation of Miscible Polymer Blends

L. Stoeber, E. M. Pearce*, and T. K. Kwei

Herman F. Mark Polymer Research Institute and Department of Chemistry, Polytechnic University, Brooklyn, NY 11201

Thermal oxidation of miscible polymer blends at 80 °C, 110 °C, and 140 °C were studied and compared with the photooxidation at 30 °C. Pure polystyrene (PS)/ poly(vinyl methyl ether) (PVME) blends, modified PS/PVME, the addition of an antioxidant to PS/PVME blends, and other miscible polymer blend systems are discussed. The oxygen uptake by PS was negligible, whereas PVME oxidized rapidly. During thermal and photooxidation the induction period was lengthened by the presence of PS in the blend. The steady-state rate of oxidation of the blend was strongly influenced by the segmental mobility of the blend, and this mobility also governed the kinetics and morphology of phase separation. The molecular weight of PVME decreased more slowly as the PS content in the blend increased. It is believed that the reaction between PVME radicals and PS resulted in less reactive PS radicals, and this phenomenon retarded oxidation. In thermal oxidation of PS/PVME blends, phase separation occurred at the end of the induction period. The steady-state rate of oxidation was proportional to the PVME content in the blend. Similarities were found in other miscible polymer blend systems as in poly(methyl methacrylate)/poly(ethylene oxide) blends. The addition of an antioxidant to PS/PVME blends lowered the rate of oxidation and lengthened the induction period. Phase separation occurred at the end of the induction period and caused a redistribution of the antioxidant between the different phases.

T HE RAPIDLY DEVELOPING FIELD OF POLYMER BLENDING is becoming increasingly important as a means to modify and extend desirable polymer properties to meet commercial needs. Polymer blends like the "impact" polystyrenes (PS), which possess outstanding properties without the brittleness of common PS, have attracted a great deal of scientific interest. Most of the

*Corresponding author

0065–2393/96/0249–0503$16.25/0

polymer blending research to date has been focused on heterogeneous polymer blends, where one polymer is dispersed into the continuous phase of another polymer. The result is a heterogeneous blend in which the molecules are not intimately mixed on a molecular level. A fundamental aspect of this research is the question of miscibility and its control, because even in heterogeneous blends it is common practice to add a graft or block copolymer as a compatibilizer to modify interfacial characteristics. More than 500 miscible or partially miscible blends are already known and that number is increasing steadily (1–5). The molecular interaction in miscible blends is on a much more intimate scale than that of heterogeneous blends. The mixing of the two different molecules is done at the segmental level and the resulting blends exhibit single-phase behavior. This interaction results in the appearance of entirely new chemical properties, unlike those of the separate components. These new sets of properties are often of great interest and require scientific explanation.

Although homogeneous and heterogeneous polymer blends have become an important class of materials, our knowledge of their chemical stability is inadequate. There is information scattered in the literature about thermal degradation of heterogeneous polymer blends (6) and thermal or photooxidation of rubber-modified PS (7–9). But the oxidation of miscible polymer blends has received only occasional attention.

Thermal and photooxidation studies of miscible blends have been a major interest at the Herman F. Mark Polymer Research Institute over the last decade. Our investigation involved thermal and photo oxygen uptake measurements of polymer blends and the measurements of chemical changes and molecular weights of the component polymers. A number of miscible polymer blends, including PS/poly(vinyl methyl ether) (PVME) (10–13), low molecular weight polystyrene (LPS)/PVME (10, 11), 4-hydroxy-modified polystyrenes (M1PS, M2PS, and M7PS)/PVME (11), hexafluoroisopropanol-modified styrene–styrene copolymer (FPS)/PVME (11), poly(methyl methacrylate) (PMMA)/poly(ethylene oxide) (PEO) (14), poly(vinylidene fluoride) (PVDF)/poly(vinyl acetate) (PVAc) (15), poly(4-hydroxystyrene) (PHOST)/poly(vinyl pyrrolidone) (PVP) (16), or PEO (17), were studied.

Miscibility Enhancement Through Hydrogen Bonding

Hydrogen bonding between two dissimilar polymers can lead to miscibility or complexation. Miscibility enhancement is therefore often obtained through the blending of two polymers, one containing a hydrogen donor group and the other a hydrogen-acceptor group (18–20). The fact that the formation of hydrogen bonds can lead to miscibility raises the question of the minimum number of hydrogen bonds necessary to obtain miscibility (21) and the effect on the thermally induced phase separation. These cloud points (22) are believed to be sensitive to the magnitude of polymer–polymer interaction (23).

The prediction was borne out in the work by Pearce, Kwei, and co-workers (*24–26*). Copolymers were synthesized to contain small amounts of strong donors, such as a series of styrene copolymers containing varying amounts of *p*-(hexafluoro-2-hydroxyl isopropyl) styrene as the comonomer unit. Miscible blends containing different functional groups to increase miscibility have been used for thermal and photooxidation studies (Figure 1). The phase diagrams (Figure 2) of PS/PVME blends were studied extensively by many researchers (*27–29*).

For thermal oxidation, Park and co-workers (*11, 30*) used 1, 2, and 7 mole% *p*-hydroxy-modified PS (M1PS, M2PS, and M7PS, respectively) with PVME. As the degree of modification increases, the phase separation temperatures of the blends become higher. The M7PS/PVME blends do not show phase separation below 300 °C.

The incorporation of the even stronger hydrogen-bonding donor of 2.5 mole% *p*-(hexafluoro-2-hydroxylisopropyl) (HHIS) group into the styrene copolymer raised the phase separation temperature above 300 °C. These blends stay miscible even after PVME has undergone some oxidation. As shown in earlier work (*18*), the incorporation of only 0.1 mole% *p*-(hexafluoro-2-hydroxyl-isopropyl) group into the styrene copolymer raised the phase separation temperature minimum up to 160 °C; at 0.6 mole% HHIS, the temperature increased to 195 °C (Figure 3). The use of preoxidized PVME, however, lowered the phase separation temperature of the blends drastically (Figure 4).

Thermal Oxidation of Miscible Polymer Blends

Thermal and photooxidation of polymer blends are complicated processes involving multicomponent systems simultaneously undergoing various free-radical reactions. In heterogeneous blends the interactions between radicals take place at the interface, whereas in miscible blends the chances of interpolymer (radical–polymer and radical–radical) reactions are increased if the polymer does not undergo phase separation during oxidation.

Park et al. (*10*) showed in earlier studies of the thermal oxidation of miscible blends of PS/PVME that the induction period of the PVME oxidation was prolonged as the weight fraction of PS in the blend increased (Figures 5 and 6). However, the steady-state oxidation rates are almost proportional to the PVME contents in the blends.

The opaqueness of the blends after the induction period shows that all the blends have undergone phase separation after minor chemical changes during the induction period of oxidation. The steady-state region of oxidation is therefore almost proportional to the PVME content of the blend. The same results were found in all PS/PVME blends measured at 140 °C. These blends undergo phase separation before oxidation. The induction periods as well as the steady-state region of oxidation are proportional to the PVME content of

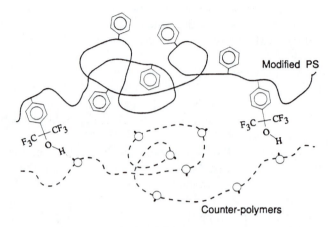

Figure 1. Schematic diagram of hydrogen bonding between modified polystyrene and counter polymers. (Reproduced with permission from reference 18. Copyright 1992 Elsevier Science Publishers.)

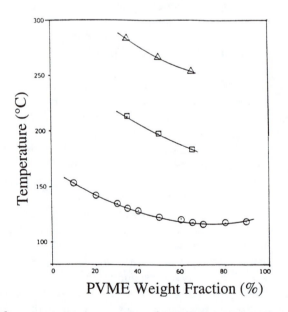

Figure 2. Phase separation temperatures of PS/PVME, M1PS/PVME and M2PS/PVME blends; (○) PS/PVME; (□) M1PS/PVME; (△) M2PS/PVME blends. (Reproduced with permission from reference 11. Copyright 1990 John Wiley & Sons, Inc.)

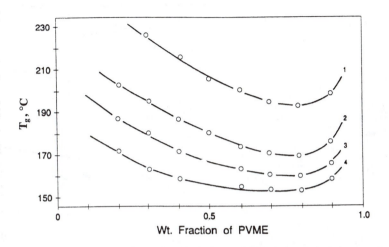

Figure 3. Cloud points of the various modified PSs and PVME systems at the heating rate of 2 °C/min: (1) 0.64 MPS, (2) 0.22 MPS, (3) 0.096 MPS, and (4) pure PS. (Reproduced with permission from reference 18. Copyright 1992 Elsevier Science Publishers.)

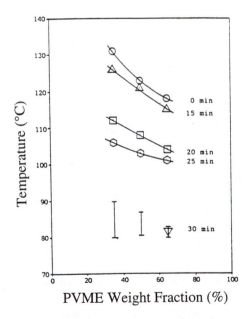

Figure 4. Cloud point curves of blends of PS with preoxidized PVME. (Reproduced from reference 10. Copyright 1990 American Chemical Society.)

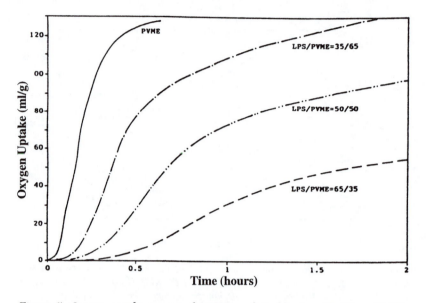

Figure 5. Oxygen uptake curves of PVME and its blends with PS at 80 °C. (Reproduced from reference 10. Copyright 1990 American Chemical Society.)

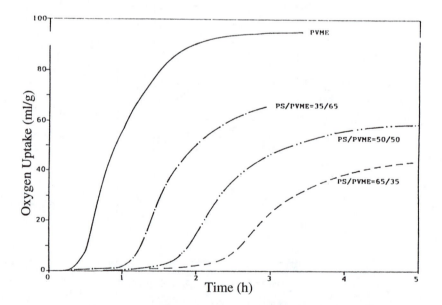

Figure 6. Oxygen uptake curves of PVME and its blends with PS at 110 °C. (Reproduced from reference 10. Copyright 1990 American Chemical Society.)

the blends (Figure 7). After the steady-state region the oxygen absorption decelerates and reaches a saturated region where the oxygen absorptions are also proportional to the PVME content in the blends.

In contrast to the PS/PVME blends, the LPS/PVME blends and the hexafluoroisopropanol modified styrene–styrene copolymer (FPS)/PVME blends stayed miscible even during oxidation (Figures 8 and 9; Table I). The induction periods of FPS/PVME and LPS/PVME blends were longer than those of the PS/PVME blends, and their oxidation rates were lower. The steady-state oxidation rates were not proportional to the PVME content of the blend because of the single-phase characteristic throughout the oxidation. The LPS/PVME blends had longer induction periods and lower oxidation rates than the FPS/PVME blends. The miscibility difference on the molecular level is believed to be responsible for that difference. The LPS polymer chain is believed to stay near the oxidized PVME chain, whereas long sequences of styrene units in FPS are thermodynamically unfavored to mix with the oxidized PVME. The strong hydrogen bonds are believed to keep the two polymers in a macroscopically single phase. The different polymer blends are likely to have, therefore, a different scale of homogeneity (Figure 10).

Photooxidation of Miscible Polymer Blends

The oxygen uptake of polymer blends during UV radiation provides major evidence for photooxidation. Although the mechanism of thermal- and photooxidation have great similarities, photooxidation was carried out at a lower temperature. The significantly reduced segment mobility at lower temperatures was expected to have an effect on the rate of reaction and the phase separation behavior. The departures from the results obtained in thermal oxidation provide important clues to the understanding of mechanism of the oxidation process in miscible polymer blends.

Chien and co-workers (*13, 31*) results of the oxygen uptake during UV-radiation of PS, PVME, and their blends are shown in Figure 11. PS absorbed less than 0.5 mL of oxygen/g in a period of 22 days, which could be regarded as negligible in the context of our investigation. On the other hand PVME consumed, after a slow beginning, approximately 8.7 mL/g/day after 3.1 days. The total consumption was about 30 mL/g in 6 days. The induction period increased with the PS content in a regular manner. The oxidation rate instead undergoes a tremendous decrease between 30 and 50% PS in the blend. The rates of the thermal oxidation of the same polymer blends are proportional to PVME content, and the departure between the thermal and photoprocesses provide important clues in the understanding of the mechanism involved (Figures 12 and 13).

The films, containing 70% by weight of PVME, turned slightly translucent after photooxidation for about 5 days. After 11 days those films showed two

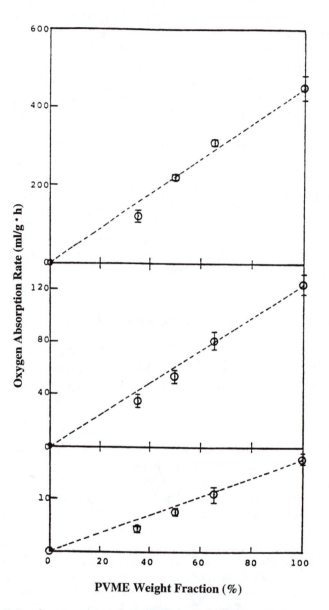

Figure 7. Steady-state oxygen absorption rates of PS/PVME blends: bottom, 80 °C; middle, 110 °C; top, 140°C. (Reproduced from reference 10. Copyright 1990 American Chemical Society.)

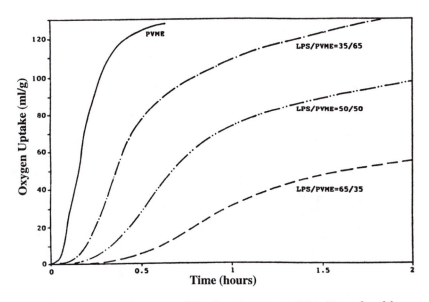

Figure 8. Oxygen uptake curves of blends with LPS at 140 °C. (Reproduced from reference 10. Copyright 1990 American Chemical Society.)

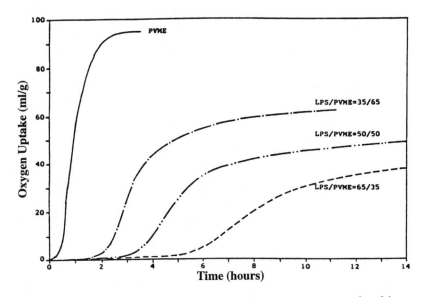

Figure 9. Oxygen uptake curves of blends with LPS at 110 °C. (Reproduced from reference 10. Copyright 1990 American Chemical Society.)

Table I. Induction Periods and Oxidation Rates of PVME and Its Blends

PS (wt%)	T (°C)	Induction Period (h)						Oxidation Rate (mL/(g·h))					
		PVME	PS	LPS	FPS	MIPS	M2PS	PVME	PS	LPS	FPS	MIPS	M2PS
0	110	0.50						124					
35	110		1.2	2.3	1.5	6.8	16.8		81.5	31.9	37.5	18.8	4.4
50	110		1.8	3.5	2.0	7.0	10.4		54.8	17.1	18.3	7.0	3.9
65	110		2.3	5.7	3.6	6.3	7.5		35.0	8.5	8.4	3.8	2.8
0	140	0.05						498					
35	140		0.05[a]	0.15	0.17	0.93	1.20		313	239	215	208	78
50	140		0.05[a]	0.28	0.23	1.23	1.53		237	130	108	56	27
65	140		0.05[a]	0.45	0.28	1.45	1.72		132	58	45	20	10

NOTE: Values for induction period and oxidation rate were measured and calculated three times for each blend. Accuracy of these values is about ±5%.

[a] Phase separation occurred before oxidation.

SOURCE: Reproduced with permission from reference 11. Copyright 1990 John Wiley & Sons, Inc.

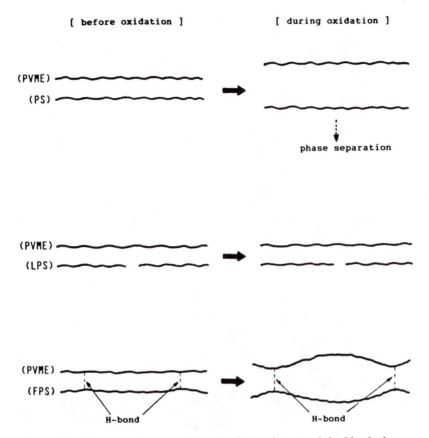

Figure 10. Schematic diagram of the miscibility changes of the blends during oxidation. (Reproduced with permission from reference 11. Copyright 1990 John Wiley & Sons, Inc.)

glass transition temperatures (T_g). Similar to the blends observed in thermal oxidation at 110 °C, the blends at 30 °C underwent phase separation. The difference is that phase separation occurred after a higher value of oxygen consumption, namely, 7 mL/g. At 30 °C a more extensive chemical change of PVME is required, due to a higher miscibility of the oxidized blend at low temperature. This demand was verified by measuring the cloud-point temperatures of three blends that had been photooxidized. The 70% PVME blend photooxidized for 3 days was still homogeneous at room temperature, whereas its cloud-point temperature was 110 °C (Figure 14).

The 50 and 30% PVME blends had even longer induction periods and slower oxidation rates than expected. The T_g values for the 70, 50, and 30% PVME blends were about –19 °C, –7 °C, and 15 °C, respectively. The lower segment mobility of the 30% blend, indicated by its high T_g, is reflected readily

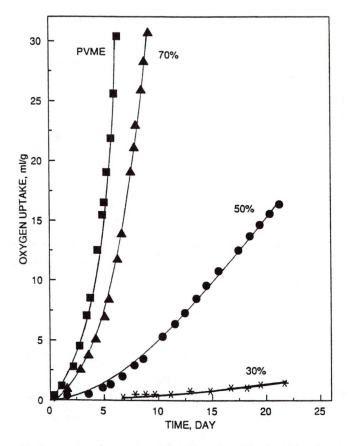

Figure 11. Oxygen uptake curves of PVME and its blends. (Reproduced with permission from reference 13. Copyright 1991 John Wiley & Sons, Inc.).

in its longer NMR spin-lattice relaxation time (32) at the temperature of the photooxidation. Therefore, the kinetics and the morphology of phase separation, as well as the oxidation rates, are strongly influenced by this change. The cross-propagation between the PVME radicals and PS is still operative, as long as both phases obtain a substantial amount of PS. The less reactive PS radical is responsible for the decrease in the rate of oxidation.

Antioxidant

Kim and co-workers (12, 33) used octadecyl 3,5-di-*tert*-butyl-4-hydroxyhydro-cinnamate to show the effect of a hindered phenol as an antioxidant (34, 35) on the thermal oxidation of miscible blends (Figures 15 and 16). The anti-oxidant-containing polystyrene (AOPS)/PVME blends showed similar charac-

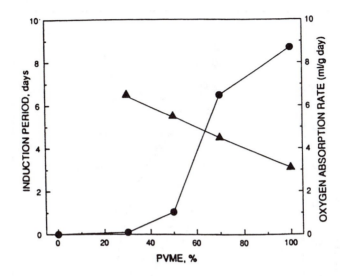

Figure 12. Induction periods (▲) and steady-state oxygen absorption rates (●) of PVME photooxidized PVME and its blends with PS. (Reproduced with permission from reference 13. Copyright 1991 John Wiley & Sons, Inc.)

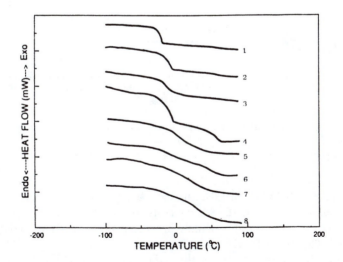

Figure 13. Differential scanning calorimetry curves of photooxidized PVME and its blends with PS: 1, PVME as prepared; 2, PVME oxidized for 11 days; 3, 70% PVME as prepared; 4, 70% PVME oxidized for 11 days; 5, 50% PVME as prepared; 6, 50% PVME oxidized for 11 days; 7, 30% PVME as prepared; 8, 30% PVME oxidized for 11 days. (Reproduced with permission from reference 13. Copyright 1991 John Wiley & Sons, Inc.)

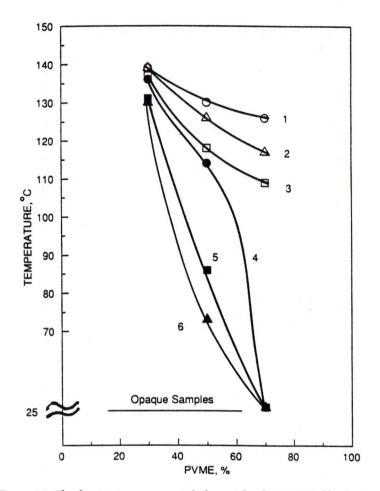

Figure 14. Cloud point temperatures of photooxidized PS/PVME blends: 1, as prepared, 2–6, oxidized for 1, 2, 4, 5.5, and 7.5 days, respectively. (Reproduced with permission from reference 13. Copyright 1991 John Wiley & Sons, Inc.)

teristics to the PS/PVME blends. The induction periods lengthened, the rates of oxidation decreased as the PS contents in the blends increased, and the saturation values were approximately proportional to the PVME contents in the blend. The AOPS/PVME blends phase separated as well as the PS/PVME blends shortly before the end of the induction period. However, all AOPS/PVME blends showed longer induction periods and slower rates of oxidation to its corresponding antioxidant-free blends, as expected (Figure 17, Table II).

Although all blends containing antioxidants showed the expected effects of antioxidants, there were some unexpected deviations. As shown in Table II, blends containing up to about 40% PS in the presence of antioxidant

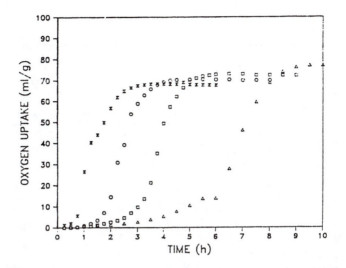

Figure 15. Oxygen uptake curves of PS/PVME blends containing 10 wt% PS with various concentrations of Irganox 1076 at 110 °C. AO contents are as follows: X, no antioxidant; ○, 0.026 wt%; □, 0.052 wt%; △, 0.102 wt%. (Reproduced with permission from reference 12. Copyright 1992 Gordon and Breach Science Publishers S.A.)

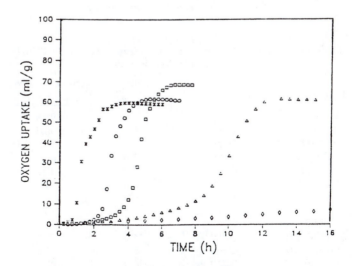

Figure 16. Oxygen uptake curves of PS/PVME blends containing 20 wt% PS with various concentrations of Irganox 1076 at 110 °C. AO contents are as follows: X, no antioxidant; ○, 0.026 wt%; □, 0.052 wt %; △, 0.102 wt%; ◇, 0.204 wt%. (Reproduced with permission from reference 12. Copyright 1992 Gordon and Breach Science Publishers S.A.)

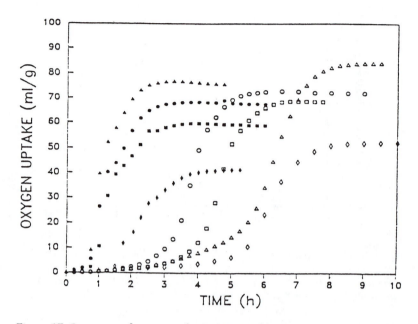

Figure 17. Oxygen uptake curves of AOPS/PVME blends at 110 °C. PS contents are as follows: △, 0 wt%; ○, 10 wt%; □, 20 wt%; ◇, 40 wt%. Filled symbols represent the blends without Irganox 1076.

Table II. Induction Periods and Oxidation Rates of PS/PVME and AOPS/PVME Blends

Sample	Composition (%)	Induction Period (h)	Oxidation Rate (mL/(g • h)]
PVME	100	0.6	128.5
PS/PVME	10/90	0.7	99.3
PS/PVME	20/80	0.9	83.9
PS/PVME	40/60	1.6	65.7
PS/PVME	60/40	3.2	36.1
PS/PVME	70/30	3.5	14.4
PS/PVME	80/20	6.5	10.3
AOPVME	100	5.2	43.1
AOPS/PVME	10/90	3.2	63.6
AOPS/PVME	20/80	4.0	54.8
AOPS/PVME	40/60	5.1	27.6
AOPS/PVME	60/40	36.0	5.1
AOPS/PVME	70/30	67.2	2.2

SOURCE: Reproduced with permission from reference 12. Copyright 1992 Gordon and Breach Science Publishers S.A.

showed shorter induction periods and higher oxidation rates than PVME in the presence of the same amount of antioxidants. Upon further increase of the PS content of the blends to more than 60%, the induction periods increased drastically and the oxidation rates decreased.

The results for the 10, 20, and 40 wt% PS blends were unexpected, because the presence of PS in blends containing no antioxidants was believed to act as an inhibitor in lengthening the induction period. Phase separation in AOPS/PVME blends occurred after a similar amount of oxygen, 2 mL/g, was absorbed as in the case of the blends containing no antioxidants. Blends containing up to about 40% PS phase separated relatively easily because the experimental temperature was quite close to the phase-separation temperature. Small changes in the chemical structure of PVME alter the miscibility of the blends, and this change is believed to lower the solubility of the antioxidant in PVME. The antioxidant is believed to redistribute to the PS phase during the initial stages of oxidative phase separation. The resulting PVME-rich phase contains, therefore, less antioxidant than the nominal amount and undergoes relatively fast autoxidation.

Activation Energy of Oxidation

Arrhenius plots of the steady-state oxidation rates as well as of the induction periods were made to show the differences in the activation energies of the blends before and after phase separation. The Arrhenius plots of the steady-state oxidation rates of the PS/PVME blends and PVME have almost the same slopes (Figure 18). This result is again an indication that in the steady-state region of oxidation the activation energy of oxidation is almost the same. This result also suggests that the oxidation of PVME proceeds independently of PS, because phase separation has already occurred (Figure 19).

Although the addition of LPS (or FPS) to PVME generally raises the activation energy of oxidation, no further raise results from increasing the concentration of LPS (or FPS) in the blends. The Arrhenius plot shows that all blends have a higher activation energy of oxidation, but the value is the same for all blends. This similarity indicates that neither FPS nor LPS took part in the oxidation process directly, but their addition generally increased the miscibility of the blends and resulted in higher activation energies, lower oxidation rates, and longer induction periods of the blends (Figures 20 and 21).

The Arrhenius plots of the oxidation rates of the phenol group containing PS/PVME blends show that the blends have different activation energies depending on the concentrations of the phenol groups. A higher concentration of phenol groups lowers the activation energies. The phenol groups are directly involved in the oxidation process. As a result, the presence of the phenol

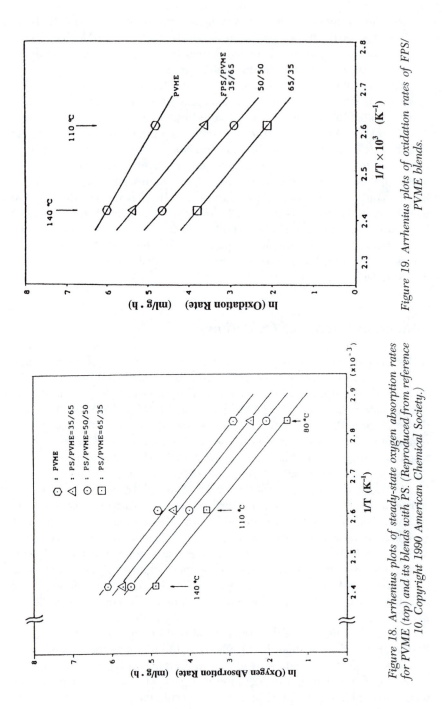

Figure 19. Arrhenius plots of oxidation rates of FPS/PVME blends.

Figure 18. Arrhenius plots of steady-state oxygen absorption rates for PVME (top) and its blends with PS. (Reproduced from reference 10. Copyright 1990 American Chemical Society.)

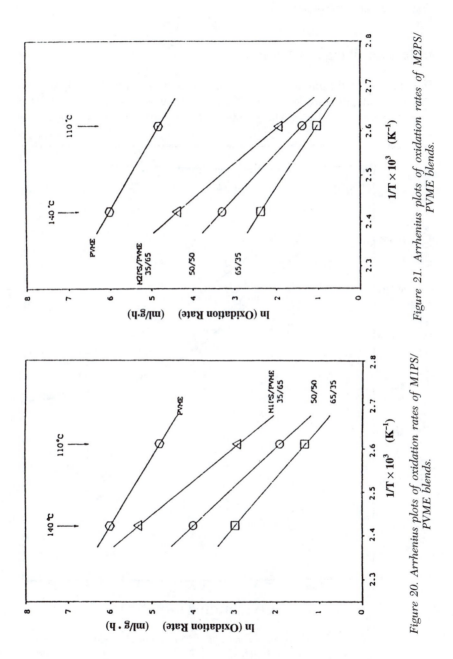

Figure 21. Arrhenius plots of oxidation rates of M2PS/PVME blends.

Figure 20. Arrhenius plots of oxidation rates of M1PS/PVME blends.

groups in PS has two contributions: an improvement of the miscibility of the blends, and an antioxidant effect.

Molecular Weight Changes

The changes of the number-average molecular weight for the different PS/ PVME blends after oxidation at 110 °C are shown for different periods of time (Figures 22 and 23). The number-average molecular weight of PVME

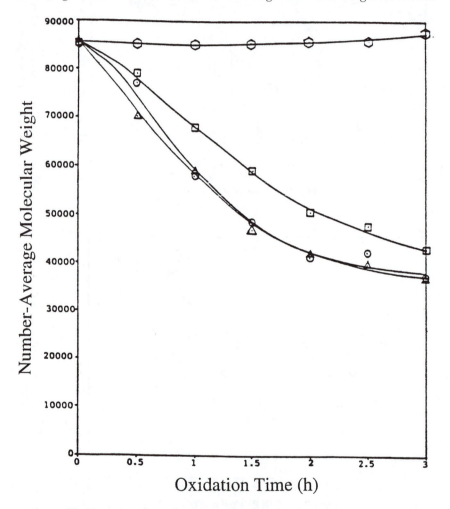

Figure 22. Changes in the molecular weight of PS as a function of oxidation time at 110 °C. PS contents are 100, 65, 50 and 30 % from the top to the bottom curves. (Reproduced from reference 10. Copyright 1990 American Chemical Society).

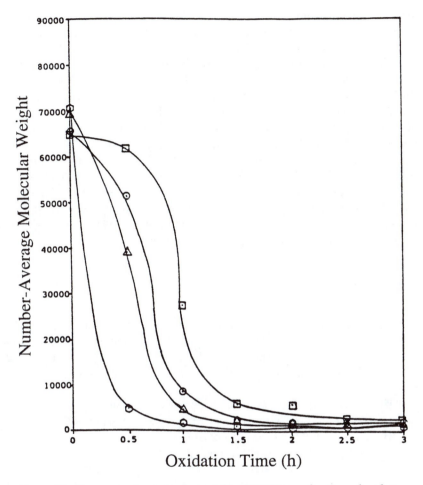

Figure 23. Changes in the molecular weight of PVME as a function of oxidation time at 110 °C (values below 2000 are inaccurate). PVME contents are as follows:
○, *100%;* △, *65%;* ○, *50%;* □, *35%. (Reproduced from reference 10. Copyright 1990 American Chemical Society).*

decreased rapidly during oxidation. A 15-fold decrease in the molecular weight was observed throughout the induction period. PS showed almost no change, or even a slight increase, of molecular weight after oxidation. However, a twofold decrease was found in the case of PS in the blends, and this decrease is an average of one chain scission per polystyrene molecule (i.e., one scission per 820 repeat units). The continuous change of PS molecular weight can be explained through a reaction between the PVME peroxy radicals and PS. The resulting PS radical is less reactive, as demonstrated by a prolonged induction period of the blend. Eventually the PS radical undergoes chain scission that leads to a decrease in molecular weight. The decrease in molecular weight even happens after phase separation, which is thought to be due to boundary-region contacts between domains.

The antioxidant-containing PS/PVME blends also showed a significant decrease in molecular weight. However, during the induction period the stabilized PVME exhibited little change in molecular weight (Figures 24 and 25).

The PS molecular weight almost did not change during an extensive oxidation for 15 days at 110 °C. The decrease in molecular weight occurred in 20 and 40% PS blends shortly after the induction periods.

It is apparent that PS is directly involved in the oxidation process in both the AO-free and the AO-containing blends. The presence of the antioxidant and of PS lengthened the induction period. It is believed that the PVME radical underwent a reaction with PS after the antioxidant was used up in the induction period, and this reaction led to a more stable PS radical that retards the oxidation process and undergoes chain scission.

Chemical Changes

The structural change of PVME as the component that undergoes oxidation is accompanied by the formation of carbonyl (32), double bonds (36), and hydroxy (hydroperoxide) groups (10). IR spectroscopy (37–40) was used by Chien et al. (13) and Park et al. (10) to show these changes (Figures 26 and 27).

The chemical changes are visible during the oxidation. The vibrational peaks of CH, CH_2, and CH_3 reduce, whereas growing intensities of $-OH/-OOH$, carbonyl, and C=C are of notice (Figures 28 and 29).

Carbonyl groups are formed through β-scission of alkoxy radicals. A close relationship between the change of the number-average molecular weight and the concentration of carbonyl groups was observed by Iring et al. (41) during the oxidation of PE. The changes of the $-OH/-OOH$ concentration of the oxidation of other polymers (42, 43) seem to be in general agreement with the results found for the oxidation of PVME. The accepted interpretation is that the $-OOH$ group increases during the induction period of the oxidation. As soon as the autoacceleration starts, the peroxides decompose into initiating radicals, a phenomenon that explains the downshift of the curve. However, deviations from this explanation are seen in the case of PVME. The maximum value in the case of PVME is 30 min, which is still within the induction period of PVME, and autoacceleration has also begun. The similar temporal evolution of $-OH/-OOH$ and C=O for PS/PVME blends and PVME during oxidation is thought to be due to the preferential presence of PVME at the surface (44, 45) in the blends. The lower surface energy component preferentially moves to the surface and dominates surface behavior.

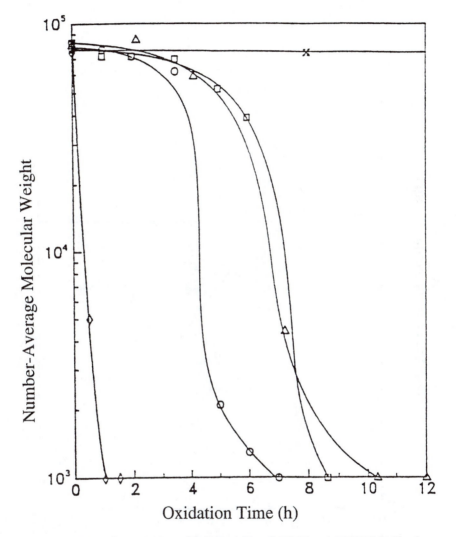

Figure 24. Number-average molecular weights of PVME in AOPS/PVME blends after oxidation. AO concentration, 0.05%. PVME contents: △, 40%; ○, 80%; □, 60%; X, 40%. Uninhibited PVME represented by ◇. (Reproduced with permission from reference 12. Copyright 1992 Gordon and Breach Science Publishers S.A.)

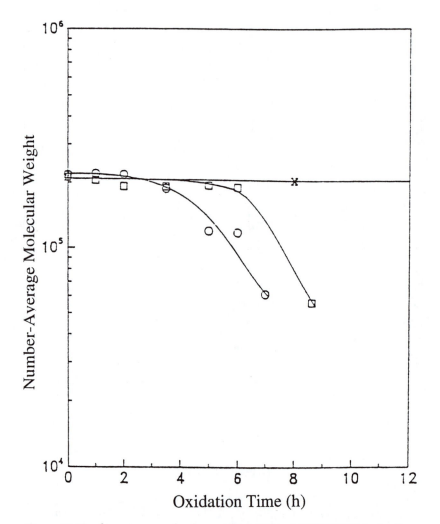

Figure 25. Number-average molecular weights of PS in AOPS/PVME blends after oxidation. AO concentration, 0.05%. PVME contents: X, 40%; box, 60%; ○, 80%. (Reproduced with permission from reference 12. Copyright 1992 Gordon and Breach Science Publishers S.A.)

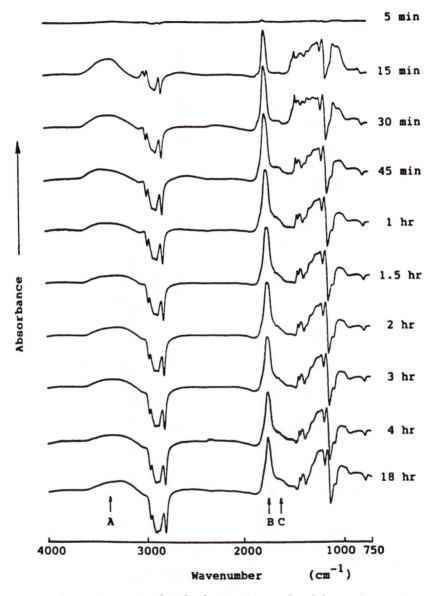

Figure 26. FTIR spectra of oxidized PVME. (Reproduced from reference 10. Copyright 1990, American Chemical Society.)

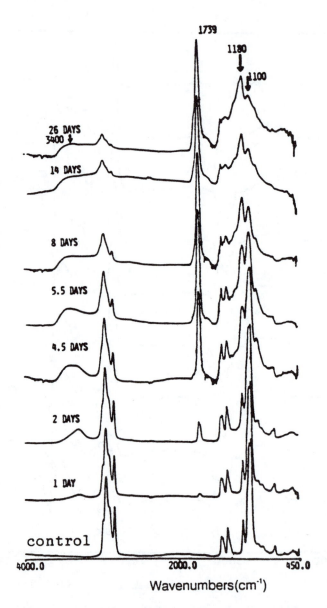

Figure 27. FTIR spectra of photooxidized PVME. (Reproduced with permission from reference 13. Copyright 1991, Copyright 1991 John Wiley & Sons, Inc.)

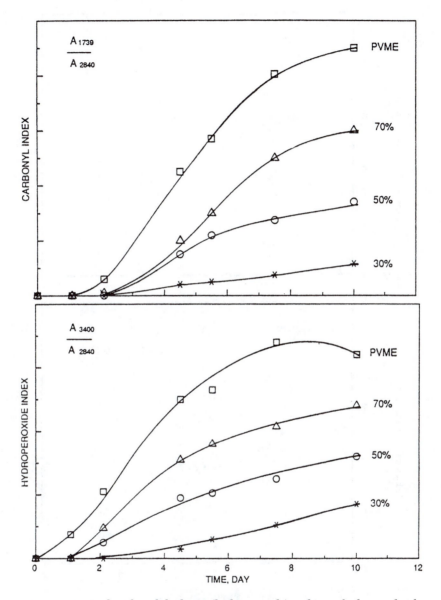

Figure 28. Carbonyl and hydroxy (hydroperoxide) indices of photooxidized polymers. (Reproduced with permission from reference 13. Copyright 1991 John Wiley & Sons, Inc.)

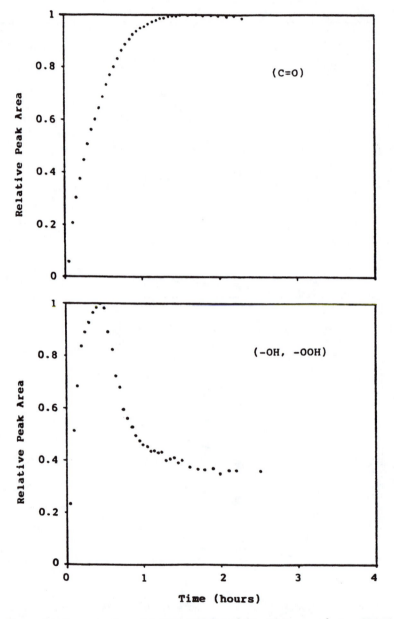

Figure 29. Concentration of (–OH/–OOH) and (C=O) groups during PVME oxidation at 110 °C. (Reproduced from reference 10. Copyright 1990 American Chemical Society.)

Other Miscible Systems

Similar results were found by Makhija et al. (*14*) on miscible blends of PMMA and PEO. PMMA consumed an insignificant amount of oxygen at 140 °C after 7 days, whereas PEO oxidized rapidly. PMMA showed a stabilizing effect on the oxidation of PEO in the blend. As the content of PMMA in the blend increased, the rates of oxidation were reduced. The blends stayed miscible during oxidation, and their oxidation rates were less than the simple mathematical averages of the values for each component (Table III, Figures 30 and 31).

Zhu et al. (*16*) found an intermolecular hydrogen-bonding effect on the thermostability of poly(*p*-hydroxystyrene) in miscible blends with poly(vinyl pyrolidone) and poly(ethyl oxazoline). A loss in weight of about 6% of poly-(*p*-hydroxystyrene) at 200–250 °C was found that was caused by cross-linking reactions involving hydroxyl groups. Fourier transform IR (FTIR) studies showed the appearance of ether groups. In the miscible blends this phenomenon vanished (Figure 32).

Table III. Steady-State Oxidation Rates for PEO and Its Blends

Compound	Conc. (%)	140 °C	120 °C	95 °C	70 °C
PEO	100	158	101	42	3.2[a]
PEO/PMMA	80/20	128	62.8	16[a]	
PEO/PMMA	50/50	59	30	8[a]	

NOTE: Values are in mL/(g • h).
SOURCE: Reproduced with permission from reference 14. Copyright 1992 John Wiley & Sons, Inc.
[a]Values were not used for calculation of activation energy.

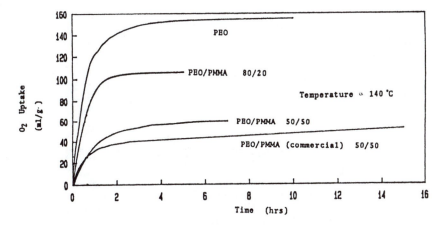

Figure 30. Oxidation curves for PEO and its blends at 140 °C. (Reproduced with permission from reference 14. Copyright 1992 John Wiley & Sons, Inc.).

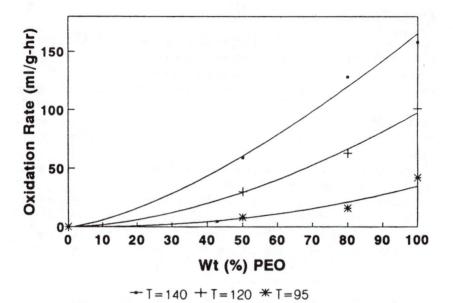

Figure 31. Oxygen absorption rates of PEO/PMMA blends as a function of PEO content. (Reproduced with permission from reference 14. Copyright 1992 John Wiley & Sons, Inc.).

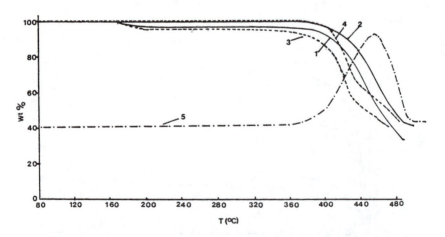

Figure 32. TGA curves of blends (normalized to PHS weight loss; heating rate 10 °C/min in nitrogen). 1, PHS-PVP blend (1:1 w/w); 2, PHS-PVP blend (1:2 w/ w); 3, PHS-PEO$_x$ blend (1: 0.8 w/w); 4, PHS-PEO$_x$ blend (1:4 w/w); 5, derivative curve of sample 2. (Reproduced with permission from reference 16. Copyright 1989).

Conclusions

A number of conclusions can be drawn from our experimental results obtained from thermal oxidation for PS/PVME blends. The presence of PS prolonged the induction period of PVME oxidation. Phase separation occurred shortly before the end of the induction period. After the reaction of about one oxygen molecule per 100 repeating units of PVME, phase separation occurred. As a result of the phase separation and due to minor chemical changes of PVME, the steady-state rate of oxidation was proportional to the PVME content of the blend. The molecular weight of PVME decreased more slowly as the PS content increased. During the homogeneous state in the induction period a reaction between PVME radicals and PS took place, most likely through the abstraction of the tertiary hydrogen atoms of PS. The resulting PS radical was less reactive in chain propagation and caused the retardation of the oxidation reaction. PS, which did not undergo molecular weight changes during oxidation by itself, experienced chain scission in the blends.

Blends of low molecular weight polystyrene and PVME, remained homogeneous throughout the oxidation. As a consequence the induction periods were longer and the oxidation rates slower than in the corresponding PS blends. Therefore, low molecular weight PS was involved in the oxidation process during both periods.

The improved miscibility of styrene copolymers containing hydrogen-bond donors were more effective than PS homopolymer in prolonging the induction period and lowering the rate of oxidation in the steady-state region of thermal oxidation of PVME. The superior retardation effect of the copolymer allowed for more extensive cross-propagation/cross-termination reactions. Phenol groups in the copolymer contributed to miscibility and acted as antioxidants.

The incorporation of an antioxidant in miscible blends prolonged the induction period and lowered the oxidation rate. At the end of the induction period, the blends phase separated and the antioxidant redistributed itself between the two phases. The molecular weight of PVME remained nearly unchanged during the induction period due the inhibitory effect of the antioxidant. A rapid decrease of the molecular weight by a factor of 80 after the induction period was observed. PS also underwent chain scission after the induction period.

The experimental results obtained from photooxidation for PS/PVME blends were similar to those obtained from thermal oxidation. The difference that was found can be ascribed to lower segment mobility in these blends that results in a slower rate of the propagation reaction and of phase separation.

References

1. Robeson, L. M. In *Contemporary Topics in Polymer Science;* Culverston, B. L., Ed.; Plenum: New York, 1990; Vol. 6, p 177.

2. *Polymer Blends;* Paul, D. R.; Newman, S., Eds.; Academic: New York, 1978.
3. Olabisi, O.; Robeson, L. M.; Shaw, M. T. *Polymer–Polymer Miscibility;* Academic: New York, 1979.
4. *Multiphase Polymers;* Cooper, S. L.; Estes, G. M., Eds.; Advances in Chemistry Series No. 176; American Chemical Society: Washington, DC, 1979.
5. *Polymer Science and Technology;* Klemper, D.; Frisch, K. C., Eds.; Plenum: New York, 1977; Vol. 10.
6. McNeill, I. C. In *Developments in Polymer Degradation;* Grassie, N., Ed.; Applied Science: London, England, 1977; Vol. 1.
7. Scott, G. In *Stabilization and Degradation of Polymers;* Allara, D. L.; Hawkins, W. L.; Advances in Chemistry Series No. 169; American Chemical Society: Washington, DC, 1978.
8. (a) Ghaffar, A.; Scott, G. *Eur. Polym. J.* **1976,** *12,* 615; (b) Ghaffar, A.; Scott, A.; Scott, G. *Eur. Polym. J.* **1977,** *13,* 83.
9. Kaplan, M. L.; Kelleher, P. G. *J. Polym. Sci. Polym. Chem. Ed.* **1970,** *8,* 3163.
10. Park, H.; Pearce, E. M.; Kwei, T. K. *Macromolecules* **1990,** *23,* 434.
11. Park, H.; Pearce, E. M.; Starnes, W. H., Jr.; Kwei, T. K. *J. Polym. Sci. Polym. Chem.* **1990,** *28,* 1079.
12. Kim, J. H.; Pearce, E. M.; Kwei, T. K. *Chem. Eng. Comm.* **1992,** *116,* 105.
13. Chien, Y. Y.; Pearce, E. M.; Kwei, T. K. *J. Polym. Sci. Polym. Chem.* **1991,** *29,* 849.
14. Makhija, S.; Pearce, E. M.; Kwei, T. K. *J. Polym. Sci. Polym. Chem.* **1992,** *30,* 2693.
15. He, L. Y.; Pearce, E. M.; Kwei, T. K. In *International Symposium on Polymers for Advanced Technologies;* Lewin, M., Ed.; VCH Publishers: New York, 1988; p 713.
16. Zhu, K. J.; Kwei, T. K.; Pearce, E. M. *J. Appl. Polym. Sci.* **1989,** *37,* 573.
17. George, A. Master's Thesis, Polytechnic University, 1993.
18. Pearce, E. M.; Kwei, T. K. In *Polymer Solutions, Blends, and Interfaces;* Noda, I.; Rubingh, D. N., Eds.; Elsevier Science, 1992.
19. Pearce, E. M.; Kwei, T. K.; Min, B. Y. *J. Macromol. Sci. Chem.* **1984,** *A21,* 1181.
20. Kwei, T. K.; Pearce, E. M.; Ren, F.; Chen, J. P. *J. Polym. Sci. Polym. Phys. Ed.* **1986,** *24,* 1597.
21. Djadoun, S.; Goldberg, R. N.; Morawetz, H. *Macromolecules* **1977,** *10,* 1015.
22. Bank, M.; Leffingwell, J.; Thies, C. *J. Polym. Sci. Part A* **1972,** *10,* 1097.
23. McMaster, L. P. *Macromolecules* **1973,** *6,* 760.
24. Ting, S. P.; Pearce, E. M.; Kwei, T. K. *J. Polym. Sci. Polym. Lett. Ed.* **1980,** *18,* 201.
25. Ting, S. P.; Bulkin, B. J.; Pearce, E. M.; Kwei, T. K. *J. Polym. Sci. Polym. Chem. Ed.* **1981,** *19,* 1451.
26. Pearce, E. M.; Kwei, T. K.; Min, B. Y. *J. Polym. Sci. Chem.* **1984,** *A21,* 1181.
27. Nishi, T.; Wang, T. T.; Kwei, T. K. *Macromolecules* **1975,** *8,* 227.
28. Davis, D. D.; Kwei, T. K. *J. Polym. Sci. Polym. Phys. Ed.* **1980,** *18,* 2337.
29. Nishi, T.; Kwei, T. K. *Polymer* **1975,** *16,* 285.
30. Park, H., Ph.D. Dissertation, Polytechnic University, Brooklyn, New York, 1989.
31. Chien, Y. Y., Ph.D. Dissertation, Polytechnic University, Brooklyn, New York, 1990.
32. Kwei, T. K.; Nishi, T.; Roberts, R. F. *Macromolecules* **1974,** *7,* 667.
33. Kim, J. H., Ph.D. Dissertation, Polytechnic University, Brooklyn, New York, 1991.
34. Shelton, J. R. In *Polymer Stabilization;* Hawkins, W. L., Ed.; Wiley-Interscience: New York, 1972.

35. Horng, P. L.; Klemchuk, P. P. In *Polymer Stabilization and Degradation;* Klemchuk, P. P., Ed.; ACS Symposium Series 280; American Chemical Society: Washington, DC, 1978.

36. Burnett, J. D.; Miller, R. M. J.; Willis, M. A. *J. Polym. Sci.* **1955**, *15*, 92.

37. Bellamy, J. *The Infrared Spectra of Complex Molecules*, 3rd ed.; Chapman and Hall: London, 1975.

38. Domke, W. D.; Steinke, H. *J. Polym. Sci. Polym. Chem. Ed.* **1986**, *24*, 2701.

39. Luongo, J. P. *J. Polym. Sci.* **1960**, *42*, 139.

40. Rugg, F. M.; Smith, J. J.; Bacon, R. C. *J. Polym. Sci.* **1954**, *13*, 535.

41. Iring, M.; Kelen, T.; Tudos, F.; Laszlo-Hedvig, Z. In *Degradation and Stabilization of Polyolefins;* Sedlacek, B.; Overberger, C. G.; Mark, H. F., Fox, T. G., Eds.; Wiley: New York, 1976; p 89.

42. Hawkins, W. L. In *Polymer Degradation and Stabilization;* Springer-Verlag: Berlin, 1984.

43. *Aspects of Degradation and Stabilization of Polymers;* Jellinek, H. H. G., Ed.; Elsevier Scientific: Amsterdam, Netherlands, 1978.

44. Pan, D. H. K.; Prest, W. M., Jr. *J. Appl. Phys.* **1985**, *58*, 2861.

45. Bhatia, Q. S.; Pan, D. H.; Koberstein, J. T. *Macromolecules* **1988**, *21*, 2166.

RECEIVED for review December 16, 1993. ACCEPTED revised manuscript February 17, 1995.

Influence of Titanium Dioxide Pigments on Thermal and Photochemical Oxidation and Stabilization of Polyolefin Films

Norman S. Allen and Hassan Katami

Department of Chemistry, Faculty of Science and Engineering, Manchester Polytechnic, Chester Street, Manchester M1 5GD, United Kingdom

The thermooxidative (oven aging) and photooxidative degradation of low-density polyethylene (LDPE) film materials (100-μm thick) containing nine types of titanium dioxide pigments were studied by IR spectroscopy and hydroperoxide analysis as a function of temperature, and apparent activation energies were calculated. The role of the sensitizing or stabilizing effects of the pigment on the rate of polymer oxidation was more important during photooxidation. The behavior of the pigments as catalysts for the thermooxidative breakdown of the polymer was confirmed by hydroperoxide analysis. The complex interactions of titanium dioxide pigments with a phenolic antioxidant and a hindered piperidine stabilizer were also examined in polypropylene film. Stabilizer effects are discussed in terms of additive adsorption onto the pigment particle surface versus the photocatalytic effect of the pigment.

PIGMENTS ARE WIDELY USED FOR THE COLORATION of thermoplastics in many commercial applications. Although they are used primarily to impart color to the polymer they can nevertheless have a marked influence on the thermal and photochemical stability of the polymer material (*1*). For example, by absorbing or screening light energy they can exhibit a protective effect, or they may be photoactive and sensitize the photochemical breakdown of the polymer. With regard to these effects there are four principal factors that can influence the photostability of a pigmented-polymer system (*2*):

0065–2393/96/0249–0537$12.00/0

- the intrinsic chemical and physical nature of the polymer itself (1)
- the environment in which the system is used
- the chemical and physical nature of the pigment and its concentration (3–6)
- the presence of antioxidants and light stabilizers as well as other additives (7, 8)

The ability of pigments to catalyze the photooxidation of polymer systems has received much attention in terms of the mechanistic behavior of the pigments. In this regard much of the information originates from work carried out on TiO_2 pigments in both polymers and model systems (1, 2). To date there are three mechanisms of the photosensitized oxidation of polymers by TiO_2, and for that matter other white pigments such as ZnO:

1. The formation of an oxygen radical anion by electron transfer from photoexcited TiO_2 to molecular oxygen (9). A recent modification of this scheme involves a process of ion-annihilation to form singlet oxygen, which then attacks any unsaturation in the polymer (10).

$$TiO_2 + O_2 \xrightarrow{h\upsilon} (TiO_2^{+\cdot} + O_2^{-\cdot})(I)$$
$$(I) \rightarrow TiO_2 + {}^1O_2 \text{ (ion-annihilation)}$$
$$(I) + H_2O \rightarrow TiO_2 + HO^{\cdot} + HO_2^{\cdot}$$
$$2HO_2^{\cdot} \rightarrow H_2O_2 + O_2$$
$$RCH_2 = CHR' + {}^1O_2 \rightarrow RCH=CHCH(OOH)R'$$

2. Formation of reactive hydroxyl radicals by electron transfer from water catalyzed by photoexcited TiO_2 (11). The Ti^{3+} ions are reoxidized back to Ti^{4+} ions to start the cycle over again.

$$H_2O \xrightarrow[TiO_2]{h\upsilon} H^+ + e' \text{ (aqu)} + {}^{\cdot}OH$$
$$[Ti^{4+}] + e' \rightarrow [Ti^{3+}]$$
$$[Ti^{3+}] + O_2 \rightarrow [Ti^{4+}]$$

3. Irradiation of TiO_2 creates an exciton (p) that reacts with the surface hydroxyl groups to form a hydroxyl radical (12). Oxygen anions are also produced and are adsorbed on the surface of the pigment particle. They produce active perhydroxyl radicals.

$$TiO_2 \xrightarrow{h\upsilon} e' + (p)$$
$$OH^- + (p) \rightarrow HO^{\cdot}$$
$$Ti^{4+} + e' \rightarrow Ti^{3+}$$

$$Ti^{3+} + O_2 \rightarrow \quad [Ti^{4+}\text{--}O^{2-}] \text{ adsorbed}$$

$$[Ti^{4+}\text{--}O^{2-}] \text{ adsorbed} + H_2O \rightarrow \quad Ti^{4+} + HO^- + HO_2^{\cdot}$$

In this chapter we attempt to address a number of the previously mentioned factors related to the influence of titanium dioxide pigments on the thermal and photooxidative degradation of low-density polyethylene (LDPE). The first factor related to the inherent stability of the polymer is addressed to some degree with regard to the importance of the pigment type on the formation of hydroperoxide and carbonyl groups, whereas the second factor, which relates to environmental effects, is interrelated in terms of the influence of temperature during thermal and photochemical aging processes. The third factor related to pigment structure is examined on a wide basis through a study of nine different pigment types. The fourth factor on stabilization is examined for polypropylene film in terms of the interactions of the titanium dioxide pigments with a hindered-piperidine light stabilizer and a hindered phenolic antioxidant.

Experimental

Materials. Polymer (LDPE) film samples (100-μm thick) were prepared and supplied by Tioxide (UK) Ltd., Stockton-on-Tees, U.K. Masterbatches were prepared at a pigment loading of 50% by weight in an MFI 20 LDPE, with an addition of 0.5% w/w (on the weight of pigment) zinc stearate by using a Banbury laboratory internal mixer. The masterbatches were then reduced to a melt–flow index (MFI) 7 grade of LDPE and films were prepared on a film tower. The final concentrations of the titanium dioxide pigments used were eventually reduced to 0.5% w/w by using the film tower. The codes used for the different pigment types studied are shown in Table I together with their various properties. The durability is defined commercially in terms of their overall weatherability.

The stabilized polypropylene film samples were prepared by using a Brabender Plasticorder (Duisburg, Germany) and processing for 10 min at 190 °C by using 0.1% w/w each of a hindered piperidine light stabilizer, bis(2,2,6,6-tetra-

Table I. Properties of Titanium Dioxide Pigments

Sample	Crystalline Modification	Coated	Crystal Size	Photodurability
A	Anatase	No	Fine	Poor
B	Anatase	Light	Fine	Poor
C	Rutile	Light	Fine	Medium
D	Rutile	Light	Fine	Medium
E	Rutile	Light	Medium	Good
F	Rutile	Medium	Fine	Very good
G	Rutile	Medium	Medium	Very good
H	Rutile	Heavy	Medium	Excellent
I	Rutile	No	Fine	Poor

NOTE: All samples underwent organic treament.

methyl-4-piperidinyl)sebacate (Tinuvin 770), and a hindered phenolic antioxidant, pentaerythrityltetra-β-(4-hydroxy-3,5-ditertiary-butylphenyl)propionate   (Irganox 1010), both supplied by Ciba-Geigy (UK) Ltd., Manchester. The anatase pigments used were samples A and B, whereas the rutile pigments were samples I and G for uncoated and coated grades, respectively, at a 1% w/w concentration. Films (200-μm thick) were then compression molded at 200 °C for 1 min.

Thermal and Photooxidation.   Polymer films were exposed in a SEPAP Sairem 30:24 unit (designed by the Universite Blaise-Pascal Clermont-Ferrand II, France) using four 500-W high-pressure Hg/W fluorescent lamps (wavelengths > 300 nm). The temperature of light exposure was carried out at 50, 60, 70, 80, and 90 °C. The 400-W medium-pressure Hg lamps normally used in this light exposure unit were replaced by the fluorescent lamps. Polymer film samples were also aged in hot air ovens over the same temperature range.

The rates of photooxidation of the polymer films were monitored by measuring the buildup in the concentration of the nonvolatile carbonylic oxidation products absorbing at 1710 cm^{-1} in the IR region. IR spectra were recorded by using a Mattson Alpha-Centauri Fourier transform IR (FTIR) instrument. Oxidation rates were determined via the well-established carbonyl index (C.I.) method (1):

$$C.I. = Log_{10}(I_0/I_t)/D \times 100$$

where D is film thickness in μm, I_0 is the initial light intensity at 1710 cm^{-1}, and I_t is the transmitted light intensity at 1710 cm^{-1}.

From the C.I. rate curves, rate constants were obtained from tangential fits over the C.I. range 0.1–0.2 with the aid of an SE Apple Macintosh system via the use of Cricket Graph software (Computer Associates International Ltd.).

Hydroperoxide Measurement.   Hydroperoxide determinations were carried out by using the iodiometric method of analysis (10). Polymer film samples (1 g) were cut into small pieces and placed into a pear-shaped 100-cm^3 flask containing sodium iodide (0.1 g) (BDH Ltd., U.K), 9.5 cm^3 of Analar-grade propan-2-ol (BDH Ltd.), and 0.5 cm^3 of glacial acetic acid (BDH Ltd.). The combination was then refluxed vigorously for 30 min with a control solution that did not contain polymer. The yellow I^{3-} was then measured spectrophotometrically at 380 nm (wavelength maximum at 357 nm) (Perkin-Elmer Lambda 7 spectrometer) to avoid any additive absorbances, and a calibration curve was set up using 70% w/w cumene hydroperoxide as a standard (Aldrich Chemical Company, U.K). The data were cross-checked using FTIR spectroscopy (11).

Results and Discussion

Thermal Oxidation.   The effects of thermal oxidative aging on the unpigmented and pigmented LDPE film samples show a number of interesting features. Rates of C.I. versus oven-aging time in hours, particularly at the higher temperatures, show a strong overlap for many of the plots. A typical example is shown in Figure 1 for the 90 °C data. Typical temperature effects on oven aging are also illustrated in Figure 2 for sample G. For an easier

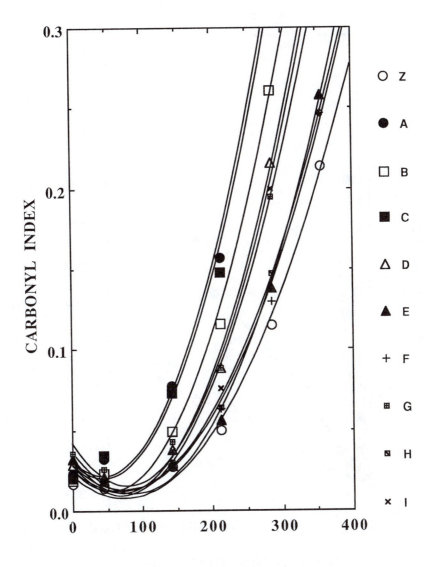

Figure 1. C.I. versus oven-aging time (h) for LDPE films (100-μm thick) at 90 °C containing no TiO₂ (Z; ○) and 0.5% w/w of pigments. For key, see Table I.

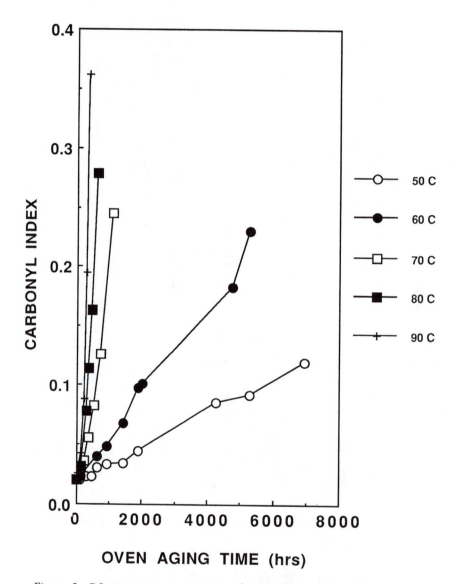

Figure 2. C.I. versus oven-aging time (h) for LDPE films (100-μm thick) containing pigment G (0.5% w/w) at various temperatures (°C).

comparison of the data, Table II compares the times to 0.1 C.I. for all the aged LDPE films; this time is often taken as the time to embrittlement for this polymer.

The first and most interesting feature is the observation that all the pigment types sensitize the thermal oxidative aging of the LDPE at all temperatures. The initial decrease in carbonyl may be due to a catalyzed decomposition of the initial carbonyl groups present in the polymer films. The second point is that the rate of oxidation accelerates with increasing temperature, as would be expected, and becomes autocatalytic above 70 °C. Thus, at 50 °C the order in thermal stabilization is control film (Z) >> E, F, H, and I >> B and G >> D >> C and A. At higher temperatures the order is very similar, although the differences between the pigments are reduced. The orders of stability are seen to be reasonably similar over the entire temperature range, and the uncoated anatase behaves as the most powerful sensitizer. Apart from pigment G all the durable, coated, rutile forms are less active as sensitizers, and two of the lightly coated, fine crystal rutile grades are the more active types (C and D).

The relationship between temperature and pigment type is compared in more detail in Figure 3 as a plot of time to 0.1 C.I. versus oven-aging temperature. This data shows that as the temperature of aging is increased the pigment type becomes less important in terms of its catalytic activity on polymer oxidation.

Hydroperoxide species are known to be important in the initiation and propagation steps of LDPE oxidation reactions. Figure 4 shows a plot of the hydroperoxide concentrations versus oven-aging time in hours at 90 °C for the various unpigmented and pigmented polymer film samples. The actual data values are given in Table III. Again, all the pigments sensitize the formation of hydroperoxide groups during aging of the polymer compared with the control sample without pigment. Furthermore, from the initial values before oven aging, all the pigmented polymer samples exhibit higher hydroperoxide concentrations than that of the control film. Thus, during the processing operation

Table II. Time (h) to 0.1 C.I. for Pigmented LDPE Films during Oven Aging

Sample	50 °C	60 °C	70 °C	80 °C	90 °C
Z (control)	9000	3400	830	383	264
A	4260	1800	402	330	165
B	6200	2570	511	340	193
C	4300	1690	509	332	170
D	5200	2260	649	350	212
E	7800	2745	653	440	250
F	7600	2957	680	460	248
G	6020	2120	611	383	217
H	7380	2720	740	400	244
I	7300	2957	739	380	220

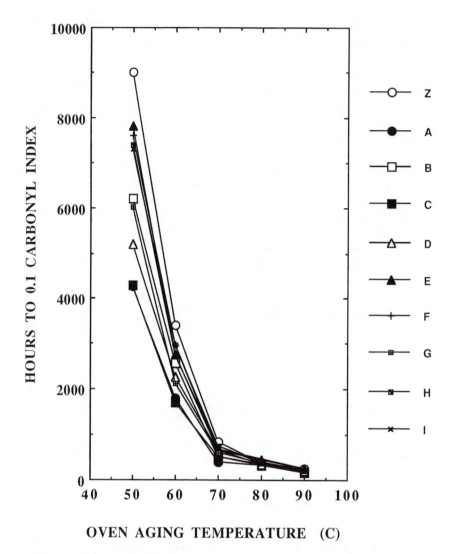

Figure 3. Time to 0.1 C.I. versus oven-aging temperature for LDPE films (100-μm thick) containing no TiO₂ (Z; ○) and 0.5% w/w of pigments. For key, see Table I.

in the Banbury mixer the titanium dioxide pigments are catalyzing the formation of hydroperoxide groups, and both the uncoated anatase and lightly coated rutile pigments are the most active types.

From the initial data in Table III, pigment activity follows the order I > A > D > B > C > G > E > F > H. Again, the coated, durable, rutile types are the least active in this regard, followed by the fine crystal grades. Pigment

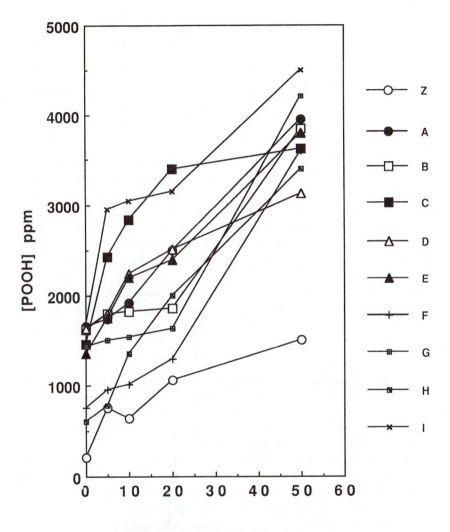

Figure 4. Hydroperoxide concentration (ppm) versus oven-aging time (h) for LDPE films (100-μm thick) at 90 °C containing no TiO$_2$ (Z; ○) and 0.5% w/w of pigments. For key, see Table I.

Table III. Hydroperoxide Concentrations (μg/g) for Pigmented LDPE Films during Oven Aging at 90 °C

Sample	0 h	5 h	10 h	20 h	50 h
Z (control)	210	760	640	1068	1504
A	1660	1742	1920	2500	3950
B	1620	1796	1824	1860	3840
C	1455	2428	2840	3400	3616
D	1633	1802	2240	2508	3128
E	1350	1750	2200	2400	3800
F	760	956	1015	1300	3600
G	1440	1505	1540	1640	4200
H	607	780	1350	1995	3400
I	1676	2953	3053	3150	4550

I is essentially classed as an uncoated fine-crystal grade of rutile and is more active than anatase in terms of hydroperoxide formation. The coated anatase pigment is much less active. Thus, the nature of the coating treatment appears to be quite important in terms of actual contact between the surface of the titanium dioxide particles and the polymer matrix. However, there are major differences among the rates of C.I. formation and those related to hydroperoxide formation.

Hydroperoxides behave as potential initiators and intermediates during oxidation of the polymer. Depending on the temperature, they will be catalytically decomposed by the titanium dioxide pigments at rates depending on the pigment type.

Photooxidation.   The photooxidation results are different and show a number of interesting and related trends. Figure 5 shows the rates of carbonyl formation versus irradiation time at 50 °C in the SEPAP unit. The effects of temperature for pigment G are shown in Figure 6. Actual data values are shown in Table IV for all the pigment types.

All the rates of carbonyl formation are autocatalytic irrespective of the temperature of exposure. Also, pigment activity increases with increasing temperature of light exposure from 50 °C to 90 °C. In the 90 °C case all the pigments, with the exception of H, operate as photosensitizers of carbonyl formation in LDPE. Obviously, during photothermal aging active hydroperoxide and carbonyl groups are formed in-situ, and they can accentuate the photooxidation rate of the polymer. Figure 7 shows a plot of the time to 0.1 C.I. for the polymer films versus exposure temperature from 50 °C to 90 °C. Differences between pigment types are reduced with increasing temperature, but not to the same degree as with thermal aging. At 50 °C the order in pigment activity in terms of ability to stabilize the polymer is H > E > F > G > D > C > Control, whereas the order for sensitization pigments is B > I > A. The uncoated pigments are the most active, followed by the fine-crystal

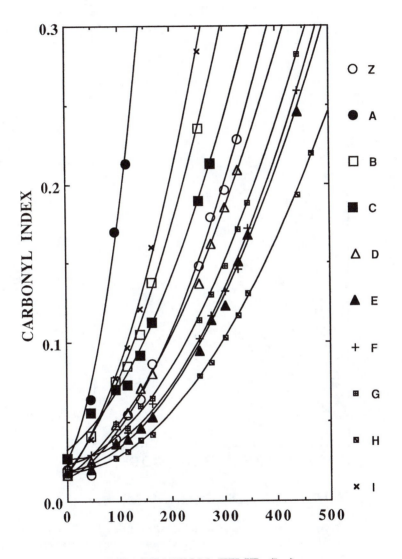

Figure 5. *C.I. versus irradiation time (h) in SEPAP for LDPE films (100-μm thick) at 50 °C containing no TiO$_2$ (Z; ○) and 0.5% w/w of pigments. For key, see Table I.*

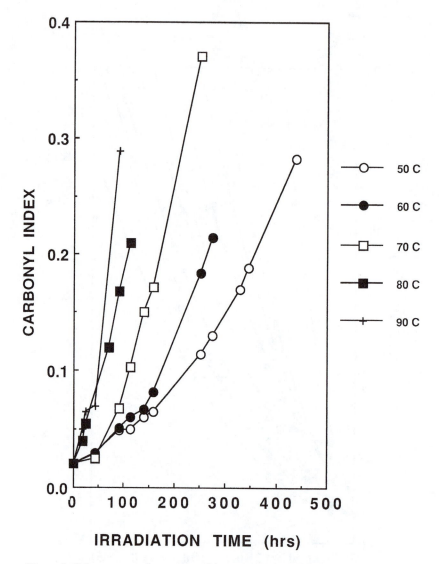

Figure 6. C.I. versus irradiation time (h) in SEPAP for LDPE film (100-μm thick) containing pigment G (0.5% w/w) at various temperatures.

grade, rutile types and then the more heavily coated durable grades. At this temperature the anatase is seen to be more active than the uncoated rutile grade I.

At 90 °C the order in pigment activity in terms of stability is H > Control > F > E > G > D > C and B > I > A. Again the order in activity is similar; the super-durable rutile grade is the more effective stabilizer, and the un-

Table IV. Time (h) to 0.1 C.I. for Pigmented LDPE Films during Irradiation

Sample	50 °C	60 °C	70 °C	80 °C	90 °C
Z (control)	190	177	112	90	86
A	63	46	24	12	7
B	126	100	65	35	36
C	145	127	80	30	36
D	193	156	109	48	40
E	254	233	140	67	60
F	249	156	150	105	75
G	226	175	123	59	48
H	296	223	128	119	90
I	116	82	63	16	12

coated and coated anatase and lightly coated fine-crystal, rutile types are the most active pigments. From this data and that shown in Table III and Figure 4, the photooxidative behavior of titanium dioxide pigments could be related to hydroperoxide formation during processing and thermal oxidation. The light stabilizing effects are evidently due to the pigments operating as UV screeners as well as absorbers. However, there will be a competitive effect with regard to the ability of the pigment to sensitize the thermal oxidation of the polymer.

Activation Energies. From the C.I. rate curves, rate constants were obtained. Arrhenius plots were obtained from the data, from which apparent activation energies (kJ/mole) were determined for carbonyl formation. These C.I. values are shown in Table V, and the activation energies are plotted in Figure 8 against the film sample letter. These values will be subject to some degree of error because the oxidation reaction rates are assumed to be first order and therefore are not absolute and are only relative. We also assume that the pigment is well dispersed and that for a 100-μm film the oxidation is considered homogeneous. From Figure 8, the activation energies for thermal oxidation are significantly higher than those for photooxidation.

Overall, during thermal oxidation the activation energies for the pigmented films are lower than that of the unpigmented control (191.86 kJ/mole), and this finding is consistent with the data showing that all the pigments operate as thermal sensitizers. For photooxidation, the activation energies are very similar to that of the control (31.02 kJ/mole). The only exception is sample D. The slightly higher values for the pigments are indicative of the fact that on photooxidation they operate more as stabilizers and that temperature is more important in controlling the overall rate of oxidation of the polymer.

Stabilization of Polypropylene Film. The data in Tables V and VI compare the embrittlement times (0.06 C.I.) for the stabilized polypropylene films with anatase and rutile pigments, respectively. Notably, in poly-

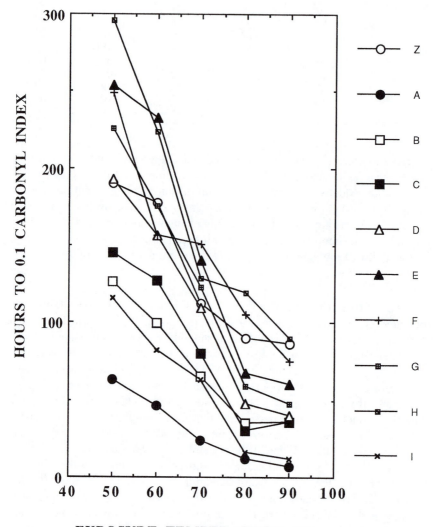

Figure 7. Time to 0.1 C.I. versus irradiation exposure temperature for LDPE films (100-μm thick) containing no TiO₂ (Z; ○) and 0.5% w/w of pigments. For key, see Table I.

Table V. Interactions of Anatase with Stabilizers in Photooxidation of Polypropylene Film

Additives	Time (h) to 0.06 C.I.
None	35
Irganox 1010	55
Tinuvin 770	500
Irganox 1010 + Tinuvin 770	455
Uncoated (UN) Anatase	10
Coated (CO) Anatase	15
(UN)Anatase + Irganox 1010	45
(CO)Anatase + Irganox 1010	50
(UN)Anatase + Tinuvin 770	70
(CO)Anatase + Tinuvin 770	110
(UN)Anatase + Irganox 1010 + Tinuvin 770	40
(CO)Anatase + Irganox 1010 + Tinuvin 770	95

NOTE: Film was 100-μm thick and was processed in a Brabender Plasticorder at 190 °C for 10 min. Pigments were 1% by weight; Irganox 101 and Tinuvin 770 were 0.1% by weight.

propylene, hindered phenolic antioxidants antagonize the light stabilizing effect of hindered-piperidine light stabilizers, and the effect is dependent on the structure of the two stabilizer types (*1, 12*). This effect is why polypropylene was chosen for study instead of polyethylene, because polyethylene gives variable effects (*13*). In the case of the anatase pigment, stabilization of the polymer is not effective when compared with the control films. Thus, even in the presence of a hindered-piperidine light stabilizer, the anatase behaves as a powerful photosensitizer

The coated type has only a slightly greater protective effect than the uncoated type. Also, both anatase pigments enhance the antagonistic effect between the hindered-phenolic antioxidant and the hindered-piperidine stabilizer. On the other hand, the rutile pigments synergize effectively with both stabilizers, and the coated grade is more effective in each case (Table VI). However, the antagonism between the two stabilizers is enhanced in the presence of rutile pigments, and the effect is more pronounced in the presence of the uncoated pigment. The enhanced antagonism may be associated with the ability of the pigments to adsorb the stabilizers onto the pigment particle surfaces and thereby enhance their interaction through the photocatalytic oxidation mechanism.

Conclusions

Our results indicate that for thermal oxidative degradation over the temperature range 50–90 °C, all the titanium dioxide pigments behave as thermal (catalytic) sensitizers. The nature of the pigment appears to control the rate of oxidation at lower temperatures, and uncoated and coated anatase and un-

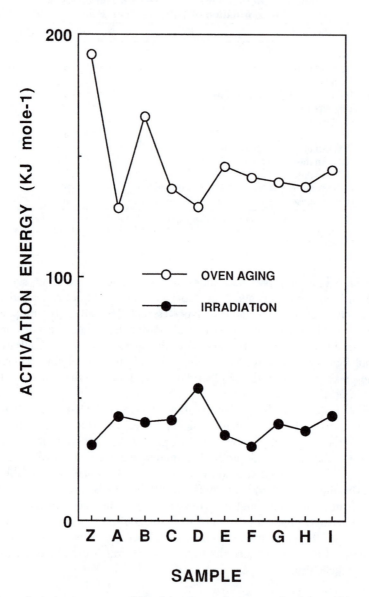

Figure 8. Activation energy (kJ/mole) versus polymer sample during (○) oven aging and (●) irradiation in the SEPAP.

**Table VI. Interactions of Rutile with Stabilizers in
Photooxidation of Polypropylene Film**

Additives	Time (h) to 0.06 C.I.
None	35
Irganox 1010	55
Tinuvin 770	500
Irganox 1010 + Tinuvin 770	455
Uncoated (UN) Anatase	15
Coated (CO) Anatase	20
(UN)Anatase + Irganox 1010	180
(CO)Anatase + Irganox 1010	265
(UN)Anatase + Tinuvin 770	580
(CO)Anatase + Tinuvin 770	850
(UN)Anatase + Irganox 1010 + Tinuvin 770	420
(CO)Anatase + Irganox 1010 + Tinuvin 770	740

NOTE: Film was 100-μm thick and was processed in a Brabender Plasticorder at 190 °C for 10 min. Pigments were 1% by weight; Irganox 101 and Tinuvin 770 were 0.1% by weight.

coated fine-crystal rutile types are the most active. The coated fine-crystal rutile grades are less active in promoting oxidation of the polymer and are followed by the least active, coated, durable, rutile grades. The rates of carbonyl formation are also autocatalytic above 70 °C and are less dependent on the pigment type as the temperature is increased.

All the pigments operate as thermal aging sensitizers for hydroperoxide formation during processing. The trends in activity on photooxidation are similar to those for thermal oxidative aging, except that the rates of carbonyl formation are autocatalytic over the whole temperature range studied, and some of the pigments behave as stabilizers. The stabilizing effect of many of the pigments, however, is converted into a sensitizing effect as the temperature is increased during irradiation. The role of the sensitizing or stabilizing effects of the pigment on the rate of polymer oxidation is more important during photooxidation. From the rates of carbonyl formation, activation energies were determined and were significantly higher for thermal aging (128.40–191.86 kJ/mole) than for photooxidation (30.79–43.08 kJ/mole). For thermal aging, all the pigments lowered the activation energy for carbonyl formation, whereas for photooxidation many of the pigments had little overall effect apart from a small increase. However, these values were taken in a carbonyl region where only minor differences existed in rates of polymer oxidation. The actual times taken for the polymer samples to achieve a given C.I. value will be different and therefore not related to the activation energies obtained here.

The rates of thermal and photooxidative aging of LDPE are interrelated with the catalytic formation of hydroperoxides by the titanium dioxide pigments during processing, and the pigments play an important role in the photothermal oxidation of the polymer. This effect is, in turn, dependent on the crystal size and structure of the titanium dioxide pigment used and the nature of its surface treatment.

In terms of the light stabilization of polypropylene, the use of a hindered-piperidine light stabilizer, Tinuvin 770, is ineffective against the photosensitizing effect of anatase. In the case of rutile pigments, strong synergism is observed with the antioxidant and light stabilizer. The antagonism between the antioxidant and light stabilizer is dependent on the nature of the surface treatment on the rutile particle surface.

Acknowledgments

We would like to thank Tioxide (U.K) Ltd., Stockton-on-Tees, United Kingdom, for partial financial support for Hassan Katami. We also thank R. E. Day, C. Watson, and J. Lawson of the Research Laboratories for helpful discussions throughout.

References

1. Allen, N. S. In *Degradation and Stabilisation of Polyolefins;* Allen, N. S., Ed.; Elsevier Science: London, England, 1983; Chapter 8, p 337.
2. Allen, N. S.; McKellar, J. F. *Photochemistry of Dyed and Pigmented Polymers;* Applied Science: London, England, 1980; p 247.
3. Kaempf, G.; Papenroth, W.; Holm, R. *J. Paint Technol.* **1974,** *46,* 56.
4. Allen, N. S.; McKellar, J. F.; Wilson, D. *J. Photochem.* **1977,** *7,* 319.
5. Allen, N. S.; Bullen, D. J.; McKellar, J. F. *J. Mater. Sci.* **1979,** *14,* 1941.
6. Day, R. E. *Polym. Degrad. Stab.* **1990,** *29,* 73.
7. Allen, N. S.; McKellar, J. F.; Wood, D. G. M. *J. Polym. Sci. Polym. Chem. Ed.* **1975,** *13,* 2319.
8. Allen, N. S.; Gardette, J. L.; Lemaire, J. *Dyes Pigments* **1982,** *3,* 295.
9. Carlsson, D. J.; Garton, A.; Wiles, D. M. *Macromolecules* **1976,** *9,* 695.
10. Carlsson, D. J.; Wiles, D. M. *Macromolecules* **1969,** *2,* 597.
11. Carlsson, D. J.; Lacoste, J. *Polym. Degrad. Stab.* **1991,** *32,* 377.
12. Allen, N. S.; McKellar, J. F. *Plast. Rubber Mater. Appl.* **1979,** *4,* 170.
13. Allen, N. S.; Hamidi, A.; Williams, D. A. R.; Loffeleman, F. F.; MacDonald, P.; Susui, P. *Plast. Rubber Proc. Appl.* **1986,** *6,* 109.

RECEIVED for review January 26, 1994. ACCEPTED revised manuscript January 10, 1995.

LIFETIME PREDICTION

The growing emphasis on quality and reliability of products has resulted in increased interest in the ability to predict the aging behaviors and useful lifetimes of polymers in application environments. A reliable lifetime prediction capability would be highly desirable both for assuring the durability of commercialized products, as well as for selecting among new materials or formulations under current development and evaluation, for use in long-term applications. Unfortunately, simple accelerated aging tests have developed a reputation for being unreliable. Problems in extrapolating short-term laboratory aging results to application environments and timescales can arise because of changes in degradation mechanism; examples include such phenomena as temperature-dependent physical transitions in materials or oxygen-diffusion-limited effects under accelerated conditions.

The final section of this book reviews advances in understanding differences between accelerated and "real world" aging processes in polymers, often making use of sensitive techniques for monitoring degradation. The improvements in fundamental understanding of aging processes under different stress levels are being incorporated in the development of advanced aging methods. In addition, mathematical models of key degradation processes are under development, which are applicable both to short-term degradation (e.g., accelerated aging tests, or melt processing) and to long-term aging applications.

Prediction of Elastomer Lifetimes from Accelerated Thermal-Aging Experiments

Kenneth T. Gillen, Roger L. Clough, and Jonathan Wise

Sandia National Laboratories, Albuquerque, NM 87185

We describe a study aimed at validating the Arrhenius lifetime prediction methodology. Ultimate tensile measurements were made on three elastomers after elevated temperature exposures. Although tensile elongation could be analyzed using the Arrhenius approach, tensile strength could not. Modulus profiles resolved this inconsistency by showing that complex, diffusion-limited oxidation effects (involving surface hardening) were present. The surface (equilibrium) modulus values were correlated with elongation and indicated that elongation is Arrhenius because cracks initiated at the hardened surface and then immediately propagated through the material. Tensile strength, on the other hand, is non-Arrhenius because it depends on the entire material cross section. We also introduce a methodology based on monitoring oxygen consumption rates that allows us, for the first time, to test the Arrhenius extrapolation assumption.

BECAUSE ELASTOMERIC MATERIALS are commonly used for long-term applications (years to decades) at room (ambient) or at moderately elevated temperature conditions, it is important to be able to predict their lifetimes. A common approach involves accelerating the chemical reactions underlying the degradation by aging at several elevated temperatures and monitoring the degradation through changes in ultimate tensile properties [elongation and tensile strength (TS)]. These accelerated thermal-aging results are generally extrapolated to use-temperature conditions by using the Arrhenius methodology (1). This method is based on the observation that the temperature dependence of the rate of an individual chemical reaction is typically

0065–2393/96/0249–0557$12.00/0

proportional to $\exp(-E_a/RT)$, where E_a is the Arrhenius activation energy, R is the ideal gas constant, and T is the absolute temperature.

In general, the aging of a polymer can be described by a series of chemical reactions, each assumed to have Arrhenius behavior. Kinetic analysis of these reactions results in a steady-state rate expression with an Arrhenius temperature dependence, where E_a now represents the effective activation energy for the mix of reactions underlying the degradation. If this relative mix of reactions remains unchanged throughout the temperature range under analysis, a linear relationship will exist between the logarithm of the time to a certain amount of material property change and $1/T$. The value of E_a is then obtained from the slope of the line. If, on the other hand, the relative mix of degradation reactions changes with changes in T, the effective E_a would be expected to change, and this change would lead to curvature in the Arrhenius plot.

Although, at first glance, Arrhenius behavior seems to be valid in many (though not all) instances, closer examination leads to some troubling concerns. For instance, ultimate tensile elongation results are often used to "confirm" Arrhenius behavior, even though the ultimate TS data available from the same mechanical-property testing (typically not reported) are often non-Arrhenius. In addition, many workers use only a single point or a few selected points from each temperature curve for their Arrhenius analysis, thereby eliminating much of their data and significantly depreciating the value of any conclusions.

When elastomers age in air environments, the chemical reactions dominating the long-term degradation usually involve the oxygen dissolved in the material (2). When attempts are made to accelerate these reactions by using elevated temperature (e.g., in air-circulating ovens), complications caused by diffusion-limited oxidation (DLO) typically enter (3–7). DLO can occur whenever the rate of oxygen consumption within the material is greater than the rate at which it can be resupplied by diffusion from the surrounding air atmosphere. This effect results in heterogeneously oxidized materials (equilibrium oxidation occurs at the sample surfaces and reduced or nonexistent oxidation occurs in interior regions). Because this physical phenomenon depends on both temperature and geometry (i.e., sample thickness), understanding its significance in accelerated simulations is critical to confident predictive extrapolations. A second common problem with the Arrhenius approach is confirming the assumption that the value of E_a derived under the accelerated conditions remains constant at lower (extrapolated) temperatures.

In an attempt to address these concerns, we have been critically examining the Arrhenius approach to better understand its capabilities and limitations. Our first paper (8) on this subject concentrated on the importance and mechanism of DLO effects for thermal aging of neoprene and styrene–butadiene rubber (SBR) elastomers. This chapter describes a detailed study of a nitrile rubber in which we apply the Arrhenius approach to ultimate tensile-property data from accelerated experiments and show that the complex results can be rationalized by understanding the underlying DLO effects. We also describe some preliminary results in which we used sensitive oxygen-consumption measurements to determine whether the E_a derived from accelerated conditions remains constant at the low temperatures appropriate to the extrapolation region. A future paper (9) will provide a more detailed discussion of this work.

Experimental

Compression-molded sheets (~2.0-mm thick) of a typical commercial nitrile rubber formulation (100 parts Hycar 1052 resin, 5 parts per hundred [pph] zinc oxide, 1 pph stearic acid, 1.5 pph 2246 [hindered phenol] antioxidant, 65 pph N774 carbon black, 15 pph Hycar 1312, 1.5 pph sulfur, and 1.5 pph 2,2'-dithiobis[benzothiazole]) and a typical commercial neoprene rubber formulation (100 parts Neoprene GN, 4 pph magnesia, 0.5 pph stearic acid, 5 pph zinc oxide, 60 pph hard clay, and 2 pph 2246 antioxidant) were obtained from Burke Rubber Co. The SBR was a proprietary material obtained in sheets (150 × 150 × 2.2 mm) from Parker Seal Corporation. The copolymer had a styrene:butadiene monomer ratio of 23:77 and number-average molecular weight ~425,000. The material was cross-linked by using a sulfur cure and contained 37% by weight carbon black and a hindered phenol stabilizer.

Strips approximately 12-mm wide and 150-mm long were cut from the sheets and aged in air-circulating ovens (± 1 °C stability). Tensile testing (12.7-cm/min strain rate; 5.1-cm initial jaw separation) was performed using an Instron Model 1130 testing machine equipped with pneumatic grips and having an extensometer clamped to the sample. For each aging time at a given temperature, three samples of the nitrile and neoprene materials typically were tensile tested; for the SBR material, a single sample was tested under each aging condition. Modulus profiles were obtained on sample cross sections by using our modulus profiling instrument, which was described in detail elsewhere (10). This instrument allows us to obtain quantitative values of the inverse tensile compliance, D^{-1}, a quantity closely related to the tensile modulus, with a resolution of ~50 µm. Tensile and modulus (D^{-1}) values for the unaged materials are given in Table I.

Oxygen consumption measurements were performed by sealing known amounts of the material with 16 cm Hg of O_2 in glass containers of known volume. To avoid DLO artifacts and therefore to assure that homogeneous oxidation occurred during the measurements, the material was cut into sufficiently thin pieces (5–7). The containers were thermally aged for time periods chosen to consume ~40% of the O_2 (to make the average partial pressure of O_2 during aging approximately equal to ambient conditions in Albuquerque, NM). The remaining O_2 content was determined using gas chromatography (7).

Results and Discussion

Nitrile Rubber.   Figure 1 shows normalized elongation results (normalized elongation is e/e_0, where e_0 is initial elongation value) at the indicated aging temperatures for the nitrile rubber. Each data point typically represents the average result from three identically aged samples. Estimated experimen-

Table I. Initial Mechanical Properties

Material	e_0 (%)	TS_0 (MPa)	D^{-1} (MPa)
Nitrile	570 ± 30	15.7 ± 0.7	4.3 ± 0.2
SBR	260 ± 30	12.8 ± 1	4.95 ± 0.2
Neoprene	620 ± 50	15.2 ± 1	7.5 ± 0.3

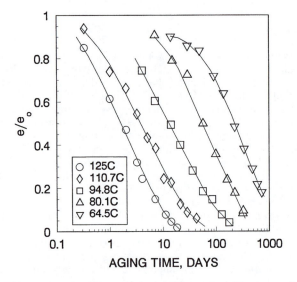

Figure 1. Ultimate tensile elongation (e) of the nitrile rubber normalized to its unaged value (e_0) versus aging time in air at the indicated temperatures.

tal uncertainties range from about ±0.05 at high values of e/e_0 to about ±0.015 at low values. Smooth curves drawn through these data are used to construct conventional Arrhenius plots at 3 levels of damage (e/e_0 = 0.75, 0.5, and 0.25). These plots (Figure 2) are linear, have identical slopes (from which E_a = 22 ± 2 kcal/mol is calculated), and therefore confirm Arrhenius behavior. Having determined E_a from the processed data, we applied the principle of time–temperature superposition (6, 11, 12) to shift all of the unprocessed raw data from Figure 1 to a selected reference temperature, T_{ref}. This process is accomplished by multiplying the times appropriate to experiments at each temperature, T, by a shift factor a_T:

$$a_T = \exp\left[\frac{E_a}{R}\left(\frac{1}{T_{ref}} - \frac{1}{T}\right)\right] \tag{1}$$

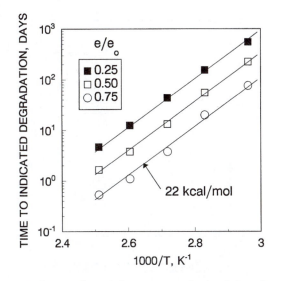

Figure 2. Arrhenius plots of elongation results for the nitrile rubber.

Using E_a = 22 kcal/mol, we obtained excellent superposition, as shown for T_{ref} = 50 °C in Figure 3. However, use of the same shift factor for the normalized TS data did not result in superposition (Figure 4). Indeed, the TS data could never be superposed because the results dropped in the later stages at high temperatures but increased or dropped less at other temperatures. Such behavior indicates fundamental temperature-dependent changes in the degradation mechanism that are contrary to the assumptions of the Arrhenius methodology.

Therefore, we are left with a dilemma as to why the Arrhenius approach works well for the elongation but fails for the TS. This dilemma can be resolved through the use of modulus profiling data. Our modulus profiling apparatus (10) allows us to quantitatively map modulus values across the cross section of degraded samples with a spatial resolution of approximately 50 μm. Representative modulus profiles for the nitrile rubber at selected aging times and temperatures are shown in Figure 5. At the highest temperature (125 °C), heterogeneity in the modulus is evident at the earliest aging times and becomes quite pronounced later. This effect is caused by DLO in which the rate of oxygen consumption within the material is greater than the rate at which it can be replenished by diffusion from the surrounding air atmosphere (5–8). For aging experiments at lower temperatures (Figure 5), the importance

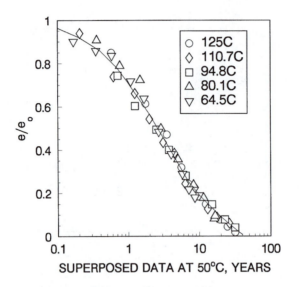

Figure 3. Time–temperature superposition of the e/e₀ data from Figure 1 obtained by using $E_a = 22$ kcal/mol.

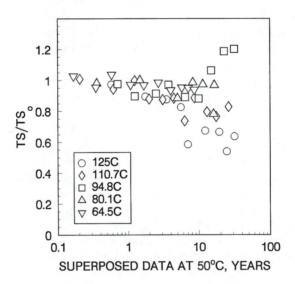

Figure 4. Time–temperature superposition of the normalized TS data for the nitrile rubber obtained by using $E_a = 22$ kcal/mol.

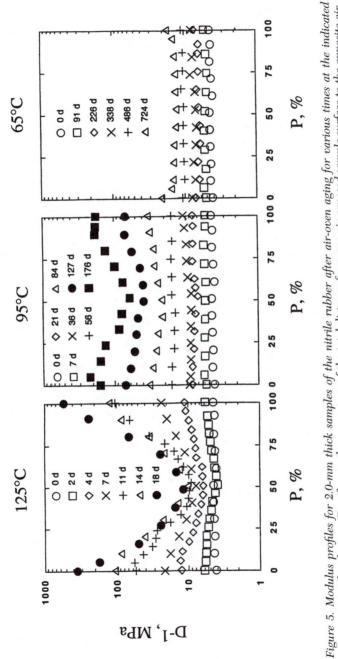

Figure 5. Modulus profiles for 2.0-mm thick samples of the nitrile rubber after air-oven aging for various times at the indicated temperatures. The abscissa, P, refers to the percentage of the total distance from one air-exposed sample surface to the opposite air-exposed surface.

This change occurs because the oxygen consumption rate decreases more rapidly ($E_a = 22$ kcal/mol) with decreasing temperature than does the oxygen diffusion rate ($E_a \approx 9$ kcal/mol). Later in the degradation, however, hardening (modulus increases) leads to a significant reduction in the oxygen diffusion rate (8) and results in the delayed appearance of heterogeneity caused again by DLO effects. Given this complexity, it is now easy to rationalize the TS results shown in Figure 4: TS is a property that depends on the force at break *integrated* over the cross section of the material, and the nitrile rubber samples aged at different temperatures clearly experience very different degrees of degradation in their interior regions.

The modulus values at the sample *surfaces*, however, are not affected by DLO anomalies. Figure 6 shows surface modulus values of the nitrile rubber versus time and temperature; these values correspond to the modulus change expected under equilibrium air (i.e., non-DLO) conditions. Following the procedure used for the elongation results in Figures 1–3, we shifted these surface modulus values to a 50 °C reference temperature by using an Arrhenius shift factor chosen to achieve the best superposition. The best superposition occurred with $E_a = 22 \pm 2$ kcal/mol (Figure 7). Because this value represents the E_a appropriate to the underlying oxidative degradation reactions, and because it is exactly the same as that found earlier for the elongation, it is clear why Arrhenius behavior occurred for the elongation. When a material is tensile tested, cracks can be expected to initiate first at the hardened, oxidized surface; if such cracks quickly propagate through the remainder of the material's cross section, the elongation value will be determined by the oxidative hardening occurring at the surface. A further confirmation that hardening of the surfaces determines the elongation results can be seen in Figure 8, which shows the experimental correlation for these properties. This correlation indicates that severe mechanical degradation ($e/e_0 \approx 0.1$) will occur when the surface modulus value increases by approximately one order of magnitude.

The correlation between modulus and elongation is anticipated in most theories of rubber elasticity (13, 14). For both Gaussian and non-Gaussian statistical treatments, the extension ratio at break, λ, is predicted to be proportional to $(M_c)^{0.5}$, where M_c represents the molecular weight between crosslinks. In addition, the statistical theories predict that the low strain modulus is directly proportional to the $(M_c)^{-1}$. This relationship implies that λ, defined as

$$\lambda = 1 + 0.01e \tag{2}$$

where e is the elongation in percent, should be directly proportional to the inverse square root of the modulus.

A test of this prediction for the nitrile data is shown in Figure 9. The results qualitatively follow the predicted correlation but quantitatively follow

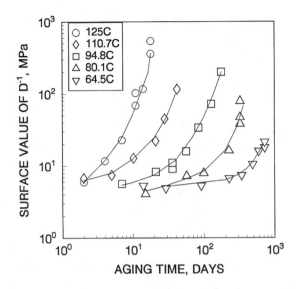

Figure 6. *Surface modulus values for the nitrile rubber versus aging time and temperature.*

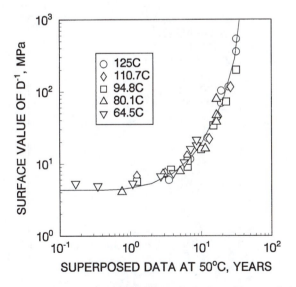

Figure 7. *Time–temperature superposition of nitrile rubber surface modulus values obtained by using $E_a = 22$ kcal/mol.*

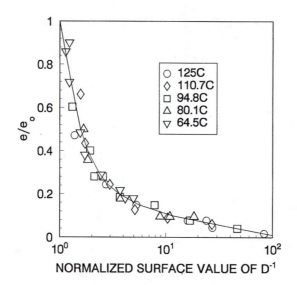

Figure 8. *Normalized elongation (e/e₀) plotted versus the normalized surface modulus value for the nitrile rubber at the indicated temperatures.*

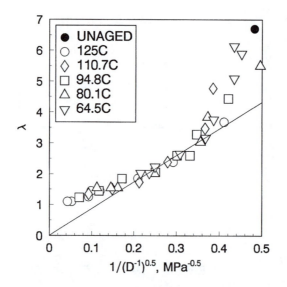

Figure 9. *Extension ratio, λ, plotted versus the inverse square root of the surface modulus for the nitrile rubber at the indicated temperatures.*

only from $\lambda \approx$ 1.5–3. There are numerous reasons for the quantitative failure of the statistical theories. First, the nitrile rubber is a filled material. In filled materials, the mechanical properties will depend on the type and amount of filler; its size, shape, agglomeration behavior; and the chemical nature of its surfaces. These characteristics make comparisons with theory much more difficult (*14*). Second, besides oxidative aging leading to a reduction in M_c, the incorporation of polar species containing oxygen may also modify the mechanical properties. Oxidation processes could also influence the chemical nature of the filler particle surfaces and thereby affect the adhesion of filler to elastomer and hence the mechanical properties. Finally, as λ approaches unity, modulus values begin to rise very rapidly and imply that the material is approaching its glassy state where rubber elasticity theory is inappropriate.

SBR and Neoprene Results.   Our earlier study of the SBR and neoprene materials first used modulus profiling results to show that complex time- and temperature-dependent heterogeneous aging effects occurred during accelerated air-oven aging exposures. By incorporating oxygen permeation and consumption measurements, plus antioxidant assay techniques, we then showed that DLO dominated the observed heterogeneities (*8*). Figures 10 and 11 show modulus profile results for SBR and neoprene, respectively, at three temperatures. Similar to the analysis done in Figures 6 and 7 for the nitrile rubber, we plotted the surface modulus values versus time and temperature and found the E_a that gives the best superposition of the results. For the SBR, this value occurred at $E_a \approx$ 24 $\pm$ 2 kcal/mol; the resulting superposition at a reference temperature of 50 °C is shown in Figure 12. For the neoprene, the best superposition occurred for $E_a \approx$ 21.5 $\pm$ 2 kcal/mol; the superposed results are shown in Figure 13.

Figures 10 and 11 show that, similar to the nitrile rubber, dramatic surface hardening occurs for the SBR and neoprene materials under air-oven aging conditions. We would therefore expect cracks to initiate at the surfaces during tensile testing. If these cracks immediately propagate through the material, the ultimate tensile elongation values should time–temperature superpose with the same value of E_a as the surface modulus values (24 kcal/mol for SBR and 21.5 kcal/mol for the neoprene). This hypothesis is confirmed by the superposed elongation results shown in Figures 14 and 15. By comparing the results from Figures 12 and 14 and from Figures 13 and 15, it is seen that severe mechanical degradation ($e/e_0 \approx$ 0.1) occurred when the surface modulus value increased by factors of about 5 and about 8, respectively, for the SBR and neoprene materials.

Although scatter in the TS data for the SBR material is too large to make definitive conclusions, poor superposition occurred when the neoprene TS data was shifted by using a 21.5 kcal/mol E_a, as seen in Figure 16. This result, of course, was anticipated for reasons similar to those discussed earlier for the nitrile TS results.

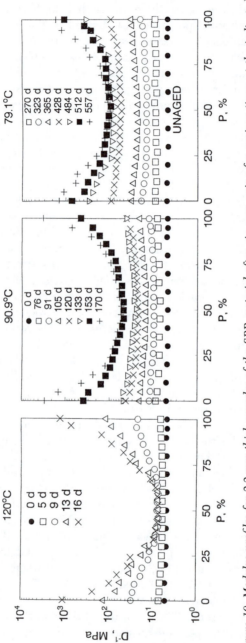

Figure 10. Modulus profiles for 2.2-mm thick samples of the SBR material after air-oven aging for various times at the indicated temperatures.

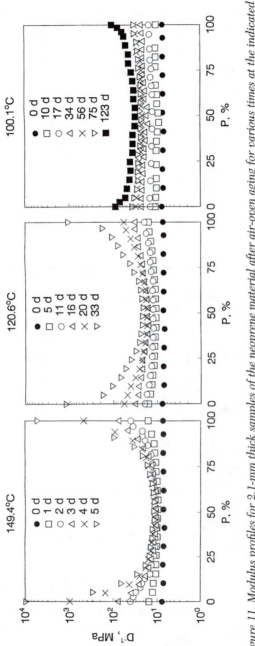

Figure 11. Modulus profiles for 2.1-mm thick samples of the neoprene material after air-oven aging for various times at the indicated temperatures.

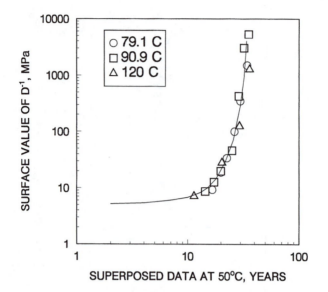

Figure 12. Time–temperature superposition of SBR surface modulus values obtained by using $E_a = 24$ *kcal/mol.*

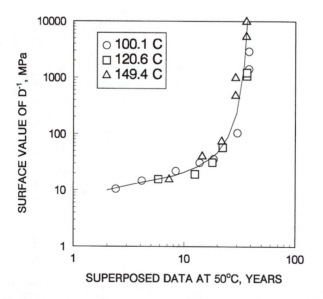

Figure 13. Time–temperature superposition of neoprene rubber surface modulus values obtained by using $E_a = 21.5$ *kcal/mol.*

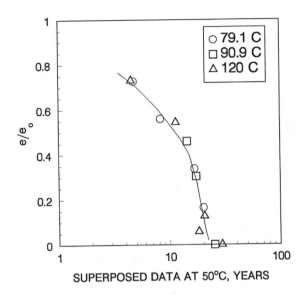

Figure 14. Time–temperature superposition of SBR-normalized ultimate tensile elongation data (e/e_0) obtained by using $E_a = 24$ kcal/mol.

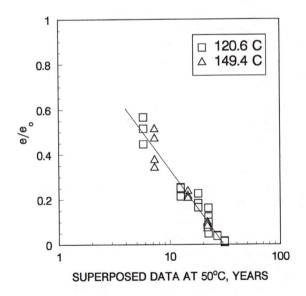

Figure 15. Time–temperature superposition of neoprene-normalized ultimate tensile elongation data (e/e_0) obtained by using $E_a = 21.5$ kcal/mol.

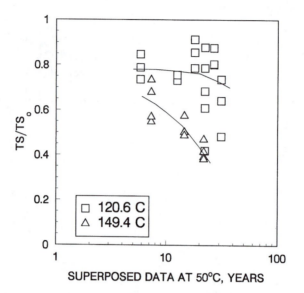

Figure 16. Time–temperature superposition of neoprene-normalized ultimate TS data (TS/TS0) obtained by using E_a = 21.5 kcal/mol.

Oxygen Consumption Measurements.   From a combination of the surface modulus and tensile elongation results for the nitrile, SBR, and neoprene materials, we concluded that the mechanical degradation under accelerated thermal aging conditions was dominated by oxidation reactions with E_a values of 22, 24, and 21.5 kcal/mol, respectively. The Arrhenius methodology assumes that these values of E_a can be used to extrapolate the accelerated results to lower-temperature use conditions and thereby allow predictions to be made at much longer times. The only method for rigorously confirming this assumption is to follow the degradation at the lower-use temperature until the end of the useful life of the material. Because this approach may take many years or decades, however, it is impractical.

An alternative approach would be to monitor the early stages of the degradation at the lower temperatures (perhaps even below the use temperature) by using an ultrasensitive technique closely correlated to the degradation mechanism. Because we have already determined that oxidation dominates the mechanical degradation of these three materials, one promising method would be to monitor oxygen consumption rates. Our approach involves taking measurements at several temperatures that encompass the lower part of the range used for the nitrile mechanical-property degradation measurements (95 °C to 65 °C) and extend into the extrapolation region (down to ambient temperature of 23 °C). If the consumption measurements are a valid means of testing the extrapolation assumption (i.e., if they are sensitive to the same oxidation reactions responsible for mechanical degradation), then time–temperature superposition of the consumption results should occur with an E_a similar to those found for the surface modulus and mechanical property results. Because oxygen consumption measurements represent macroscopic averages across the sample cross section, they are potentially prone to the same DLO artifacts noted for TS earlier. For this reason, we carried out our oxygen consumption experiments under conditions (thin enough samples) guaranteed to yield homogeneous (constant) oxidation (7).

Figure 17 shows the oxygen consumption results for the nitrile rubber. As usual, we used the time–temperature superposition approach to find values of E_a that yielded reasonable superposition. In this case, an $E_a = 20 \pm 2$ kcal/mol resulted in excellent superposition, as shown at $T_{ref} = 23$ °C in Figure 18. This result is equal, within the small experimental uncertainties, with the E_a results found for surface modulus and elongation (22 ± 2 kcal/mol). Therefore, it offers strong evidence for the validity of the extrapolation assumption in this case. We are currently generating similar results for a number of additional materials to further test the applicability of this approach.

Conclusions

In thermooxidative aging environments, most elastomers harden (i.e., their modulus increases) due to oxidation. For typical accelerated thermal-aging conditions, DLO effects are important, and these effects can be conveniently monitored by using modulus profiling techniques. By applying time–temper-

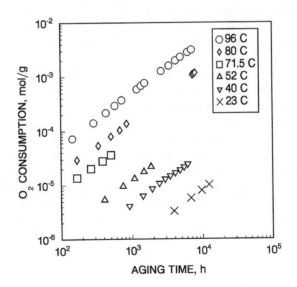

Figure 17. Oxygen consumption results for the nitrile rubber versus aging time at the indicated temperatures.

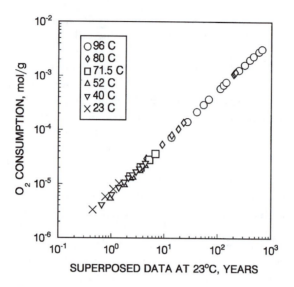

Figure 18. Time–temperature superposition of the nitrile rubber oxygen consumption results shown in Figure 17 obtained by using E_a = 20 *kcal/mol.*

ature superposition to the surface modulus results obtained at various temperatures for three elastomeric materials (a nitrile rubber, an SBR, and a neoprene rubber), we were able to determine the Arrhenius E_a for the oxidation reactions underlying the hardening (22, 24, and 21.5 kcal/mol, respectively). Excellent time–temperature superposition is achieved when the same E_a values are used to shift the ultimate tensile elongation data, and this result implies that the hardening at the surfaces determines changes in this mechanical property. Because cracks should initiate at the hardened surface of a material, surface cracks, once initiated, must rapidly propagate through the cross section of the material and lead to mechanical failure. In contrast, changes in TS at break (a property determined by the integrated strength over the sample cross section) are non-Arrhenius, as would be expected given the spatially nonuniform degradation seen at different temperatures.

When Arrhenius behavior is found for mechanical property data, it is often necessary to extrapolate the results to lower temperatures, assuming that the same E_a is appropriate in the extrapolation region. A method based on oxygen consumption measurements is introduced to allow verification of this assumption. Results on the nitrile rubber sample are consistent within experimental uncertainty with this assumption. This method may have general applicability to samples dominated by oxidation processes.

Acknowledgments

This work was performed at Sandia National Laboratories, supported by the U.S. Department of Energy under Contract DE-AC04-94AL85000. We thank G. M. Malone for able technical assistance.

References

1. Dakin, T. W. *AIEE Trans.* **1948,** 67, 113–118.
2. Grassie, N.; Scott, G. *Polymer Degradation and Stabilisation;* Cambridge University: Cambridge, England, 1985; p 86.
3. van Amerongen, G. J. *Rubber Chem. Technol.* **1964,** 37, 1065–1152.
4. Cunliffe, A.; Davis, A. *Polym. Degrad. Stab.* **1982,** 4, 17–37.
5. Gillen, K. T.; Clough, R. L. In *Handbook of Polymer Science and Technology;* Cheremisinoff, N., Ed.; Dekker: New York, 1989; Vol. 2, pp 167–202.
6. Gillen, K. T.; Clough, R. L. In *Irradiation Effects on Polymers;* Clegg, D. W.; Collyer, A. A.; Eds.; Elsevier Applied Science: London, England, 1991; pp 157–223.
7. Gillen, K. T.; Clough, R. L. *Polymer* **1992,** 33, 4358–4366.
8. Clough, R. L.; Gillen, K. T. *Polym. Degrad. Stab.* **1992,** 38, 47–56.
9. Wise, J.; Gillen, K. T.; Clough, R. L. *Polym. Degrad. Stab.*, in press.
10. Gillen, K. T.; Clough, R. L.; Quintana, C. A. *Polym. Degrad. Stab.* **1987,** 17, 31–47.
11. Ferry, J. D. *Viscoelastic Properties of Polymers;* Wiley: New York, 1970; Chapter 11.
12. Gillen, K. T.; Clough, R. L. *Polym. Degrad. Stab.* **1989,** 24, 137–168.
13. Treloar, L. R. G. *The Physics of Rubber Elasticity*, 3rd ed.; Clarendon: Oxford, England, 1975.
14. Mullins, L. In *The Chemistry and Physics of Rubber-Like Substances;* Bateman, L.; Ed.; McLaren: London, England, 1963; pp 301–328.

RECEIVED for review December 16, 1993. ACCEPTED revised manuscript February 17, 1995.

Mechanisms of Photooxidation of Polyolefins: Prediction of Lifetime in Weathering Conditions

Jacques Lemaire, Jean-Luc Gardette, Jacques Lacoste, Patrick Delprat, and Daniel Vaillant

Laboratoire de Photochimie Moléculaire et Macromoléculaire, Unité de Recherche Associée, Centre National de la Recherche Scientifique 433 - Université Blaise Pascal, 63177 Aubiere France

The fate of solid synthetic polymeric materials aging outdoors could be predicted from accelerated laboratory experiments using a simulation or a mechanistic approach. In the simulation methodology, the relevance of the observed phenomena is deduced from the similarities among the physical and chemical aggressions in natural and artificial exposures. In the mechanistic approach, the relevance of the observed phenomena is controlled at the molecular level through the recognition of the chemical changes in the solid matrix. In the mechanistic approach, which considers the polymeric systems as photochemical reactors, it is essential to analyze the intermediate and final products at a very low reaction extent. The chemical analysis is mainly based on spectrophotometric or microspectrophotometric techniques coupled with specific gaseous reagents. The comparison between weathering kinetics and accelerated artificial aging kinetics should be used only when the nature and the spatial distribution of the various photoproducts are similar. As examples, the photooxidations of polypropylene and photo(bio)degradable low-density polyethylene are presented and recent results are emphasized.

THE PREDICTION OF THE LONG-TERM LIFETIME of polymeric materials, especially in outdoor conditions, is a very difficult problem. The progress of scientific knowledge in this field, whose development began no sooner than 1970, came up against a well-established empiricism lasting for more than 50 years. This tenet is still the background of more than 90% of the world activity in control, development, and even research. As soon as they were used,

0065–2393/96/0249–0577$12.50/0

polymers have shown problems of durability. In the 1950s, some experimental methods of studying the phenomena of degradation urgently had to be developed to try to solve these difficulties. Because the phenomena involved in long-term evolution looked complex, the chosen techniques could only be empirical. At that time, polymers were considered *black boxes*, where physical and chemical stresses of the environment were artificially applied with the hope of observing in the laboratory the same phenomena as in natural conditions. These phenomena were only indirectly studied from the variations of physical properties in the use conditions of these polymers (e.g., mechanical properties, aspect of surfaces, transparency, and discoloration). The tentative approach based on a *simulation* of strains was both successful and unsuccessful but without any possibility to justify each case.

In fact, the long-term behavior of polymers exposed to UV radiations, heat, oxygen, and water could be predicted in the laboratory with a good coefficient of security through experimental techniques based on scientific data obtained from fundamental research. In 1970, a fundamental approach to the phenomena of photoaging of polymers was not thought to be fruitful, because a very complex situation existed that might not be able to be simplified. The interactions of a polychromatic light, of a solid state more or less organized, and of a strong perturbation agent like oxygen were not supposed to be simple. The Laboratory of Photochemistry, University of Clermont-Ferrand, France, began to work in that particular field of fundamental research in 1972, because a bet was made on a possible stylization of these phenomena.

Some 12 years later, the experience gained from studying several thousand blends in conditions of accelerated photoaging showed that reproduction of chemical evolutions in conditions very similar to natural ones was possible in the laboratory. The exact mimesis of natural strains no longer had to be found, and it was sufficient to work in experimentally relevant conditions and to control, at a molecular scale, the similarity of the mechanisms of chemical evolution in weathering and in artificial aging conditions. An acceleration of the detrimental phenomena resulted only from an acceleration of chemical reactions, in most cases. The relevancy of the data collected could be checked throughout these studies based on SEPAP 12.24 and SEPAP 12.24H units, which were much easier to use compared with the simulation techniques (*1–4*).

Research of mechanisms of photochemical, thermal, and hydrolytic aging carried out for more than 20 years uncovered the following facts :

1. A polymer lives as a chemical reactor. Its degradation implies the appearance of generally low concentrations of chemical groups (for example oxidized groups), and this chemical evolution is responsible for the degradation of physical properties. In weathering, there are no examples of physical aging without chemical modifications.

2. The chemical evolution does not depend on the mechanical stresses. These stresses modify only the physical consequences of the chemical evolution and not the kinetics of the chemical reaction; therefore, the acceleration of the chemical evolution does not depend on the external and internal mechanical stresses. By considering the chemical evolution, laboratory data may be converted into durations of use in natural conditions on the basis of the acceleration of the chemistry (consequently, the chemical evolution is used as a base for any transfer of data).

3. The chemical evolution is a characteristic of the mechanism of evolution of any specific material. A precise formulation (polymer + filler + pigments or dyes + additives) must be characterized with a specific acceleration factor. A set of various materials or various formulations classified from artificial aging experiments cannot be transferred into use conditions without taking into account the acceleration factors that are necessarily different.

4. The acceleration of chemical evolutions is not only allowed but is a strict requirement because

    - it is impossible to extrapolate the data collected in the earlier phases of the evolution of the materials (any treatment based on homogeneous kinetics is not acceptable)
    - the material must reach a level of chemical evolution that leads to a mechanical degradation

5. On the other hand, the acceleration of the chemical events should be provoked while maintaining the relevancy of the phenomena (conditions inducing a lack of oxygen in solid polymers should be avoided, for example).

6. Any accelerated aging corresponds to an acceleration of the chemical evolution.

7. Eventually, the description of the chemical evolution must be associated with the criteria of degradation that have been selected:

    - description with products able to be observed by means of vibrational spectroscopy (IR, Raman) correlated with variations of mechanical properties
    - description with products able to be observed by means of electronic spectroscopy (e.g., UV, visible, colorimetric, and emission) correlated with variations of aspect

These basic principles led to a number of consequences.

1. When a chemical mechanism implies several processes of the same importance, there is no hope to accelerate all of them with the same

acceleration factor. The experimental tests in the laboratory misrepresent the reality.

2. When phenomena of physical transfer are superposed on the chemical evolution (i.e., oxygen diffusion or stabilizers migration), all these dynamic processes cannot be accelerated with the same factor of acceleration.

3. Only the systems where evolution is controlled by only one dynamic process can be transferred from accelerated laboratory conditions to nonaccelerated conditions of use. In fact, this case has been met fairly often in weathering, where the photooxidative process is the controlling one.

4. The migration of stabilizer additives is a major process when controlling the aging phenomena. Effects of the stabilizers should be tentatively examined, and the perturbation provoked by migrations should be evaluated.

5. On a practical point of view, it is acceptable to consider a common acceleration factor for different formulations of a basic polymer involving different additives with a similar mechanism of action.

6. Standardization and specifications should take into account the nature of the polymer under test and the nature of the formulation. The consideration of the aging conditions is also a prerequisite.

7. Devices used to simultaneously study several physical and chemical processes (simulation devices such as the Weatherometer and Xenotest) have to be considered as compromises, as a base of language whose advantage is to be common and whose disadvantage is to be only an approximation.

8. According to some Japanese works, some accelerated setups are developing based on the use of light intensities that are 3–5 times the daylight intensity. These apparatus are close to analytical devices, which means they are only able to examine the preponderant processes involved in photoaging (like a SEPAP 12.24 setup).

9. Durability studies of polymers in artificial aging (nonanalytical) and in natural aging are generally based on macroscopic criteria. These measures could be completed with analyses of the chemical evolution either of elementary layers (thickness from 5 to 40 μm) from surface to core of the samples, or of microzones on surfaces of small sizes (e.g., 10×10 μm or 10×100 μm). Profiles of the degradation products and of additives can thus be determined in thick systems. The basic technique is presently Fourier transform IR (FTIR) microspectrophotometry. However, FT-Raman microspectrophotometry will be combined with FTIR as the basic technique in the near future.

As emphasized in the previous sections, chemical analysis should allow the recognition of the evolution mechanisms in artificially accelerated conditions and throughout weathering. The required information deals with the exact chemical nature of the intermediate and final groups formed on the main polymer chains or branches and with the spatial repartition of these photoproducts in the exposed systems (films or plaques). When a common mechanism has been observed, that is, when *relevancy* is controlled, comparison between weathering kinetics and artificial aging kinetics supplies the required acceleration factor. As an example, the evolution of polymeric systems based on photostable blends of heterophasic polypropylene (PP) and on photo(bio)degradable polyethylene (PE) are described with some details.

Evolution of Heterophasic PP Systems

The oxidation of isotactic and atactic PP, initiated photochemically, thermally, or radiochemically, has been studied for years by many research groups. The general features of the oxidation mechanisms are fairly well described, although the initiation steps remain largely unknown. However, the formation of the primary radicals could have many different origins, and the analysis generally starts with the two macroradicals formed from the normal structure of PP.

A summary of most of the results reported in the literature (5–13) yields the mechanism of isotactic PP photooxidation represented in Scheme I. This mechanism indicates the formation and conversion of the main intermediate photoproducts (associated tertiary hydroperoxides, chain-end, and chain ketones) and the formation of the final oxidation products that accumulate in the matrix (e.g., tertiary alcohols). Although PP appears as the polymer whose photooxidation is the best understood, several questions remain unanswered.

- What is the exact origin of the acid groups formed?
- What is the relative importance of the two β-scission processes of the alkoxy radicals? (In 1973, the formation of methylated and chain-end ketones was shown to be favored on model hydroperoxides (*14*).

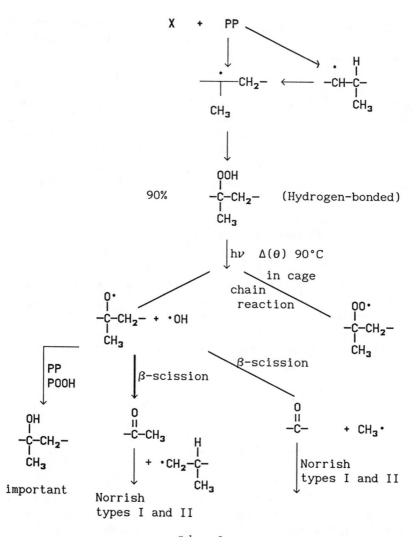

Scheme I.

- Why are photooxidation and thermooxidation stoichiometries so similar, even though the Norrish type I and type II photoprocesses should significantly convert the ketonic intermediates?
- What is the exact assignment of the 1740 cm^{-1} band that is observed in the IR spectrum of photooxidized PP and has been frequently understood as IR absorption of ester groups?

A few more recent works on ethylene–propylene copolymers (EPR) have not provided more information on these specific questions (15, 16). The development of heterophasic PP, that is, blends of isotactic PP as major constituents and EPR noddles, prompts new effort to understand the involved problems of PP photooxidation. Some original results are reported in this chapter.

Himont Spheripol PP and Hoechst heterophasic PP have been photooxidized in the form of thin films (100 μm) or thick plates in a SEPAP 12.24 photoaging unit. FTIR analysis was generally carried out by using a transmission technique with the film and photoacoustic detection to analyze the most superficial layers of the thick plates (a 10-μm superficial layer was indeed analyzed). Films were photooxidized as homogeneous reactors, and photoproduct profiles were observed in thick unpigmented or pigmented plates (pigmented with TiO_2 or carbon black).

The assignment of complex absorption massives in IR spectra of oxidized samples was considerably eased by the derivatization techniques first proposed by Carlsson and co-workers (*17*). SF_4 treatment of photooxidized heterophasic PP films converted the acidic groups formed in acid fluorides absorbing at 1840–1841 cm^{-1} and not at 1848 cm^{-1}. A series of low molecular weight carboxylic acids of various structures was introduced into a heterophasic PP matrix and submitted to SF_4 treatment. The 1840 cm^{-1} absorption band could only be assigned to an α-methylated acid:

Such a carboxylic structure could not be formed from the acyl groups resulting from Norrish type I processes on the intermediate ketone group. As explained in the next sections, a new route for the acid formation should be proposed. On the basis of pulsed 300-MHz ^{13}C NMR techniques, the analysis in the solid state on heterophasic PP thermooxidized at 140 °C up to a large oxidation extent was tentatively carried out. Chain-end ketones whose signal appears at 206 ppm largely prevailed over chain ketones whose signal appears at 216.2 cm^{-1}. Vinylidene groups absorbing at 112 ppm (in J-modulated echo spectrum) were also observed among more conventional oxidized groups (e.g., acid, hydroperoxide, and tertiary alcohols).

In the β-scission of alkoxy radicals, which accounts for the favored formation of chain-end ketones, a macro-alkyl radical is simultaneously produced (*see* Scheme II). This macro-alkyl radical, before rearrangement into vinylidene groups or into a tertiary radical, could be oxidized into a primary hydroperoxide, then into an aldehyde, and finally into an acidic group that is α-methylated (*see* Scheme III). A similar radical could be formed in the Norrish type I process observed from the chain-end ketones or from the macroketones (*see* Scheme IV).

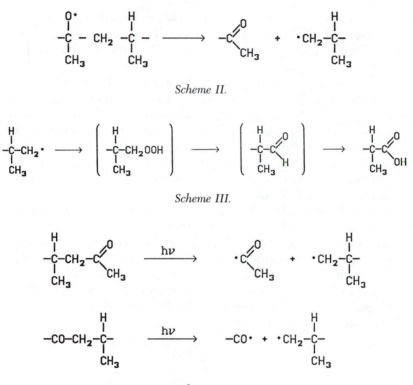

Scheme II.

Scheme III.

Scheme IV.

These routes should not be considered as important because the stoichi-ometries of photo- and thermooxidation are very similar, as shown by the IR spectra and by the rates of production of the various oxidized products (Figures 1 and 2).

The 1740 cm^{-1} shoulder that appears in the thermo- or photooxidation of isotactic or heterophasic PP was reassigned to an acidic group that would be hydrogen-bonded to a vicinal hydroperoxide.

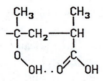

The conversion of hydroperoxides through chemical (NO or SO$_2$) or physical (photolysis or thermolysis) posttreatments or the conversion of acidic groups

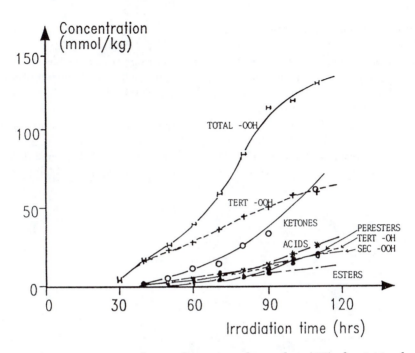

Figure 1. Kinetic accumulation of isotactic polypropylene (iPP) photoinitiated products (SEPAP 12.24 at 60 °C).

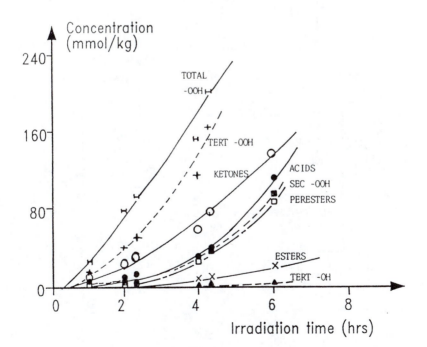

Figure 2. Kinetic accumulation of iPP thermally initiated oxidation products (100 °C) after elimination of induction period.

through SF_4 or NH_3 neutralization induces the parallel disappearance of this absorption at 1740 cm^{-1}. These observations are not consistent with the conventional assignment of the 1740 cm^{-1} band to ester groups. From the various observations reported in the previous section, the main photooxidation route of heterophasic PP could be represented by Scheme V. The chemical evolution of heterophasic PP could be described by the variations of the concentration of the acidic groups, which could be considered critical photoproducts in the sense of the mechanistic approach. However, part of the acidic groups formed corresponds to low molecular weight compounds. The formation of acetic acid and α-methylated levulinic acid has been shown by completing the

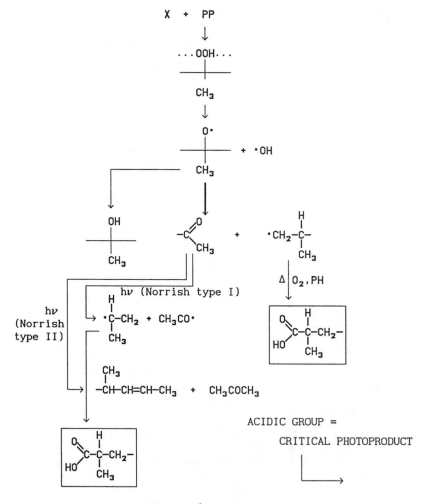

Scheme V.

IR analysis of the solid sample and identifying the volatile photoproducts based on IR and high-performance liquid chromatography analysis. A loss of these photoproducts by migration out of the sample can occur during exposure.

Therefore, IR analysis of the solid polymer film gives an underestimation of the carboxylic acids concentration. The total concentration of ketonic compounds is also underestimated by analysis of the polymeric sample because an important loss of acetone is monitored during irradiation. The notable loss of oxidation photoproducts explains why a drastic decrease of mechanical properties is monitored, whereas only a weak concentration of carbonyl species can be detected by measuring the oxidation of the solid sample (*see* Figure 3).

The photooxidation *mechanism* of unstabilized heterophasic PP accounts for all the analytical and kinetic aspects of the oxidation carried out in a SE-PAP 12.24 unit, that is, in the conditions of moderate, accelerated artificial photoaging. The relevancy to weathering of the observed phenomena could be assumed through the comparison of both mechanisms, and the only acceptable differences are in the kinetic parameters. No important modifications in the relative concentrations of intermediate and final photoproducts should

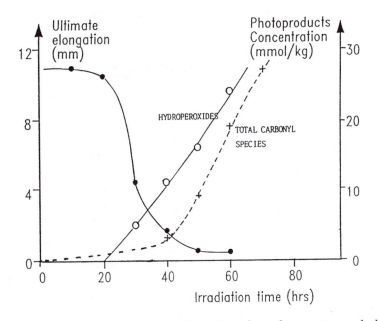

Figure 3. Correlation between the loss of mechanical properties and the appearance of oxidation products (hydroperoxides estimated by iodometric and total carbonyl species by IR spectrometry).

be observed for the main oxidation route and especially for the final critical photoproduct. Within that strict limitation, the acceleration factor for the main route could be determined from the rates of accumulation of critical photoproduct in accelerated conditions and throughout weathering.

As an example, an FTIR analysis was performed on a Spheripol 100-μm film and on plaque submitted to weathering in Clermont-Ferrand, France. The observed FTIR spectra were compared with the corresponding spectra for the same samples exposed in a SEPAP 12.24 unit. The series of FTIR spectra obtained with the film samples are compared in Figures 4 and 5 in the range of hydroxylated and carbonylated group absorption. The evolution of these spectra are identical, especially when considering the carbonyl massive. The photoacoustic (PAS)–FTIR spectra of the superficial layers of the plaque exposed in both conditions are very similar too.

A device named SEPAP 50.24 has been designed at the University of Clermont-Ferrand to study the very photostable polymeric materials in the conditions of ultra-acceleration. Ultra-acceleration is only due to higher light

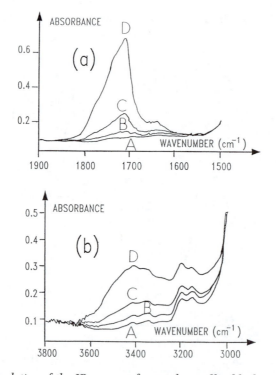

Figure 4. Evolution of the IR spectra of a copolymer film [thickness (e) = 100 μm] on natural exposure in Clermont-Ferrand: (a), carbonyl vibration region; (b), hydroxyl vibration region; A, 0 days; B, 66 days; C, 95 days; and D, 160 days.

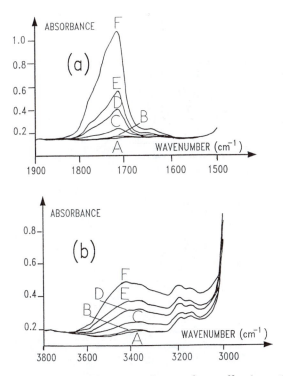

Figure 5. Evolution of the IR spectra of a copolymer film (e = 100 μm) on irradiation at λ ≥ 300 nm and 60 °C: (a), carbonyl vibration region; (b), hydroxyl vibration region; A, 0 h; B, 25 h; C, 50 h; D, 72 h; E, 90 h; and F, 156 h.

intensity and higher temperature compared to SEPAP 12.24; the spectral distribution of the filtered light is unvarying. When exposed in this unit, an unstabilized heterophasic film is photooxidized with a stoichiometry that is somewhat different. As shown on Figure 6, the main maximum of the carbonyl massive appears to be around 1735 cm^{-1} and not 1713 cm^{-1}, and absorptions at 3077, 1640, and 910 cm^{-1} are favored. These absorptions should be assigned to the build up of vinyl groups that reveal a photooxidation of ethylene segments of the EPR noddles, a process that is relatively more important than in weathering conditions. The ultra-acceleration technique is, in this case, unacceptable. The same series of experiments was reported with heterophasic PP stabilized with 1000, 2500, and 5000 ppm of a low molecular weight hindered amine light stabilizer (HALS), denoted HALS-1, or with 1000, 2500, and 5000 ppm of high molecular weight HALS (HALS-2). The moderate acceleration and ultra-acceleration have been compared.

Figure 7 represents the various FTIR spectra observed with a film sample exposed in SEPAP 12.24 in the range of interest. The consumption of the HALS stabilizer at 1740 cm^{-1} was observed before the development of any

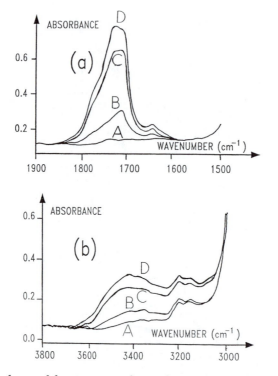

Figure 6. Evolution of the IR spectra of a copolymer film (e = 100 μm) on ultra-accelerated irradiation at λ ≥ 300 nm and 70 °C: (a), carbonyl vibration region; (b), hydroxyl vibration; A, 0 h; B, 18 h; C, 39 h; and D, 43 h.

significant oxidation. The carbonyl massive developed with a main maximum around 1735 cm^{-1} and a parallel development in the hydroxyl range are then observed. Exactly the same variations are observed with the two different HALS and when the film samples are exposed in natural conditions in Clermont-Ferrand.

In Figure 8, the variations of the absorbance at 1713 cm^{-1} (acid groups) are compared for the various samples exposed in SEPAP 12.24. The observed photostability order is very consistent with the results obtained in weathering. With the stabilized samples, the same stoichiometries are observed in the ultra-accelerated unit and in SEPAP 12.24. The variations of the absorbance at 1713 cm^{-1} of the various films, represented in Figure 9, could be compared with the variations presented in Figure 8. The two series of results are very consistent.

As we mentioned in a previous section, the coexistence of two dynamic processes could be an inherent difficulty. Figures 10 and 11 compare the PAS–FTIR spectra corresponding to the evolution of the most superficial layers of two PP plaques containing 2500 ppm of a migrating HALS (Figure 10) and

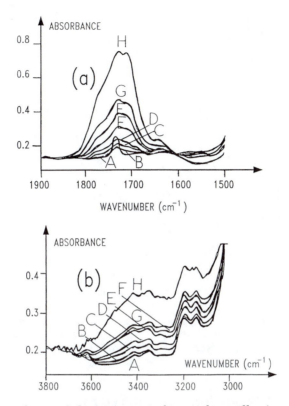

Figure 7. Evolution of the IR spectra of a copolymer film (e = 100 μm) containing 0.5% of HALS 1 on irradiation at λ ≥ 300 nm and 60 °C: (a), carbonyl vibration region; (b), hydroxyl vibration region; A, 0 h; B, 1074 h; C, 1626 h; D, 1925 h; E, 2000 h; F, 2089 h; G, 2237 h; and H, 2331 h.

2500 ppm of a nonmigrating HALS (Figure 11). The spectra of Figure 10 are identical to those observed with film samples in any conditions, whereas the spectra of Figure 11 are quite different; in the superficial layers, a compound accumulated with a band peaking at 1732 cm⁻¹. Interestingly enough, the carbonyl compound absorbing at 1732 cm⁻¹ could result from a grafting of the HALS additive (converted into nitroxy radical) to an oxidized macroradical (which could be a ketyl radical). This observation is only possible in an ultra-acceleration unit in which the loss of volatile photoproducts is kept low due to the reduced time scale. When two dynamic processes coexist, that is, a photooxidation process and a diffusion process, it is important to check that the chemical evolution is controlled during the whole lifetime of the material by the same process in weathering and in accelerated artificial conditions. When the accelerated conditions favor the noncontrolling process, irrelevant results are obtained.

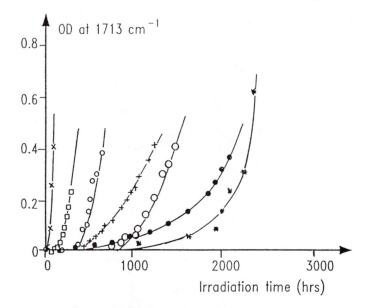

Figure 8. Evolution of optical density at 1713 cm⁻¹ with irradiation times of films (e = 100 µm) on irradiation at λ ≥ 300 nm and 60 °C: X, no additive; o, 0.1% HALS 1; O, 0.25% HALS 1; ☀, 0.5% HALS 1; □, 0.1% HALS 2; +, 0.25% HALS 2; and ●, 0.5% HALS 2.

Evolution of Photo(Bio)Degradable PE

In a recent European Brite-Euram Contract (Brite-Euram Contract BREU 170, Proposal BE 3120-89, Université Blaise Pascal, Clermont-Fd, France, and Aston University, United Kingdom), we had to evaluate the bio-assimilation properties of highly photooxidized PE systems. The control of the abiotic degradation of a series of *photodegradable* systems enabled us to compare weathering, artificial nonaccelerated photoaging (in the UV cabinet of Aston University), moderately accelerated photoaging (in SEPAP 12.24), and ultra-accelerated photoaging (in SEPAP 50.24). The comparison was made on the basis of FTIR spectroscopy and on the basis of gel permeation chromatography (GPC). When the photostabilization of a polymeric material has to be evaluated, the FTIR analysis of the matrix is the more informative because the oxidation occurs at a very low extent, most photoproducts are nonvolatile, and cross-linking could be important only in the very earliest phases. When the ultimate fate of a polymeric material is characterized after heavy fragmentation, GPC is the most informative. The photodegradable PEs examined were

- a photolytic system based on a copolymer ethylene–CO (1% CO) (Film A₂)

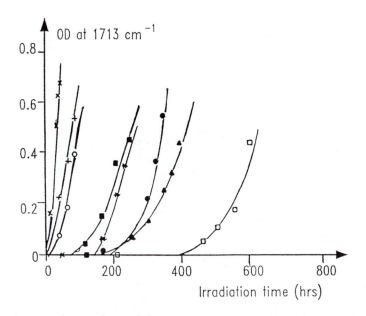

Figure 9. Evolution of optical density at 1713 cm⁻¹ with irradiation times of films (e = 100 μm) on ultra-accelerated irradiation at λ ≥ 300 nm and 60 °C: X, no additive; o, 0.1% HALS 1; *, 0.25% HALS 1; □, 0.5% HALS 1; +, 0.1% HALS 2; ■, 0.25% HALS 2; ▲, 0.5% HALS 2; and ●, 0.25% HALS 1 + 0.25% HALS 2.

- a nonpigmented PE photosensitized by iron carboxylate (Film B_1)
- a TiO$_2$-pigmented PE photosensitized by iron carboxylate (Film C_2)
- a nonpigmented PE photosensitized by iron and nickel dithiocarbamates (Scott–Gilead system) (*18*) (Film B_5)

The photooxidation was followed by the simultaneous determination of the variations of the absorbance at 1715 cm⁻¹ (by measuring the concentration of acid groups considered as the critical photoproduct in polyethylene matrices) and the variations of the average weight-average molecular weight, $\overline{M}_w$. The variations of $\overline{M}_w$ versus the extent of the photooxidation are shown on Figure 12. Large decreases of $\overline{M}_w$ from 3 million to 4,000 or 36,000 were observed, and demonstrated that chain scissions largely prevailed on crosslinking. The photolytic copolymer (Film A_2) initially presented the fastest decrease; however, afterward the decrease in $\overline{M}_w$ due to photooxidation of fragments proceeded slower than in other systems.

A more extensive photooxidation was carried out in the SEPAP 12.24 with C_2 films. After an exposure equivalent to 2 years of weathering, $\overline{M}_w$ decreased to around 2,000, whereas the absorbance at 1715 cm⁻¹ was around 2 for a 45-μm film. The results obtained in weathering conditions and in the UV

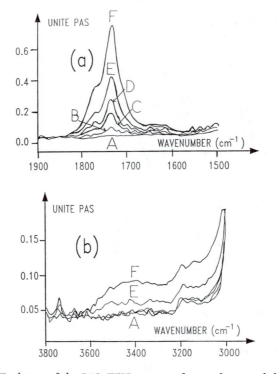

Figure 10. Evolution of the PAS–FTIR spectra of a copolymerized thick plaque containing 0.25% migrating HALS 1 on ultra-accelerated irradiation at λ ≥ 300 nm and 70 °C: (a), carbonyl vibration region; (b), hydroxyl vibration region; A, 0 h; B, 154 h; C, 461 h; D, 610 h; E, 774 h; and F, 938 h.

cabinet of Aston University (i.e., in nonaccelerated conditions) for the B_5 and C_2 systems are very consistent with the data obtained in SEPAP 12.24 (Table I).

The reduction of $\overline{M}_w$ of the C_2 system with the oxidation extent was compared throughout weathering, moderate accelerated photoaging (SEPAP 12.24), and ultra-accelerated photoaging (SEPAP 50.24). As shown in Figure 13, the decrease in $\overline{M}_w$ is very similar in the three conditions. If the decrease in $\overline{M}_w$ is slightly antagonized by cross-linking, that cross-linking is not favored in ultra-accelerated conditions.

Conclusions

The recognition of an evolution mechanism throughout weathering and artificial aging allows a fairly efficient control of relevancy. When the simulation techniques were proposed, acceleration was not accepted to avoid the observation of irrelevant phenomena in laboratory conditions. On the other hand,

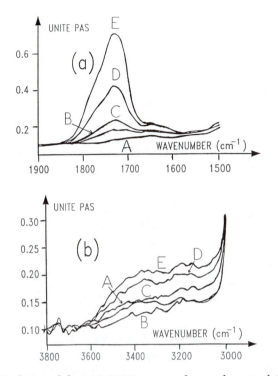

Figure 11. Evolution of the PAS–FTIR spectra of a copolymerized thick plaque containing 0.25% nonmigrating HALS 1 on ultra-accelerated irradiation at λ ≥ 300 nm and 70 °C: (a), carbonyl vibration region; (b), hydroxyl vibration region; A, 0 h; B, 154 h; C, 461 h; D, 610 h; E, 774 h; and F, 938 h.

an aging matrix should not be considered as a homogeneous (photo)reactor at the molecular scale. By using conventional kinetics, an expression of the evolution rate of the matrix cannot be obtained, even approximately; and the data collected at the initial time, in nonaccelerated conditions, cannot be extrapolated to the real lifetime of the polymeric material. Thus, experimental acceleration should appear as a fundamental necessity and its limitation should be controlled. Because identification of the evolution mechanism is essentially based on the nature of the intermediate and final photoproduct, like in any phenomenological approach, the following techniques should be used:

- FTIR and Raman spectroscopies for matrices converted into homogeneous reactors
- micro-FTIR and micro-Raman spectroscopies for examining the elementary layers of matrices that present some photoproduct profiles
- GPC for molecular weight determinations

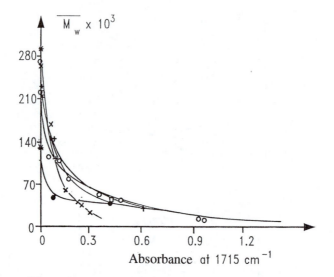

Figure 12. $\overline{M}_w$ variations of B_1 (X), A_2 (✳), C_2 (o), C_3 (□), and B_5 (+) films irradiated in SEPAP 12.24 unit at 60 °C (e = 40 μm).

Table I. Results for B_5 and C_2 Film Systems

Conditions	Film	Thickness (μm)	ΔA at 1715 cm⁻¹	$\overline{M}_w$
Clermont-Ferrand	B_5	34	—	229,500
			0.142	239,300
			0.173	208,800
			0.305	105,100
			0.428	63,600
	C_2	45	—	271,000
			0.133	125,000
			0.249	76,700
			0.630	27,100
UV cabinet, Aston	B_5	35	2.13	4,500
University	C_2	45	2.07	8.500

Coupling with chemical dosages or with gaseous-specific in situ reagents (like SF_4, NO, and NH_3) is advantageous in many circumstances. Acceleration could be used every time a single (even complex) dynamic process controls the evolution of the matrix. The acceleration level should be controlled by the invariance of this prevalent mechanism. When deviations are pointed out from experimental results, the data collected in laboratory conditions could not be directly transferred in the conditions of the material's real life. However, the data could be useful to design new accelerated conditions if deviations are accounted for. The frequent reasons for deviations are generally either bi-

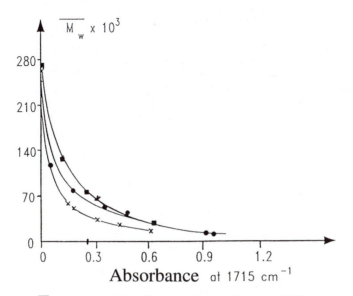

Figure 13. $\overline{M}_w$ variations of C_2 film according to ■, *natural UV-exposure;* ●, *accelerated photoaging in SEPAP 12.24 at 60 °C; and* X, *ultra-accelerated photoaging in SEPAP 50.24 at 80 °C.*

molecular reactions between intermediate species, whose concentrations in real conditions are too low to interact, or diffusional processes of reactants (e.g., O_2, H_2O, or stabilizers) that appear to be too slow in accelerated conditions.

References

1. Ly, T.; Sallet, D.; Lemaire, J. *Macromolecules* **1981,** *15,* 1437.
2. Lemaire, J.; Arnaud, R.; Gardette, J. L. *Rev. Gén. Caoutch. Plast.* **1981,** *613,* 87.
3. Ginhac, J. M.; Arnaud, R.; Lemaire, J. *Makromol. Chem.* **1981,** *182,* 1229.
4. Penot, G.; Arnaud, R.; Lemaire, J. *Makromol. Chem.* **1981,** *183,* 2731.
5. Adams, J. H. *J. Polym. Sci. A* **1970,** *1(8),* 1279.
6. Adams, J. H.; Goodrich, J. E. *J. Polym. Sci. A* **1970,** *1(8),* 1269.
7. Chien, J. C. W.; Vandenberg, E. J.; Jabloner, H. *J. Polym. Sci. A* **1968,** *1(6),* 381.
8. Niki, E.; Decker, C.; Mayo, F. R. *J. Polym. Sci. Polym. Chem. Ed.* **1973,** *11,* 2813.
9. Carlsson, D. J.; Wiles, E. J. *Macromolecules* **1969,** *2(6),* 597.
10. Carlsson, D. J.; Wiles, E. J. *Macromolecules* **1969,** *2(6),* 587.
11. Carlsson, D. J.; Wiles, E. J. J. *Macromol. Sci. Rev. Macromol. Chem.* **1976,** *C14(1),* 65.
12. Adams, J. H. *J. Polym. Sci. A* **1980,** *1(8),* 1077.
13. Carlsson, D. J.; Chmela, S.; Lacoste, J. *Macromolecules* **1990,** *23,* 4934.
14. Mill, T.; Richardson, H.; Mayo, F. R. *J. Polym. Sci. Polym. Chem. Ed.* **1973,** *11,* 2899.
15. Lacoste, J.; Singh, R. P.; Boussand, J.; Arnaud, R. *J. Polym. Sci. Polym. Chem. Ed.* **1987,** *25,* 2799.

16. Singh, R. P.; Lacoste, J.; Arnaud, R.; Lemaire, J. *Polym. Degrad. Stab.* **1988,** *20,* 49.
17. Carlsson, D. J.; Dobbin, C. J. R.; Jensen, J. P. T.; Wiles, D. M. In *Polymer Stabilization and Degradation;* Klemchuk, P. P., Ed.; ACS Symposium Series 280; American Chemical Society: Washington, DC, 1985; pp 359–371.
18. Gilead, D.; Scott, G. *Dev. Polym. Stab.* **1982,** *5,* 71.

RECEIVED for review January 26, 1993. ACCEPTED revised manuscript December 9, 1994.

Degradation Profiles of Thick High-Density Polyethylene Samples after Outdoor and Artificial Weathering

J. C. M. de Bruijn[1]

Delft University of Technology, Faculty of Industrial Design Engineering, Laboratory for Mechanical Reliability, Leeghwaterstraat 35, 2628 CB Delft, The Netherlands

The depth and degree of degradation in the surface layer of polymers differs for different processing and exposure conditions and possibly causes differences in service life. Injection and compression-molded high-density polyethylene samples were artificially weathered in a Xenotest 1200 and a weatherometer and were exposed outdoors in Miami, Florida, and Delft, The Netherlands. Degradation profiles were measured using a specially designed microfoil tensile test, a Fourier transform IR spectrometer, and a density gradient column. Degradation profiles of samples weathered in the Xenotest, weatherometer, and outdoors in Florida all showed a constant degradation depth with exposure time and a horizontal plateau near the surface. In contrast, the samples exposed outdoors in Delft showed a concave-shaped profile and an increasing degradation depth with exposure time.

WEATHERING OFTEN LEADS to brittle failure of otherwise ductile plastic products and may strongly reduce the service life of plastic products used or stored outdoors such as bottle crates (*1*) and pipes (*2, 3*). This brittle failure is often caused by the initial fracture of the outer surface layers.

In products with a large wall thickness, the influence of weathering is often limited to a surface layer, either due to limited oxygen diffusion or to limited UV penetration (*4–7*). Although the depth of this layer may be small (e.g., 0.5 mm) compared with the complete wall thickness (e.g., 4 mm), it can

[1]Current address: KEMA Nederland B.V., KEMA Inspection Technology, Plastics and Rubber, PO Box 9035, 6800 ET, Arnhem, The Netherlands

0065–2393/96/0249–0599$12.50/0

cause brittle fracture of the product (6, 8) analogous to a brittle surface layer on a ductile polymer. Despite the increasing amount of literature on the oxidation profiles in weathered samples, it is not yet clear how the mechanical-failure behavior depends on the oxidation profile (9, 10).

This research was concentrated on the shape of the degradation profiles in different exposure conditions, either natural or artificial. The relation between degradation profiles and failure behavior has been dealt with before (11). The objective of the overall research was to find the critical degradation profile that accounted for the failure of complete samples (11). The exposure time to failure was defined as t_{fail}, and all tests were performed close to the time to failure.

Experimental

Material. The material was a non-UV-stabilized, nonpigmented, high-density polyethylene (HDPE) supplied by DSM. It had a narrow molecular weight distribution and a melt flow index of 0.8 g/min (ISO 1133). A phenolic antioxidant (<500 ppm) was added for processing stabilization, and the material also contained standard film additives.

Processing. A number of samples were injection molded on an Arburg injection-molding machine. From a number of the samples the outer 100 ± 15 μm was removed by using a microtome (Reichert-Jung Autocut 2040) operated at room temperature. This procedure was done to reveal the effects of the less crystalline and possibly more orientated skin layer on photooxidation and failure behavior. A number of samples were compression molded to study a homogenous morphology free from the complicating factors due to the injection-molding process. The dimensions of the samples were approximately 6 × 4 × 50 mm³ and were in accordance with ISO 179/2D (Charpy impact strength determination).

Exposure Conditions. *Xenotest 1200.* The exposure in the Xenotest 1200 U (Original Hanau Quarzlampen GmbH) took place at The Netherlands Organization (TNO) for Applied Scientific Research, TNO Plastics and Rubber Research Institute. Most conditions of this artificial weathering were in accordance with DIN 53 387 (ISO 4892). The samples had a stainless steel backing and the lamps were filtered using three filters. One-third of the circumference had no UV filter (which allowed radiation wavelengths below 300 nm), and the remainder was filtered with UV Spezialglas. The samples were exposed in Wendelauf (one-half of the time the samples were turned toward the lamp; the other half, opposite to the lamp), which yields an average total UV radiation (TUVR) of 49.5 W/m² (300–400 nm; Figure 1), whereas the maximum TUVR (when turned toward the lamps) was 90 ± 5 W/m² (300–400 nm). The black panel temperature was 45 ± 5 °C.

Weatherometer. The exposure in the Weather-Ometer (WOM) weathering machine (Atlas Electric Devices Company) took place at DSM Research, Department for Polymer Developments. The samples were exposed without backing, and two borosilicate filters with watercooling in between were used for filtering. The TUVRs were 32.2 W/m² (300–400 nm, Figure 1) and 24.08 W/m² (295–385

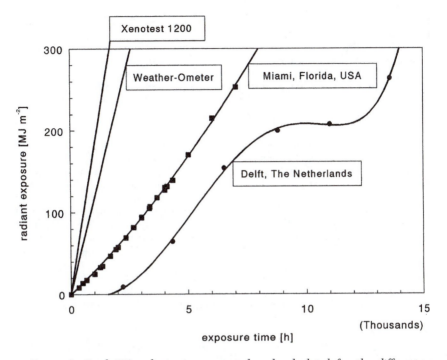

Figure 1. Total UV radiation as measured and calculated for the different exposure conditions.

nm) (0.28 W/m² at 340 nm; bandwidth, 1 nm). The black panel temperature was 50 °C. The samples were continuously turned toward the lamp.

Outdoor Exposure: Miami, Florida.   A number of samples were exposed in a subtropical climate on a site of the South Florida Test Service Inc. in Miami, Florida. The exposure started on 10 December 1990 and lasted 26 weeks (4368 h) (period A). A second set of samples was exposed starting 4 September 1991. These samples were exposed up to 42 weeks (7056 h) (period B).

Global radiation and TUVR (295–385 nm) were recorded at the test site (Figure 1). The TUVR varied between 3.7 and 5.3% of the global radiation.

The irradiance according to the average optimum daylight is 0.3 W/m² at 340 nm (bandwidth, 1 nm) and 30 W/m² in the range from 300 to 400 nm, whereas there is no irradiance below 300 nm (*12*). On average, the temperature varies yearly between 15 and 32 °C. During periods A and B the maximum average day temperature was approximately 27 °C. The relative humidity varied between 35 and 95% and the yearly rainfall is approximately 1200 mm. The samples were directly exposed to the weather at an angle of 45° to the south and without backing.

Outdoor Exposure: Delft.   A number of samples were exposed in a European climate at the TNO site in Delft, The Netherlands. The exposure field was situated a few hundred meters from a busy highway. No special care was taken to shield the samples from exhaust fumes. The exposure started on 20 November 1990 and finished on 20 August 1992.

The TUVR (300–400 nm) was calculated from the global radiation, which was measured at the test site by using translation factors (13) (Figure 1).

Clearly, the data show a waveform due to the seasonal variations (Figure 1). The maximum average day temperature during the exposure period was approximately 14 °C. The samples were directly exposed to the weather at an angle of 45° to the south and with an anodized aluminum film backing.

Sample Preparation.   To be able to obtain microfoil tensile tests (MFTT), Fourier transform IR (FTIR), and density results with respect to the thickness of the samples, films with a thickness of 10 μm were microtomed from the sample. To avoid the influence of degradation of the side edges, approximately 1 mm on each side of the samples was removed using the microtome with a steel knife operated at room temperature. To avoid fracture in the clamps in the MFTT system, two side grooves with a depth of approximately 1 mm were introduced opposite to each other in the middle of the sample by using a chisel with radius of 0.3 mm. The microtoming was performed at -120 ± 15 °C because preliminary results showed that the cutting of the films at room temperature caused deformation of the material, which made it impossible to study the material's morphology. Therefore, special care was taken to cool the knife and sample to a temperature around the glass transition temperature of the amorphous phase of PE.

Microfoil Tensile Testing.   A very simple but effective way to measure the mechanical strength across the thickness of an exposed sample is to perform tensile tests on films microtomed from the sample. An MFTT system was designed and built in our own laboratory as previously described (14).

The temperature was kept at 23 ± 2 °C. Clamp separation speeds of 0.3 mm/s for the injection-molded samples and 0.03 mm/s for the compression-molded samples were used. The lower speed for the films cut from the compression-molded samples was necessary because of their more brittle behavior. After the tensile test the thickness of the remaining film parts were measured using a micrometer. An MFTT force displacement plot was determined, and from this plot the nominal strain (ε_{nom}) was calculated:

$$\varepsilon_{nom} = \frac{ds}{2r_{notch}} \tag{1}$$

where ds is the displacement until fracture, and r_{notch} is the notch radius measured by using an optical microscope.

The notch diameter is taken as a reference, because there is no defined initial length for tensile bars with a large fillet radius. A notch was used instead of a large fillet radius because it was not possible to obtain the fillet radius without damaging the embrittled surface layer (14). The use of a notch instead of a large fillet radius, however, hampers the relation between the MFTT results and the results from macroscopic tensile test bars. The nominal strain defined in this way can only be used as a relative measure.

FTIR Measurements.   The chemical changes in the material were measured with an FTIR spectrometer Polaris (Mattson Instruments, Inc.) and analyzed with ICON software. One hundred scans were averaged and a resolution of 4 cm^{-1} was used. The microtomed films of 10-μm thickness were pressed between KBr (powder) to avoid interference. Similar to Furneaux et al. (*4*), a carbonyl and vinyl index were defined by dividing the absorbance measured at 1712 and 909 cm^{-1}, respectively, by the absorbance of a reference peak at 1368 cm^{-1}.

Density Measurements.   The density of weathered samples can increase because of chemicrystallization (*15, 16*), the increase in polar groups (*17*), oxygen uptake (*16*), or the loss of volatile products (*15*). The density was measured very accurately by means of a density gradient column (Davenport 6-columns density apparatus) operated at 23 ± 0.1 °C. A mixture of 2-propanol and distilled water was used to obtain a gradient column with a range of approximately 0.05 g/mL. Previous research (*18*) showed that these liquids do not affect PE.

From microtomed samples removed from a given depth below the original front surface, at least 3 films were first wetted in 2-propanol (low density) and then in distilled water (high density), each for at least half an hour. By following this procedure, the films sank quickly and did not contain air bubbles once released in the column. The films were left for at least 20 h to stabilize before their density was determined.

Results

Although seven different combinations of the exposure conditions and ways of processing were tested, only a few, typical results will be shown.

Microfoil Tensile Test.   *Injection-Molded Samples Exposed in Weatherometer.* As shown in Figure 2, the depth over which the nominal strain changes compared with the nominal strain of films taken from the unexposed sample was approximately 750 μm and was independent of exposure time (≤1300 h). Even at 800 h, just before the first samples started to fracture, the films taken near the surface (0–200 μm) had a nominal strain comparable with that of the unexposed counterpart. However, the nominal strain between 200–700 μm was considerably lower. After 800-h exposure (just after t_{fail}), the nominal strain decreased somewhat near the surface (<100 μm) but still remained high compared with films taken at a depth up to 700 μm. After a relatively long exposure time (1300 h), the nominal strains from films taken from the outer surface (550 μm) were more or less homogeneously degraded.

The nominal tensile stress was less sensitive to degradation than the nominal strain. After 800 h (before t_{fail}) no changes were seen in the degradation profile, whereas after 1300 h the nominal tensile stress was drastically reduced at the outer 500 μm.

Compression Molded Samples Exposed in Weatherometer.   The nominal strain of the compression-molded samples was considerably lower than that of the injection-molded samples, even though a slower clamp separation speed was used. The nominal strain and the nominal tensile stress decreased

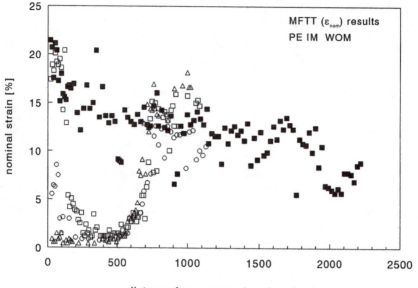

Figure 2. Nominal strain results (MFTT) of injection-molded samples weathered in the WOM. Key: ■, *unexposed;* □, *800-h exposure;* ○, *991-h exposure; and* △, *1300-h exposure.*

more or less homogeneously; a less degradation-sensitive skin, as in the injection-molded samples, was not observed. The degradation depth was limited to approximately 600 μm, which was less than the limit for the injection-molded samples (750 μm).

FTIR Spectroscopy of Injection Molded Samples Exposed in Weatherometer. Similar to the MFTT result, FTIR analysis determined an initial strong increase in the vinyl index underneath the surface (300–600 μm), but after relatively long exposure times there was also an increase near the surface (Figure 3). The vinyl results show a clear trend in contrast to the results of the carbonyl index (not shown here), which were difficult to interpret due to scatter (11).

Density Gradient Column. *Injection-Molded Samples Exposed in Weatherometer.* As was the case with the MFTT and FTIR results, the change in density (Figure 4) was restricted to approximately 750 μm independent of exposure time (<1300 h). The density increased steadily near the surface (<750 μm) when the exposure time increased. Because of the strong density increase near the surface of the sample, where the largest change on weathering takes place, it is illustrative to use an alternative presentation. In

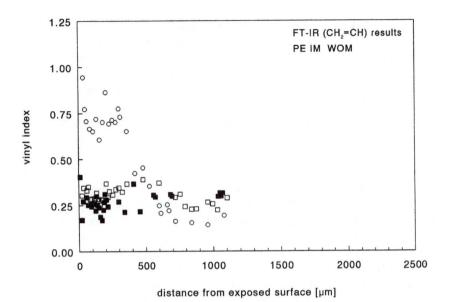

Figure 3. Vinyl index results (FTIR) of injection-molded samples exposed in the WOM. Key: ■, unexposed; □, 800-h exposure; and ○, 1300-h exposure.

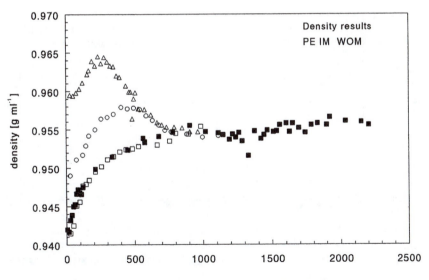

Figure 4. Density gradient column results of injection-molded samples exposed in the WOM. Key: ■, unexposed; □, 283-h exposure; ○, 991-h exposure; and △, 1300-h exposure.

Figure 5 the density increase, that is, the difference between the density of the exposed samples and the original sample, is plotted. The density increase shows a plateau whose depth decreases at longer exposure times.

A strong density gradient occurred through the sample depth due to change in crystallinity. The crystallinity calculated according to the formula and values of Moy and Kamal (*19*) yielded densities of the unexposed sample of approximately 63% at the surface and approximately 72% in the middle of the sample.

Compression-Molded Samples Exposed in Weatherometer. Figure 6 shows that the density and therefore the crystallinity of the compression-molded material were higher compared with those of the injection-molded material. The density increased uniformly with respect to the sample thickness without showing the presence of a less degradation-sensitive outer layer or skin. Similar to the MFTT measurements, the depth at which the density after exposure deviated from the density before exposure was approximately 600 µm. As expected, the crystallinity varied in the compression-molded samples considerable less than in the injection-molded samples. However, the crystallinity was higher for the compression-molded samples: 73% at the surface and 76% in the middle of the sample.

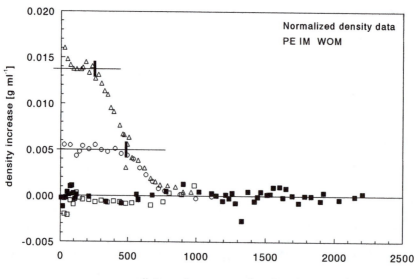

Figure 5. Normalized density data of injection-molded samples exposed in the WOM. Key: ■, unexposed; □, 283-h exposure; ○, 991-h exposure; and △, 1300-h exposure.

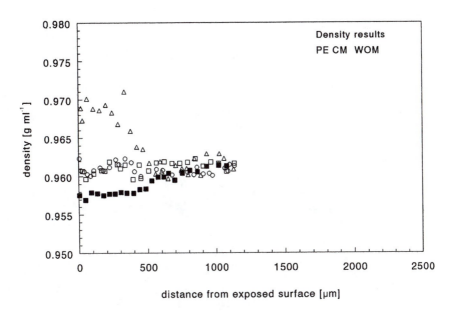

Figure 6. Density of compression-molded samples exposed in the WOM. Key: ■, *unexposed;* □, *336-h exposure;* ○, *621-h exposure; and* △, *1097-h exposure.*

Injection-Molded Samples Exposed Outdoors in Delft. Figure 7 shows that the density increase reduced consistently from the surface toward the deeper layers and that there was no constant degradation depth. In contrast to the other exposure conditions, there was an increasing depth of approximately 300 μm after one year and approximately 850 μm after 1.5 years (13,152 h). The whole sample seemed more or less completely degraded after 1 year and 9 months (15,312 h).

In the next section the relation among the different experimental techniques, the depth of degradation, and the presence of a less degradation-sensitive skin will be discussed.

Discussion

Less Degradation-Sensitive Skin.   On the basis of experimental results found in the literature, the extent of degradation is expected to decrease from the surface toward the middle (*4, 6*). However, the MFTT, FTIR, and density results show that around t_{fail}, the changes in properties (i.e., nominal strain or vinyl index) measured in the outer 100–200-μm layer of injection-molded samples exposed in the Xenotest, WOM, and Florida outdoors were less or at the most equal to the changes in properties deeper into the sample. After longer exposure times this effect disappeared.

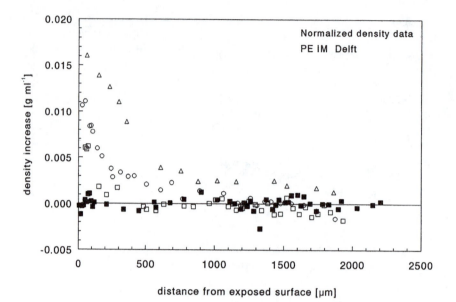

Figure 7. Normalized density data of injection-molded samples exposed outdoors in Delft. Key: ■, *unexposed;* □, *8760-h exposure;* ○, *13,152-h exposure; and* △, *15,312-h exposure.*

The injection-molded samples exposed in Delft and the compression-molded samples exposed in either the WOM or outdoors in Florida did not show a less degradation-sensitive skin. The skin of injection-molded samples without an outer 100 μm was somewhat more degradation sensitive than the skin of the virgin injection-molded samples; however, this difference was still less than expected.

On the basis of these results we concluded that the less degradation-sensitive skin must have been caused by the use of the injection-molding process but also depended on the exposure conditions. The settings of the injection-molding machine were those normally used for the injection molding of Charpy test samples. Although the effect of varying processing conditions on the degradation profile was not studied, it is quite likely that a less degradation-sensitive skin will also be present in injection-molded products in practice. A number of explanations can be given for this phenomenon.

- A possible reason found with MFTT was the lower crystallinity and thus the higher amount of tie molecules. Because of the higher tie molecule density the toughness was kept intact for a longer time than the toughness of the underlying higher-crystalline material (*20, 21*). However, because the less degradation-sensitive skin was also seen in the FTIR and density measurements, this explanation is not that likely.

- The less degradation-sensitive skin may be caused by the presence of more chromophores in the bulk compared to the skin. This increased concentration may be due to the oxidation of the hot bulk material directly after the samples are released from the mold.
- The stable skin might be either a consequence of surface cross-linking promoted by oxidative degradation or surface deposition (blooming) of antioxidant or other additive.

No satisfying answer was found for the less degradation-sensitive skin, although the oxygen diffusion into the material after the sample was released from the mold was the most likely reason. Further research is needed to confirm this assumption. However, three-point bending tests on exposed samples demonstrated that the presence of the skin had no significant effect on the time to failure (*11*).

Relation among Experimental Techniques.   The most pronounced results were obtained with the MFTT and the density measurements and not with the more familiar FTIR measurements. For both the unexposed injection and unexposed compression-molded samples the density steadily increased from the surface toward the middle of the sample, whereas the nominal strain density decreased. The density and thus the crystallinity were higher in the compression-molded samples, whereas the nominal strain was considerably lower. The nominal tensile stress and vinyl index were more or less homogeneously distributed over the sample depth. However, the carbonyl index did not show a homogeneous distribution (*11*).

Generally, there was reasonable agreement among the different exposure conditions regarding the change in nominal strain, vinyl index, and density versus the distance from the exposed surface. A decrease in the nominal strain corresponded to an increase in the vinyl index and in the density. Only in the Delft outdoor exposure samples did the nominal strain show considerable changes at large depths before a change in density was observed. Generally, the nominal tensile stress only started to show significant changes (a decrease) around t_{fail}, whereas the carbonyl index showed too much scatter to be able to reveal a general trend.

The scatter of the carbonyl and vinyl index measurements, compared with the MFTT and the density measurements, was caused by the low amount of carbonyl and vinyl groups formed during oxidation. The larger amount of scatter in the carbonyl index compared with the vinyl index may be due to the use of thin films (10 µm) and the presence of KBr, which caused a water peak near the carbonyl peak. Although the data show much scatter, the increase in the amount of carbonyl appeared in some cases to be less than the increase in the amount of vinyl, whereas according to the literature vinyl is only expected to be formed after the formation of a carbonyl group (*22*). The lower increase in carbonyl may be caused by a Norrish type II reaction, which only slowly yields carbonyl groups that directly decompose by a fast Norrish type II reaction and lead to a vinyl group and volatile acetone. No changes in macromolecular carbonyl concentration occur (Verdu, J., ENSAM, personal communication). A second alternative for the lower increase in carbonyl compared with vinyl is given by a reaction of the macromolecule with oxygen that leads directly to two vinyl groups and a hydroperoxide (23). Variability in the carbonyl measurements could also be a consequence of antioxidant diffusions to the surface.

Shape of Degradation Profiles.   The degradation profiles of samples weathered in the Xenotest, WOM, and outdoors in Florida all showed a constant degradation depth with exposure time and a horizontal plateau near the surface, which was previously found by a number of researchers (1, 4, 10, 24). This constant degradation was in contrast with the samples exposed in Delft, which showed a concave degradation profile and an increasing degradation depth with exposure time. Because of the importance of understanding the mechanism behind artificial and outdoor weathering for the interpretation of the data, both types of degradation profiles are discussed. Different authors have suggested several causes for a limited degradation depth (4, 25). However, the two most likely causes for a constant degradation depth are decreasing UV intensity and restriction of oxygen diffusion. Most authors (4, 10, 26) concluded that the limited degradation depth is, at least for non-UV-stabilized materials, due to a low rate of oxygen diffusion compared with a high rate of oxidation. In the following sections, the profiles that were experimentally found will be explained by using the theory of oxygen diffusion.

The diffusion of oxygen into a solid polymer is a relatively slow process, and beyond a certain thickness it can become the rate-controlling process (4, 27). This characteristic means that the photooxidation occurs heterogeneously with respect to the thickness of the specimen and that the process occurs primarily in the surface layers that are accessible to oxygen. To study the effect of oxygen diffusion, most authors start from Fick's second law modified by a term that describes the rate, r, at which the reactant (in this case oxygen) is consumed. This rate is a function of the reactant concentration, C, as a function of time, t, place, x, and D, the oxygen diffusion coefficient.

Often a steady state is assumed; that is, the rate of oxygen consumption is exactly matched by the oxygen supply by diffusion, and thus the oxygen concentration profile does not change with time (9, 10). This modification yields eq 2:

$$D \frac{d^2 C(x)}{dx^2} - rC(x) = 0 \qquad (2)$$

The value of $C(x)$ can be determined if D and the reaction rate, $r(C(x))$, are known.

Oxygen Diffusion Coefficient. Although D can depend on the state of the aging process (i.e., time), the reactant concentration (*10*), and the morphology (orientation and crystallinity) of the polymer (*28*), the diffusion coefficient is still often assumed to be a constant (independent of depth, time, or concentration). A PE with a density of 0.964 g/mL has an oxygen diffusion coefficient of 1.7×10^{-11} m²/s at 25 °C (*29, 30*). The value of D depends on the temperature (T) according to eq 3:

$$D = D_0 \exp \left(\frac{-E_d}{RT} \right) \tag{3}$$

where D_0 is a preexponential factor for oxygen diffusion, E_d is the activation energy for oxygen diffusion, and R is the gas constant.

Oxidation Rate. The value of $r(C(x))$ can be approximated from an analysis of the reaction kinetics to predict trends (*9, 10, 28*). A simple approximation is given by eq 4:

$$r(C(x)) = \frac{\alpha C(x)}{\beta C(x) + 1} \tag{4}$$

where α and β are constants. A calculation of the oxidation profile from eq 4 can be made by using numerical methods. In this chapter the profiles will be explained qualitatively by using the two extreme cases in which the profile can easily be solved analytically. In a future paper a more advanced analytical approximate of the solution of these equations is given (*31*).

For high oxygen concentrations eq 4 leads to a reaction rate of α/β; the system then behaves as a zero-order reaction and is independent of oxygen concentration. For low oxygen concentrations or low values of β, eq 4 leads to a reaction rate of αC; the system then behaves as a first-order reaction.

A number of oxidation experiments have determined that oxidation at atmospheric pressure occurs according to a zero-order rate until oxygen depletion starts to set in (toward the middle of the sample) and oxidation occurs at a first-order rate. Both regimes are separated by a region with a mixed kinetic behavior (*10, 26*). The reaction rate can probably never be completely explained by a zero-order dependence on the oxygen concentration, because it also depends on UV intensity (i.e., pseudo zero order).

Extent of Degradation. The reaction rate can be determined from the concentration profile, and because of the simple steady-state approach the reaction rate is constant in time. Therefore, the conversion or extent of reaction at point x is given by eq 5 (*9*):

$$Q(x) = \int_0^{t_{exp}} r(C(x))dt = r(C(x))t_{exp} \tag{5}$$

where t_{exp} is exposure time and Q is the conversion or extent of reaction. When this conversion is divided by the conversion at the surface, Q_0, a reduced conversion is obtained.

Several authors (4, 11, 18, 28, 32, 33) found that the extent of degradation, in contrast to the steady-state assumption, is not a linear function of the exposure time. From their experiments it follows that the rate of oxygen absorption and the rate of the formation of chemical reaction products increases with exposure time due to the expected autocatalytic effect. Consequently, the extent of degradation is increasing more than proportionally with the exposure time. However, even though the steady-state assumption is no longer valid, the experimentally determined degradation profile was still reasonably well fitted with numerical solutions of eq 2 for the photooxidation of PE by Furneaux et al. (4) and Zabara (34) and for the radiochemical oxidation of LDPE by Papet et al. (33) and Gillen and co-workers (18, 24).

Degradation Profile of Samples Exposed Outdoors in Delft.
Degradation profiles of samples exposed outdoors in Delft are concave, and the degradation depth increases with exposure time. This shape is the opposite of the degradation profiles of samples exposed in the other conditions.

The concave degradation profile, in terms of oxygen diffusion, must be caused by a first-order dependence between oxygen concentration and reaction rate. However, if the degradation profile is explained with first-order kinetics, the oxygen concentration must be below the critical oxygen concentration; therefore, the oxygen must diffuse in slowly or be used up quickly to enable first-order kinetics. Compared with other exposure conditions, both intensity and temperature are lower in Delft. However, if one looks at the activation energy of oxygen diffusion (approximately 37 kJ/mole; ref. 30) and the activation energy of the overall photooxidation rate constant for photooxidation (approximately 56 kJ/mole; ref. 35), the rate of oxygen consumption must be relatively low and the oxygen diffusion relatively high compared with the other conditions. Therefore, first-order kinetics is not to be expected.

However, the activation energy of photooxidation is lower at the beginning of photooxidation (36), and this discrepancy implies that the reaction rate remains high while the diffusion coefficient decreases compared with the other exposure conditions. Therefore, first-order conditions are more likely to prevail sooner because of the low ratio of oxygen diffusion over reaction rate. The oxygen diffusion coefficient may be even lower than expected because of a temperature decrease due to air pollution (caused by the nearby highway) in Delft, or because the average maximum daily temperature is used, whereas the actual temperature was considerably lower during most of the exposure period.

If one assumes that oxygen diffusion limits the degradation process, then an increasing degradation depth can be expected by taking into account the dark periods in Delft. Because of the relatively long dark periods in Delft compared with the other exposure conditions, oxygen may diffuse in the sample to deep layers and cause oxidation during the light periods.

Length of Horizontal Plateau. The conversion according to a first-order rate leads to degradation profiles with only a concave shape, whereas the conversion according to a zero-order rate leads to a shouldered profile with a horizontal plateau (Figure 5). The length of the horizontal plateau (LOP) can be calculated. A plateau is formed if the oxygen concentration is above the critical oxygen concentration (C_c); that is, the concentration at which the reaction rate becomes concentration-dependent. Consequently, LOP ends when $C(x)$ becomes equal to C_c (11):

$$LOP_{1,2} = \frac{L}{2} \pm \sqrt{\left(\frac{L}{2}\right)^2 - \frac{2D(C_0 - C_c)}{r_0}} \qquad (6)$$

Equation 6 shows that an increase in intensity and thus in r_0 leads to a decrease of LOP. This decrease is in agreement with experimental findings (10). On the other hand an increase in the diffusion coefficient leads to an increase in the LOP. The high UV intensities used in artificial weathering machines probably lead to an LOP considerably lower than the LOP obtained in outdoor aging.

Large oxygen concentrations yield a horizontal plateau and thus lead to zero-order kinetics. In the degradation profiles of the samples exposed in the Xenotest, the WOM, and outdoors in Florida, the degradation depth remained constant with exposure time and the degradation showed a reasonable plateau. The presence of a plateau is a typical phenomenon that occurs when the reaction rate in the surface layers is independent of the oxygen concentration; therefore, zero-order conditions prevail near the surface. The oxygen concentration decreases toward the middle. When it becomes less than the critical concentration, the reaction rate tends more toward a first-order dependence. This phenomenon caused the concave shape of the profile at the end of the plateau (Figure 5).

The LOP in the degradation profiles could be established for samples exposed in the Xenotest, WOM, and outdoors in Florida from the experimental results such as those shown in Figure 5. It is possible to calculate r_0 from the LOP after rewriting eq 5 and taking an arbitrary value for C_c:

$$r_0 = \frac{2D(C_c - C_0)}{LOP(LOP - L)} \qquad (7)$$

where r_0 is the zero-order rate constant, D is the diffusion coefficient, C_0 is the initial oxygen concentration in the outer surface of the sample, and L is the thickness of the sample. For all calculations a C_0 of 8.4×10^{-8} mole/g at the polymer surface was used. The diffusion coefficient was based on the air temperature measured in the different exposure conditions and calculated according to eq 4. For the air temperature in the outdoor exposure sites of Florida and Delft, the average maximum day temperature was used (35). The calculated oxygen diffusion coefficient was divided by an arbitrary factor of 2 for the compression-molded samples to account for the higher crystallinity. For C_c an arbitrary value of 1.7×10^{-8} mole/g was used; this value represents 20% of the oxygen concentration at the surface and more or less corresponds to values found in the literature (33).

The calculated values of the front surface reaction rate, r_0, are plotted against the exposure time in Figure 8. Clearly, the reaction rate increased with increasing exposure time due to autooxidation caused by the accumulation of oxidation products. The increase in r_0 was based on the assumption that the change in the LOP was caused solely by an increase in r_0 and not by

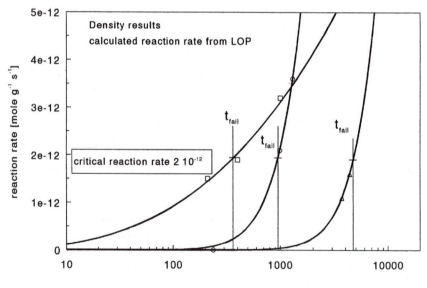

Figure 8. Zero-order reaction rate calculated on the basis of the length of the horizontal plateau in the degradation profiles showing autoacceleration. Key: □, Xenotest; ○, WOM; and △, Florida.

a decrease of the solubility or the oxygen diffusion coefficient during oxidation. The fact that the r_0 calculated according to eq 6 increased with time is in contradiction with the assumption of a steady-state reaction. This statement implies that eq 6 is not valid. However, even though the values may not be quantitatively correct, there is a clear trend toward an increased reaction rate.

The order of magnitude of the reaction rate obtained from the calculations based on the LOP can be verified by considering the measured carbonyl and vinyl concentration. As discussed in reference 11, the increase in the concentration of carbonyl was approximately 1×10^{-5} mole/g around t_{fail}, which was 360 h (or 1.3×10^6 s). The amount of carbonyl formed is assumed to be proportional to half the amount of oxygen absorbed (9). The oxidation rate then becomes $1 \times 10^{-5} /(2 \times 1.3 \times 10^6) = 3.85 \times 10^{-12}$ mole/(g • s), which is of the same order of magnitude as the values calculated (Figure 8).

Degradation Depth. The exposed samples show a limited degradation depth: that is, the maximum depth at which the degradation profile differs from the profile of an unexposed sample. However, the samples exposed outdoors in Delft show an increase in degradation depth. The degradation depth is an interesting measure, because it may determine the time to failure (6, 8). When the different exposure conditions are compared, the only variation seems to be the depth at which changes take place and the slope at which the degree of degradation decreases from the plateau or from the surface toward the middle. The results of the different measurements were averaged and summarized. For injection molded samples, the depths of the Xenotest, WOM, and Florida exposure, respectively, were 500, 750, and 900 μm. For the injection molded samples without the 100-μm skin, the average degradation depth from the Xenotest was 450 μm. The WOM and Florida exposure conditions for compression molded samples yielded degradation depth values of 600 and 700 μm, respectively.

Generally, the degradation depth increased, whereas the slope decreased going from Xenotest to WOM to Florida outdoor exposure. Injection-molded samples without the outer 100-μm layer had a lower degradation depth compared with the complete injection-molded samples. Compression-molded samples with more homogeneity but higher crystallinity (73–76%) with respect to the depth in the sample showed a small degradation depth compared with the injection-molded samples. The injection-molded samples had a lower overall crystallinity ($\pm 70\%$), and the crystallinity increased about 10% going from the surface (63%) toward the middle of the sample (72%).

According to the theory of oxygen-limited photooxidation, the degradation depth, a_i, depends on D and r according to eq 8:

$$a_i \propto \sqrt{\frac{D}{r}} \qquad (8)$$

The reaction rate for photooxidation is expected to show a similar temperature dependence and is also a function of the UV intensity:

$$r = k_0 \exp\left(\frac{-E_a}{RT}\right) I^\alpha \tag{9}$$

where k_0 and a are constants, E_a is the activation energy, R is the gas constant, and I is the UV intensity. Combination of eqs 3, 8, and 9 yields eq 10:

$$a_i \propto \left(\frac{D_0}{k_0}\right)^{1/2} \exp\left(\frac{E_a - E_d}{2RT}\right) I^{-\alpha/2} \tag{10}$$

According to the literature E_a for photooxidation is approximately 56 kJ/mole (35). The intensity is determined from the cumulative radiant exposure at the time to failure divided by the time to failure. When the degradation depth was calculated according to eq 10 by using the corrections for the temperature for the oxygen diffusion coefficient and the reaction rate, it showed a rather poor fit. From further analysis of the data, the best fit was obtained when the influence of the small temperature differences on the reaction rate in the various exposure conditions was neglected. The temperature range involved, however, was too narrow (27–35 °C) to judge the temperature dependence properly.

When the degradation depth is divided by the square root of the (temperature-dependent) oxygen diffusion coefficient a corrected degradation depth (a_i^*) follows:

$$a_i^* = \frac{a_i}{\sqrt{D}} = cI^{-\alpha/2} \tag{11}$$

where c is a constant. The very simple formula yields good results considering the assumptions made and the experimental difficulties involved. The following assumptions make the reliability of the prediction questionable:

- The radiant exposure in Florida was given between 295–385 nm in contrast to the exposure conditions in the Xenotest and WOM (300–400 nm).
- The spectrum of the Xenotest contained wavelengths below 300 nm.
- The intensity was calculated roughly by dividing the total radiant exposure at the time to failure by the time to failure, which yielded less reliable values for outdoor exposure conditions.
- The temperature used was not the sample temperature, whereas in case of outdoor exposure the average maximum air temperature was used.

Interestingly, α was approximately 0.8, which was expected on the basis of values given in the literature. The value of α should lie between 0.5 and 1 for non-UV- and UV-stabilized materials (28). In the beginning of the process the value of α equals approximately 1 if the small amount of antioxidant is still present and acts as a photostabilizer. However, during exposure α is expected to decrease to 0.5; therefore, a value of α somewhere between 0.5 and 1 is reasonable. The relation between the data probably improved when the influence of the temperature on the reaction rate was not taken into account because the activation energy in the beginning of photooxidation was lower (36). Values given in the literature are, in most cases, determined at embrittlement. In addition, the temperature differences between the different exposure conditions are too small to draw reliable conclusions.

The degradation depth and the LOP of injection-molded samples were significantly larger than the degradation depth and LOP of compression-molded samples weathered in the WOM and outdoors in Florida. According to the theory on oxygen diffusion, a higher rate of oxidation and a lower oxygen diffusion can account for a smaller oxidation depth. The lower coefficient of diffusion most likely caused the smaller depth because the diffusion coefficient decreases with increasing crystallinity.

Only a small difference occurred in the degradation depth of injection-molded samples with and without a 100-μm skin layer exposed in the Xenotest 1200. The oxygen diffusion coefficient of the complete samples was probably high at the surface and decreased toward the middle (because of the increasing crystallinity). By removing the outer 100 μm the diffusion becomes relatively more difficult and the degradation depth is likely to decrease. The skin effect in the complete samples may cause the rate of oxidation to be relatively higher, and this effect again leads to a lower degradation depth.

Conclusions

The most sensitive results with the least amount of scatter were obtained with the MFTT and the density measurements and not with the more familiar FTIR measurements. Generally, there was a reasonable agreement among the different exposure conditions regarding the change in nominal strain, vinyl index, and density versus the distance from the exposed surface. A decrease in the nominal strain corresponded to an increase in the vinyl index and in the density. Generally, the nominal tensile stress only started to show significant changes (decrease) around t_{fail}, whereas the carbonyl index showed too much scatter to be able to reveal a general trend.

The presence of a less degradation-sensitive skin was clearly seen in the MFTT measurements and, to some extent, in FTIR and density measurements. This less degradation-sensitive skin was only seen in injection-molded samples and not in the compression-molded samples. After longer exposure time the less degradation-sensitive skin disappeared, but unlike the other exposure conditions the samples exposed outdoors in Delft showed most degradation near the skin.

The exposed samples showed a limited degradation depth, which is the maximum depth at which the degradation profile differs from the profile of an unexposed sample. However, the samples exposed outdoors in Delft showed an increase in degradation depth. When the different exposure conditions were compared, the only variation seemed to be the depth at which changes took place and the slope at which the degree of degradation decreased from the plateau or from the surface toward the middle. Generally, the degradation depth increased, whereas the slope decreased going from Xenotest to WOM to Florida outdoor exposure. Injection-molded samples without the outer 100-μm layer had a lower degradation depth compared with the complete injection-molded samples.

A zero-order reaction-rate dependence of the oxygen concentration can explain qualitatively the degradation profile in the injection-molded samples in the Xenotest, WOM, and Florida outdoor exposure. Degradation profiles showed a horizontal plateau and the degradation depth remained constant. The degradation profiles of the samples exposed in Delft exhibited no plateau or only a very small horizontal one. Instead, they had a concave shape and showed an increasing degradation depth with increasing exposure time. The cause may be that the reaction rate was first order due to the low oxygen diffusion. The oxygen diffusion may have been very low due to the low temperature and perhaps to air pollution. On the other hand, the reaction rate in the beginning of the oxidation process probably was not very temperature dependent and was not lower in Delft compared with the other exposure conditions. The increasing degradation depth can be explained by the relatively long dark periods that enabled oxygen diffusion to deeper layers.

The degradation depth is proportional to the square root of the oxygen diffusion coefficient (which decreases with increasing crystallinity) and inversely proportional to the square root of the oxidation rate. By fitting the experimental results, the temperature dependence of the reaction rate appeared to be small. However, several assumptions had to be made, and the temperature of the experiments only varied to a small extent. Therefore, the validity of the equation must be verified for a larger range of temperatures.

The length of the horizontal plateau of the degradation profile decreased with exposure time and indicated an increasing reaction rate with exposure time. This result corresponds to a progressive increase in the degree of reaction as measured as the total increase of the density over the front surface of the sample.

The smaller degradation depth of a compression-molded sample compared with an injection-molded sample was most likely due to the lower oxygen diffusion coefficient caused by a higher crystallinity.

References

1. Kulshreshtha, A. K.; Kaushik, V. K.; Pandey, G. C.; Chakrapani, S.; Sharma, Y. N. *J. Appl. Polym. Sci.* **1989**, *37*, 669–679.
2. Birch, M. W.; Williams, J. G.; Marshall, G. P. In *Proceedings of the 3rd International Conference on Deformation, Yield, and Fracture in Polymers;* Churchill College, Cambridge, 29 March–1 April 1976; pp 6.1–6.7.
3. Qureshi, F. S.; Hamid, S. H.; Maadhah, A. G.; Amin, M. B. *Polym. Plast. Technol. Eng.* **1989**, *28*, 663–670.
4. Furneaux, G. C.; Ledbury, K. J.; Davis, A. *Polym. Degrad. Stab.* **1980–1981**, *3*, 431–442.
5. Bykov, Ye V.; Bystritskaya, Ye. V.; Karpukhin, O. N. *Vysokomol. (Soedin., Ser. A)* **1987**, *29(7)*, 1347–1352.
6. Schoolenberg, G. E., Ph.D. Thesis, Delft University of Technology, Delft, 1988.
7. Schoolenberg, G. E.; Vink, P. *Polymer* **1991**, *32(3)*, 432–437.
8. Rolland., L., Ph.D. Thesis, Illinois Institute of Technology, Chicago, IL, 1983.
9. Cunliffe, A. V.; Davis, A. *Polym. Degrad. Stab.* **1982**, *4*, 17–37.
10. Audouin, L.; Langlois V.; Verdu, J.; de Bruijn, J. C. M. *J. Mater. Sci.* **1994**, *29*, 569–583.
11. de Bruijn, J. C. M., Ph.D. Thesis, Delft University of Technology, Delft University Press: Delft, 1992.
12. Atlas, Bulletin No. 1360, 1986.
13. McTigue, F. H.; Blumberg, M. *Appl. Polym. Symp.* **1967**, *4*, 175–188.
14. de Bruijn, J. C. M.; Meijer, H. D. F. *Rev. Sci. Instrum.* **1991**, *62(6)*, 1620–1624.
15. Winslow, F. H. *Makromol. Chem. Suppl.* **1979**, *2*, 27.
16. Titus, J. B. *The Weathering of Polyolefins;* Report 32; Plastic Technical Evaluation Center: Dover, NJ, 1968.
17. Reich, L.; Stivala, S. S. *Elements of Polymer Degradation;* McGraw-Hill: New York, 1971.
18. Gillen, K. T.; Clough, R. L.; Dhooge, N. J. *Polymer* **1986**, *27*, 225–232.
19. Moy, F. H.; Kamal, M. R. *Polym. Eng. Sci.* **1980**, *20*, 957–964.
20. Torikai, A.; Shirakawa, H.; Nagaya, S.; Fueki, K. *J. Appl. Polym. Sci.* **1990**, *40*, 1637–1646.
21. Tidjani, A.; Arnaud, R.; DaSilva, A. *J. Appl. Polym. Sci.* **1993**, *47*, 211–216.
22. Scott, G. In *Mechanism of Photodegradation and Stabilization of Polyolefins: Ultraviolet Light Induced Reactions with Polymers;* Labana, S. S., Ed.; ACS Symposium Series 25; American Chemical Society: Washington, DC, 1976; pp 340–366.
23. Gijsman, P.; Hennekens, J.; Tummers, D. *J. Polym. Degrad. Stab.* **1993**, *39*, 225–233.
24. Gillen, K.T.; Clough, R. L. In *Handbook of Polymer Science and Technology: Performance Properties of Plastics and Elastomers;* Cheremisinoff, N. P., Ed.; Dekker: New York, 1989; Vol. 2, Chapter 6, pp 167–202.
25. Clough, R. L.; Gillen, K.T. *Polym. Degrad. Stab.* **1992**, *38*, 47–56.
26. Dalinkevich, A. A.; Kiryushkin, S. G.; Shlyapnikov, Yu. A. *Int. Polym. Sci. Tech.* **1992**, *19*, T/85–T/92.
27. Davis, A.; Sims, D. *Weathering of Polymers;* Elsevier Applied Science: London, 1983.
28. Vink, P. In *Degradation and Stabilisation of Polyolefins;* Allen, N. S., Ed.; Elsevier Applied Science: London, 1983; Chapter 5, pp 213–246.
29. Michaels, A. S.; Bixler, H. J. *J. Polym. Sci.* **1961**, *50*, 413–439.

30. Stannet, V. In *Diffusion in Polymers;* Crank, J.; Park, G. S., Eds.; Academic: London, 1968; Chapter 2, pp 41–73.
31. Jansen, K. M. B. *Polym. Eng. Sci.,* in press.
32. Verdu, J. *Vieillissement des Plastiques;* AFNOR Technique: Paris, 1984.
33. Papet, G.; Audouin-Jirackova, L.; Verdu, J. *Radiat. Phys. Chem.* **1989,** *33(4),* 329–335.
34. Zabara, M.Ya. *Rapra* **1971,** *42c11– 93T,* 40–42.
35. Vincent, J. A. J. M.; Jansen, J. M. A.; Nijsten, J. J. H. In *Proceedings of the 4th Annual International Conference on Advances in the Stabilization and Degradation of Polymers;* Lucerne, Switzerland, 2-4 June 1982.
36. Huvet, A.; Philippe, J.; Verdu, J. *Eur. Polym. J.* **1978,** *14,* 709–713.

RECEIVED for review January 26, 1994. ACCEPTED revised manuscript September 22, 1994.

Comparison of UV Degradation of Polyethylene in Accelerated Test and Sunlight

Pieter Gijsman, Jan Hennekens, and Koen Janssen

DSM Research BV, PO Box 18, 6160 MD Geleen, The Netherlands

Studies of aging of polyethylene in an accelerated (Xenon) test and outdoors in Geleen, The Netherlands, showed unexpected differences at the same degree of oxygen uptake. In outdoor weathering, about twice the oxygen uptake was necessary to give the same drop of the elongation at break and to form the same amount of carbonyl groups and unsaturation as in the accelerated weathering. The results may be explained by assuming different contributions to oxygen uptake from initiation by charge–transfer complexes and from the propagation reaction for accelerated and outdoor weathering. A possible improved accelerated test is discussed.

$\mathbf{P}$OLYOLEFINS DEGRADE UNDER THE INFLUENCE of UV-light, which can lead to failure of articles in outdoor applications (*1, 2*). Determination of this UV-stability is a problem. In general, outdoor aging is too slow to be useful in the development of stabilizer formulations or for quality control. This flaw led to the development of several accelerated weathering tests (e.g., Weather-OMeter, Xenontester, Suntester, UVCON, QUV, and SEPAP; ref. 2). Most of these accelerated weathering devices show poor correlation between the stabilities measured with them and those measured outdoors (*3–6*). Recent Fourier transform IR (FTIR) studies (*7*) suggested that this lack of correlation may be due to differences in degradation mechanisms in accelerated and outdoor testing.

The most direct way to study the oxidation of polymers is to determine the rate of oxygen uptake by the polymer, but this method is experimentally difficult. In many studies it is done by measuring the drop in pressure in a closed system (*8, 9*). In such experiments the drop of the pressure is assumed to correspond quantitatively to the consumption of oxygen. This assumption can lead to errors if gaseous oxidation products are formed. The problem can

0065–2393/96/0249–0621$12.00/0

be overcome if the amount of oxygen left is determined after each degradation period (10, 11).

This chapter describes a study of the photooxidation of polyethylene (PE) in an accelerated (Xenon) test and in outdoor weathering in Geleen, The Netherlands. The degradation was followed by measuring the true oxygen uptake by determination of the reduction of the amount of oxygen in the gas phase, along with changes in the FTIR spectra and the mechanical properties. The combination of these techniques led to new insights.

Experimental

Degradation studies used 150-μm thick, blown, low-density PE [M_w 91,000; number of branches (CH_3) per 1000 carbons = 20, amount of unsaturation ($C=C$) per 10^5 carbons = 55] films, designated PE I and PE II. Processing was done in the presence of a stabilizer, which was then extracted with refluxing chloroform. PE I was extracted for 25 h and PE II for 150 h.

Both types of aging test were performed in a closed Durethan glass system with flat windows facing the light source. This type of glass is transparent to light with a wavelength above 290 nm. All experiments were done with air containing 0.83% helium to check possible leakages.

Accelerated weathering was done in a Suntester (Hanau; filtered xenon lamp; intensity at 340 nm, 0.3 W/m²). The temperature was measured continuously inside the sample cells and was always in the 30–40 °C range. Sunlight exposures were done in Geleen, The Netherlands, by using plaques facing south at an angle of 45°. Outdoor weathering was started on 29 January 1989 for PE I and on 14 February 1989 for PE II. The temperature both outdoors and in the closed system was measured continuously and varied with the season (Figure 1).

Oxygen uptake was determined by periodic gas chromatographic analysis of the gas phase by using the method previously described (11). Chemical changes were recorded by FTIR analysis. Absorptions were calculated as the difference between the peak absorption and the absorption at a baseline. For the absorptions at 1712 and 1642 cm⁻¹, the baseline was drawn between 1840 and 1600 cm⁻¹; for the absorption at 908 cm⁻¹, the baseline was drawn between 950 and 860 cm⁻¹. In addition to the chemical changes, the impairment of the mechanical properties was determined by using the elongation at break as a criterion (in absolute percentages).

Results

Accelerated and sunlight weathering of the polymers resulted in an impairment of the mechanical properties (Figure 2). For the accelerated weathering the drop of the elongation at break began after 1500–2000 h exposure; within 3000 h the polymers became totally brittle. For the samples weathered outdoors the reduction of the elongation at break started after 12,000 h, and after 20,000 h exposure all polymers were brittle. The time until the elongation at break dropped to 50% of its original value was 6–8 times longer for the outdoor weathering than for the accelerated test.

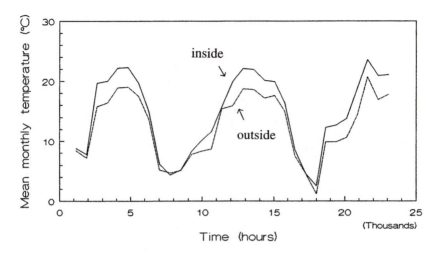

Figure 1. *Mean monthly outdoor temperature as a function of exposure time inside (solid line) and outside (broken line) the closed system.*

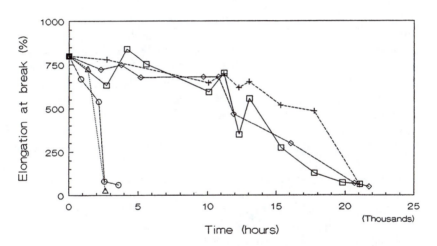

Figure 2. *Elongation at break (%) as a function of exposure time (h) for accelerated weathering of PE I (△) and PE II (○) and for outdoor weathering of PE I (□ and +) and PE II (◇).*

During accelerated weathering the oxygen uptake started immediately and was almost linear with time (Figure 3). During the first 1000 h the oxygen uptake was comparable for the two PEs, but after 1000 h a small deviation was found. After 3000–4000 h of accelerated weathering the polymers had an oxygen uptake of almost 1 mol/kg.

During outdoor weathering the oxygen uptake curves showed totally different behavior (Figure 3). Exposure of these samples started in January and

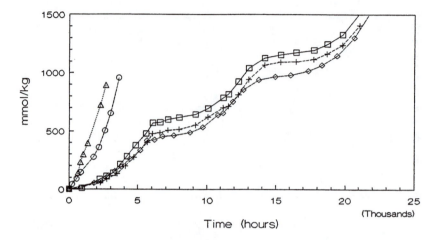

Figure 3. Oxygen uptake (mmol/kg) as a function of exposure time (h) for accelerated weathering of PE I (△) and PE II (○) and the outdoor weathering of PE I (□ and +) and PE II (◇).

February and the samples began to take up oxygen in the early spring. During the summer the rate of oxidation became constant. It slowed down at the end of the summer. During autumn and winter there was almost no oxygen uptake and in the early spring a second increase of the oxidation rate took place. After 22,000 h of outdoor weathering the total oxygen uptake was about 1.5 mol/kg for all three samples. The reproducibility of the outdoor weathering data was good.

Changes in the IR spectra were also recorded during the degradation. For accelerated weathering the increase of the carbonyl absorption (1712 cm^{-1}) was almost linear in time, and it reached 0.8 after about 3000 h of degradation (Figure 4). In outdoor weathering the development of the carbonyl absorption is different (Figure 4). However, the shape of the curves is comparable with the oxygen uptake curves (Figure 3). After 20,000 h the carbonyl absorption was approximately 0.5.

Changes in the amount of unsaturation during weathering were also recorded. The changes of the concentration of total unsaturation (absorption at 1642 cm^{-1}) and of end unsaturation (absorption at 908 cm^{-1}) are shown in Figures 5 and 6, respectively. The two curves show the same form as the corresponding plots for carbonyl absorption.

As expected, accelerated weathering was faster than outdoor weathering. However, for outdoor weathering during the summer period the rate of oxygen uptake was only 2.5 times slower than for the accelerated weathering. The mean rate of oxygen uptake for three summer periods was 0.2 mmol/(kg·h), and the oxygen uptake rate during the accelerated weathering was 0.5 mmol/(kg·h). The acceleration factor is even smaller if we take into account

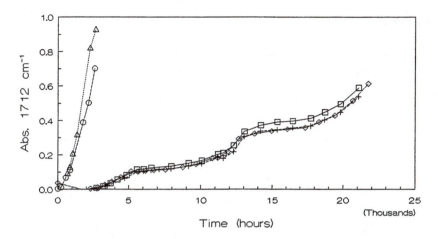

Figure 4. Carbonyl absorption at 1712 cm⁻¹ as a function of exposure time (h) for accelerated weathering of PE I (△) and PE II (○) and the outdoor weathering of PE I (□ and +) and PE II (◇).

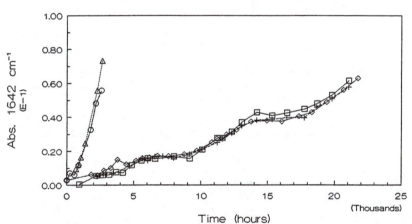

Figure 5. Concentration of unsaturation (absorbance, 1642 cm⁻¹) as a function of exposure time (h) for accelerated weathering of PE I (△) and PE II (○) and the outdoor weathering of PE I (□ and +) and PE II (◇).

that the outdoor degradation only took place during part of the day, whereas the accelerated degradation was continuous. If outdoor degradation only occurred for 10 h per day, then the outdoor oxidation rate in this period was comparable to that in accelerated weathering.

The rates of formation of carbonyl and of end unsaturation during accelerated and outdoor summer weathering differed more than the oxygen uptake rates. For the accelerated test the rate of formation of carbonyl groups was 7.5 times higher than the mean formation rate over three summers of outdoor weathering. Similarly, end unsaturation was produced 10 times faster in accelerated weathering than in sunlight.

The impairment of the mechanical properties was expected to be related directly to the oxidation of the polymer. Nevertheless, we found a difference between accelerated and sunlight weathering in the relationship between oxygen absorption and the loss of elongation at break (Figure 7). For the ac-

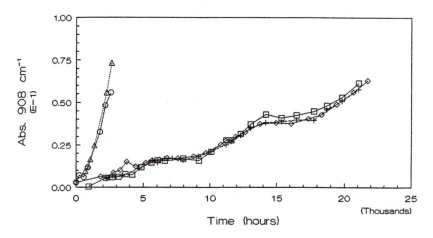

Figure 6. End unsaturation (absorbance, 908 cm⁻¹) as a function of exposure
time (h) for accelerated weathering of PE I (△) and PE II (○) and the outdoor
weathering of PE I (□ and +) and PE II (◇).

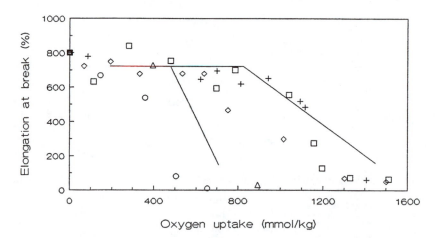

Figure 7. Elongation at break (%) vs. oxygen uptake (mmol/kg) for accelerated
weathering of PE I (△) and PE II (○) and the outdoor weathering of PE I (□
and +) and PE II (◇).

celerated weathering the drop of the elongation at break started at an oxygen uptake of about 350 mmol/kg, whereas for the sunlight weathering an oxygen uptake of 700 mmol/kg was observed before any significant loss in elongation at break was measured. It is interesting to compare the relations between oxygen uptake and the formation of the different products for the accelerated and the sunlight weathering. Accelerated tests can be reliable only if these relations are the same for both conditions.

The IR spectra for samples having an oxygen uptake between 350 and 450 mmol/kg are shown in Figures 8 and 9. Even though the samples had taken up about the same amount of oxygen for the accelerated and sunlight weathering, the differences are remarkable. The oxygen uptake led to larger changes of the IR spectra for the accelerated weathering than for the outdoor weathering.

The relationship between carbonyl formation and the oxygen reacted with the polymer depended on the kind of exposure. The conversion of oxygen into carbonyl groups was higher for the accelerated than for the outdoor weathering (Figure 10). The same effect was found for the relation between the oxygen uptake and the IR absorptions at 1642 and 908 cm^{-1} (Figures 11 and 12, respectively). These figures reveal the difference in degradation chemistry during accelerated and outdoor weathering.

Because mechanical property measurements are destructive, a large amount of polymer had to be exposed to provide enough material to allow determination of the time until mechanical failure. In literature studies it is common to assume a relationship between the amount of carbonyl groups formed and the changes in mechanical properties (3). The carbonyl groups can be measured using a nondestructive method such as IR so that only a small amount of polymer has to be exposed to determine the stability. This method has led to an increased use of the rate of change of the IR absorption at 1712 cm^{-1} to determine the stability of a polymer. A plot of the elongation at break versus the carbonyl absorption (1712 cm^{-1}) shows that this relationship is not as universal as expected and depends on the kind of exposure (Figure 13). During outdoor exposure an increase of the carbonyl absorption led to a larger drop of the elongation at break than in the accelerated exposure. The relationship between the amount of end unsaturation (IR absorption at 908 cm^{-1}) and the elongation at break is much less dependent on the kind of exposure (Figure 14). Thus, it is better to use the IR absorption at 908 cm^{-1} to determine the stability of a polymer instead of the change of the IR absorption at 1712 cm^{-1}.

The relationships between different IR absorptions developing during weathering also depend on the kind of weathering. The relationships between IR absorptions at 908 and 1712 cm^{-1} are plotted in Figure 15. The ratio of end unsaturation to carbonyl groups is higher for the outdoor than for the accelerated weathering.

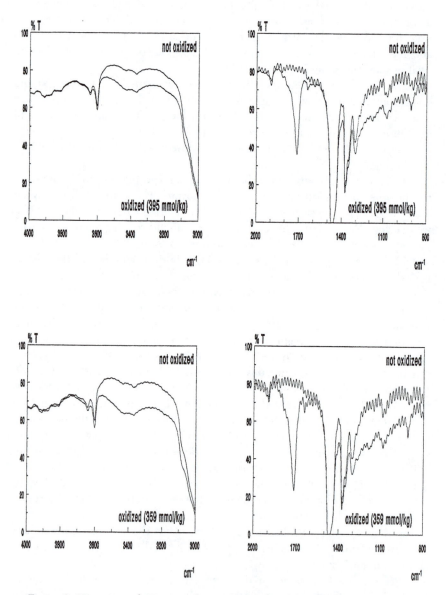

Figure 8. IR spectra of PE exposed to accelerated weathering (PE I and PE II) before and after taking up 395 mmol/kg (PE I, top) and 359 mmol/kg (PE II, bottom) of oxygen.

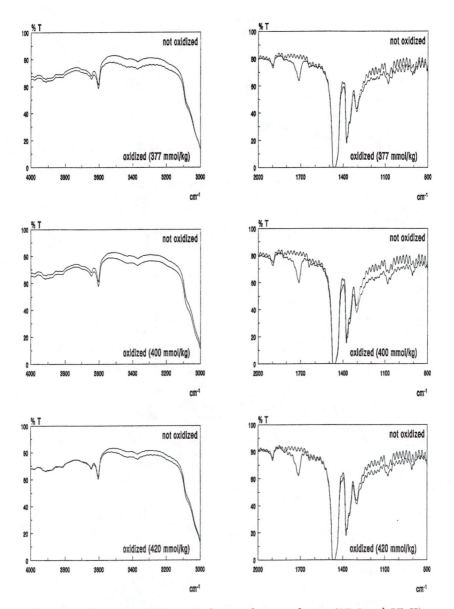

Figure 9. IR spectra of PE exposed to outdoor weathering (PE I and PE II) before and after taking up 377 mmol/kg (PE I, top), 400 mmol/kg (PE I, middle), and 420 mmol/kg (PE II, bottom) of oxygen.

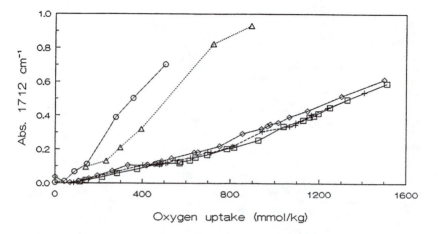

Figure 10. Carbonyl absorption at 1712 cm⁻¹ vs. oxygen uptake (mmol/kg) for accelerated weathering of PE I (△) and PE II (○) and the outdoor weathering of PE I (□ and +) and PE II (◇).

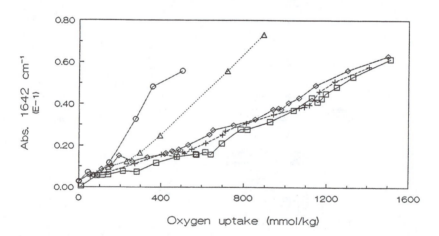

Figure 11. Unsaturation (absorbance, 1642 cm⁻¹) vs. oxygen uptake (mmol/kg) for accelerated weathering of PE I (△) and PE II (○) and the outdoor weathering of PE I (□ and +) and PE II (◇).

Discussion

Environmental factors such as light intensity, spectral distribution, and temperature have an influence on the degradation rate during exposure of a polymer. Differences in these factors can lead to differences in oxygen uptake, changes of the IR spectra, and impairment of the mechanical properties. A good correlation between an accelerated test and outdoor aging will be found

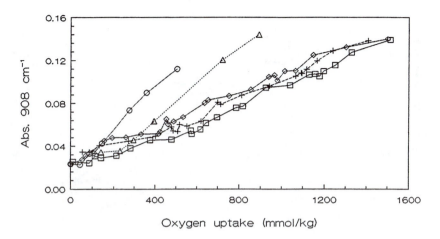

Figure 12. End unsaturation (absorbance, 908 cm⁻¹) vs. oxygen uptake (mmol/ kg) for accelerated weathering of PE I (△) and PE II (○) and the outdoor weathering of PE I (□ and +) and PE II (◇).

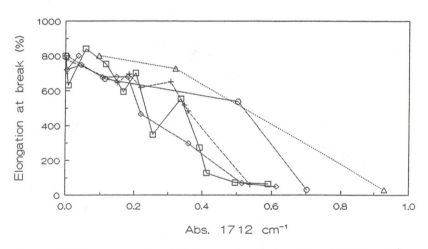

Figure 13. Elongation at break (%) vs. carbonyl absorption at 1712 cm⁻¹ for accelerated weathering of PE I (△) and PE II (○) and the outdoor weathering of PE I (□ and +) and PE II (◇).

only if the different degradation-determining factors are accelerated in the same way.

   The basic mechanism of the photodegradation of PE is well known. However, there is still discussion about the relative importance of the different reactions. Several publications (*12–14*) suggested that the initiating capacities of added ketones in PE and of hydroperoxides formed by thermal oxidation

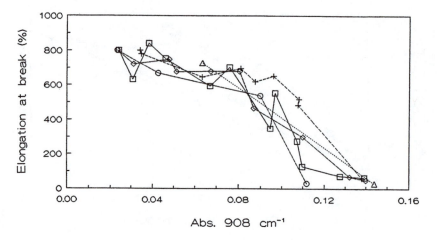

*Figure 14. Elongation at break (%) vs. end unsaturation (absorbance, 908 cm⁻¹)
for accelerated weathering of PE I (△) and PE II (○) and the outdoor weathering
of PE I (□ and +) and PE II (◇).*

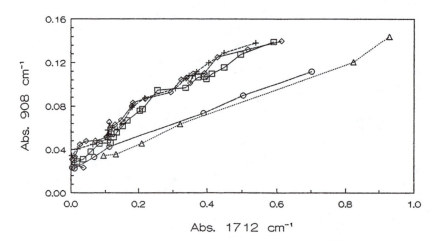

*Figure 15. End unsaturation (absorbance, 908 cm⁻¹) vs. carbonyl absorption
(1712 cm⁻¹) for accelerated weathering of PE I (△) and PE II (○) and the
outdoor weathering of PE I (□ and +) and PE II (◇).*

are low. The suggestion was made that hydroperoxides in PE do not initiate photooxidation, because they decompose without forming radicals.

Gugumus (*15–17*) postulated an alternative mechanism of radical formation. He suggested that the main source of new radicals is hydrogen peroxide, which is formed from a charge–transfer complex (CTC) of the polymer with oxygen. Reaction of the CTC leads to the formation of a *trans*-vinylene group and hydrogen peroxide. Hydrogen peroxide can initiate oxidation by thermal or photochemical decomposition. Other CTCs are also possible, and end unsaturation or cross-links can be formed instead of *trans*-vinylene groups (*11*).

Absorptions due to unsaturation are certainly found in the IR spectra of our exposed polymers. The absorption at 908 cm^{-1} is due to end unsaturation, whereas that at 1642 cm^{-1} is also due to unsaturation (*9, 18*). Unfortunately, the IR spectra were not clear enough to allow detection of the absorption due to *trans*-vinylene groups. Besides unsaturation, CTCs might also lead to cross-linking (*11*), which can not be detected with IR.

In an earlier study (*11*) comparing the accelerated weathering of unstabilized and hindered amine light stabilized (HALS) PE, we found that the major part of the oxygen taken up by HALS PE was converted into gaseous products, whereas the degradation of unstabilized PE resulted in many different products. These results could be explained by proposing that the oxidation is initiated by reaction of oxygen–polymer CTCs that results in water and radicals. In the case of unstabilized PE these radicals initiate the autoxidation, whereas for the HALS polymer the radicals are trapped by the stabilizer and do not initiate oxidation.

For unstabilized PE this mechanism would mean that part of the oxygen is converted via the CTC into water, but most is consumed in the normal oxidation. For the stabilized PE the radicals formed in the CTC reaction are trapped, and the normal oxidation does not start and the major part of the oxygen is consumed in the CTC reaction and yields water (*11*).

Similar to the accelerated weathering of the unstabilized and HALS PE, accelerated and outdoor weathering of unstabilized PE gave different oxidation products. For the accelerated weathering of the HALS PE and for the outdoor weathering of the unstabilized PE, the oxygen uptake led only to small changes of the IR spectra and the mechanical properties. For accelerated weathering of the HALS PE, this phenomenon was explained by assuming that the main part of the oxygen was converted into water through a CTC initiation mechanism. This mechanism might also explain the differences between the results for the accelerated and the outdoor weathering of the unstabilized PE. Differences between the accelerated and outdoor weathering probably are due to a higher oxygen conversion into water through a CTC for the outdoor weathering and a higher conversion of the oxygen in the normal oxidation process during accelerated weathering.

The ratio of the amounts of end unsaturation to carbonyl groups formed was higher for outdoor than for accelerated weathering (Figure 15). This result might be due to a difference in the rate of formation of end unsaturation through the Norrish type II reaction or to the CTC initiation reaction. At the same oxygen uptake the concentration of carbonyl groups was lower for the outdoor than for the accelerated weathering.

Norrish type II reactions of aldehydes and methyl ketones can result in gaseous products and cause a drop in the concentration of carbonyl groups in the polymer, whereas for all other ketones the Norrish type II reaction does not cause a drop in the concentration of carbonyl groups. A higher contribution of the Norrish type II reaction during outdoor weathering would result in gaseous products that would not be detected and might explain the higher ratio of unsaturation to carbonyl groups. However, a higher efficiency of the Norrish type II reaction should also cause a faster decline of the mechanical properties. We found that oxygen uptake led to a faster decline of the mechanical properties in accelerated as opposed to outdoor weathering. Thus, a higher contribution from the Norrish type II reaction in outdoor weathering does not explain our results.

The other possible reason for a higher ratio of end unsaturation to carbonyl groups when comparing outdoor and accelerated weathering is that the initiation by a CTC is more efficient in outdoor weathering. This phenomenon might be due to the difference in temperature during degradation. Because the stability of a CTC increases with decreasing temperature (19), the lower temperature during outdoor weathering (the mean monthly temperature was between 0 and 25 °C) in comparison with accelerated weathering (between 30 and 40 °C) might be the reason for a higher efficiency of CTC initiation in outdoor weathering.

The major part of the end unsaturation in accelerated aging appeared to be due to the Norrish type II reaction, whereas in outdoor aging the major part was due to CTC initiation. Initiation by a CTC can be attributed to several reactions (11). In sunlight weathering a lower amount of end unsaturation is formed for a given level of reacted oxygen than in the case of accelerated weathering. This difference means that most of the CTCs result in other products (*trans*-vinylene or cross-linking).

The oxygen uptake in outdoor and in accelerated weathering was probably due to different reactions. Our results can be explained by assuming that in accelerated weathering most of the oxygen was consumed via the propagation reaction and resulted in ketones and chain scission. In outdoor weathering the major part of the oxygen absorbed was converted through a CTC into water, and only a minor part was consumed in the propagation.

Because the outdoor degradation in summer took place during only part of the day (e.g., 10 h), the oxygen uptake in this period was as fast as in accelerated weathering. Thus, an accelerated test that simulated our summer sunlight weathering might show about the same oxygen uptake rate as our accelerated test. This alternative test would show smaller rates for the formation of carbonyl groups and end unsaturation but would give a better correlation with outdoor weathering.

The differences between accelerated and outdoor weathering are probably due to a difference in aging temperature. Thus, a decrease of the temperature during accelerated weathering might lead to an accelerated weathering test that gives a good correlation with outdoor weathering.

A better correlation between accelerated and outdoor weathering will probably be found if conditions can be established so that more oxygen is consumed through the CTC initiation mechanism during accelerated weathering. The stability of a CTC of oxygen with a polymer depends on temperature and the oxygen pressure. A decrease of the temperature and an increase of the oxygen pressure will result in a higher CTC stability. Thus, decreasing the temperature or increasing the oxygen pressure should lead to more initiation through the CTC. Under these conditions, an accelerated weathering test would probably correlate better with outdoor weathering.

Conclusions

Accelerated degradation was faster than outdoor degradation and led to a more rapid drop of the elongation at break, higher oxygen uptake, and faster changes of the IR spectra. The outdoor aging mainly took place in spring and summer; in autumn and winter the polymers did not degrade at all. Thus, part of the acceleration was attributed to the fact that outdoors the polymers degraded only during part of the year. Comparison of the rate of degradation in accelerated and summer outdoor weathering showed that the acceleration factor depends on the degradation parameter measured. For oxygen uptake the acceleration factor was only 2.5, which is small especially because outdoors the polymers only degrade during part of the day. For the carbonyl and end unsaturation formation rates the acceleration factors are larger (*7–10*).

The conversion of oxygen into other products depends on the kind of exposure. A higher oxygen uptake is necessary to bring about the same drop of the elongation at break and to get the same changes of the carbonyl and end unsaturation absorptions in the IR spectra for outdoor compared with accelerated weathering. Thus, in the case of outdoor weathering the oxygen uptake must lead to different oxidation products than in the case of accelerated weathering.

The differences between accelerated and outdoor weathering are probably due to a change of the mechanism leading to oxygen uptake. We suggest that in accelerated weathering most of the oxygen is consumed through the propagation reaction and gives all expected products, whereas in outdoor weathering most of the oxygen is consumed by an initiation reaction caused by a CTC of oxygen and the polymer and yields water. The higher conversion of oxygen via a CTC during outdoor weathering is probably due to a higher stability of these complexes at low temperatures. A decrease of the temperature or an increase of the oxygen pressure during accelerated weathering should lead to more initiation through a CTC and thus a better correlation with outdoor weathering.

References

1. Gugumus, F. In *Plastic Additives Handbook*; Gächter, R.; Müller, H., Eds.; Hanser Publishers: Munich, Germany, 1990; pp 129–262.
2. Rabek, J. F. In *Photostabilization of Polymers;* Elsevier: London, 1990.
3. Gugumus, F. In *Developments in Polymer Stabilisation;* Scott, G., Ed.; Applied Science: London, 1987; Vol. 8, pp 239–289.
4. Vincent, J. A. J. M.; Jansen, J. M. A.; Nijsten, J. J. H. *Proceedings of the International Conference on Advances in the Stabilization and Controlled Degradation of Polymers;* Lucerne, Switzerland, 1982.
5. Laus, T. *Plast. Rubbers Mater. Appl.* **1977,** *2,* 77.
6. Kingsnorth, D. J.; Wood, D. G. M. *The Weathering of Plastics and Rubber;* PRI Symposium; 1976; p E2.
7. Titjani, A.; Arnaud, R. *Polym. Deg. Stab.* **1993,** *39,* 285.
8. Bigger, S. W.; Delatycki, O. *J. Polym. Sci. Part A: Polym. Chem.* **1987,** *25,* 3311.
9. Foster, G. N. In *Oxidation Inhibition in Organic Materials;* Pospíšil, J.; Klemchuk, P. P., Eds.; CRC: Boca Raton, FL, 1990; Vol. 2, pp 314–331.
10. Rose, J.; Mayo, F. R. *Macromolecules* **1982,** *15,* 948.
11. Gijsman, P.; Hennekens, J.; Tummers, D. *Polym. Degad. Stab.* **1993,** *39,* 225.
12. Ginhac, J. M.; Gardette, J. R.; Arnaud, R.; Lemaire, J. *Makromol. Chem.* **1981,** *182,* 1017.
13. Arnaud, R.; Moison, J.-Y; Lemaire, J. *Macromolecules* **1984,** *17,* 332.
14. Gugumus, F. *Makromol. Chem. Macromol. Symp.* **1989,** *25,* 1.
15. Gugumus, F. *Makromol. Chem. Macromol. Symp.* **1989,** 27.
16. Gugumus, F. *Angew. Makromol. Chem.* **1990,** *176/177,* 241.
17. Gugumus, F. *Polym. Degrad. Stab.* **1991,** *34,* 205.
18. Adams, J. H. *J. Polym. Sci. Part A: Polym. Chem.* **1970,** *8,* 1077.
19. Chien, J. C. W. *J. Phys. Chem.* **1965,** *69,* 4317.

RECEIVED for review May 14, 1994. ACCEPTED revised manuscript May 31, 1995.

Measurement of Radical Yields To Assess Radiation Resistance in Engineering Thermoplastics

Kirstin Heiland[1], David J. T. Hill[2]*, Jefferson L. Hopewell[2], David A. Lewis[3], James H. O'Donnell[2], Peter J. Pomery[2], and Andrew K. Whittaker[4]

[1]Faculty of Science, Griffith University, Nathan, Queensland 4111, Australia
[2]Polymer Materials and Radiation Group, Department of Chemistry, The University of Queensland, Brisbane, Queensland 4072, Australia
[3]Watson Research Center, IBM, Yorktown Heights, NY 10598
[4]The Centre for Magnetic Resonance, The University of Queensland, Brisbane, Queensland 4072, Australia

Radiation chemical yields for radicals were assessed for a variety of engineering thermoplastics following γ-irradiation under vacuum at 77 K and at a low dose rate. On the basis of these radical yields the radiation resistance of the polymers increased in the following order: poly(phenylene oxide), polyamide, poly(arylene ether sulfone), poly(arylene ether phosphine oxide), polyimide, and poly(arylene ether ketone). This order was similar to that found by other workers based on measurements of the tensile strength of the polymers following electron-beam irradiation of a high dose at a high dose rate.

INCORPORATION OF AROMATIC UNITS into a polymer chain imparts a resistance to degradation by high-energy radiation to the polymer (1). The aromatic units are capable of degrading absorbed energy to heat through their manifold of vibrational energy states, and these units can also scavenge small radical species such as hydrogen atoms or methyl radicals to prevent them for participating in abstraction reactions. These reactions may lead to further breakdown of the polymer chain.

Highly aromatic polymers with phenylene groups in the backbone of the polymer chain would thus be expected to exhibit a resistance to degradation by high-energy radiation such as γ-radiation and electron beams. These polymers generally have high glass-transition and melting temperatures (2). These

* Corresponding author

characteristics make the polymers suitable for use in high temperature environments, particularly in environments that are beyond the useful range for most other thermoplastics. The phenylene groups also give the chains greater rigidity; this class of polymers is characterized by having a high modulus and a high tensile strength (2–4).

Polymers that are strong and resistant to high thermal and ionizing radiation environments can find uses in a wide variety of applications, such as in nuclear facilities, satellites, and other aerospace structures. Thus, there is considerable interest in assessment of the radiation resistance of these polymers both in air and in vacuum. In many of the applications for these materials, an appropriate balance must be reached between the desirable polymer properties and the requirement that the materials must be processible. This need for processibility has resulted in the incorporation of heteroatoms in the polymer chains, and particularly in the incorporation of oxygen in the form of arylene ether groups, such as those found in the engineering thermoplastics Kapton and polyetheretherketone (PEEK).

Comparison of the radiation sensitivity of various polymers that exhibit a resistance to high-energy radiation poses a problem. In their practical applications the materials are generally subjected to irradiation at relatively low dose rates; to study the behavior of these polymers under the conditions of their use, very long irradiation times are required to induce significant and observable changes in many of their properties (3, 4). Some of these properties, such as tensile strength, chemical composition, and molecular weight, can often be difficult to measure for these materials. To overcome this problem, some researchers have resorted to accelerated degradation studies using high dose rates, such as those available through the use of electron beams (4). However, radiation chemistry may be dose-rate dependent (5), and at very high dose rates significant changes in the temperature of the irradiated material occur, in some cases ≥ 25 °C.

We found (6) that electron spin resonance (ESR) spectroscopy can provide a valuable means of assessing the radiation resistance of materials, even when the radiation dose rate is low and the materials are relatively insensitive to radiation damage. When high-energy radiation is absorbed by a polymer, radical anions and cations are formed along with excited states and neutral radical species (6, 7). The primary radical cations that are formed undergo further reactions and are not usually observed even at liquid-nitrogen temperatures. However, the radical anions and neutral radical intermediates are more stable and can be thermally trapped at 77 K. Previous studies demonstrated (6, 8) that there is usually a good correlation among the trapped radical yields measured at low temperature and many of the other property changes that take place in a polymer on irradiation. Thus, studies of radiation chemical yields for radicals can provide a measure of the radiation damage to a material and provide a means of assessing the relative radiation resistance of a family of polymers.

Because of the improved sensitivity of modern ESR spectrometers, it is possible to measure, with acceptable accuracy, radical concentrations as low as 10^{14} radicals per gram of polymer. Therefore, the radiation resistance of a polymer can be assessed at low absorbed dose (typically 10^{-4} to 10^{-3} times that required for tensile strength measurements) and hence at low dose rates.

In this chapter we review our results from studies of the radiation chemistry of a range of radiation-resistant, highly aromatic thermoplastics. We assess the radiation resistance of the materials by ESR studies at low temperature by using low dose rates and relatively short irradiation times. We also compare our results with those obtained by other techniques that require much higher dose rates or much longer irradiation times.

Experimental

Polymer samples of approximately 50 mg were placed in Spectrasil quartz ESR tubes and evacuated for approximately 24 h at $>10^{-2}$ Pa. The temperature of the samples was then raised gradually to approximately 20 K above their glass-transition temperatures to ensure that any absorbed oxygen or residual solvent was completely removed. The tubes were then sealed under vacuum before radiolysis.

Irradiations were carried out by using an Atomic Energy of Canada Limited (AECL) Gammacell at the University of Queensland or the ^{60}Co facility at the Australian Nuclear Science and Technology Organization. The irradiation dose rates for each source were determined by Fricke dosimetry at room temperature and ranged from 1 to 5 kGy/h. Irradiations at 77 K were carried out in a Dewar flask, and appropriate allowances were made for attenuation of the radiation dose rate by the liquid nitrogen. Absorbed doses for the polymers were obtained by correcting for mass–energy absorption coefficients.

The ESR spectra were obtained by using a Bruker ER200D X-band spectrometer that was fitted with a provision for improvement of the spectrum signal-to-noise ratio by data accumulation. The radical concentrations were obtained by use of a Varian pitch standard reference. Care was taken to avoid microwave power saturation of the sample during spectral acquisition, and spectra were routinely obtained at approximately 2 μW.

Polymers

The range of engineering thermoplastics studied in this work include the proprietary polyimides Kapton (DuPont) and Ultem (General Electric), polyamides Kevlar and Nomex (DuPont), poly(phenylene oxide) (PPO) (General Electric), poly(arylene ether ether ketone)s, PEEK, PEEK ketone (PEEKK) (Hoechst), poly(arylene ether sulfone)s, Udel (ICI), synthetic polymers (*9*, *10*), and synthetic poly(arylene ether phosphine oxide)s (*9*, *11*). The chemical structures of these polymers are given in Tables I and II.

Results and Discussion

The ESR spectra observed after radiolysis of the polymers at 77 K are characterized by singlets centered in the region $g = 2.003$–2.004, where g is

Table I. Chemical Structures of Proprietary Polymers

Polymer	Structure
PEEK	
PEEKK	
Kapton	
Ultem	
Udel	
Nomex	
Kevlar	
PPO	

the spectroscopic splitting factor (Figure 1). An exception to this general rule was PPO, which was characterized by a singlet with superimposed fine structure. This fine structure was accounted for by the formation of benzyl-type radicals arising from the loss of a hydrogen atom from the methyl groups in the polymer (12). The sharp peaks observable near the center in these spectra arise from the formation of paramagnetic species in the Spectrasil quartz tubes during radiolysis. The contribution to the area of the spectra arising from these

Table II. Chemical Structures of Synthetic Polymers

Polymer	Structure

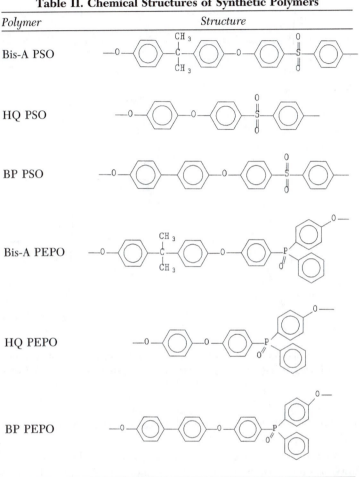

quartz signals were subtracted when the polymer-radical concentrations were calculated.

The observed singlet spectra are composite spectra made up of contributions from three or more components (*3, 10, 11, 13*). This conclusion was based on the use of microwave power saturation of the ESR spectra, photobleaching, and thermal annealing studies of the irradiated samples. For the polymers, the major components of the spectra were assigned to anion radicals and neutral phenoxyl, phenyl, and cyclohexadienyl radicals in varying proportions. Importantly, no evidence for the formation of main-chain scission radicals, methyl radicals, or hydrogen abstraction radicals was determined by

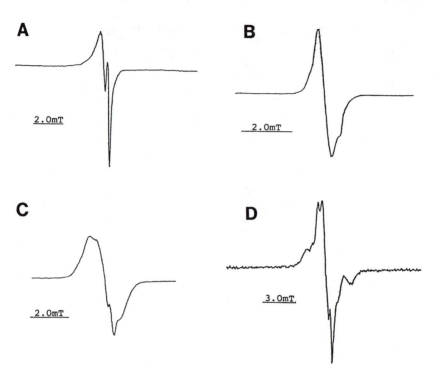

Figure 1. ESR spectra obtained at 77 K after radiolysis at 77 K of A, Udel (20 kGy); B, PEEK (10 kGy); C, poly(bisphenyl-A phosphine oxide) (8 kGy); and D, PPO (15 kGy).

radiation chemistry at the isopropylidine linkage in those polymers containing bisphenyl-A as a component (3, 13).

Plots of the radical concentration versus absorbed dose for several representative polymers irradiated at 77 K are shown in Figure 2. In every case the plots show evidence of dose saturation, even at relatively low doses. This saturation effect was attributed (11) to saturation of the anion radicals, but dose saturation is also observed for the polymers if they are irradiated at ambient temperature (Figure 3). This observation suggests that dose saturation of the neutral radical intermediates may also play a role, because the anion radicals are thermally unstable at temperatures above about 150–200 K (10–12), and these anion radicals will not be present following radiolysis at ambient temperature.

At 77 K the motions of the polymer chains will be very restricted, so that radical recombination reactions will be limited to those that can occur within the cage. Thus, the radical concentration versus dose plots in Figure 2 provide a reliable means by which to assess the relative radiation resistance of the polymers. Those polymers with the greatest resistance to radiation damage

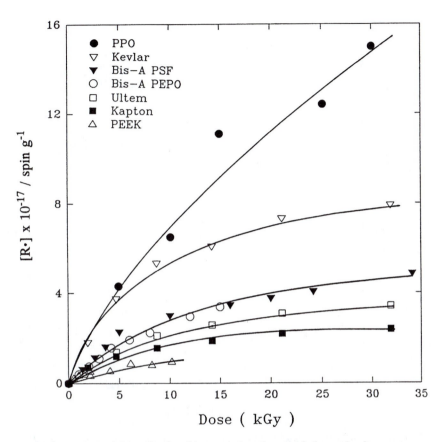

Figure 2. Plots of the radical yield versus dose for γ-radiolysis of polymers at a dose rate of 1–5 kGy/h at 77 K.

will have the lowest radical concentrations following irradiation at 77 K. However, at higher temperatures some chain motions can take place, and these motions can lead to radical decay reactions. Evidence for this phenomenon is demonstrated by the lower radical yields observed following annealing from 77 K to higher temperatures or following radiolysis of poly(bisphenyl-A sulfone) and poly(bisphenyl-A phosphine oxide) at ambient temperature (Figure 3).

On the basis of the information provided in Figure 2, the polymers were ranked in order of their resistance to high-energy radiation under vacuum. The ranking is presented in Table III. Also given in Table III are values of the radical concentrations for each polymer obtained for an absorbed dose of 10 kGy. Irradiation of the polymers to this dose, which can be achieved with reasonably short irradiation times at a low dose rate, is sufficient to establish the relative ranking of a polymer in the series.

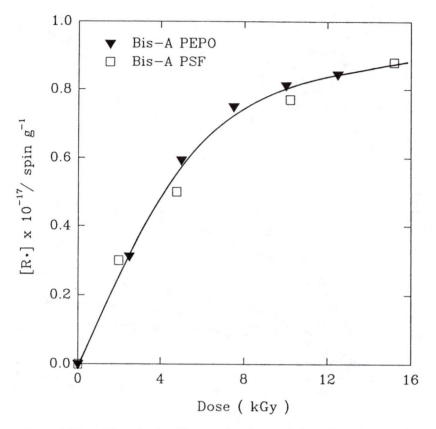

Figure 3. Plots of the radical yield versus dose for γ-radiolysis of poly(bisphenyl-A sulfone) (□) and poly(bisphenyl-A phosphine oxide) (▼) at a dose rate of 1–5 kGy/h at 303 K.

This ranking can be compared with that reported by Sasuga et al. (*4*), who studied the tensile properties of a range of engineering thermoplastics following electron-beam irradiation at ambient temperature, dose rate of 5 × 10³ Gy/s, and doses up to 120 MGy for Kapton. The ranking proposed by Sasuga et al. (*4*) was polyimide > polyetherketone > polyamide > polyetherimide > polyarylate > polysulfone > PPO. The ranking obtained on the basis of our ESR studies differs slightly from this order in that the polyamide was found to produce significantly more radicals at 77 K than the polysulfone and polyetherimide, but the polyimides and polyetherketones remain ranked higher than the polysulfones and PPO. The higher ranking of the polyamide by Sasuga et al. could result from secondary radiation chemistry of these polymers, such as cross-linking, at the higher temperature used in the electron beam study.

**Table III. Ranking of Polymers
by Radical Yields Measured
after Irradiation at 77 K**

Polymer Ranking	Radical Yield (10^{20} spin/kg)
PEEK	1.0
PEEKK	1.1
Kapton	1.7
Ultem	2.2
Bis-A PEPO	2.5
Udel	2.8
Nomex	5.6
Kevlar	5.6
PPO	5.7

NOTE: Radical yields are those for an absorbed dose of 10 kGy and a dose rate of 1–5 kGy/h. Experimental error in radical yields is approximately 5%.

Sasuga and Hagiwara (*14*) studied the radiolysis of engineering thermoplastics in air and in oxygen. They irradiated the polymers at ambient temperature to high dose, up to 30 MGy in air and 12 MGy in oxygen, and at a dose rate of 10 kGy/h. In general, irradiation in air or oxygen increases the yield of scission compared to that observed for irradiation in vacuum and thus increases the yield of radicals. On the basis of tensile measurements, they reported (*14*) that the relative radiation resistance of the polymers in an oxygen atmosphere was similar to that observed in their previous electron-beam study: radiation resistance of the aromatic polymers increased from poly(sulfone) to polyester to polyamide to PEEK.

This order of radiation resistance is similar to that observed in our present study, except that the polyamide is ranked higher than the polysulfone. Thus, although a relatively high concentration of radicals are formed on the γ-radiolysis of the polyamides at 77 K, the tensile properties observed after γ-irradiation in air at ambient temperature do not reflect this feature of their radiation chemistry. This observation suggests that the secondary radiation chemistry of the polyamides is different from that of the other polymers, as indicated previously.

The ESR measurement of radical yields at 77 K can also provide useful information about the radiation resistance of a group of closely related polymers, such as the poly(arylene ether sulfone)s or the poly(arylene ether phosphine oxide)s. The radiation chemical yields at 77 K, called G-values, for formation of radicals in the biphenyl, bisphenyl-A, and hydroquinone poly(sulfone)s at low dose are given in Table IV. These data indicate that the G-values are similar for the three copolymers but that the biphenyl polymer is slightly more radiation resistant than the bisphenyl-A and hydroquinone

Table IV. Comparison of Low-Dose Radiation Chemical Yields for Radicals at 77 K and Sulfur Dioxide at 303 K for a Series of Poly(arylene ether sulfone)s

Polymer	$G(R^{\cdot})$	$G(SO_2)$
BP PSO	0.51	0.06
Bis-A PSO	0.57	0.15
HQ PSO	0.56	0.14

NOTE: G values are numbers of events per 16 aJ of absorbed energy. Experimental errors in G values are approximately 5%.

polymers. Interestingly, this order is similar to that observed for the formation of gaseous sulfur dioxide assessed by gas chromatography. Sulfur dioxide requires a much higher absorbed dose to allow accurate quantitation of the product yield (Table IV). This observation is also consistent with that of Sasuga and co-workers (4, 14).

The radical yields for the three poly(phosphine oxide)s were the same at low dose [$G(R^{\cdot})$ 0.58]. These chemical yields for radical formation are similar in magnitude to that observed for the corresponding poly(sulfone)s at this temperature. However, the radical yields for PEEK at 77 K were significantly lower than those for poly(hydroquinone sulfone) and poly(hydroquinone phosphine oxide) (Figure 4).

In comparing the relative radiation resistance of the poly(sulfone)s and the poly(phosphine oxide)s, our studies indicated that the nature of the radiation chemistry in the two polymers is different, even though the radiation chemical yields for radical formation are very similar for the two polymers. In the sulfone polymers, the radiation chemistry occurred at both the sulfone and ether groups in the polymer chain (3); whereas in the phosphine oxide polymers the bonds to phosphorus are stable toward radiation degradation, and scission occurs at the ether unit of the polymer chain (13). Thus, subtle differences in the radiation chemistry of these two polymer families exist.

The higher radiation resistance of PEEK compared with the poly(sulfone)s and poly(phosphine oxide)s could be due to the semicrystalline nature of PEEK. However, the studies carried out by Sasuga and co-workers (4, 14), in which the effect of crystallinity on the radiation resistance of PEEK was investigated, showed that the semicrystalline nature of the polymer, while playing an important role, does not completely account for the higher radiation resistance observed for this polymer. The carbonyl group seems to confer a radiation resistance greater than that for the corresponding sulfone and phosphine oxide units.

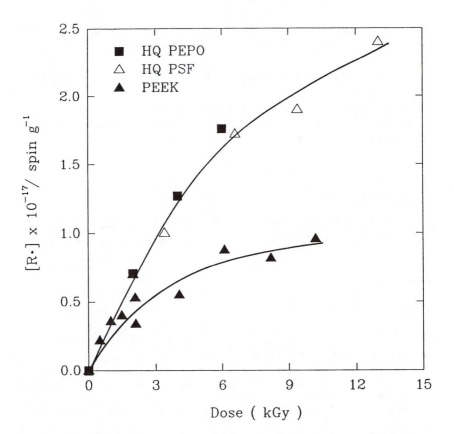

Figure 4. Plots of the radical yield versus dose for γ-radiolysis of poly(hydroquinone sulfone) (△), poly(hydroquinone phosphine oxide) (■), and PEEK (▲) at a dose rate of 1–5 kGy/h at 77 K.

Summary

The radiation resistance of a series of engineering thermoplastics were investigated at low dose by using γ-radiation at 77 K. The radical yields measured by ESR spectroscopy indicated that the radiation resistance of the polymers increased in the order PPO, polyamide, poly(arylene ether phosphine oxide) and poly(arylene ether sulfone), polyimide, poly(arylene ether ketone). The radiation chemical yields for the three arylene ethers in the poly(arylene ether sulfone)s and poly(arylene ether phosphine oxide)s were very similar in magnitude, but the biphenyl group in the polysulfone series yielded a slightly smaller value of G(R·) than that observed for the hydroquinone and bisphenyl-A polymers. This order of relative radiation resistance differs from that observed by other workers based on tensile strength measurements following electron-beam irradiation at high dose rates: The radiation resistance of the polyamide was ranked lower than that reported from the electron-beam study. This difference could be due to

- thermal heating of the samples subjected to electron-beam irradiation at high dose rates
- the high absorbed doses used in the electron-beam study
- differences in the secondary radiation chemistry of the different polymers.

Acknowledgments

We thank the Australian Research Council and the Australian Institute of Nuclear Science and Engineering for financial support for this research.

References

1. O'Donnell, J. H.; Sangster, D. F. *Principles of Radiation Chemistry;* Edward Arnold: London, 1970.
2. Smith, C. D.; Gunyor, A.; Keister, K. M.; Marand, H. A.; McGrath, J. E. *Polym. Prepr. (Am. Chem. Soc., Div. Polym. Chem.)* **1991,** *32,* 93–94.
3. Lewis, D. L.; O'Donnell, J. H.; Hedrick, J. L.; Ward, T C.; McGrath, J. E. In *The Effects of Radiation on High-Technology Polymers;* Reichmanis, E.; O'Donnell, J. H., Eds.; ACS Symposium Series 381; American Chemical Society: Washington, DC, 1989; pp 252–261.
4. Sasuga, T.; Hayakawa, N.; Yoshida, K.; Hagiwara, M. *Polymer* **1985,** *26,* 1039–1045.
5. O'Donnell, J. H. In *The Effects of Radiation on High-Technology Polymers;* Reichmanis, E.; O'Donnell, J. H., Eds.; ACS Symposium Series 381; American Chemical Society: Washington, DC, 1989; pp 1–13.
6. Hill, D. J. T.; O'Donnell, J. H.; Pomery, P. J. In *Materials for Microlithography;* Thompson, L. F.; Willson, C. G.; Frechet, J. M. J., Eds.; ACS Symposium Series 266; American Chemical Society: Washington, DC, 1984; pp 125–149.
7. Campbell, K. T.; Hill, D. J. T.; O'Donnell, J. H.; Pomery, P. J.; Winzor, C. L. In *The Effects of Radiation on High-Technology Polymers;* Reichmanis, E.; O'Donnell, J. H., Eds.; ACS Symposium Series Series 381; American Chemical Society: Washington, DC, 1989; pp 80–94.
8. Tenney, D. R.; Slemp, W. S. In *The Effects of Radiation on High-Technology Polymers;* Reichmanis, E.; O'Donnell, J. H., Eds.; ACS Symposium Series 381; American Chemical Society: Washington, DC, 1989; pp 224–251.
9. These polymers were synthesised by Professor J. E. McGrath and Co-workers, Department of Chemistry, Virginia Polytechnic Institute and State University, Blacksburg, VA.
10. Hill, D. J. T.; Lewis, D. A.; O'Donnell, J. H.; Pomery, P. J.; Hedrick, J. L.; McGrath, J. E. *Polym. Int.* **1992,** *28,* 233–237.
11. Hill, D.J.T.; Hopewell, J.L.; O'Donnell, J.H.; Pomery, P.J.; McGrath, J. E.; Priddy, D. B.; Smith, C. D. *Polym. Degrad. Stab.* **1994,** in press.
12. Hill, D. J. T.; Hunter, D. S.; Lewis, D. A.; O'Donnell, J. H.; Pomery, P. J. *Radiat. Phys. Chem.* **1990,** *36,* 559–563.

13. Heiland, K; Hill, D. J. T.; O'Donnell, J. H.; Pomery, P. J. *Polym. Adv. Technol.* **1994,** 5, 116–121.
14. Sasuga, T.; Hagiwara, M. *Polymer* **1987,** 28, 1915–1921.

RECEIVED for review January 26, 1994. ACCEPTED revised manuscript September 28, 1994.

Thermal Scission and Cross-Linking during Polyethylene Melt Processing

Robert T. Johnston and Evelyn J. Morrison

Dow Plastics, Building B-1470-D, Freeport, TX 77541

We used temperature and time-dependent changes in vinyl and vinylidene concentration to develop a kinetic model that allows estimation of whether cross-linking or thermal scission will dominate during melt processing of major classes of polyethylenes (PE). Torque rheometer processed high-density, linear low-density-, and low -density PE exhibited losses of unsaturated groups when cross-linking was dominant and formation of unsaturated groups when thermal scission was dominant. Cross-linking was attributed to addition of alkyl radicals to olefin groups, and the activation energy for addition to vinyl groups determined was approximately 18 kJ/mol. The activation energy for β-cleavage of secondary alkyl radicals to form vinyl groups was approximately 91 kJ/mol. Reliable estimates of activation energies for reactions involving vinylidene groups could not be determined, but similar reactions were observed for vinylidenes as for vinyls. trans-Vinylene-group formation could not account for the vinyl concentration reductions observed.

BECAUSE POLYETHYLENE (PE) is one of the largest volume thermoplastic polymers yet structurally one of the simplest, understanding and controlling degradation during melt processing is the object of considerable effort in both industry and academia. PE worldwide capacity in 1992 was approximately 110 billion pounds (*1*). PE is manufactured via several different processes, including:

- high-pressure free-radical polymerized low-density polyethylene (LDPE) made in autoclave and tubular reactors
- low-pressure Ziegler-catalyzed ethylene (-olefin copolymers such as ethylene-1-butene, ethylene-1-hexene, and ethylene-1-octene copoly-

mers, collectively referred to as linear low-density polyethylene (LLDPE), which are made in both solution and gas phase reactors

- high-density polyethylene (HDPE), including Phillips-process chromium-oxide-catalyzed products and Ziegler-catalyzed solution and slurry reactor products
- new families of polyolefins based on metallocene catalysts that are under development.

PE is used to manufacture films, molded articles, fibers, wire and cable insulation, sheet, pipe, tubing, foam, and many other useful products. To meet the performance requirements of these applications, dozens of manufacturers make hundreds of different grades of PE. These grades are differentiated on the basis of molecular weight (M_w), molecular weight distribution (MWD), density, comonomer, comonomer content and distribution, short chain branch (SCB) and long chain branch (LCB) content and distribution, and additive content. These variations in polymer composition produce wide variations in performance, including variations in susceptibility to degradation and the consequences thereof.

PE degradation may occur at any stage from manufacture to final disposal, but for most PE applications the stage where degradation occurs most rapidly is during melt processing, when the polymer is exposed to severe conditions. Typical PE melt processing operations consist of extrusion at 175–325 °C (even higher in a few processes) followed by shaping in a die. The extrudate is typically molded, cast, or blown and then cooled to solidify the plastic in the desired shape. The fabrication process involves exposure to heat, oxygen, and mechanical shear. Residence time in this harsh environment can range from a few seconds to several minutes or even hours.

Simultaneous exposure to heat and oxygen leads to thermooxidative degradation. Some of the results of thermooxidation during melt processing include the formation of oxidation products such as aldehydes, ketones, and acids (2, 3). These compounds may cause off-taste, odor, and discoloration in fabricated products. Thermooxidation may also effect molecular weight changes with resultant melt viscosity changes. Viscosity increases or decreases depending on whether scission or cross-linking predominates. Cross-linking and scission can alter extrusion rates, affect melt strength, alter melt extensibility and orientation, cause melt fracture, or create gels (cross-linked network particles dispersed in film).

Some practical questions that a technologist may ask include, Why does PE sometimes cross-link and sometimes scission during melt processing? Under what conditions does each predominate? In particular, how do PEs made by different processes compare in this regard? We will consider these questions and describe the important role of unsaturation in determining the cross-linking versus scission balance as a function of processing temperature. Our approach will be to briefly review the existing literature on the role of vinyl unsaturation in PE melt degradation, and then to describe recent experimental results from our own studies of several classes of PEs.

Mechanisms

Thermooxidation.   The mechanism of PE thermooxidation has been extensively reviewed in the literature (4–6). Figure 1 shows a simplified scheme for this free-radical chain reaction. The first step of the process is the formation of alkyl radicals from the polymer (RH). Many schemes have been proposed for the initiation step, but because of the rapid rate of the chain reaction and the degenerate chain-branching kinetics, the precise nature of the initiation step is of less importance under melt processing conditions than the simple fact that it occurs. Once the alkyl radicals are formed, they may react rapidly with oxygen if available and generate alkylperoxy radicals. The peroxy radicals may then abstract hydrogen from the polymer substrate to form hydroperoxides. At the elevated temperatures characteristic of PE melt processing, hydroperoxides rapidly cleave to form alkoxy and hydroxyl radicals. These compounds may in turn rapidly abstract hydrogen from the polymer substrate to form more alkyl radicals. This process is the engine that drives a series of other reaction processes, such as free-radical disproportionations to form ketones or unsaturated vinylene groups along the polymer backbone.

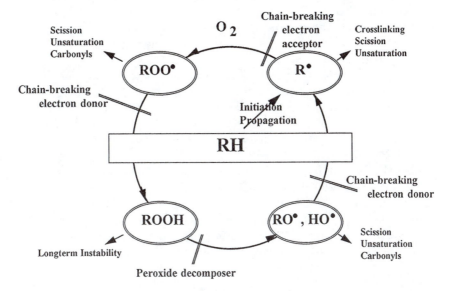

Figure 1. Mechanism of thermooxidation.

The process is disrupted by using antioxidants to break the cycle at the points shown.

Scission.   Among the subsequent reactions that alter molecular weight, there are two primary mechanisms for scission. The first, referred to in this chapter as "oxidative scission," involves β-cleavage of oxygen-centered alkoxy radicals (direct scission of peroxy radicals was also proposed (7). Oxidative scission has a relatively low activation energy (~59 kJ/mol) (8, 9). It is the dominant scission reaction at moderate temperatures or under conditions of high oxygen concentration.

The second scission mechanism, referred to in this chapter as "thermal scission", involves β-cleavage of alkyl radicals. Cleavage of a secondary alkyl radical produces a vinyl group (CH_2=CH–R) and an alkyl radical, whereas cleavage of a tertiary alkyl radical produces a vinylidene group (CH_2=CRR') and an alkyl radical. Thermal scission does not require oxygen, but because thermooxidation can generate alkyl radicals, thermal scission can occur under thermooxidative conditions. Activation energies of 84–117 kJ/mol were determined for β-cleavage of model alkyl radicals (10), so thermal scission becomes increasingly important relative to oxidative scission as temperature increases.

Cross-Linking.   PE cross-linking under thermooxidation conditions has often been attributed to alkyl radical coupling. However, this coupling is increasingly recognized to be of minor importance as compared to the addition reaction of alkyl radicals to olefinic unsaturation. Because the concentration, type, and distribution of unsaturated groups (vinyls, vinylidenes, and t-vinylenes) is dependent on PE polymerization conditions, understanding the differences between various grades of PEs with respect to melt-processing degradation behavior is not possible without an understanding of the role of olefinic unsaturation.

Studies of each of the major classes of PE have shown the importance of vinyl unsaturation in thermooxidative cross-linking during melt processing. In HDPE, cross-linking was accompanied by a reduction in vinyl concentration (11). HDPE devoid of vinyl groups did not cross-link (12, 13). Ziegler HDPEs (with low vinyl concentration) did not cross-link under extrusion conditions, whereas Phillips HDPE (with high vinyl concentration) did. Cross-linking was attributed to the addition of alkyl radicals to vinyl groups (14). LDPE cross-linking was partially attributed to addition of alkyl radicals to vinyl groups (15, 16). A kinetic model for LDPE molecular weight change due to olefin reactions was developed (17). Hydrogenated, vinyl-free LLDPE did not cross-link (18). LLDPE vinyl concentration decreased during melt processing and was accompanied by cross-linking (18, 19). A ^{13}C NMR study of melt processed ethylene-1-hexene medium-density copolymers showed that vinyl reduction was accompanied by LCB formation and provided compelling evidence for formation of "T"-type linkages between main-chain alkyl radicals and vinyl end groups (*20*). Sixty percent of the vinyl decay was attributable to LCB formation.

Even though the importance of vinyl unsaturation in the balance between cross-linking and scission is increasingly recognized, there is still misunderstanding of its importance in determining the melt degradation characteristics of different grades of PE. For example, the difference in cross-linking rates between HDPE and LLDPE was attributed to SCB rather than vinyl concentration differences. The reduced SCB content in the HDPE was hypothesized to cause "a greater possibility of intermolecular recombination in HDPE due to the smaller distances between the chains" (*19*).

Alternative explanations for the disappearance of vinyl unsaturation have been proposed. These explanations include allylic hydrogen abstraction and isomerization of the vinyl group (*19*) or fragmentation with ethylene formation (*21*). Even though these processes may occur to some extent, they cannot explain the LCB formation observed by ^{13}C NMR spectroscopy. Addition of peroxy- or alkoxyradicals to vinyl groups could compete with addition of alkyl radicals, but peroxy- or alkoxyradicals would produce either thermally unstable linkages (ROOR) or cross-links (ROR) that would not explain the LCB formation. Thus, even though other reactions of vinyl groups are probable, the addition of alkyl radicals to vinyl groups appears to be among the most important contributors to vinyl disappearance and LCB formation and cross-linking during PE melt processing.

The role of other unsaturated groups is not so clear. In a study of HDPE, Moss and Zweifel (*14*) concluded that *t*-vinylenes and vinylidene groups did not participate in the alkyl radical addition reactions in the same manner as vinyl groups did. The absence of cross-linking in polypropylene despite the formation of unsaturated groups (due to disproportionation and scission) led to the hypothesis that steric hindrance in higher substituted olefins prevented the addition reaction observed with HDPE vinyl groups (*22*). On the other hand, it is well known that higher substituted olefins in other polymers undergo free-radical addition reactions (*23, 24*). Because these observations were made in polymers having relatively high concentrations of olefins as compared to PE, similar reactions may occur in PE but are not usually detected because of the low olefin concentration.

Temperature Dependence of Cross-Linking and Scission.

Many studies (*4, 5, 12, 25–28*) of PE degradation have shown the tendency for cross-linking to dominate at lower temperatures and scission to dominate at higher temperatures. Several studies (*11, 14, 17, 28*) have shown that vinyl concentration after processing is minimized at intermediate temperatures. These results are consistent with the expected competition between vinyl decay due to cross-linking and vinyl formation due to thermal scission; at low temperatures, the rate of vinyl reaction is slow so vinyl decay is minimal, whereas at high temperatures vinyl formation becomes dominant due to thermal scission.

Gol'dberg et al. (17) developed a kinetic model for PE cross-linking and scission involving alkyl radicals under "thermal mechanical" degradation conditions. They derived activation energies of 146 and 96 kJ/mol for scission and cross-linking, respectively. These values are high compared with those obtained with model compounds; for example, studies of alkyl radical addition to model olefins have produced activation energy estimates of approximately 21–38 kJ/mol (10). Two primary concerns with respect to this work are that

- an activation energy difference between cross-linking and scission of 50 kJ/mol was assumed; therefore, constraints were placed on the parameter estimation process that was already limited by the strong correlation between Arrhenius parameters
- the model was based on molecular weight changes, but accurate molecular weight data for cross-linked or LCB-containing polymers are difficult or impossible to obtain.

Despite these limitations, this model provided an excellent qualitative description of the factors affecting the cross-linking/scission balance during PE melt processing.

In our own studies (29, 30) we extended this approach in an effort to understand the differences in melt processing behavior of several important classes of PEs. Rather than attempt to accurately measure MWD changes, we focused on the underlying changes in olefinic unsaturation (which can be measured even in polymers cross-linked well beyond the gel point). By using robust statistical tools, kinetic parameters for the changes in olefin groups were determined. The resultant quantitative description of changes in unsaturation can then be related qualitatively to viscosity changes obtained during melt processing.

Experimental

Materials. Representatives from the major classes of PEs were studied (Table I). A range of melt indices, densities, comonomers, and production processes was included. The type of unsaturated groups and their concentration varied with each sample, depending on both the polymerization process and the specific molecular design characteristics of each grade. All the samples were free of antioxidant except LLD-2, which contained 480 ppm Irganox 1076 hindered phenolic antioxidant [octadecyl 3-(3,5-di-t-butyl-4-hydroxyphenyl)propionate]. HD-1 originally contained phenolic antioxidant, but the antioxidant was removed by hexane extraction before this study.

Equipment and Procedures. For torque rheometer processing stability testing, 40.00 g of PE sample was melt-processed in a HaakeBuchler Rheocord

Table I. Description of Resins Used

Code	Manu-facturer	Description	MI	D	t-Vinyl-enes/ 1000 C	Vinyls/ 1000 C	Vinyl-idenes/ 1000 C
LLD-1	Dow	Solution Process Ziegler; Ethylene/Octene	0.92	0.920	0.04	0.30	0.07
LLD-2	Union Carbide	Gas Phase Process Ziegler; Ethylene/Butene	0.99	0.919	0.02	0.06	0.04
HD-1	Phillips Petroleum	Phillips Process HDPE	49	0.955	0.01	0.94	0.00
HD-2	Dow	Ziegler Process HDPE	0.03	0.949	0.01	0.09	0.02
LD-1	Dow	High Pressure LDPE (autoclave)	7.36	0.916	0.04	0.09	0.30
LD-2	Dow	High Pressure LDPE (tube)	1.84	0.925	0.02	0.07	0.14

NOTE: Vinylidenes corrected for underlying methyl absorbance by bromination/peak subtraction method. MI is melt index measured in g/10 min (I_2–ASTM D–1238, condition E). D is density in g/cc.

System 40 torque rheometer by using an open Rheomix 600 mixer (modified with stainless steel bushings) and roller style mixing blades. Isothermal control software maintained melt temperature within 1 °C of the set point after the initial melting and thermal equilibration period (approximately 150–300 s, depending on temperature). Samples were mixed for 60 s at a rotor speed of 60 rpm and then for 2940 additional seconds at 10 rpm (10 rpm was used to achieve oxygen deficient conditions) (*31*). Compared with studies conducted at higher revolutions per minute (*19, 32*). low revolutions-per-minute conditions may better simulate the oxygen-deficient extruder zones where gel formation sometimes occurs. In addition, low revolutions-per-minute measurements are rheologically more sensitive to viscosity changes due to cross-linking. Finally, at 10 rpm and the relatively high temperatures used, mechanodegradation may be assumed to be negligible (*33–35*). Torque data were continuously collected, and up to three small (0.5-g) samples were removed from selected experimental runs for analysis by IR spectroscopy (*28*).

Results and Discussion

Torque Data. Figures 2–9 show the torque data for the various PEs and one PE blend. As shown in Figure 2, no change in torque occurred in LLD-1 at 150 °C and 175 °C once the melt temperature equilibrated. (The difference in torque between the two temperatures is simply due to the temperature dependence of viscosity.) LLD-1 predominantly cross-linked (torque increase) between 200 °C and 263 °C. Scission dominated during the latter portion of the experiment. Above 263 °C (Figure 3), scission became increasingly important, and no torque increase was evident at ≥325 °C.

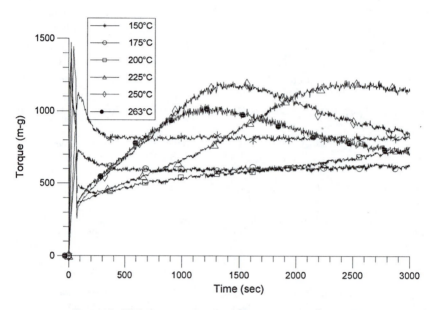

Figure 2. LLD-1 torque response between 150 and 263 °C.

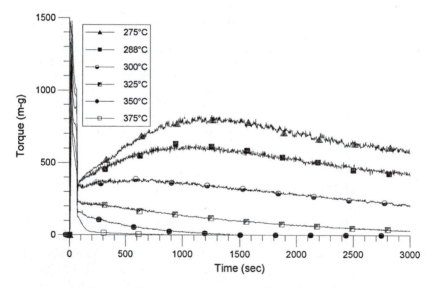

Figure 3. LLD-1 torque response between 275 and 375 °C.

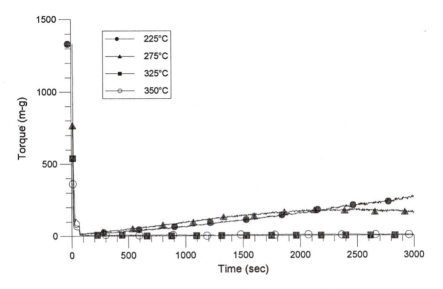

Figure 4. HD-1 torque response between 225 and 350 °C.

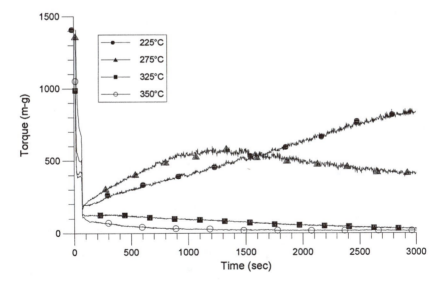

Figure 5. LLD-1/HD-1 blend torque response between 225 and 350 °C.

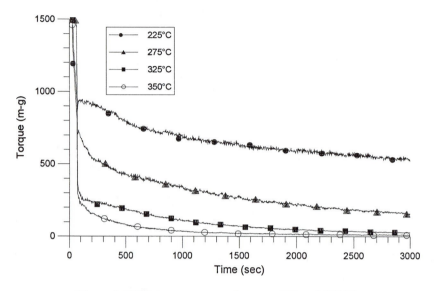

Figure 6. HD-2 torque response between 225 and 350 °C.

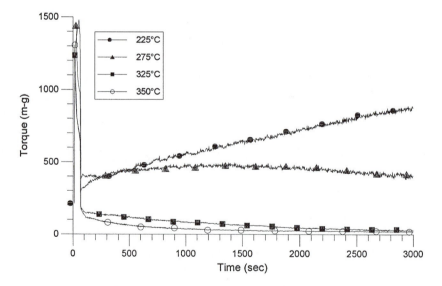

Figure 7. LLD-2 torque response between 225 and 350 °C.

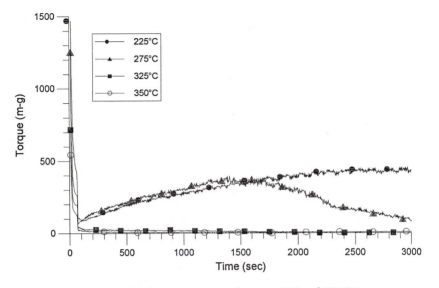

Figure 8. LD-1 torque response between 225 and 350 °C.

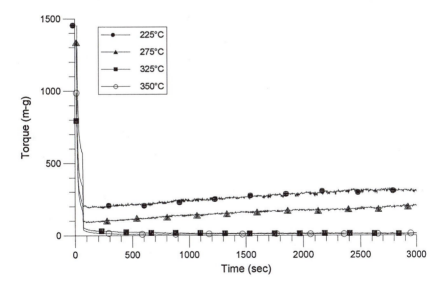

Figure 9. LD-2 torque response between 225 and 350 °C.

Figure 4 shows that HD-1 had a measurable torque increase at 225 °C and 275 °C but not at 325 °C or 350 °C, where the torque was too low to measure accurately. A 1:1 mixture of LLD-1 and HD-1 had a slight torque plateau at 325 °C but only a decrease in torque after 400 s or at 350 °C (Figure 5). HD-2 exhibited only declines in torque at all temperatures (Figure 6).

LLD-2 (Figure 7) exhibited torque increases at 225 °C and 275 °C, even though the dominance of cross-linking at 275 °C was significantly less than in LLD-1. At 325 °C and 350 °C, scission was dominant.

Cross-linking dominated at 225 °C and 275 °C in LD-1 (Figure 8), but scission dominated at 325 °C and 350 °C. LD-2 (Figure 9) had only modest rates of torque increase at 225 °C and 275 °C, and scission was dominant at ≥325 °C.

What determined the different torque results with each resin? Obviously, the absolute value of torque was determined largely by the molecular weight of the polymer. But what determined the predominant direction of torque change? Especially, what governed the transition from predominant cross-linking to predominant scission behavior?

Vinyl Reaction Model.   If the rapid, low-activation-energy process of hydrogen abstraction by primary alkyl radicals is ignored, the thermal reactions of alkyl radicals may be represented simplistically as a reversible process:

$$R^\bullet + CH_2 = CR' \sim\sim\sim \xrightarrow{k_1} R^\bullet \tag{1}$$

$$R^\bullet \xrightarrow{k_2} R^\bullet + CH_2{=}CR'' \sim\sim\sim \tag{2}$$

where k is the rate constant.

We first consider the vinyl ($R' = R'' = H$) case. If a steady-state alkyl radical concentration is assumed,

$$dV/dt = -k_{1'}V + k_{2'} \tag{3}$$

where V is vinyl concentration, and t is time. The integrated form of the rate equation is

$$V = (V_0 - k_{2'}/k_{1'})\exp(-k_{1'}t) + k_{2'}/k_{1'} \tag{4}$$

By using a modified form of the Arrhenius equation to reduce the correlation between parameters, an expression for $k_{1'}$ in terms of $E_{1'}$ and $k_{b1'}$ is

$$k_{1'} = k_{b1'}\exp[(E_{1'}/R)(1/T - 1/T_b)] \tag{5}$$

where $E_{1'}$ is the activation energy, $k_{b1'}$ is the rate at an arbitrary reference temperature T_b, generally selected at the midpoint of the data set (in this case, 548 K, 275 °C). An analogous expression for $k_{2'}$ was obtained by using $E_{2'}$ and $k_{b2'}$. These expressions for $k_{1'}$ and $k_{2'}$ were substituted into eq 3. Equation 3 was then fitted to the experimental vinyl concentrations (excluding data below 200 °C, where the steady-state assumption would be unreasonable) by using SimuSolv modeling software with a generalized reduced-gradient optimization method and optimizing $k_{b1'}$, $k_{b2'}$, $E_{1'}$ and $E_{2'}$ in the nonintegrated form of the rate equation. Statistical output showed that the model accounted for 97% of the variability in the data, and estimates for $E_{1'}$ and $E_{2'}$ were 18 and 91 kJ/ mol for cross-linking and scission, respectively. Figure 10 shows the relationship between experimental and model-predicted vinyl concentrations.

Figures 11–18 show the experimental vinyl concentrations together with plots of the integrated rate equation calculated for each temperature and initial vinyl concentration (taken as the vinyl concentration of the first sample collected). The results show a reduction in vinyl concentration at low-to-intermediate temperatures corresponding to the conditions where cross-linking predominated in the torque curves. At higher temperatures, an increase in vinyl concentration due to thermal scission was observed in most samples. An exception was HD-1 and its blend with LLD-1. In these cases, the high initial vinyl concentration resulted in lower vinyl concentrations even at the highest temperatures, even though the rate was lower than at the intermediate temperatures.

Considering the wide range of PEs and temperatures included in the analysis, this simplified model described vinyl concentration surprisingly well. The usefulness of the model lies in the qualitative understanding it can bring to the relationship among polymer structure, processing temperature, and the resultant mode of degradation. Thus, from eq 3, an equilibrium vinyl concentration ($V = V_{eq}$) exists for each temperature at which vinyl concentration does not change ($dV/dt = 0$). Because $V_{eq} = k_{2'}/k_{1'} = k_2/k_1$, the equilibrium vinyl concentration can be estimated from the model (Figure 19). Given a PE and initial vinyl concentration, a processing temperature exists at which the initial vinyl concentration (V_0) will equal V_{eq}. At that temperature, cross-linking and *thermal* scission will be in balance. Conversely, for each processing temperature, an equilibrium vinyl concentration exists that must be reached before cross-linking and thermal scission will be balanced. For example, V_{eq} = 0.25 vinyls/1000 carbons at 330 °C. Figure 12 shows that LLD-1 was almost at vinyl equilibrium at 325 °C. Also, scission became dominant at that temperature (Figure 3).

If the vinyl concentration in a given PE is greater than the V_{eq} for the processing temperature, then cross-linking will predominate over thermal scission. If the vinyl concentration is lower than V_{eq}, then thermal scission will dominate over cross-linking.

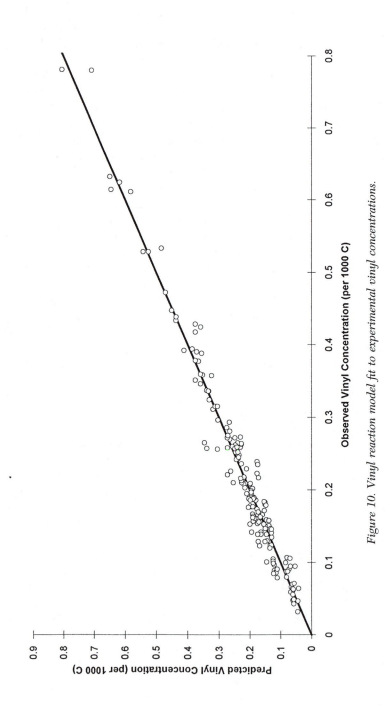

Figure 10. Vinyl reaction model fit to experimental vinyl concentrations.

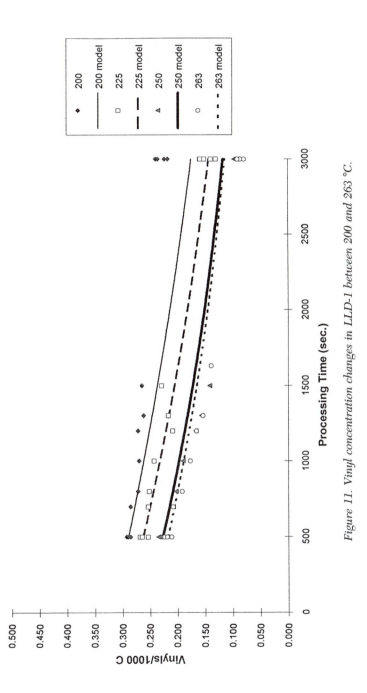

Figure 11. Vinyl concentration changes in LLD-1 between 200 and 263 °C.

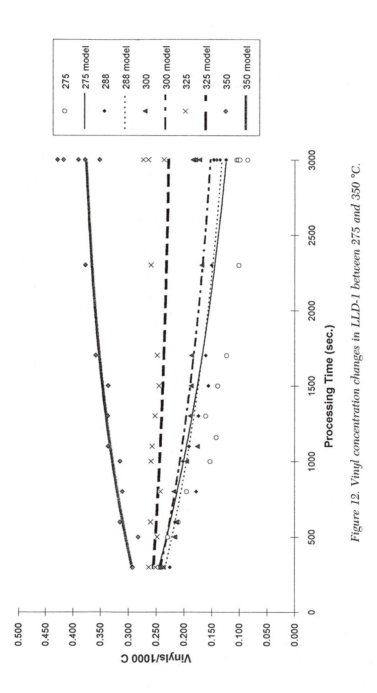

Figure 12. Vinyl concentration changes in LLD-1 between 275 and 350 °C.

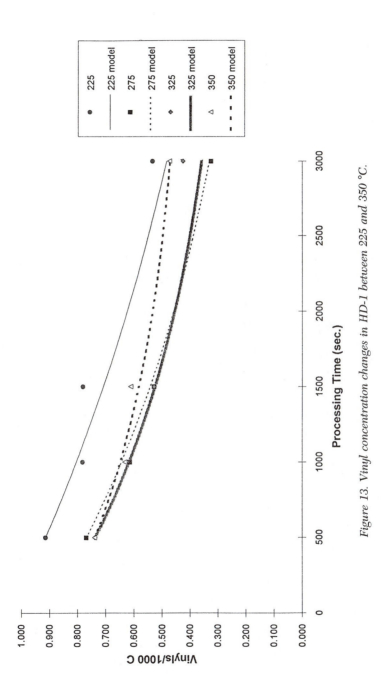

Figure 13. Vinyl concentration changes in HD-1 between 225 and 350 °C.

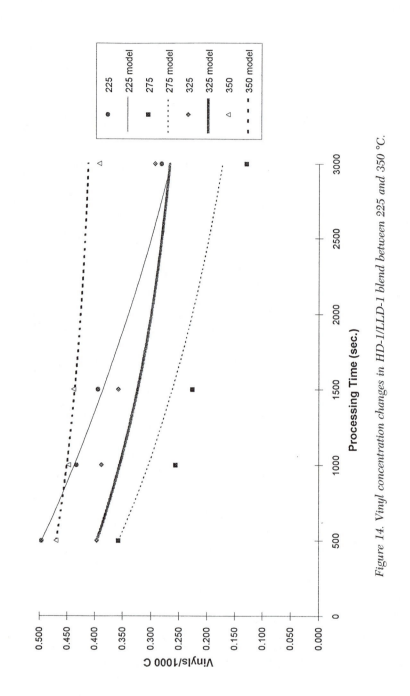

Figure 14. Vinyl concentration changes in HD-1/LLD-1 blend between 225 and 350 °C.

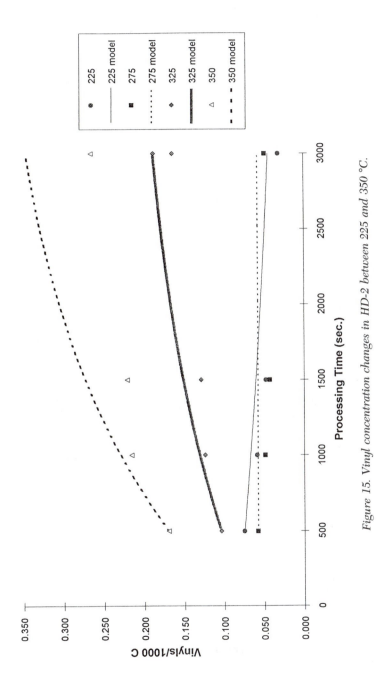

Figure 15. Vinyl concentration changes in HD-2 between 225 and 350 °C.

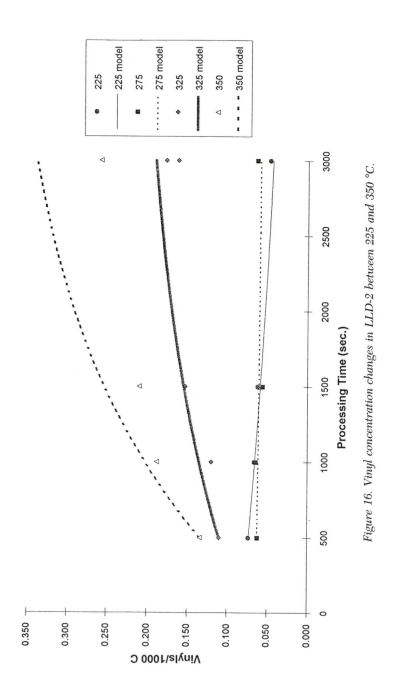

Figure 16. Vinyl concentration changes in LLD-2 between 225 and 350 °C.

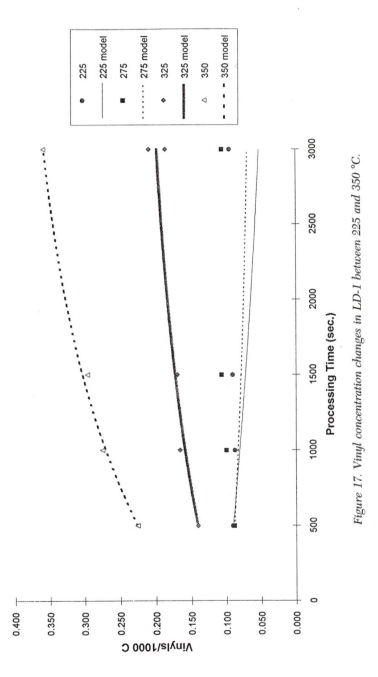

Figure 17. Vinyl concentration changes in LD-1 between 225 and 350 °C.

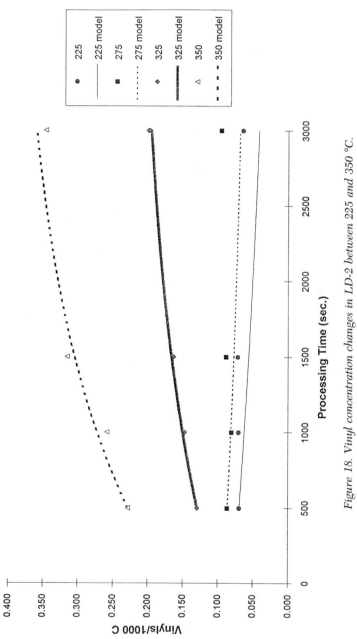

Figure 18. Vinyl concentration changes in LD-2 between 225 and 350 °C.

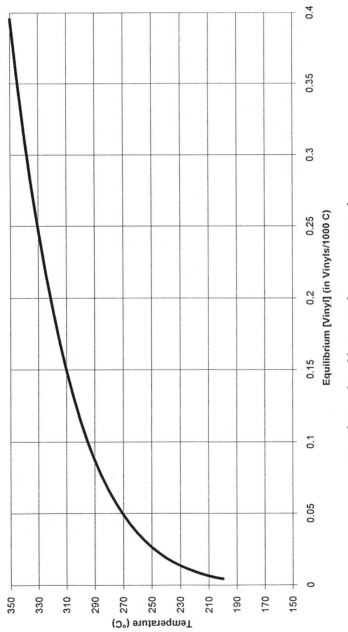

Figure 19. Relation of equilibrium vinyl concentration and temperature.

The reason the model can only qualitatively describe the relationship of olefin content and viscosity change is because total scission = thermal scission + oxidative scission, but the model only describes the thermal component. In this sense, the model may be used to estimate the upper-limit temperature at which cross-linking will be dominant. The actual transition temperature will be lower depending on the extent of oxidative scission.

Except to the extent that oxidized radical species may directly attack vinyl groups, the balance between thermal scission and cross-linking should be independent of oxygen concentration and depend only on k_1 and k_2. This relationship is suggested by the results shown in Figure 20, where the transition from predominant cross-linking to scission for LLD-1 in a nitrogen purged mixer was still near 325 °C.

Vinylidene Reaction Model. Even though much of the variation in PE melt degradation behavior as a function of temperature and PE class can be explained by the vinyl reaction model, other unsaturated moieties probably participate in cross-linking reactions and affect the cross-linking/scission balance.

Using equations analogous to those for vinyls, we attempted to model the changes in vinylidene concentration. Unfortunately, the vinylidene concentration in most of the samples was too low to avoid excessive correlation during parameter estimation. Even though the estimates were highly correlated, rea-

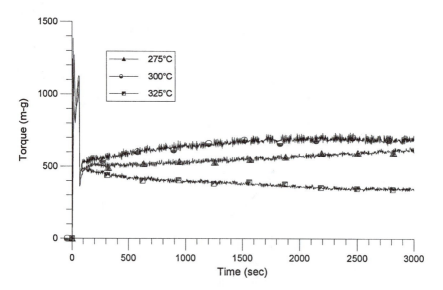

Figure 20. Torque changes in LLD-1 processed in a nitrogen purged mixer (all torque curves adjusted for temperature difference to a 250 °C basis by using 5.124 kcal/mol flow activation energy).

sonable parameter estimates could be obtained by using LD-1 alone, which had sufficiently high vinylidene content to display both decay and formation of vinylidene groups. Statistical output showed that the model accounted for 99% of the variability in the data, and estimates for $E_{1'}$ and $E_{2'}$ were 33 and 84 kJ/mol for cross-linking and scission, respectively. Figure 21 shows the changes in vinylidene concentration for LD-1. It is evident that the vinylidene group concentrations changed similarly to those for vinyl groups in LLD-1.

trans-Vinylenes and Carbonyls.   None of the samples had a high enough *t*-vinylene content to enable us to observe possible contributions to cross-linking reactions. However, we were interested in the relation of *t*-vinylene and vinyl concentration because of the possible isomerization of vinyls to *t*-vinylenes.

As shown in Figures 22–24, the *t*-vinylene formation rate increased with temperature and was highly correlated with carbonyl formation. Figure 25 shows that for the entire data set, there was a correlation between *t*-vinylenes and carbonyls but not between *t*-vinylenes and vinyls. Also plotted in Figure 25 are the calculated reacted vinyls. Even though these values increased with *t*-vinylene concentration, the correlation was poor as compared with the values for carbonyls.

Table II shows a specific example of the lack of correlation of *t*-vinylene formation and vinyl reaction. HD-1 had three times more vinyl reduction than LLD-1 at 225 °C (a temperature where vinyl formation due to thermal scission is minimal), yet it had the same rate of *t*-vinylene formation. We conclude that isomerization of vinyl groups to *t*-vinylenes cannot explain the loss in vinyl concentration. Ketonic carbonyl and *t*-vinylene formation is attributed to disproportionation of alkoxy- and alkyl radicals, respectively.

The increase in carbonyls and *t*-vinylenes with increasing temperature is indicative of increased rates of oxidation and oxidative scission. The fact that an equilibrium in vinyl or vinylidene concentration may have been achieved does not mean that degradation has stopped, or that physical property deterioration is not continuing.

Conclusions

In this study we wished to model the olefin-based reactions affecting cross-linking versus scission balance in PE melt processing. Building on previous work, we developed a simplified kinetic model by using SimuSolv modeling techniques and a broad data set. The model allows estimation of the temperature above which a given PE will show predominant scission during melt processing. The model explains much of the variability in cross-linking/scission behavior observed in commercial PEs under melt processing conditions.

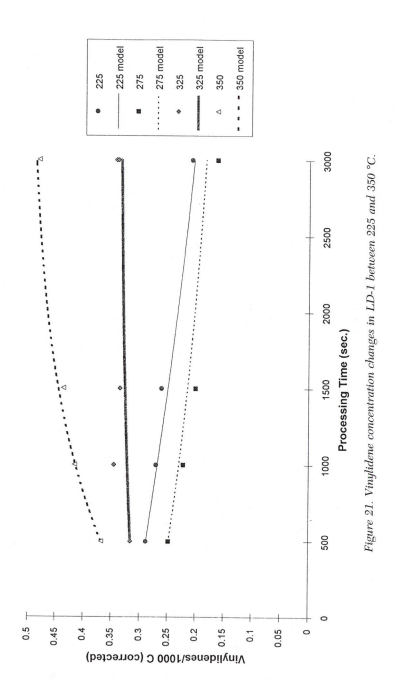

Figure 21. Vinylidene concentration changes in LD-1 between 225 and 350 °C.

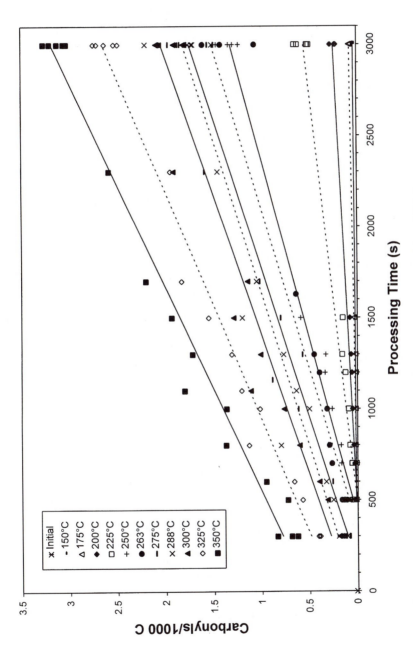

Figure 22. Carbonyl formation at 150–350 °C in LLD-1.

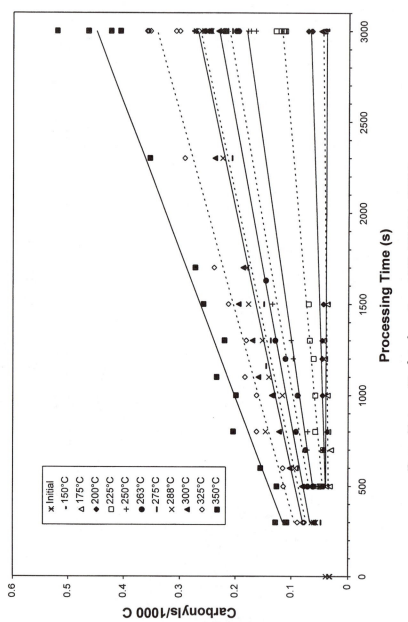

Figure 23. trans-Vinylene formation at 150–350 °C in LLD-1.

Trans-vinylenes/1000 C

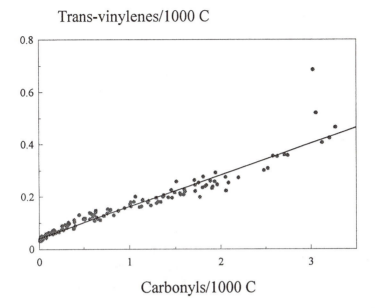

Carbonyls/1000 C

Figure 24. Correlation between trans-vinylene *and carbonyl formation in LLD-1. For straight line shown:* $y = 0.12x + 0.04$; $R^2 = 0.93$.

Even though efforts to model vinylidene reactions were limited by the data available, parameter estimates were obtained for an autoclave LDPE resin with a high vinylidene concentration. This model showed the same trends in olefin concentration as the vinyl model did, and this result contradicts the assumption that vinylidene groups are too hindered for cross-linking reactions.

Limited t-vinylene concentrations in the polymers precluded determination of a possible role of vinylene groups in cross-linking reactions. Comparisons of t-vinylene formation rates and both vinyl and carbonyl concentration changes suggest that the formation of t-vinylenes is not attributable to vinyl isomerization.

Estimation of the dominance of cross-linking versus scission in a given PE must therefore take into account vinyl concentration, vinylidene concentration, temperature, relative oxidation rate, and possibly t-vinylene concentration. If the oxidation rate is assumed to be similar for different PEs processed under the same processing conditions, then the major factor causing different PEs to degrade differently during processing will be the unsaturation content. For most classes of PEs, vinyl concentration will be the most important variable, followed by the vinylidene concentration.

In practice, oxidation rates are generally not the same for different PEs. In some cases, an increase in oxidation rate can occur due to catalyzing effects

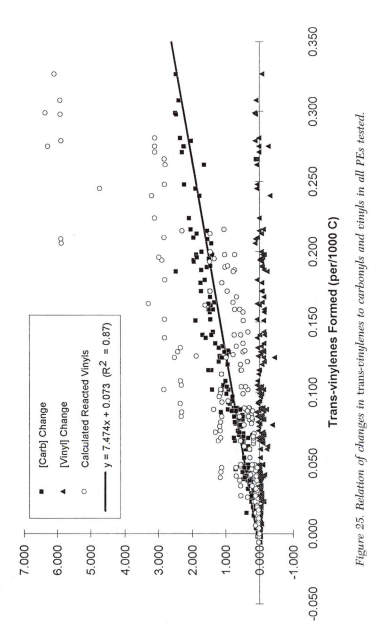

Figure 25. Relation of changes in trans-vinylenes to carbonyls and vinyls in all PEs tested.

Table II. Comparison of *t*-Vinylene and Vinyl Concentrations at 225 °C for LLD-1 and HD-1

Sample	Time(s)	Car-bonyls	t-Vinylenes	Vinyls	Carbonyls	t-Vinylenes	Vinyls
					Increase Between 500 and 3000 s		
LLD-1	500	0.03	0.05	0.27	0.00	0.00	0.00
	1500	0.15	0.07	0.23	0.12	0.02	−0.04
	3000	0.65	0.12	0.14	0.62	0.07	−0.13
HD-1	500	0.08	0.02	0.91	0.00	0.00	0.00
	1500	0.32	0.04	0.78	0.24	0.02	−0.13
	3000	0.65	0.09	0.53	0.56	0.08	−0.38

of metal impurities or polymerization catalyst residues. However, under melt processing conditions where the rate of hydroperoxide thermal decomposition is already high, catalyst residues have relatively little effect.

By far the largest determinant of oxidation rate is the antioxidant package in a given PE. By reducing the rate of overall oxidation, antioxidants reduce both cross-linking and scission. However, most antioxidants are more effective peroxy radical scavengers than alkyl radical scavengers, and they reduce oxidative scission relatively more than the alkyl-radical-dominated cross-linking and thermal scission reactions. Thus, at high temperatures, these antioxidants can shift the overall cross-linking/scission balance toward cross-linking. However, they will not cause cross-linking to dominate once the olefin concentration is at or below the equilibrium olefin concentration for the processing temperature (30). Properly stabilized, PEs with a wide range of olefin contents can be melt-processed without excessive degradation.

References

1. *Chemical Economics Handbook;* SRI International: Menlo Park, CA, 1993.
2. Hoff, A.; Jacobsson, S. *J. Appl. Polym. Sci.* **1982**, *27*, 2539.
3. Holmstrom, A.; Sorvik, E. M. *J. Polym. Sci. Polym. Chem. Ed.* **1978**, *16*, 2555.
4. Stivala, S. S.; Kimura, J.; Gabbay, S. M. In *Degradation and Stabilization of Polyolefins;* Allen, N. S., Ed.; Applied Science: London, 1983; Chapter 3, pp 63–185.
5. Kelen, T. *Polymer Degradation;* Van Nostrand Reinhold: New York, 1983; Chapter 6, pp 107–136.
6. Grassie, N.; Scott, G. *Polymer Degradation and Stabilization;* Cambridge University Press: Cambridge, England, 1985; pp 17–67, 86–118.
7. Iring, M.; Tudos, F.; Fodor, Z.; Kelen, T. *Polym. Degrad. Stab.* **1980**, *2*, 143.
8. Kerr, J. A. *Chem. Rev.* **1966**, *66*, 465.
9. Mill, T.; Hendry, D. G. In *Comprehensive Chemical Kinetics: Liquid-Phase Oxidation;* Bamford, C. H.; Tipper, C. F. H., Eds.; Elsevier: Amsterdam, Netherlands, 1980; Vol. 16, p 52.
10. Kondratiev, V. N. *Rate Constants of Gas Phase Reactions;* Academy of Sciences of the USSR; National Technical Information Society: Springfield, VA, 1972; COM-72-10014.
11. Rideal, G. R.; Padget, J. C. *J. Polym. Sci. Symp. No. 57,* **1976**, 1.

12. Holmstrom, A.; Sorvik, E. M. *J. Polym. Sci. Symp. No. 57,* **1976,** 33.
13. Witt, D. R.; Hogan, J. P. *J. Polym. Sci. Part A-1,* **1970,** *8,* 2689.
14. Moss, S.; Zweifel, H. *Polym. Degrad. Stab.* **1989,** *25,* 217.
15. Sweeting, O. J. In *The Science and Technology of Polymer Films;* Sweeting, O. J., Ed.; Wiley-Interscience: New York, 1971; Vol. 2, p 180.
16. Gol'dberg, V. M.; Yarlykov, B. V.; Paverman, N. G.; Berezina, Y. I.; Akutin M. S.; Vinogradov, G. V. *Polym. Sci. USSR,* **1979,** *20,* 2740.
17. Gol'dberg, V. M.; Zaikov, G. E. *Polym. Degrad. Stab.* **1987,** *19,* 221.
18. Johnston, R. T. *Proceedings of the SPE LLDPE RETEC;* Society of Petroleum Engineers: Richardson, TX, 1985; p 59.
19. Iring, M.; Foldes, E.; Barabas, K.; Kelen, T.; Tudos, F. *Polym. Degrad. Stab.* **1986,** *14,* 319.
20. Randall, J. C. In *Cross-linking and Scission in Polymers;* Guven, O., Ed.; Kluwer: Dordrecht, Netherlands, 1990; pp 60–62.
21. Foldes, E.; Iring, M.; Tudos, F. *Polym. Bull.* **1987,** *18,* 525.
22. Hinsken, H.; Moss, S.; Pauquet J.; Zweifel, H. *Polym. Degrad. Stab.* **1991,** *34,* 279.
23. Loan, L.D. *Pure Appl. Chem.* **1972,** *30,* 173.
24. Endstra, W. C.; Wreesmann, C. T. J. In *Elastomer Technology Handbook;* Cheremisinoff, N. P., Ed.; CRC: Boca Raton, FL, 1993; p 499.
25. Holmstrom, A.; Sorvik, E. M. *J. Appl. Polym. Sci.* **1974,** *18,* 761.
26. Holmstrom, A.; Sorvik, E. M. *J. Appl. Polym. Sci.* **1974,** *18,* 3513.
27. Drake, W. O.; Pauquet, J. R.; Todesco, R. V.; Zweifel, H. *Die Angew. Makro. Chem.* **1990,** *176/177,* 215.
28. Johnston, R. T. *Proceedings on Advances in the Stabilization and Controlled Degradation of Polymers;* Institute in Materials Science: State University of New York at New Paltz, 1986; p 57.
29. Johnston, R. T.; Morrison, E. J. *Presentation at Advances in the Stabilization and Controlled Degradation of Polymers;* Lucerne, Switzerland, 1990.
30. Johnston, R. T.; Slone, E. J. *Proceedings SPE Polyolefins RETEC VII;* Society of Petroleum Engineers: Richardson, TX, 1991; pp 207–219.
31. Trizisky, J. D. *Polym. Eng. Sci.* **1979,** *19,* 805.
32. Iring, M.; Fodor, Z.; Barabas, K.; Kelen, T.; Tudos, F. *Polym. Bull.* **1986,** *16,* 159.
33. Porter, R. S.; Casale, A. *Polym. Eng. Sci.* **1985,** *25,* 129.
34. Holmstrom, A.; Andersson, A.; Sorvik, E. M. *Polym. Eng. Sci.* **1977,** *17,* 728.
35. Hanson, D. E. *Polym. Eng. Sci.* **1969,** *9,* 405.

RECEIVED for review December 6, 1993. ACCEPTED revised manuscript September 28, 1994.

INDEXES

Author Index

Affiliation Index

Subject Index

A

I

Copy editing: Scott Hofmann-Reardon
Indexing: Colleen P. Stamm
Production: Zeki Erim, Jr.
Acquisition: Anne Wilson
Cover design: Alan Kahan